ROBUST MECHANISM DESIGN

The Role of Private Information and Higher Order Beliefs

World Scientific Series in Economic Theory
(ISSN: 2251-2071)

Series Editor: Eric Maskin *(Harvard University, USA)*

Published

Vol. 1 Equality of Opportunity: The Economics of Responsibility
by Marc Fleurbaey and François Maniquet

Vol. 2 Robust Mechanism Design: The Role of Private Information and Higher Order Beliefs
Dirk Bergemann and Stephen Morris

World Scientific Series in Economic Theory – Vol. 2

ROBUST MECHANISM DESIGN

The Role of Private Information and Higher Order Beliefs

Dirk Bergemann
Yale University, USA

Stephen Morris
Princeton University, USA

NEW JERSEY • LONDON • SINGAPORE • BEIJING • SHANGHAI • HONG KONG • TAIPEI • CHENNAI

Published by

World Scientific Publishing Co. Pte. Ltd.
5 Toh Tuck Link, Singapore 596224
USA office: 27 Warren Street, Suite 401-402, Hackensack, NJ 07601
UK office: 57 Shelton Street, Covent Garden, London WC2H 9HE

Library of Congress Cataloging-in-Publication Data
Bergemann, Dirk.
Robust mechanism design : the role of private information and higher order beliefs / by Dirk Bergemann & Stephen Morris. -- 1st ed.
p. cm. -- (World scientific series in economic theory ; 2)
Includes bibliographical references and index.
ISBN-13: 978-9814374583
ISBN-10: 981437458X
1. Robust control. 2. Prices. I. Morris, Stephen. II. Title.
TJ217.2.B47 2012
629.8'312--dc23

2012005125

British Library Cataloguing-in-Publication Data
A catalogue record for this book is available from the British Library.

In-house Editor: Alisha Nguyen

Typeset by Stallion Press
Email: enquiries@stallionpress.com

Printed in Singapore.

To our wives, Kishwar and Violeta, who gave us steadfast love and support in the decade we wrote the papers in this book, and to our children, Reza and Yasmin, and Luis, Thomas and Edward whose inspiring first decade it was.

"Equilibrium robustness in informational variables is critical, if one wants to use results from the mechanism design literature in real life applications. The papers included in the Bergemann and Morris book describe state of the art progress in this direction of research. The book is an excellent resource for established game theorists, who want to learn more about this subject; and for PhD students, who look for exciting problems to investigate."

Ehud Kalai
Kellogg School of Management, Northwestern University

"The question of the design of institutions has been at the centre of some of the most important economic theory in the past four decades. Bergemann and Morris have made seminal contributions to the understanding of how uncertainty can and should be incorporated into mechanism design, and this volume reproduces a collection of their most important work in the area. The volume will be an important reference for those working in the area and those who wish to apply the ideas in economic models."

Andrew Postlewaite
Department of Economics, University of Pennsylvania

Foreword

Good social, political, and economic decisions are normally contingent on the preferences of individuals in society over the possible options. Indeed, the very concept of economic efficiency (Pareto optimality) demands preference contingency: an option is efficient provided there is no other option that everyone prefers. Yet, unfortunately for public decision-makers, individuals' preferences are not typically publicly known, a limitation that can seriously interfere with making the right choice.

A great accomplishment of mechanism design theory has been to show that, in many cases, this limitation can be circumvented. That is, it is often possible in principle to devise a mechanism or game whose outcome in equilibrium is the same as the one that would have been chosen had preferences been public in the first place.

Even so, a good many of the mechanisms exhibited in the literature have been justifiably criticized for depending too sensitively on details that the mechanism designer or the individuals themselves are not likely to know, at least not very precisely. For example, although the literature on Bayesian mechanism design arose to handle the case where individuals do not know one another's preferences, much of it requires that they and the mechanism designer have common knowledge of the prior probability distribution from which these preferences are drawn. This implies, in particular, that if individual A somehow learns B's preferences, then he will also know what B beliefs about A's preferences are — an implication that, in practice, is often implausible.

Dirk Bergemann and Stephen Morris have made a major contribution to mechanism design by developing the idea that, when constructing a mechanism, we should carefully model an individual's type *space*: not only his possible preferences, but also his possible *beliefs* (about others' preferences, about others' beliefs about *his* preferences, etc.). Furthermore, we should ordinarily take this type space to be *larger* (allowing for a greater variety of beliefs) than is conventionally assumed. By doing so, we make the mechanism more *robust* than those of the standard literature.

I am delighted that this volume in the Economic Theory series gathers twelve important papers that Bergemann and Morris have written with each other and with other authors on robust mechanism design. The volume also provides a detailed and beautifully lucid introduction to their work through the particular example of how to allocate a single indivisible good.

Eric Maskin
Editor-in-Chief
World Scientific Series in Economic Theory

Acknowledgments

We would like to thank Eric Maskin, for inviting us to present the material in this series edited by him, for valuable conversations about the material and lastly for the role that his work on efficient auctions, Maskin (1992) and Dasgupta and Maskin (2000), played in the development of the present work. We would also like to thank co-authors Hanming Fang, Takahashi Kunimoto, Moritz Meyer-ter-Vehn, Karl Schlag and Olivier Tercieux for agreeing to have their joint work with us included in this volume and Andreas Blume, Tilman Börgers, Jacques Cremer, Moritz Meyer-ter-Vehn, Andy Postlewaite, Phil Reny and Olivier Tercieux for comments on this introduction. We had the opportunity to present the material in this introduction at a number of invited lectures, notably at Boston University, Northwestern University and the European and North American Econometric Society Meetings and a set of slides which cover and accompany this introduction can be found at http://dirkbergemann.commons.yale.edu/files/2010/12/robustmechanism design1.pdf.

Contents

Robust Mechanism Design: An Introduction

Dirk Bergemann and Stephen Morris

1 Introduction

This volume brings together a number of contributions on the theme of robust mechanism design and robust implementation that we have been working on in the past decade. This work examines the implications of relaxing the strong informational assumptions that drive much of the mechanism design literature. It collects joint work of the two of us with each other and with coauthors Hanming Fang, Moritz Meyer-ter-Vehn, Karl Schlag and Olivier Tercieux. We view our work with these co-authors as thematically closely linked to the work of the two of us included in this volume.

The objective of this introductory essay is to provide the reader with an overview of the research agenda pursued in the collected papers. The introduction selectively presents the main results of the papers, and attempts to illustrate many of them in terms of a common and canonical example, the single unit auction with interdependent values. It is our hope that the use of this example facilitates the presentation of the results and that it brings the main insights within the context of an important economic mechanism, the generalized second price auction. In addition, we include an extended discussion about the role of alternative assumptions about type spaces in our work and the literature, in order to explain the common logic of the informational robustness approach that unifies the work in this volume.

The mechanism design literature of the last thirty years has been a huge success on a number of different levels. There is a beautiful theoretical literature that has shown how a wide range of institutional design questions can be formally posed as mechanism design problems with a common structure. Elegant characterizations of optimal mechanisms have been obtained. Market design has become more important in many economic arenas, both because of new insights from theory and developments in information and computing technologies, which enable the implementation of large scale trading mechanisms. A very successful econometric literature has tested auction theory in practise.

However, there has been an unfortunate disconnect between the general theory and the applications/empirical work: mechanisms that work in theory or are optimal in some class of mechanisms often turn out to be too complicated to be used in practise. Practitioners have then often been led to argue in favor of using simpler but apparently sub-optimal mechanisms. It has been argued that the optimal mechanisms are not "robust" — i.e., they are too sensitive to fine details of the specified environment that will not be available to the designer in practise. These concerns were present at the creation of the theory and continue to be widespread today.[1] In response to the concerns, researchers have developed many attractive and influential results by imposing (in a somewhat ad hoc way) stronger solution concepts and/or simpler mechanisms motivated by robustness considerations. Our starting point is the influential concern of Wilson (1987) regarding the robustness of the game theoretic analysis to the common knowledge assumptions:

> "Game theory has a great advantage in explicitly analyzing the consequences of trading rules that presumably are really common knowledge; it is deficient to the extent it assumes other features to be common knowledge, such as one agent's probability assessment about another's preferences or information."
>
> "I foresee the progress of game theory as depending on successive reductions in the base of common knowledge required to conduct useful analyses of practical problems. Only by repeated weakening of common knowledge assumptions will the theory approximate reality."

[1]Hurwicz (1972) discussed the need for "non-parametric" mechanisms wich are independent of the distributional assumptions regarding the willingness-to-pay of the agents. Wilson (1985) states that trading rules should be "belief-free" by requiring that they "should not rely on features of the agents' common knowledge, such as their probability assessments." Dasgupta and Maskin (2000) seek "detail-free" auction rules "that are independent of the details — such as functional forms or distribution of signals — of any particular application and that work well in a broad range of circumstances."

Wilson emphasized that as analysts we are tempted to assume that too much information is common knowledge among the agents, and suggested that more robust conclusions would arise if researchers were able to relax those common knowledge assumptions. Harsanyi (1967–1968) had the original insight that relaxing common knowledge assumptions is equivalent to working with a type space which is larger if there is less common knowledge. A natural theoretical question then is to ask whether it is possible to explicitly model the robustness considerations in such a way that stronger solution concepts and/or simpler mechanisms emerge endogenously. In other words, if the optimal solution to the planner's problem is too complicated or too sensitive to be used in practice, it is presumably because the original description of the planner's problem was itself flawed. We would like to investigate if improved modelling of the planner's problem endogenously generates the "robust" features of mechanisms that researchers have been tempted to assume. Our research agenda in robust mechanism design is therefore to *first* make explicit the implicit common knowledge assumptions and then *second* to weaken them.

Thus, formally, our approach suggests asking what happens to the conventional insights in the theory of mechanism design when confronted with larger and richer type spaces with weaker requirements regarding the common knowledge between the designer and the agents. In this respect, a very important contribution is due to Neeman (2004) who showed that the small type space assumption is of special importance for the full surplus extraction results, as in Myerson (1981) and Cremer and McLean (1988). In particular, he showed that the full surplus extraction results fail to hold if agents' private information doesn't display a one-to-one relationship between each agent's beliefs about the other agents and his preferences (valuation). The extended dimensionality relative to the standard model essentially allows for a richer set of higher order beliefs.

Similarly, the analysis of the first price auction in Chapter 8 (by Hanming Fang and Morris) looks at the role of richer type spaces by allowing private values but multidimensional types. There, each bidder observes his own private valuation and a noisy signal of his opponent's private valuation. This model of private information stands in stark contrast to the standard analysis of auctions with private values, where each agent's belief about his competitor is simply assumed to coincide with the common prior. In the presence of the multidimensional private signal, it is established in Chapter 8 that the celebrated revenue equivalence result between the first and the second price auction fails to hold. With the richer type space, it is not even possible to rank the auction format with respect to their expected revenue.

2 Leading Example: Allocating a Private Good with Interdependent Values

It is the objective of this introduction to present the main themes and results of our research on robust mechanism design through a prominent example, namely the efficient allocation of a single object among a group of agents. We are considering the following classic single good allocation problem with interdependent values. There are I agents. Each agent i has a "payoff type" $\theta_i \in \Theta_i = [0,1]$. Write $\Theta = \Theta_1 \times \cdots \times \Theta_I$. Each agent i has a quasi-linear utility function and attaches monetary value $v_i : \Theta \to \mathbb{R}$ to getting the object, where the valuation function v_i has the following linear form:

$$v_i(\theta) = \theta_i + \gamma \sum_{j \neq i} \theta_j .$$

The parameter γ is a measure of the interdependence in the valuations. If $\gamma = 0$, then we have the classic private values case. If $\gamma > 0$, we have positive interdependence in values, if $\gamma < 0$, we have negative interdependence. If $\gamma = 1$, then we have a model of common values.

In this setting, a social choice function must specify the allocation of the object and the (expected) payments that agents make as a function of the payoff type profile. Thus a social choice function f can be written as $f(\theta) = (q(\theta), y(\theta))$ where the allocation rule determines the probability $q_i(\theta)$ that agent i gets the object if the type profile is θ, with $q(\theta) = (q_1(\theta), \ldots, q_I(\theta))$; and transfer function, $y(\theta) = (y_1(\theta), \ldots, y_I(\theta))$, where $y_i(\theta)$ determines the payment that agent i makes to the planner.

If $\gamma < 1$, then the socially efficient allocation is to give the object to an agent with the highest payoff type θ_i. Thus *an* efficient allocation rule is given by:

$$q_i^*(\theta) = \begin{cases} \dfrac{1}{\#\{j : \theta_j \geq \theta_k \text{ for all } k\}}, & \text{if } \theta_i \geq \theta_k \text{ for all } k; \\ 0, & \text{otherwise;} \end{cases}$$

The specific form of the tie-breaking rule, here simply assumed to be uniform by construction of $q_i^*(\theta)$, is without importance. If $\gamma = 1$, there are common values and all allocations are efficient, but the above q^* continues to form an efficient allocation rule. While the papers in this volume deal with general allocation problems — and in particular, are not restricted to quasi-linear environments — this introduction will survey our results using this example and focussing on this efficient allocation rule.

Now let us consider mechanisms for allocating the object. Suppose for the moment that we were in a private value environment, i.e., $\gamma = 0$. Then a well-known mechanism to achieve the efficient allocation is the second price sealed bid auction. Here, each player i announces a "bid" $b_i \in [0, 1]$, and the object is allocated to the highest bidder who pays the second highest bid. Each agent has a dominant strategy to bid his true payoff type θ_i and the object is allocated efficiently. The second price sealed bid mechanism is a specific instance of a Vickrey-Clarke-Groves mechanism which are known to achieve efficiency and incentive compatibility in dominant strategies for a large class of allocation problems in private value environments with quasi-linear utility.

Maskin (1992) introduced a suitable generalization of the Vickrey-Clarke-Groves mechanism to an environment with interdependent values. With interdependence, that is for $\gamma \neq 0$, the "generalized" Vickrey-Clarke-Groves mechanism asks each agent i to report, or "bid" $b_i \in [0, 1]$, but now the rule of the "generalized" second price sealed bid auction is that agent i with the highest report or "bid" wins, and pays the second highest bid plus γ times the bid of others:

$$y_i^*(b) = \left(\max_{j \neq i}\{b_j\} + \gamma \sum_{j \neq i} b_j \right) q_i^*(b).$$

We observe that if $\gamma = 0$, then the payment rule of the "generalized" second price sealed bid auction reduces to the familiar rule of the second price sealed bid auction. If agents bid "truthfully," setting their bid b_i equal to their payoff type θ_i, then the generalized second price auction leads to the realization of the social choice function $f^*(\theta) = (q^*(\theta), y^*(\theta))$.

As long as parameter of interdependence is $\gamma \leq 1$, ensuring that a single crossing property is satisfied, this social choice function is "ex post incentive compatible". That is, if an agent expected other agents to report their types truthfully, he has an incentive to report his type truthfully. Conditioning on truthtelling by the other agents, the utility of a winning bidder who tells the truth is

$$\left(\theta_i + \gamma \sum_{j \neq i} \theta_j\right) - \left(\max_{j \neq i}\{\theta_j\} + \gamma \sum_{j \neq i} \theta_j\right).$$

This expression is greater than 0 if $\theta_i > \max_{j \neq i}\{\theta_j\}$ and less than 0 if $\theta_i < \max_{j \neq i}\{\theta_j\}$.

We observe that the winning bidder cannot affect the transfer through this report; this is the VCG aspect of the generalized second price auction. Now if his payoff type is larger than the payoff type of everybody else, he would like to win the object, and thus he cannot do better than bid his true value. On the other hand, if agent i's payoff type is lower than the highest payoff type among the remaining bidders, then he would have to report a higher type to receive object, but as $\theta_i < \max_{j \neq i}\{\theta_j\}$, the resulting net utility for bidder i would be negative. If $\theta_i = \max_{j \neq i}\{\theta_j\}$, the agent would be indifferent between winning the object or not. We have thus established that the efficient allocation is implemented with ex post incentive compatibility conditions. Thus the generalized second price auction ensured that, for all beliefs and higher order beliefs, there is an equilibrium that leads to the efficient allocation.

This mechanism is "robust" in the sense that as long as there is common knowledge of the environment and payoffs as we described them, there will be an equilibrium where the efficient allocation rule is followed whatever the beliefs and higher order beliefs of the agents about the payoff types of the other agents. Ex post incentive compatibility is clearly *sufficient* for "partial robust implementation," i.e., the existence of a mechanism with the property that, whatever agents' beliefs and higher order beliefs, there is an equilibrium giving rise to the efficient allocation. In chapter 1, we study when the existence of an ex post incentive compatible direct mechanism is *necessary* for partial robust implementation. But formalizing this question is delicate, and has been the subject of some confusion in the literature. In the next section, we will discuss how the language of type spaces can be used to formalize this and other questions and to highlight some subtleties in the formalization.

3 Type Spaces

We will be interested in situations where there is common knowledge of the structure of the environment described in the previous section, but the planner may not know much about each agent's beliefs or higher order beliefs about other agents' types. Thus rather than making the usual "Bayesian" assumption that the planner knows some true common prior over $\Theta = \Theta_1 \times \cdots \times \Theta_I$, we want to be able to capture the planner's uncertainty about agents' types, and what each agent believes about other agents' types, by allowing richer type spaces.

It is important to study type spaces that are richer than Θ, because we want to allow for the possibility that two types of an agent may be identical from a payoff type perspective, but have different beliefs about, say, the payoff types of other agents. In addition, we want to allow for interim type spaces, where there are no restrictions on a type's interim belief about other agents' types. Requiring that types' interim beliefs be derived from some prior probability distribution on the type space, in other words that the type space constitutes a common prior type space, will then represent an important special case. In what follows, we will focus on finite type spaces but our results readily extend to infinite type spaces and some of the chapters in this book explicitly consider such infinite type spaces.

Agent i's *type* is $t_i \in T_i$. A type of agent i must include a description of his *payoff type*. Thus there is a function

$$\widehat{\theta}_i : T_i \to \Theta_i,$$

with $\widehat{\theta}_i(t_i)$ being agent i's payoff type when his type is t_i. A type of agent i must also include a description of his beliefs about the types of the other agents. Writing $\Delta(Z)$ for the space of probability distributions on Z, there is a function

$$\widehat{\pi}_i : T_i \to \Delta(T_{-i}),$$

with $\widehat{\pi}_i(t_i)$ being agent i's *belief type* when his type is t_i. Thus $\widehat{\pi}_i(t_i)[E]$ is the probability that type t_i of agent i assigns to other agents' types, t_{-i}, being an element of $E \subseteq T_{-i}$. We will abuse notation slightly by writing $\widehat{\pi}_i(t_i)[t_{-i}]$ for the probability that type t_i of agent i assigns to other agents having types t_{-i}. Now a *type space* is a collection

$$\mathcal{T} = (T_i, \widehat{\theta}_i, \widehat{\pi}_i)_{i=1}^{I}.$$

The standard approach in the mechanism design literature is to assume a common knowledge prior, $p \in \Delta(\Theta)$, on the set of payoff types Θ. This standard approach can be modelled in our language by identifying the set of types T_i with the payoff types Θ_i and defining beliefs by

$$\widehat{\pi}_i(\theta_i)[\theta_{-i}] \triangleq \frac{p(\theta_i, \theta_{-i})}{\sum_{\theta'_{-i} \in \Theta_{-i}} p(\theta_i, \theta'_{-i})}.$$

It is useful to distinguish two distinct, critical and strong, assumptions embedded in the standard approach. First, it is assumed that there is a unique belief type associated with each payoff type. More precisely, we will say that a type space $\mathcal{T}$ is a *payoff type space* if each $\widehat{\theta}_i$ is a bijection, so that the set of possible types is identified with the set of payoff types. While often motivated by analytic convenience, when maintained in particular applications, this assumption is often strong and unjustified. This assumption need not be paired with the common prior assumption, but it often is. Type space $\mathcal{T}$ is a *common prior type space* if there exists $\pi \in \Delta(T)$ such that

$$\sum_{t_{-i} \in T_{-i}} \pi(t_i, t_{-i}) > 0 \quad \text{for all } i \text{ and } t_i,$$

and

$$\widehat{\pi}_i(t_i)[t_{-i}] = \frac{\pi(t_i, t_{-i})}{\sum_{t'_{-i} \in T_{-i}} \pi(t_i, t'_{-i})}.$$

Thus the standard approach consists of requiring both that $\mathcal{T}$ is a payoff type space and that $\mathcal{T}$ is a common prior type space. We can think of this as the smallest type space that is used in the Bayesian analysis that embeds the payoff environment described above. The standard approach makes strong common knowledge assumptions of the type that Wilson (1987) and others have argued should be expunged from mechanism design. For example, a well known implication of the standard approach is that if the common prior p is picked generically (under Lebesgue measure), the seller is able to fully extract the agents' surplus (Myerson (1981) and Cremer and McLean (1988)). While the insight that correlation in agents' types can be exploited seems to be an economically important one, it is clear that full surplus extraction is not something which can be carried out in practise. While a number of assumptions are underlying the model of full surplus extraction,[2] Neeman (2004) highlights the role of the implausible assumption that "beliefs determine preferences" (BDP), i.e., that there is a common knowledge of a mapping that identifies a unique possible valuation associated with any given belief over others' types. The innocuous looking "genericity" assumption obtains its bite by being combined with the strong common knowledge assumptions entailed by the payoff type space restriction.

To illustrate the role of richer types spaces, let us consider an example from Chapter 8, Fang and Morris (2006). Suppose there are two agents whose valuations of the object are either low (v_l) or high (v_h), with each valuation equally likely. In addition, each agent observes a low (l) or high (h) signal which reflects the other agent's valuation with probability $q \geq \frac{1}{2}$. This situation is modelled in the language of this essay by setting $I = 2$; $\Theta_i = \{v_l, v_h\}$; $T_i = \{v_l, v_h\} \times \{l, h\}$; $\widehat{\theta}_i(\theta_i, s_i) = \theta_i$; writing $s_j \simeq \theta_i$ if $(\theta_i, s_j) = (v_l, l)$ or (v_h, h),

$$\widehat{\pi}_i((\theta_i, s_i))[(\theta_j, s_j)] = \begin{cases} q^2, & \text{if } s_i \simeq \theta_j \text{ and } s_j \simeq \theta_i; \\ q(1-q), & \text{if } s_i \simeq \theta_j \text{ but not } s_j \simeq \theta_i; \\ q(1-q), & \text{if } s_j \simeq \theta_i \text{ but not } s_i \simeq \theta_j; \\ (1-q)^2, & \text{if neither } s_i \simeq \theta_j \text{ nor } s_j \simeq \theta_i. \end{cases}$$

In this type space, there are independent private values as represented by the payoff types, but there are multidimensional types. The BDP property ("beliefs determine preferences") fails because an agent's beliefs about others' types depend only on his signal and thus reveal no information about his valuation.

At the other extreme from the payoff type space is the largest type space embedding the payoff relevant environment described above which places no restrictions on agents' beliefs or higher order beliefs about other agents' payoff types, allowing for any beliefs and higher order beliefs about payoff types. This is the universal type space of Harsanyi (1967–1968) and Mertens and Zamir (1985), allowing players to hold all possible beliefs and higher order beliefs about others' payoff types.[3] For much of this

---

[2]Robert (1991), Laffont and Martimort (2000) and Peters (2001) highlight the importance of risk neutrality and unlimitied liability, absence of collusion and absence of competition, respectively.

[3]The universal space is an infinite type space, so the language in this section must be extended appropriately to incorporate it. In the exposition here, we maintain common

book, we will study a number of classic mechanism problems allowing for all possible beliefs and higher order beliefs or, equivalently, the universal space.[4] By re-working key results in the literature under this admittedly extreme assumption, we hope to highlight the importance of informational robustness.

However, we believe that the future of work on robust mechanism design will consist of exploring type spaces which are intermediate between payoff type spaces and the universal type space. Such intermediate type spaces embody intermediate common knowledge assumptions about higher order beliefs. In the remainder of this section, we discuss examples of intermediate type spaces that are discussed in this book and in the literature.

In some strands of the implementation literature, it is explicitly or implicitly assumed that there is a true prior p over the payoff types which is common knowledge among the agents, but which the planner does not know. The complete information implementation literature can be subsumed in this specification. We can represent this as follows. The type space is $T_i \triangleq \Delta(\Theta) \times \Theta_i$, with a typical element $t_i = (p_i, \theta_i)$. The payoff type is defined in the natural way, $\widehat{\theta}_i(p_i, \theta_i) \triangleq \theta_i$. The belief type is defined on the assumption that there is common knowledge of the true prior among the agents:

$$\widehat{\pi}_i(p_i, \theta_i)[(p_j, \theta_j)_{j \neq i}] \triangleq \begin{cases} p_i(\theta_{-i}|\theta_i), & \text{if } p_j = p_i \text{ for all } j \neq i; \\ 0, & \text{otherwise.} \end{cases}$$

hoi and Kim (1999) is a representative example of a contribution that explicitly works with this class of type space in an *incomplete information* setting. Choi and Kim (1999) is also discussed in Chapter 1, where we show that in a quasi-linear environments with budget balance and two agents, we can always partially implement allocation rules on the above type spaces, even though it is not possible to partially implement on all type spaces.

certainty that each agent is certain of his own payoff type and that preferences are pinned down by a profile of payoff types. These assumptions are not present in the standard settings where universal type spaces are developed. But the standard construction can be straightforwardly adapted to incorporate these assumptions — see, e.g., the discussion in Section 2.5 of Chapter 1 and Heifetz and Neeman (2006).

[4]As discussed in Section 2.5 of Chapter 1, there is a small gap between the union of all possible type spaces and the universal space that arises from "redundant" types. We will ignore this distinction for purposes of this introductory essay.

A second classic intermediate type space is the common prior universal type space. In the universal type space, there is no requirement that agents' beliefs be derived from some common prior. But it makes sense to discuss the subset of the universal type space where a common prior assumption holds. As described formally above, a type space is a common prior type space if there is a probability measure on the type space such that the players' beliefs over other players' types are conditional beliefs under that common prior. The common prior universal type space embeds all such common prior type spaces. In this book, the results on partial implementation (in Chapter 1) do not depend on whether the common prior assumption is imposed or not, but results of full implementation (in Chapters 3, 4 and 7) do. Chapter 5 examines the implications for robust full implementation of restricting attention to common prior type spaces: the results are unchanged if there are strategic complementarities in the direct mechanism (which is true under negative interdependence in preferences, i.e., $\gamma < 0$ in the single good example), but are drastically changed if there are strategic substitutes (which happens with positive interdependence, i.e., $\gamma > 0$ in the single good example).

A third natural class of models to study is when many but not all beliefs are consistent with a given payoff type. In particular, we can assume that there is a benchmark belief corresponding to each payoff type and his true belief must be within a small neighborhood of that benchmark belief. More generally, suppose that if agent i is type θ_i, then his beliefs over the payoff types of others are contained in a set $\Psi_i(\theta_i) \subseteq \Delta(\Theta_{-i})$. A local robustness condition is the requirement that there is common knowledge that all types of all agents have beliefs over others' payoff types within such set. Thus we fix, for each agent i, $\Psi_i : \Theta_i \to 2^{\Delta(\Theta_{-i})}/\varnothing$. Now, in any type space, an agent's beliefs over others' payoff types are implicitly defined and by writing $\psi_i(t_i)$ for those beliefs, we have that:

$$\psi_i(t_i)[\theta_{-i}] = \sum_{\{t_{-i}:\widehat{\theta}_{-i}(t_{-i})=\theta_{-i}\}} \widehat{\pi}_i(t_i)[t_{-i}].$$

Now suppose we restrict attention to type spaces with the property that $\psi_i(t_i) \in \Psi_i(\widehat{\theta}_i(t_i))$ for all agents i and types t_i. If we require each payoff type to have only a single possible belief about others' payoff types (i.e., each $\Psi_i(\theta_i)$ is a singleton), this reduces to the payoff type restriction above. If we put no restrictions on beliefs (i.e., each $\Psi_i(\theta_i) = \Delta(\Theta_{-i})$), then we have the universal type space. A natural "local robustness" approach is to allow Ψ_i to consist of a benchmark belief and a small set of beliefs which are close and versions of this approach have been pursued in a number of settings. Lopomo, Rigotti, and Shannon (2009) and Jehiel, Meyer-Ter-Vehn, and Moldovanu (2010) examine local robust implementation of social choice functions. Artemov, Kunimoto, and Serrano (2010) examine locally robust (full) virtual implementation of social choice functions. In Chapter 7, we report on the effect of local robustness considerations in our work on robust virtual implementation.

These three classes of restrictions are merely representative. Other results in the literature can be understood as reflecting intermediate classes of type spaces in between payoff type spaces and the universal type space. Gizatulina and Hellwig (2010) consider all type spaces with the restriction that agents are *informationally small* in the sense of McLean and Postlewaite (2002); they show that notwithstanding a failure of the BDP property highlighted by Neeman (2004), it is possible to extract almost the full surplus in quasilinear environments. We follow Ledyard (1979) in restricting attention to *full support type spaces* in Chapter 1 (Section 4.2).

Other results in the literature can be understood as allowing richer types spaces, by allowing payoff perturbations outside the payoff type environment, but then imposing restrictions on beliefs and higher order beliefs about (perturbed) payoffs types. Type spaces which maintain *approximate common knowledge* of benchmark type spaces are studied by Chung and Ely (2003) and Aghion, Fudenberg, Holden, Kunimoto, and Tercieux (2009) as well as in Chapter 9. Oury and Tercieux (2011) can be interpreted as a study of type spaces which are *close in the product topology* to some set of benchmark type spaces. Allowing perturbations outside the specified payoff type environment are important in these results.

A final class of restrictions imposed on type spaces are those labelled "generic". As noted above, a classic argument that full surplus extraction is possible on finite type spaces relies on a restriction to "generic" common priors to ensure that the "beliefs determine preferences" property holds (Cremer and McLean (1988) and Neeman (2004)). Here, genericity is applied to finite payoff type spaces (McAfee and Reny (1992) report an extension to infinite payoff type spaces). Since the payoff type space restriction entails such strong common knowledge assumptions and the BDP property seems unnatural it is interesting to ask if the BDP property holds generically for richer type spaces. It is important to note first of all that the property will fail dramatically if we look at the (payoff type) universal type space: by construction, every combination of payoff type and beliefs about others' types are possible, and thus BDP fails. Therefore a small literature has examined whether BDP holds if we look at the common prior universal type space (the full surplus extraction question is not well posed without the common prior assumption). Unfortunately, there is no agreement or naturally compelling definition of "typical" or "generic" properties in infinite type spaces. Bergemann and Morris (2001) noted that among the (infinite) space of all finite common prior types within the universal type space, one can always perturb a BDP type by a small amount in the product topology and get a non-BDP type and conversely perturb the non-BDP type by a small amount to get back to a BDP type. For topological notions of genericity, answers depend on the topology adopted and the topological definition of genericity employed (see results in Dekel, Fudenberg, and Morris (2006), Barelli (2009), Chen and Xiong (2010), Chen and Xiong (2011) and Gizatulina and Hellwig (2011)).[5] Heifetz and Neeman (2006) report an approach based on alternative geometric and generalized measure theoretic views of genericity for infinite state spaces. We do not consider restrictions based on "genericity" notions in this book. The work on genericity is important but complements rather than substitutes for work which highlights transparently the implicit common knowledge assumptions built into type spaces (such as the BDP property) and judges the relevance of the type spaces for economic analysis based on the plausibility and relevance of those assumptions directly.

We conclude this section by emphasizing that the "payoff type" framework described above is not without loss of generality. In particular, it is assumed that all agents' utility depends only on a vector of payoff types with the property that each element of the vector is known by each agent. Put differently, it is assumed that the join of agents' information fully determines all agents' preferences. This assumption is natural for private value environments and captures important interdependent value environments, but it is restrictive. To see this, consider the single good environment where each agent i's valuation of the object is given by

$$v_i = \theta_i + \gamma \sum_{j \neq i} \theta_j. \tag{1}$$

[5]But see Chen and Xiong (2011) for a problem in the analysis of Barelli (2009).

We maintain common knowledge that each agent i knows his own payoff type θ_i. What is the content of this assumption? Summing (1) across agents gives

$$\sum_{i=1}^{I} v_i = (1 + \gamma(I-1)) \sum_{i=1}^{I} \theta_i,$$

and re-arranging then gives:

$$\sum_{j \neq i} \theta_j = \left(\frac{1}{(1 + \gamma(I-1))} \sum_{i=1}^{I} v_i \right) - \theta_i. \tag{2}$$

Substituting (2) into (1) gives

$$\begin{aligned} \theta_i &= v_i - \gamma \sum_{j \neq i} \theta_j \\ &= v_i - \gamma \left(\left(\frac{1}{(1 + \gamma(I-1))} \sum_{i=1}^{I} v_i \right) - \theta_i \right), \end{aligned}$$

which implies

$$\begin{aligned} \theta_i &= \frac{1}{1-\gamma} \left(v_i - \frac{\gamma}{(1 + \gamma(I-1))} \sum_{i=1}^{I} v_i \right) \\ &= \frac{1}{1-\gamma} \left(\left(1 - \frac{\gamma}{(1 + \gamma(I-1))} \right) v_i + \frac{\gamma}{(1 + \gamma(I-1))} \sum_{j \neq i} v_j \right). \end{aligned} \tag{3}$$

Thus common knowledge of the payoff type environment implicitly entails the extreme sounding assumption that there is common knowledge that each agent i knows a particular linear combination of the agents' values, as expressed in (3).

We nonetheless maintain the payoff type environment throughout the work in this book because we are focussed on classical questions about implementing social choice functions (and correspondences) which would be impossible if knowing the join of agents' information is not sufficient to implement the social choice functions. In Bergemann, Morris, and Takahashi (2010), we introduce a language for characterizing interdependent types in terms of revealed preference in strategic settings. This richer language can be used to explore settings beyond the payoff type environment.

4 Robust Foundations for Dominant and Ex Post Incentive Compatibility

In chapter 1 (Bergemann and Morris (2005)), we ask whether a planner can design a mechanism with the property that for any belief and higher order beliefs that the agents may have, there exists a Bayesian equilibrium of the corresponding incomplete information game where an acceptable outcome is chosen. If we can find such a mechanism, then we say that we have a solution to the robust mechanism design problem. The construction of an ex post incentive compatible mechanism that delivers an acceptable outcome is clearly sufficient, but is it necessary? We call this the ex post equivalence question.

In the special case of private values, ex post incentive compatibility reduces to dominant strategies incentive compatibility. There has been an extended debate, going back to the very beginnings of the development of mechanism design, about whether dominant strategies incentive compatibility should be required or whether Bayesian incentive compatibility is sufficient. Scholars have long pointed out that — as a practical matter — the planner was unlikely to know the "true prior" over the type space. Therefore, it would be desirable to have a mechanism which was going to work independent of the prior. For a private value environment, Dasgupta, Hammond, and Maskin (1979), Ledyard (1978) and Ledyard (1979) observed that if a direct mechanism was going to implement a social choice correspondence for every prior on a fixed type space, then there must be dominant strategies implementation. Other scholars pointed out that if the planner did not know the prior (and the agents do) then we should not restrict attention to direct mechanisms. Rather, we should allow the mechanism to elicit reports of the true prior from the agents. After all, since this information is *non-exclusive* in the sense of Postlewaite and Schmeidler (1986), this elicitation will not lead to any incentive problems. A formal application of this folk argument appears in the work of Choi and Kim (1999).

Chapter 1 provides a resolution of this debate by carefully formalizing — using the type space language above — what is and is not being assumed about what is common knowledge about beliefs. This leads to a more nuanced answer to the prior debate about the necessity of dominant strategies incentive compatibility, as well as the extension to an environment with interdependent values. In particular, we show that under some circumstances, even if the planner is able to let the mechanism depend on the agents' beliefs and higher order beliefs (and thus elicit any knowledge that agents may have about priors on a fixed type space), it is still true that ex post incentive compatibility is necessary for Bayesian implementation for all possible beliefs. This is true if the planner is trying to implement a social choice correspondence which is "separable," a property that is automatically satisfied by social choice *functions.* But for some multi-valued social choice correspondences, it is impossible to identify an ex post incentive compatible selection from a social choice correspondence; but nonetheless, it is possible to find a mechanism with an acceptable equilibrium on any type space. We can illustrate both of these points with the single good allocation example.

Let us first consider the case of a social choice function $f(\theta) = (q(\theta),\ y(\theta))$ specifying the allocation and transfers in our single good environment. For a given (large) type space $\mathcal{T}$ and a given social choice function f, interim incentive compatibility on a type space $\mathcal{T}$ requires that:

$$\sum_{t_{-i}\in T_{-i}}\left[\left(\widehat{\theta}_i(t_i)+\gamma\sum_{j\neq i}\widehat{\theta}_j(t_j)\right)q_i(\widehat{\theta}(t))+y_i(\widehat{\theta}(t))\right]\widehat{\pi}_i(t_i)[t_{-i}]$$

$$\geq \sum_{t_{-i}\in T_{-i}}\left[\left(\widehat{\theta}_i(t_i)+\gamma\sum_{j\neq i}\widehat{\theta}_j(t_j)\right)q_i(\widehat{\theta}(t_i',t_{-i}))\right.$$

$$\left.+\ y_i(\widehat{\theta}(t_i',t_{-i}))\right]\widehat{\pi}_i(t_i)[t_{-i}]$$

for all i, $t\in T$ and $t_i'\in T_i$.

We refer here to "interim" rather than "Bayesian" incentive compatibility to emphasize that the beliefs of agent i, $\widehat{\pi}_i(t_i)[t_{-i}]$, are interim beliefs (without the necessity of a common prior). Now, intuitively, the larger the type space of each agent, the more incentive constraints there are to satisfy, and the harder it becomes to implement a given social choice function. As we consider large type space, that is as we move from the smallest type space, the payoff type space, to the largest type space, the universal type space, the incentive problems become successively more difficult.

It is then natural to ask whether there is a "belief free" solution concept that can guarantee that a reporting strategy profile of the agents remains an equilibrium for all possible beliefs and higher order beliefs. A social choice function $f(\theta) = (q(\theta), y(\theta))$ is ex post incentive compatible if, for all i, $\theta \in \Theta$, $\theta_i' \in \Theta_i$:

$$\left(\theta_i + \gamma \sum_{j \neq i} \theta_j\right) q_i(\theta) + y_i(\theta) \geq \left(\theta_i + \gamma \sum_{j \neq i} \theta_{-j}\right) q_i(\theta_i', \theta_{-i}) + y_i(\theta_i', \theta_{-i})$$

Under "ex post incentive compatibility" each payoff type of each agent has an incentive to tell the truth *if* he expects all other agents to tell the truth (whatever his beliefs about others' payoff types). Now, given the above definitions, it is apparent that a sufficient condition for robust truthful implementation is that there exists an allocation rule as a function of agents' payoff types that is "ex post incentive compatible," i.e., in a payoff type direct mechanism, each agent has an incentive to announce his type truthfully whatever his beliefs about others' payoff types. We show in Chapter 1 that a social choice function f is interim incentive compatible on every type space $\mathcal{T}$ if and only if f is ex post incentive compatible.

The above discussion applied to social choice *functions*. Does it extend to social choice correspondences, where multiple outcomes are acceptable for the planner for any given profile of payoff types? Suppose that the planner wanted to implement an allocation rule q but did not care about transfers — i.e., the usual setting in which efficient allocations are studied. Then we would allow for more general payment rules $\widetilde{y} = (\widetilde{y}_1, \ldots, \widetilde{y}_I)$ that could depend on agents' beliefs and higher order beliefs, with each $\widetilde{y}_i : T \to \mathbb{R}$. Thus we would ask whether for a fixed allocation rule q, could we find for every type space $\mathcal{T}$ payment rules $(\widetilde{y}_1, \ldots, \widetilde{y}_I)$ such that the incentive compatibility condition

$$\sum_{t_{-i} \in T_{-i}} \left[\left(\widehat{\theta}_i(t_i) + \gamma \sum_{j \neq i} \widehat{\theta}_j(t_j)\right) q_i(\widehat{\theta}(t_i, t_{-i})) + \widetilde{y}_i(t_i, t_{-i})\right] \widehat{\pi}_i(t_i)[t_{-i}]$$

$$\geq \sum_{t_{-i} \in T_{-i}} \left[\left(\widehat{\theta}_i(t_i) + \gamma \sum_{j \neq i} \widehat{\theta}_j(t_j)\right) q_i(\widehat{\theta}(t_i', t_{-i})) + \widetilde{y}_i(t_i', t_{-i})\right] \widehat{\pi}_i(t_i)[t_{-i}],$$

for all i, $t \in T$ and $t_i' \in T_i$. By allowing the transfers to depend on the beliefs and higher order beliefs we weaken the incentive constraints.

Now, the criticism of the classical justification of dominant strategies discussed above argued that Dasgupta, Hammond, and Maskin (1979), Ledyard (1978) and Ledyard (1979) were flawed because they did not allow transfers to depend on beliefs. However, in this single good environment, it turns out that allowing transfers to depend on higher order beliefs does not help. In fact, ex post equivalence continues to hold in this environment and holds more generally in quasi-linear environments where a planner has a unique acceptable outcome (not specifying transfers) but does not care about transfers. Such a correspondence is a leading of example of what we call a "separable" correspondence.

In view of these results, the notion of ex post equilibrium may be viewed as incorporating concern for robustness to beliefs and higher-order beliefs. This "ex post equivalence" result also suggest that the robustness requirement imposes a striking simplicity on the implementing mechanism. The language of large, and larger, type spaces would suggest that we have to solve successively more difficult incentive problems. After all, as we demand robustness with respect to some or all beliefs and higher-order beliefs, the number of incentive constraints are increasing. But we make the problem more difficult, we eventually have to solve the incentive constraints at every profile θ exactly, without reference to any expectation over payoff profiles. Thus, while the incentive constraints per se are demanding, the set of constraints reduces and hence the solution becomes substantially easier to compute as we only need to verify the incentive constraints at the exact payoff type profiles θ rather than the much larger set of possible types t.

But ex post equivalence does not hold in general in the case of general correspondences. In Chapter 1, we give some abstract examples to make this point. In particular, we describe a private values example with the feature that dominant strategies implementation is impossible but interim implementation is possible on any type space, and this seems to be the first example in the literature noting this possibility. The example points to the fact that interim incentive compatibility can occur for all type spaces, using mechanisms that elicit and respond to the beliefs of the agents, even if ex post incentive compatibility is impossible. Here, let us report an interdependent values example due to Jehiel, Moldovanu, Meyer-Ter-Vehn, and Zame (2006) which makes the same point in the single good allocation problem.

Suppose now that the payoff type of agent i is given by $\theta_i = (\theta_{i1}, \theta_{i2}) \in [0,1]^2$ and that the value of the object to agent i is then

$$v_i(\theta) = \theta_{i1} + \gamma \sum_{j \neq i} \theta_{j1} + \varepsilon \prod_{j=1}^{I} \theta_{j2}. \tag{4}$$

with $\varepsilon > 0$. In the two agent case where $\gamma = 0$, this example was analyzed by Jehiel *et al.* (2006). In this case, the only ex post incentive compatible social choice functions are trivial ones where the allocation of the object is independent of all agents' types. Under the assumption that the object must always be allocated to one of the two agents, this example thus illustrates the general result of Jehiel *et al.* (2006) that in generic quasi-linear environments with interdependent values and multidimensional types, ex post implementation of non-trivial social choice functions is impossible.[6] But it is straightforward to see that an almost efficient allocation of the object can be robustly implemented, since if the object is sold by a second price auction, each agent will have an incentive to bid within ε of θ_{i1}. This observation can be extended to interdependent values if interdependence is not too large, with $0 < \gamma < \frac{1}{I-1}$; in this case, then the generalized second price auction would implement the correspondence of almost efficient allocations. We postpone until our discussion of Chapters 3 and 9 an explanation of why we need $\gamma < \frac{1}{I-1}$ and how much this argument generalizes.

An important class of economic environments where our separability condition fails are quasi-linear environments where transfers are required to be budget balanced. We show in Chapter 1 that ex post equivalence holds nonetheless in some special cases: if there are two agents (Proposition 2) or if each agent has at most two types (Proposition 6). The latter result highlights the importance of allowing rich type spaces: example 3 shows that we can have partial robust implementation on all payoff type spaces but not on the universal type space. This environment is important because it includes the classic public good problem. In general ex post equivalence fails in this environment (a detailed example was presented in a working paper version of Chapter 3, Bergemann and Morris (2005)). Thus there is not a general equivalence between dominant strategy implementation and robust implementation for public good problems.

---

[6]If there are more than two agents, or if the object is not allocated to either of two agents, then agents are assumed indifferent between outcomes which violates the genericity condition in the impossibility result of Jehiel *et al.* (2006). Bikhchandani (2006) discusses non-trivial ex post incentive compatible allocations that arise if the object need not be allocated to any agent.

See Jehiel and Moldovanu (2001) for analysis of how multidimensional types already limit the possibility of efficient in standard Bayesian settings and Eso and Maskin (2002) and Jehiel, Meyer-Ter-Vehn, and Moldovanu (2008) for more on settings with non-trivial ex post implementation with multidimensional type spaces in environments failing genericity condition of Jehiel *et al.* (2006).

One implication of our results in Chapter 1 is that we can distinguish settings where a restriction to dominant strategies equilibrium (under private values) or ex post equilibrium in mechanism design problems can or cannot be justified by informational robustness arguments. Thus Dasgupta and Maskin (2000) and Perry and Reny (2002) use ex post equilibrium as a solution concept in studying efficient auctions with interdependent values. This is equivalent to robust partial implementation.

Our analysis in Chapter 1 is limited to asking whether a fixed social choice correspondence — mapping payoff type profiles to sets of possible allocations — can or cannot be robustly partially implemented. Thus we focus on a "yes or no" question. Many of the most interesting questions involve asking what happens when we consider what is the best mechanism for the universal type space when we are interested in a finer objective, and a number of recent papers have addressed this question. Chung and Ely (2007) consider the objective of revenue maximization for the seller of a single object (under the seller's beliefs about agents' valuations), allowing all possible beliefs and higher order beliefs of the agents, and show conditions under which the seller cannot do better than using a dominant strategy mechanism. The best mechanism from the point of view of the seller would generally allow many outcomes for any given profile of payoff type profiles, and will not in general be separable, and thus the results of Chapter 1 do not apply. Smith (2010) and Börgers and Smith (2011) study the classic problems of public good provision and general social choice with rich private preferences (i.e., the Gibbard-Satterthwaite question) respectively. They identify simple mechanisms that perform better than dominant strategy mechanisms — in the sense of providing weakly better outcomes on all type spaces and strictly better outcomes on some type spaces — for each of these two problems. Yamashita (2011) identifies a mechanism that performs better than any dominant strategy mechanism in the classic bilateral trading problem (the notion of robustness is different from that considered in Chapter 1 but similar results would hold with our notion of robustness). Finally, Bierbrauer and Hellwig (2011) combine the informational robustness approach studied here with a requirement that the social objective be collusion-proof and then obtain restrictions on the social choice function which satisfy both desiderata.

An interesting question for further analysis is the extent to which the results in Chapter 1 continue to hold for more local versions of robustness. Lopomo, Rigotti, and Shannon (2009) identify settings where local robust implementation of a social choice function is equivalent to ex post implementation. Jehiel, Meyer-Ter-Vehn, and Moldovanu (2010) give examples illustrating when this equivalence doesn't hold, but nonetheless show that local robust implementation is a very strong and, in particular, generically impossible with multidimensional payoff types.

5 Full Implementation

All of the above results are phrased in terms of incentive compatibility, and by use of the revelation principle, are therefore statements about the existence of a truthtelling equilibrium in the direct mechanism. The construction of the truthtelling equilibrium of course presumes that when we verify the truthtelling constraint of agent i that the other agents are telling the truth as well. This does not address — let alone exclude — the possibility of other equilibria in the direct mechanism; equilibria in which the agents are not telling the truth, and importantly, in which the social choice function is not realized.

As private information may enable the agents to coordinate behavior in many different ways, the designer has to be concerned with the fact that there may exist equilibrium behavior by the agents which does not realize his objective. The notion of *full implementation*, in contrast to *truthful* or *partial implementation*, addresses this by requiring that every equilibrium in the mechanism attains the social objective.[7]

In Chapter 2, we restrict attention to the solution concept of ex post equilibrium, and ask what conditions are required for full ex post implementation i.e., all ex post equilibria to deliver outcomes in the social choice correspondence? In Chapter 3, we will move on to ask when is it possible interim implement a social choice correspondence for all possible higher order beliefs. In general, the latter is a more stringent requirement. We say

[7]There is a large literature in economic theory — much of it building on the work of Maskin (1999) — devoted to the problem of full implementation: When is it the case that there is a mechanism such that every equilibrium in this mechanism is consistent with a given social choice correspondence? While elegant characterizations of implementability were developed, the "augmented" mechanisms required to achieve positive results were complex and seemed particularly implausible. While the possibility of multiple equilibria does seem to be a relevant one in practical mechanism design problems, particularly in the form of collusion and shill bidding, the theoretical literature so far has not developed practical insights, with a few recent exceptions such as Ausubel and Milgrom (2005) and Yokoo, Sakurai, and Matsubara (2004).

that a social choice correspondence that is interim implementable for all possible type spaces is *robustly implementable.*

5.1 *Ex Post Implementation*

Chapter 1 required — for any beliefs and higher order beliefs — *an* equilibrium that delivered the right outcome. This required ex post incentive compatibility or — equivalently — that truth-telling is an ex post equilibrium of the "direct" mechanism where agents just report their payoff types. Now, in Chapter 2, we ask: if we take ex post equilibrium as the primitive solution concept, when can we design a mechanism such that, not only does an ex post equilibrium deliver the right outcome, but also *every* ex post equilibrium delivers the right outcome. Thus there is *full* implementation under the solution concept of ex post equilibrium — and we call this *ex post implementation.* We show that — in addition to ex post incentive compatibility — an ex post monotonicity condition is necessary and almost sufficient. The ex post monotonicity condition neither implies nor is implied by Maskin monotonicity (necessary and almost sufficient for implementation under complete information). By "almost sufficient", we mean sufficient in economic environments and after an additional no veto condition also sufficient in general environments.

In a direct mechanism, such as the generalized second price auction, undesirable behavior by agent i is easiest interpreted as a *misreport* or *deception* θ'. In a direct revelation mechanism, if agents misreport θ' rather than truthfully report θ, then the resulting social outcome is given by $f(\theta')$ rather than $f(\theta)$. The notion of ex post monotonicity guarantees that *(i)* a whistle-blower (among the agents) will alert the principal of deceptive reporting θ' by receiving a reward and *(ii)* a whistle-blower will not falsely report a deception.

The social choice function $f = (q, y)$ satisfies ex post monotonicity if for every θ, θ' with $f(\theta) \neq f(\theta')$, there exist i, $\widehat{q}_i \in [0,1]$ and $\widehat{y}_i \in \mathbb{R}$ such that

$$\left(\theta_i + \gamma \sum_{j \neq i} \theta_j\right) \widehat{q}_i + \widehat{y}_i > \left(\theta_i + \gamma \sum_{j \neq i} \theta_j\right) q_i(\theta_i', \theta_{-i}') + y_i(\theta_i', \theta_{-i}'),$$

while

$$\left(\theta_i'' + \gamma \sum_{j \neq i} \theta_j'\right) q_i(\theta_i'', \theta_{-i}') + y_i(\theta_i'', \theta_{-i}') \geq \left(\theta_i'' + \gamma \sum_{j \neq i} \theta_j'\right) \widehat{q}_i + \widehat{y}_i$$

for all θ_i''.

Proposition 3 in Chapter 2 then establishes that the social choice function implied by the generalized second price auction satisfies the ex post monotonicity condition. Moreover, due to the quasi-linearity of the utility function, it also represents an economic environment, and hence can be fully implemented in an ex post equilibrium, provided there are three or more bidders. In fact, with interdependent values, or $\gamma \neq 0$, the implementation can be achieved in the direct mechanism itself and does not need to make use of an augmented mechanism. In other words, the direct mechanism is shown to have a unique ex post equilibrium if $\gamma \neq 0$. This three or more player result contrasts with the observation of Birulin (2003) that, with only two players, there are a continua of undominated ex post equilibria in the direct mechanism of the single good allocation problem.

5.2 *Robust Implementation in the Direct Mechanism*

But can the planner design a mechanism with the property that for any beliefs and higher order beliefs that the agents may have, *every* equilibrium has the property that an acceptable outcome is chosen? We call this "robust implementation" and investigate the possibility of robust implementation in Chapters 3 and 4. We should immediately emphasize that the question of robust implementation is *not* the same as the ex post implementation question analyzed in Chapter 2: to rule out bad equilibria in Chapter 2, it was enough to make sure you could not construct a "bad" ex post equilibrium; for robust implementation, we must rule out bad Bayesian, or interim, equilibria on all type spaces. In Chapter 3, we consider a well-behaved environment with payoff type spaces represented by intervals of the real line and "aggregator single crossing" preferences. In this environment, we give a "contraction property" — equivalent to not too much interdependence in types — and show that if *strict* ex post incentive compatibility and the contraction property hold, then robust implementation is possible in the direct mechanism. If either fails, robust implementation is impossible in *any* mechanism.

To describe the results in more detail we return to the (generalized) second price auction. We start with the private value environment, where it is well-known that the second price auction has many equilibria in which the agents do not tell the truth, and in consequence the allocation is not guaranteed to be efficient. The reason is that truthtelling is only a weak best response and hence just a dominant strategy, but not a strictly dominant strategy. The good news is that we can easily modify the original auction so that truthful bidding becomes a strictly dominant strategy. Fix $\varepsilon > 0$. Now, with probability $1 - \varepsilon$, let us allocate the object to the highest bidder and have him pay the second highest bid. With the complementary probability ε, let us randomly and uniformly pick an agent, and allocate the object to that agent with probability b_i, a probability that is proportional to his bid. Thus the ε-allocation rule (parameterized by ε) is defined by:

$$q_i^{**}(\theta) \triangleq (1 - \varepsilon)q_i^{*}(\theta) + \varepsilon q_i(\theta), \tag{5}$$

with

$$q_i(\theta) \triangleq \frac{\theta_i}{I}.$$

This *modified* generalized second price auction is supported by an associated set of (expected) transfers conditional on the reported type profile θ:

$$y_i^{**}(\theta) = \frac{\varepsilon}{2I}\theta_i^2 + (1 - \varepsilon)\left(\max_{j \neq i}\{\theta_j\}\right) q_i^{*}(\theta). \tag{6}$$

The transfer rule $y_i^{**}(\theta)$ supports truthtelling as an equilibrium in strictly dominant strategies, that is $b_i = \theta_i$ forms a *strictly* dominant strategy in this mechanism. The strictness is established by making the allocation responsive to the bid of agent i even if agent i is not the highest bidder. It follows that whatever agent i's beliefs or higher order beliefs about θ_{-i} are, he will have a *strictly* dominant strategy to set $b_i = \theta_i$. In our language, for any $\varepsilon > 0$, we can guarantee the *robust implementation* of the almost efficient, or $\varepsilon-$efficient allocation rule q^{**}.[8]

Now consider the case of interdependent values $\gamma \neq 0$. We can modify the generalized second price sealed bid auction to turn the *ex post* equilibrium into a strict *ex post* equilibrium, just as we modified the second price sealed bid auction. We construct the following allocation rule $q_i^{**}(\theta)$. With

[8] In related modifications of the second price auction in a private value environment, Plum (1992) considers a convex combination of a first-price and a second-price auction (with a small weight on the former) and Blume and Heidhues (2004) introduce a small reserve price in the second price auction. Either of these modifications render the equilibrium outcome unique, but in contrast to the present formulation, these modifications do not strengthen truth-telling from a weakly dominant to a strictly dominant strategy.

probability $1-\varepsilon$, we have the winning bidder i pay:

$$\max_{j\neq i}\{\theta_j\}+\gamma\sum_{j\neq i}\theta_j,$$

and with probability ε, we randomly and uniformly pick an agent, and allocate the object to that agent with probability b_i, a probability that is proportional to his bid. In the event that agent i is assigned the object, he then pays:

$$\frac{1}{2}\theta_i+\gamma\sum_{j\neq i}\theta_j.$$

Now, in this modification of the generalized second price auction, the associated transfers can be written as, in a generalization of (6):

$$y_i^{**}(\theta)=\frac{\varepsilon}{2I}\theta_i^2+\frac{\varepsilon\theta_i}{I}\left(\gamma\sum_{j\neq i}\theta_j\right)+(1-\varepsilon)\left(\max_{j\neq i}\{\theta_i\}+\gamma\sum_{j\neq i}\theta_j\right)q_i^*(\theta). \quad (7)$$

The social choice function in this modified generalized second price auction is given by a pair of allocation and transfer functions: $f^{**}(\theta)=(q^{**}(\theta),y^{**}(\theta))$. The net utility of agent i, given a true payoff profile θ and reported payoff profile θ', is explicitly given by:

$$\left(\theta_i+\gamma\sum_{j\neq i}\theta_j\right)\left(\frac{\varepsilon}{I}\theta_i'+(1-\varepsilon)q_i^*(\theta')\right)$$
$$-\frac{\varepsilon}{2I}\theta_i'^2-\frac{\varepsilon\gamma\theta_i'}{I}\sum_{j\neq i}\theta_j'-(1-\varepsilon)\left(\max_{j\neq i}\{\theta_j'\}+\gamma\sum_{j\neq i}\theta_j'\right)q_i^*(\theta').$$

The net utility function is a linear combination of the efficient allocation rule and the proportional allocation rule. It is straightforward to compute the best response of each agent i, given a point belief about the payoff type profile θ and reported profile θ'_{-i} of the remaining agents. The best response is linear in the true valuation and in the size of the misrepresentation $(\theta_j-\theta_j')$, downwards or upwards, of the other agents:

$$\theta_i'=\theta_i+\gamma\sum_{j\neq i}(\theta_j-\theta_j'). \quad (8)$$

From here, it follows that the reports of agent i and agent j are strategic substitutes if $\gamma>0$ and strategic complements if $\gamma<0$. For example, with

$\gamma > 0$, if agent j increases his report, then in response agent i optimally chooses to lower his report.

From (8), we can conclude that truthtelling indeed forms a strict ex post equilibrium. But even though we have a strict ex post incentive compatible mechanism, we cannot guarantee the robust implementation of q^{**}. In fact, we shall now show that the direct mechanism robustly implements the efficient outcome if and only if the interdependence is moderate,[9] or

$$|\gamma| < \frac{1}{I-1}.$$

Moreover, no mechanism, whether it is the direct mechanism or an augmented mechanism is able to robustly implement the efficient outcome if the interdependence is too large, or:

$$|\gamma| \geq \frac{1}{I-1}.$$

This necessary and sufficient condition for robust implementation should be compared with the necessary and sufficient condition for robust partial implementation, which we earlier showed to require the single crossing condition, namely

$$\gamma \leq 1.$$

As we analyzed truthtelling in the direct mechanism for all possible beliefs and higher-order beliefs, all we had to do was to guarantee the incentives to reveal the private information, agent by agent, while presuming truthtelling by other agents. Now, as we seek robust implementation, we cannot suppose the truthtelling behavior of the other agents but rather have to guarantee it. We shall obtain this guarantee by identifying restrictions on the rational behavior of each agent, and then use these restrictions to inductively obtain further restrictions. More formally, we shall analyze the outcome of the mechanism under rationalizability with incomplete information. An action, which in the direct mechanism, simply constitutes a reported payoff type, is called incomplete information rationalizable if it survives the

[9]The importance of this moderate interdependence condition arose earlier in the work of Chung and Ely (2001) who showed that it was sufficient for implementing the efficient outcome in the unperturbed generalized second price auction under iterated deletion of weakly dominated strategies.

process of iterative elimination of dominated strategies. As rationalizability with complete information, rationalizability under incomplete information defines an inductive process: first suppose that every payoff type θ_i could send any message m_i; then, second, delete those messages m_i that are not a best response to some conjecture over pairs of payoff type and messages (θ_{-i}, m_{-i}) of the opponents that have not yet been deleted. The inductive procedure is then to repeat the second step until convergence is achieved.

We observe that the notion of incomplete information rationalizability is belief free as the candidate action needs only to be a best response to some beliefs about the other agents actions and payoff types. We can focus on the notion of incomplete information rationalizability because of the following epistemic result: a message m_i can be sent by an agent with payoff type θ_i in an interim equilibrium on some type space if and only if m_i is "incomplete information rationalizable" for payoff type θ_i. The equivalence between robust and rationalizable implementation is an incomplete generalization of Brandenburger and Dekel (1987) and can be seen as a special case of the incomplete information results of Battigalli and Siniscalchi (2003). It illustrates a general point well-known from the literature on epistemic foundations of game theory: that equilibrium solution concepts only have bite if we make strong assumptions about type spaces, i.e., we assume small type spaces where the common prior assumption holds.

We now describe the inductive argument for rationalizability in the direct mechanism for the single-unit auction. For concreteness, we shall assume here positive interdependence, $\gamma > 0$, but all the relevant arguments go through with negative interdependence, after suitably reversing the signs. In the direct mechanism a message m_i is simply a reported payoff type θ'_i. Each agent i has some conjecture about the other agents true type profile θ_{-i} and their reported type profile θ'_{-i}. We denote such a conjecture by λ_i:

$$\lambda_i(\theta_{-i}, \theta'_{-i}) \in \Delta(\Theta_{-i} \times \Theta_{-i}).$$

We can then ask what is the set of reports that agent i might send for some conjecture $\lambda_i(\theta_{-i}, \theta'_{-i})$ over his opponents' types θ_{-i} and reports θ'_{-i} in the $k-$th step of the inductive procedure. We denote this set by $\beta_i^k(\theta_i)$. We restrict the conjectures $\lambda_i(\theta_{-i}, \theta'_{-i})$ of agent i in step k to be of the form that type θ_j can only be conjectured to send message θ'_j if it was rationalizable at step $k-1$, i.e., if $\theta'_j \in \beta_i^{k-1}(\theta_j)$.

We initialize the inductive process at step $k = 0$ by allowing all possible reports $\beta_i^0(\theta_i) = [0, 1]$. In the context of the almost efficient allocation rule $f^{**}(\theta) = (q^{**}(\theta), y^{**}(\theta))$ and the associated ex post compatible transfer $y_i^{**}(\theta)$, the expected payoff of agent i is quadratic in his report θ_i'. It follows that the best response of agent i to a probability one conjecture about his opponents true type and reported type profiles to be $(\theta_{-i}, \theta_{-i}')$, is given by the linear best response θ_i':

$$\theta_i' = \theta_i + \gamma \sum_{j \neq i} (\theta_j - \theta_j'). \tag{9}$$

Thus if he expects the other agents to underreport their type, i.e., $\theta_j - \theta_j' > 0$, then the best response of agent i is to correct this by overreporting his type. We notice that the best response has a self-correcting property. With the correction induced by the reported type θ_i' in (9), the reported valuation of agent i actually equals his true valuation:

$$\theta_i' + \gamma \sum_{j \neq i} \theta_j' = \theta_i + \gamma \sum_{j \neq i} (\theta_j - \theta_j') + \gamma \sum_{j \neq i} \theta_j' = \theta_i + \gamma \sum_{j \neq i} \theta_j.$$

The best response (9) of agent i only corrects the valuation of agent i, the other reported valuations continue to differ from the true valuations under the best response of agent i. The linear best response property then leads to a set of best responses $\beta_i^k(\theta_i)$ in step k, which can be characterized in terms of a lower and upper bound:

$$\beta_i^k(\theta_i) = [\underline{\beta}_i^k(\theta_i), \bar{\beta}_i^k(\theta_i)].$$

With the inductive procedure, the bounds $\{\underline{\beta}_i^k(\theta_i), \bar{\beta}_i^k(\theta_i)\}$ in step k are determined by the restrictions identified in round $k - 1$:

$$\{(\theta_{-i}, \theta_{-i}') : \theta_j' \in \beta_j^{k-1}(\theta_j), \forall j \neq i\}.$$

The upper bound $\bar{\beta}^k(\theta_i)$ identifies the largest rationalizable report by agent i with payoff type θ_i. It is obtained by identifying a feasible point conjecture at which the sum of underreports of the other agents, $\sum_{j \neq i} (\theta_j - \theta_j')$, is maximized:

$$\bar{\beta}^k(\theta_i) = \theta_i + \gamma \max_{\{(\theta_{-i}', \theta_{-i}) : \theta_j' \in \beta_j^{k-1}(\theta_j), \forall j \neq i\}} \left\{ \sum_{j \neq i} (\theta_j - \theta_j') \right\}.$$

The largest rationalizable report for agent i, given his payoff type θ_i, arises under the conjecture that the remaining agents maximally underreport relative to their true payoff type. But the lowest reported type of payoff type θ_j is given by the lower bound obtained in the preceding step $k-1$, and thus using the lower bound $\underline{\beta}_j^{k-1}(\theta_j)$ from step $k-1$ explicitly, we get:

$$\bar{\beta}^k(\theta_i) = \theta_i + \gamma \max_{\theta_{-i} \in \Theta_{-i}} \left\{ \sum_{j \neq i} (\theta_j - \underline{\beta}_j^{k-1}(\theta_j)) \right\}.$$

Similarly, the lowest possible report of payoff type θ_i, the "maximal" underreport, emerges from the point conjecture that the remaining agents are "maximally" overreporting relative to their true type, thus:

$$\underline{\beta}_i^k(\theta_i) = \theta_i + \gamma \max_{\theta_{-i} \in \Theta_{-i}} \left\{ \sum_{j \neq i} (\theta_j - \bar{\beta}_j^{k-1}(\theta_j)) \right\}.$$

Given the compactness of the payoff type set, in fact $\Theta_i = [0, 1]$, we obtain explicit expressions for the lower and upper bounds. In step $k = 1$, the conjectures about the other players are unrestricted, and so for every j:

$$\max_{\theta_j \in \Theta_j} (\theta_j - \underline{\beta}_j^0(\theta_j)) = 1 - 0 = 1,$$

and hence

$$\bar{\beta}^1(\theta_i) = \theta_i + \gamma(I - 1),$$

and more generally we find that in step k the upper bound is given by

$$\bar{\beta}^k(\theta_i) = \theta_i + (\gamma(I - 1))^k, \tag{10}$$

and likewise the recursion for the lower bound yields:

$$\underline{\beta}^k(\theta_i) = \theta_i - (\gamma(I - 1))^k. \tag{11}$$

We thus find that a reported payoff type θ_i', different from the true type θ_i, can be eliminated for sufficiently large k from the best response set, or

$$\theta_i' \neq \theta_i \Rightarrow \theta_i' \notin \beta^k(\theta_i),$$

provided that:

$$|\gamma|(I - 1) < 1 \Leftrightarrow |\gamma| < \frac{1}{I - 1}. \tag{12}$$

We then have a sufficient condition for robust implementation, which requires that the interdependence among the agents is only moderate in this sense of the above inequality. The next question then is whether the above sufficient condition is also a necessary condition for robust implementation. Indeed, suppose that the parameter of interdependence, γ, were larger than the inequality (12) requires, or:

$$\gamma \geq \frac{1}{I-1}.$$

We can use the richness of the possible type space $\mathcal{T}$ to identify specific types, in particular specific belief types, under which the interim expected valuations of any two payoff types θ_i and θ_i', with $\theta_i \neq \theta_i'$, are indistinguishable. Thus suppose that each payoff type θ_i is convinced, i.e., has the point conjecture that the payoff type θ_j of agent j is given by:

$$\theta_j \triangleq \frac{1}{2} + \frac{1}{\gamma(I-1)}\left(\frac{1}{2} - \theta_i\right), \quad \forall j.$$

If we now compute the interim expected value of the object for i under the above belief, we find that the interim expected value of the object for agent i is in fact independent of θ_i:

$$\theta_i + \gamma(I-1)\left(\frac{1}{2} + \frac{1}{\gamma(I-1)}\left(\frac{1}{2} - \theta_i\right)\right) = \frac{1}{2}(1 + \gamma(I-1)).$$

It then follows immediately that the payoff types cannot be distinguished in the direct mechanism, as each payoff type θ_i assigns the same expected value to the object given his private information. We say that the payoff types are *indistinguishable*, and in fact they are indistinguishable in any, direct or indirect, mechanism. We have thus established that in the single unit auction, robust implementation is possible (using the modified generalized VCG mechanism) if

$$|\gamma| < \frac{1}{I-1}, \tag{13}$$

and conversely that robust implementation is impossible (in *any* mechanism) if

$$|\gamma| \geq \frac{1}{I-1}.$$

This result has to be contrasted with robust incentive compatibility condition, namely the ex post incentive compatibility, which required (only) that $\gamma < 1$.

In Chapter 3, we generalize the property of moderate interdependence (13) and refer to it more generally as a "contraction property", as it is suggested by the contraction like property of the lower and upper bounds, (10) and (11), respectively. We assume that preferences are single crossing with respect to a one dimensional aggregator of agents' types. A "deception" specifies for each payoff type of each agent, a set of payoff types that might be misreported. Our contraction property requires that for any deception, there is at least one misreport of one type of one "whistleblowing" agent for whom the misreports of others will not reverse the sign of the impact of the whistleblower's misreport on his preferences. The robust implementation result that we established above in the context of the single unit auction can now be stated for the general environment as follows. Robust implementation is possible in the *direct* (or any augmented) mechanism if and only if strict ex post incentive compatibility and the contraction property hold.

A noteworthy aspect of the above result is that the strict separation between possibility and impossibility not only holds for the direct mechanism but for any other, possibly augmented mechanism. To wit, the literature on implementation frequently uses "augmented mechanism" to obtain sufficient conditions for implementation. Here, the robustness requirement implies that augmented mechanisms, relative to the simple mechanism in the form of the direct mechanism, lose their force. Hence, the more stringent requirements of robust implementation reduce the role of complex and overly sensitive mechanisms.

The above analysis also demonstrates that while robust implementation is a strong requirement, it is weaker than dominant strategy implementation. After all, in the environment with interdependent values, a dominant strategy equilibrium does not even exist, nonetheless truthtelling in the direct mechanism is an ex post equilibrium, and as we showed is indeed the unique incomplete information rationalizable outcome.

As we saw in the example of the single unit auction, the "contraction property" had a natural interpretation in a linear valuation environment. This interpretation remains, even in a *nonlinear utility environment*, provided that the aggregator remains linear. For example, a linear aggregator for each agent i might be of the form:

$$h_i(\theta) = \theta_i + \sum_{j \neq i} \gamma_{ij} \theta_j,$$

where each weight γ_{ij} measures the importance of payoff type j for preference of agent i. In the case of the linear aggregator, we can form an interaction matrix based on the weights γ_{ij} across all agent pairs i and j:

$$\Gamma \triangleq \begin{bmatrix} 0 & |\gamma_{12}| & \cdots & |\gamma_{1I}| \\ |\gamma_{21}| & 0 & & \vdots \\ \vdots & & \ddots & |\gamma_{I-1I}| \\ |\gamma_{I1}| & \cdots & |\gamma_{II-1}| & 0 \end{bmatrix}.$$

We can then give a generalized version of the moderate interdependence condition in terms of the interaction matrix Γ. In Chapter 3, we show that the interaction matrix has the contraction property if and only if largest eigenvalue of the interaction matrix is less than 1.

5.3 *The Robustness of Robust Implementation*

In Chapter 9, Moritz Meyer-ter-Vehn and Morris show that if there is a approximate common knowledge that we are in an environment close to a strict version of that of Chapter 3 (i.e., with one dimensional interdependent values under an aggregator function and a uniformly strict contraction property, and uniformly strict ex post incentive compatibility), then the social choice correspondence consisting of almost efficient allocations can be robustly implemented.

This result can be illustrated by the two dimensional perturbation of the single good allocation problem we discussed in Section 4 of this introduction. Thus suppose again that the payoff type of agent i is given by $\theta_i = (\theta_{i1}, \theta_{i2}) \in [0,1]^2$ and that the value of the object to agent i is, as earlier in (4):

$$v_i(\theta) = \theta_{i1} + \gamma \sum_{j \neq i} \theta_{j1} + \varepsilon \prod_{j=1}^{I} \theta_{j2}$$

with $\varepsilon > 0$ and $\gamma < \frac{1}{I-1}$. It is an implication of the lower hemicontinuity of rationalizable outcomes that in the modified generalized second price auction of chapter 3, types $(\theta_{i1}, \theta_{i2})$ will have an incentive to bid something in the neighborhood of θ_{i1}. The social choice correspondence of almost efficient allocations of the private good is therefore almost robustly (fully) implemented.

While Chapter 9 delivers a robust full implementation result — by generalizing arguments in Chapter 3 — the purpose of Chapter 9 is only to deliver a partial implementation result. This raises the question of whether it is possible to get partial robust implementation of the almost efficient allocations (without full robust implementation) without the moderate interdependence condition of $\gamma < \frac{1}{I-1}$. While the argument presented in Chapter 9 relies directly on the arguments of Chapter 3, there is an important connection between partial and full implementation identified by Oury and Tercieux (2011), which might indicate that there is a strong link between partial and full implementation. They show that requiring continuous, but partial, implementation in complete or incomplete information settings implies the necessity of full implementation.

5.4 *Robust Implementation in the General Mechanism*

Chapter 3 restricted attention to a class of well-behaved environments. In contrast, in Chapter 4, we characterize robust implementation in general environments with general mechanisms. By robust implementation we mean that *every* equilibrium on *every* type space $\mathcal{T}$ generates outcomes consistent with the social choice function f. As we seek to identify necessary and sufficient conditions for robust implementation, conceptually there are (at least) two approaches to obtain the conditions. One approach would be to simply look at the interim implementation conditions for every possible type space $\mathcal{T}$ and then try to characterize the intersection or union of these conditions for all type spaces. But in Chapter 4, we focus our analysis on a second, more elegant, approach. We first establish an equivalence between robust and *rationalizable implementation* and then derive the conditions for robust implementation as an implication of rationalizable implementation. The advantage of the second approach is that after establishing the equivalence, we do not need to argue in terms of large type spaces, but rather derive the results from a novel argument using the iterative deletion procedure associated with rationalizability. This equivalence was already used in Chapter 3, but for the arguments in Chapter 4 we allow for general (perhaps infinite and non-compact mechanisms), and thus new versions of the equivalence results must be developed.

As suggested by the analysis in the direct mechanism, ex post incentive compatibility and a robust monotonicity condition are necessary and almost sufficient for robust implementation. And, in the aggregator single crossing environment of Chapter 3, robust monotonicity is equivalent to the contraction property.

5.5 *Rationalizable Implementation*

In Chapters 3 and 4, we establish necessary and sufficient conditions for "robust implementation" in *environments with incomplete information.* In particular, we showed that a social choice function f can be interim (or Bayesian) equilibrium implemented for all possible beliefs and higher order beliefs if and only if f is implementable under an incomplete information version of rationalizability. These results prompted us to refine and further develop the rationalizability arguments in *environments with complete information.* In Chapter 10, we (together with Olivier Tercieux) establish stronger necessary and sufficient conditions than in the incomplete information environment and show that these conditions are almost equivalent to the Nash equilibrium implementation conditions when the social choice function is responsive (a social choice function is responsive if distinct states imply distinct social choices). With respect to the necessary conditions, we strengthen the monotonicity condition, due to Maskin (1999), from a weak inequality to a strict inequality.

Writing the strict Maskin monotonicity condition in the context of the single good example, we say that a social choice function f satisfies strict Maskin *monotonicity* if $f(\theta) \neq f(\theta')$ implies that for some $i, \widehat{q}_i$ and $\widehat{y}_i$,

$$\left(\theta'_i + \gamma \sum_{j \neq i} \theta'_{-j}\right) \widehat{q}_i + \widehat{y}_i > \left(\theta'_i + \gamma \sum_{j \neq i} \theta'_{-j}\right) q_i(\theta) + y_i(\theta),$$

and

$$\left(\theta_i + \gamma \sum_{j \neq i} \theta_{-j}\right) q_i(\theta) + y_i(\theta) > \left(\theta_i + \gamma \sum_{j \neq i} \theta_{-j}\right) \widehat{q}_i + \widehat{y}_i.$$

The latter condition requires that if the socially desired alternatives differ in state θ and θ', then there must exist an agent i and a reward allocation $(\widehat{q}_i, \widehat{y}_i)$ such that if the true state were θ' and agent i were to expect the other agents to claim that the state is θ, i could be offered a reward $(\widehat{q}_i, \widehat{y}_i)$ that would give him a strict incentive to "report" the deviation of the other agents, but that the reward y would not tempt him if the true state were in fact θ. The strengthening of the monotonicity condition, commonly referred to as Maskin monotonicity, is that we require that the reward y gives agent i a strict incentive to "report truthfully" if the true state were θ. In the single good example, the efficient allocation rule $f^*(\theta) = (q^*(\theta), y^*(\theta))$ fails Maskin monotonicity and thus strict Maskin monotonicity, because the allocation is on the boundary, but nearly efficient rules such as $f^{**}(\theta) = (q^{**}(\theta), y^{**}(\theta))$, defined earlier by (5) and (6), are both Maskin monotonic and strict Maskin monotonic.

Given that we are stating the result in terms of a social choice function, rather than a social choice correspondence, the notion of full implementation is akin to requiring that the game (generated by the mechanism) has a unique equilibrium (outcome). The implementation results in Chapter 10 then suggest that sufficient conditions to get a unique rationalizable outcome are similar to those required for a unique Nash equilibrium outcome, provided that the social choice function is responsive. This is noteworthy as the necessary and almost sufficient condition of Maskin monotonicity is much weaker than the well-known conditions under which there are close relationships between the uniqueness of Nash equilibrium and the uniqueness of the rationalizable outcomes, such as supermodular or concave games. The present results indicate the strength of the implementation approach to reduce the number of equilibria. By using infinite message spaces and stochastic allocations, we strengthen the positive implementation results under Nash equilibrium to the weaker solution concept of rationalizability.

The techniques by which we identify necessary and almost sufficient "monotonicity" conditions for robust implementation under incomplete information in Chapter 4 and rationalizable implementation under complete information in Chapter 10 can be extended to identify necessary and almost sufficient monotonicity conditions for implementation in rationalizable strategies in standard incomplete information environments. These conditions are related to but stronger than the Bayesian monotonicity conditions identified by Postlewaite and Schmeidler (1986) and Jackson (1991) for equilibrium implementation under incomplete information. These conditions are developed and used in Oury and Tercieux (2011).

5.6 *The Role of the Common Prior*

In the presentation of the results thus far, we did not place any restrictions on the agents' beliefs and higher order beliefs. In Chapter 5, we investigate the impact of restricting attention to common prior type spaces.

We recall that in the single unit auction model, the best response of agent i was of the linear form:

$$\theta_i' = \theta_i + \gamma \sum_{j \neq i} (\theta_j - \theta_j'). \tag{14}$$

If $\gamma < 0$, the negative informational interdependence gives rise to strategic complementarity in the reporting of the payoff types in the direct mechanism. Conversely, positive informational interdependence in agents' types, or $\gamma > 0$ gives rise to strategic substitutability in direct mechanism.

The relationship between the informational interdependence and the nature of the strategic interaction then allows us to offer sharp predictions on the role of the common prior. With strategic complements, we know that in games of complete information, there is no gap between Nash equilibrium and rationalizable actions in the sense that there are multiple equilibria if and only if there are multiple rationalizable actions. This is a well-known result which appeared prominently in Milgrom and Roberts (1990). Now, with the linear best response property (14), this result remains true with the appropriate solution concepts for games with incomplete information. In particular, given the restriction to a common prior type space, the behavior under incomplete information rationalizability is equivalent to behavior in the incomplete information correlated equilibrium. In other words, there is a unique Bayes Nash equilibrium if and only if there is unique incomplete information rationalizable outcome. Thus, provided that we are considering mechanism with strategic complementarities, whether or not we restrict attention to common prior type spaces makes no difference, and in particular the contraction property continues to play the same role as a necessary and sufficient condition for robust implementation, as described earlier.

On the other hand, if we consider environments that give rise to strategic substitutability in the direct mechanism, then the presence of a common prior facilitates the robust implementation. Here, it is possible to robustly implement the allocation problem, even if the contraction property fails. In particular, in the single unit auction model we can allow the parameter of interdependence γ to satisfy:

$$\frac{1}{I-1} < \gamma < 1,$$

and still guarantee robust implementation in the direct mechanism if we restrict attention to type spaces satisfying the common prior assumption. This leads to the following result in Chapter 5. If the reports are strategic complements, then robust implementation with a common prior implies robust implementation without a common prior. If the reports are strategic substitutes, then robust implementation with a common prior fails to imply robust implementation without a common prior.

5.7 *Dynamic Mechanisms*

All the results discussed so far have dealt with static mechanisms. In Chapter 6, we analyze the modified generalized second price auction in a dynamic mechanism. We consider the ascending (or English) auction in a *complete information* environment. We ask whether the sequential mechanism offers advantages relative to the static, direct revelation, mechanism in terms of achieving robust implementation. The advantage of the sequential mechanism is the ability to reveal and communicate private information in the course of the mechanism. The revelation of private information can decrease the uncertainty faced by the bidders and ultimately improve the final allocation offered by the mechanism. In auctions, the source of the uncertainty can either be payoff uncertainty (uncertainty about payoff relevant information) or strategic uncertainty (uncertainty about the bids of the other agents). We show that the efficient outcome is fully implemented even when

$$\frac{1}{I-1} < \gamma < 1.$$

Recall that in this setting, we know that full robust implementation (with incomplete information) is not possible under any mechanism and that full implementation does not occur in this direct mechanism even with complete information. Thus we show that in at least some settings, sequential refinements help achieve full implementation.

This result is in the spirit of the classical results of Moore and Repullo (1988) showing the possibility of full implementation of social choice functions even when Maskin monotonicity fails, if subgame perfection is used as a solution concept within the dynamic mechanism. Aghion, Fudenberg,

Holden, Kunimoto, and Tercieux (2009) show that full implementation is no longer possible, even under subgame perfection, if the mechanism is required to work also for types close to complete information.

More closely related to our work in Chapter 6, Mueller (2009) and Penta (2011) examine the robustness of dynamic mechanisms in environments with incomplete information. The results are sensitive to the sequential refinement used in this context, with Mueller (2009) obtaining very permissive results with a stronger refinement and Penta (2011) getting less permissive results with a weaker refinement. Our approach uses a version of Penta's weaker refinement, but results in Penta (2011) suggest that our positive results in Chapter 6 do rely heavily on the complete information assumption.

5.8 *Virtual Implementation*

In complete as well as in incomplete information settings, the relaxation from "exact" implementation to "virtual" implementation leads to a significant weakening of the necessary conditions for implementation. Virtual implementation, as initially defined by Matsushima (1988) and Abreu and Sen (1991), requires that the social choice function arises with probability arbitrarily close to 1, but not necessarily equal to 1. In Chapter 7, we characterize robust virtual implementation in general environments with well-behaved, finite or compact, mechanisms. We show that *ex post* incentive compatibility and a robust measurability condition are necessary and almost sufficient for robust virtual implementation. Robust measurability can also be naturally interpreted as a restriction on the amount of interdependence of agents' types. But it neither implies nor is implied by robust monotonicity. However, in the aggregator environment of Chapter 3, robust measurability and robust monotonicity are both equivalent to the contraction property and the only impact of relaxing "exact" to "virtual" robust implementation is the relaxation from *strict* ex post incentive compatibility in Chapter 3 to *weak* ex post incentive compatibility.

With respect to our leading example, the single unit auction, the transition from the generalized second price auction to the modified second price auction, can now be interpreted as the virtual implementation of the generalized second price auction. After all, in the modified generalized second price auction, the allocation of the generalized second price auction is only chosen with probability $1-\varepsilon$, for some $\varepsilon > 0$. The key result in Chapter 7 is a characterization of when two payoff types are *strategically distinguishable* in the sense that they can be guaranteed to behave differently in some mechanism. The condition of robust measurability now requires that *strategically indistinguishable* types are treated the same by the social choice function.

We now provide an exact characterization of strategic distinguishability in the context of the single-unit auction. If we have sets of payoff types, Ψ_1 and Ψ_2, of agents 1 and 2, respectively, we say that the set Ψ_2 separates the set Ψ_1 if knowing agent 1's preferences and knowing that agent 1 is sure that agent 2's type is in Ψ_2, we can rule out at least some payoff type of agent 1 in Ψ_1. Now consider an iterative process where we start, for each agent, with all subsets of his payoff type set, namely the power set of Θ_1, 2^{Θ_1}, and — at each stage — delete subsets of payoff types that are separated by every remaining subset of types of his opponents. A pair of types are said to be *pairwise inseparable* if the set consisting of that pair of types survives this process. We show that two types are strategically indistinguishable if and only if they are pairwise inseparable.

If there are private values and every payoff type is value distinguished, then every pair of types will be pairwise separable and thus strategically distinguishable. Thus strategic indistinguishability arises only when the degree of interdependence in preferences is large. We can illustrate this within the context of our single-unit auction example. As the utility function $u_i(\cdot)$ is linear in the monetary transfer for all types and all agents, the separability must come from different valuations of the object. For given type set profile Ψ_{-i} of all agents but i, we can identify the set of possible (expected) valuations of agent i with type θ_i by writing:

$$\begin{aligned} V_i(\theta_i, \Psi_{-i}) \\ &= \left\{ v_i \in \mathbb{R}_+ \,\middle|\, \exists \lambda_i \in \Delta(\Psi_{-i}) \text{ s.t. } v_i = \theta_i + \gamma \sum_{\theta_{-i} \in \Psi_{-i}} \lambda_i(\theta_{-i}) \sum_{j \neq i} \theta_j \right\} \\ &= \left[\theta_i + \gamma \sum_{j \neq i} \min \Psi_j,\ \theta_i + \gamma \sum_{j \neq i} \max \Psi_j \right], \end{aligned} \tag{15}$$

where we write with minor abuse of notation, $\min \Psi_j$ and $\max \Psi_j$ to identify the smallest and largest real number in the set Ψ_j, respectively. Now we say that Ψ_{-i} separates Ψ_i if and only if

$$\bigcap_{\theta_i \in \Psi_i} V_i(\theta_i, \Psi_{-i}) = \varnothing.$$

By the linearity of the valuation, this is equivalent to requiring that

$$V_i(\max \Psi_i, \Psi_{-i}) \cap V_i(\min \Psi_i, \Psi_{-i}) = \varnothing.$$

By (15), this will hold if and only if

$$\max \Psi_i + \gamma \sum_{j \neq i} \min \Psi_j > \min \Psi_i + \gamma \sum_{j \neq i} \max \Psi_j.$$

We can rewrite the inequality as

$$\max \Psi_i - \min \Psi_i > \gamma \sum_{j \neq i} (\max \Psi_j - \min \Psi_j).$$

Thus Ψ_{-i} separates Ψ_i if and only if the difference between the smallest and the largest element in the set Ψ_i is larger than the weighted sum of the differences of the smallest and the largest element in the remaining sets Ψ_j for all $j \neq i$. Conversely, Ψ_{-i} does not separate Ψ_i if the above inequality is reversed, i.e.,

$$\max \Psi_i - \min \Psi_i \leq \gamma \sum_{j \neq i} (\max \Psi_j - \min \Psi_j). \tag{16}$$

We write Ξ_i^k for the k-th level inseparable sets of player i, and we have:

$$\Xi_i^0 = 2^{\Theta_i},$$

and define an inductive process by:

$$\Xi_i^{k+1} = \{\Psi_i \in \Xi_i^k | \Psi_{-i} \text{ does not separate } \Psi_i, \text{ for some } \Psi_{-i} \in \Xi_{-i}^k\},$$

and a (finite) limit type set profile is defined by:

$$\Xi_i^* = \bigcap_{k \geq 0} \Xi_i^k.$$

Now, we can identify the k-th level inseparable set for the single unit auction example as follows. By (16), we have

$$\Xi_i^k = \left\{ \Psi_i \in \Xi_i^k \,\middle|\, \max \Psi_i - \min \Psi_i \leq \gamma \sum_{j \neq i} \max_{\Psi_j \in \Xi_j^k} (\max \Psi_j - \min \Psi_j) \right\},$$

Now by induction, we have that

$$\Xi_i^{k+1} = \{\Psi_i | \max \Psi_i - \min \Psi_i \leq (\gamma(I-1))^k\}.$$

Thus if $\gamma(I-1) < 1$, Ξ_i^* consists of singletons, $\Xi_i^* = (\{\theta_i\})_{\theta_i \in [0,1]}$, while if $\gamma(I-1) \geq 1$, Ξ_i^* consists of all subsets, $\Xi_i^* = 2^{[0,1]}$. In consequence, we find that if $\gamma < \frac{1}{I-1}$, so that interdependence is not too large, every distinct pair of types are pairwise separable. If on the other hand, $\gamma \geq \frac{1}{I-1}$, then every pair of payoff types are pairwise inseparable.

While our sufficiency argument for robust virtual implementation builds on Abreu and Matsushima (1992), the interpretation of our results ends up being rather different. In a standard Bayesian setting, the measurability condition of Abreu and Matsushima (1992) is arguably a weak technical requirement. As a result, the "bottom line" of the virtual implementation literature has been that full implementation, i.e., getting rid of undesirable equilibria, does not impose any substantive constraints beyond incentive compatibility, i.e., the existence of desirable equilibria. By requiring the more demanding, but more plausible, robust formulation of incomplete information, we end up with a condition that is substantive (imposing significantly more structure in interdependent value environments than incentive compatibility) and easily interpretable.

A conclusion that emerges from the Chapters 3, 4 and 7, and that we developed here within the single good example, is that we keep ending up with the same moderate interdependence condition, $\gamma < \frac{1}{I-1}$, as a necessary and sufficient condition for full implementation. In general, though, the robust monotonicity condition of Chapter 4 (and its contraction property version in Chapter 3) neither implies nor is implied by robust measurability, as we show by examples in Chapter 7. Kunimoto and Serrano (2010) present a detailed discussion of these conditions and develop an argument as to why robust measurability should be seen as the weaker of the independent conditions. Artemov, Kunimoto, and Serrano (2010) characterize an analogue of robust measurability under intermediate robustness conditions and argue that it is a weak condition.

In Section 6.3 of Chapter 7, we do briefly re-consider the single good example under a local robustness condition: suppose that in the single good example, each agent puts probability mass $1-\delta$ on a uniform distribution over the payoff types of other agents, but that δ probability may be allocated to any beliefs. Thus if $\delta = 0$, we have a standard payoff type space with independent types and if $\delta = 1$ we have the universal type space that is the focus of this book. We show that virtual implementation is possible if and only if $\delta\gamma < \frac{1}{I-1}$. In this sense, at least within the single good example, the path to global robustness through local robustness is smooth.

6 Open Issues

In most of the work discussed, we defined the allocation problem in terms of social choice function or correspondence, which specified for each profile of payoff types θ a specific allocation. Importantly, the social choice function was defined independent of the beliefs of the agents and/or the principal. While this specification accommodates many allocation problems, in particular the socially efficient allocation, it cannot represent others, such as revenue maximizing allocations. Here, the allocation rule typically depends on the beliefs of the principal or the agents, as the optimal allocation relies on trading off outcomes across different states, where the trade-offs have to be evaluated with the likelihood of each state, and hence requires the use of beliefs. In Chapter 11, Karl Schlag and Bergemann suggest a possible approach to analyze revenue maximization problems in the absence of prior beliefs. They consider the classic monopoly problem of a seller who offers a homogenous good to buyers with privately known valuations. In the absence of a (common) prior, we require the seller to minimize his expected regret through an optimal pricing policy. The resulting pricing policy hedges against the uncertainty with respect to the true distribution through a uniquely determined randomized pricing policy. And while the resulting mixed strategy can be interpreted as the optimal pricing rule against a specific prior distribution, a random pricing policy is never the uniquely optimal policy given a known prior. In fact, against a known prior, there always exists an optimal pricing rule that is deterministic. In Chapter 12, Karl Schlag and Bergemann consider the problem of optimal pricing when the seller has *some* prior information. In this version of the problem the seller knows that demand will be in a *small* neighborhood of a *given* model distribution. We characterized the optimal pricing policy under two distinct, but related, decision criteria with multiple priors: (*i*) maximin expected utility and (*ii*) minimax expected regret. The resulting model can be interpreted as a locally robust version of the classic problem of optimal monopoly pricing.

A second, and related, limitation of the present work is that we were mostly concerned with "global" notions of robustness. We allowed for any beliefs and higher order beliefs consistent with the existing model. It would be of interest to look at "local" notions of robustness, where more limited perturbation of the types and information structures are considered. For example, in ongoing work, Bergemann and Morris (2011b), we consider games with incomplete information and ask what predictions can be made with the knowledge of a common prior over the payoff relevant states, and importantly in the absence of any additional information about the private information, the type space, of the agents. Thus, we consider a common prior about the relevant state, but are agnostic with respect to the beliefs and higher-order beliefs of the agents. In Bergemann and Morris (2011b), we use the structure of quadratic payoffs, and hence linear best response to analyze the set of possible equilibrium distributions in terms of moment restrictions. In Bergemann and Morris (2011a), we develop the associated equilibrium concept, which we refer to as Bayes correlated equilibrium for general finite action, finite agent games with incomplete information and establish how the equilibrium set depends on and changes with the private information of the agents.

A similar approach would seem to have promise in the realm of mechanism design as well. For example, in a first price auction, one might attempt to find the set of possible equilibrium bid distributions that are generated by all information structures consistent with a given common prior over valuations. Likewise, one might investigate, the nature of the optimal auction when the principal has only limited information about the nature of the private information of the agents.

References

Abreu, D., and H. Matsushima (1992): "Virtual Implementation in Iteratively Undominated Strategies: Incomplete Information," Discussion paper, Princeton University and University of Tokyo.

Abreu, D., and A. Sen (1991): "Virtual Implementation in Nash Equilibrium," *Econometrica*, 59, 997–1021.

Aghion, P., D. Fudenberg, R. Holden, T. Kunimoto, and O. Tercieux (2009): "Subgame-Perfect Implementation under Value Perturbations and the Hold-Up Problem," Discussion paper, Harvard University.

Artemov, G., T. Kunimoto, and R. Serrano (2010): "Robust Virtual Implementation with Incomplete Information: Towards a Reinterpretation of the Wilson Doctrine," Discussion paper, University of Melbourne, McGill University and Brown University.

Ausubel, L. M., and P. Milgrom (2005): "The Lovely But Lonely Vickrey Auction," in *Combinatorial Auctions*, ed. by P. Cramton, R. Steinberg, and Y. Shoham. MIT Press, Cambridge, MA.

Barelli, P. (2009): "On the Genericity of Full Surplus Extraction in Mechanism Design," *Journal of Economic Theory*, 144, 1320–1332.

Battigalli, P., and M. Siniscalchi (2003): "Rationalization and Incomplete Information," *Advances in Theoretical Economics*, 3, Article 3.

Bergemann, D., and S. Morris (2001): "Robust Mechanism Design," Discussion paper, available at http://www.princeton.edu/smorris/pdfs/robustmechanism2001.pdf, Yale University.

——— (2005): "Robust Mechanism Design," *Econometrica*, 73, 1771–1813.

——— (2011a): "Correlated Equilibrium in Games of Incomplete Information," Discussion paper, Cowles Foundation for Research in Economics, Yale University and Economic Theory Center, Princeton University.

——— (2011b): "Robust Predictions in Games of Incomplete Information," Discussion paper 1821, Cowles Foundation for Research in Economics, Yale University, and Economic Theory Center, Princeton University.

Bergemann, D., S. Morris, and S. Takahashi (2010): "Interdependent Preferences and Strategic Distinguishability," Discussion paper CFDP 1772 and ETC 008-2011, Cowles Foundation for Research in Economics, Yale University and Economic Theory Center, Princeton University.

Bierbrauer, F., and M. Hellwig (2011): "Mechanism Design and Voting for Public-Good Provision," Discussion paper, University of Cologne and Max Planck Institute for Research on Collective Goods.

Bikhchandani, S. (2006): "*Ex Post* Implementation in Environments with Private Goods," *Theoretical Economics*, 1, 369–393.

Birulin, O. (2003): "Inefficient *Ex Post* Equilibria in Efficient Auctions," *Economic Theory*, 22, 675–683.

Blume, A., and P. Heidhues (2004): "All Equilibria of the Vickrey Auction," *Journal of Economic Theory*, 114, 170–177.

Börgers, T., and D. Smith (2011): "Robust Mechanism Design and Dominant Strategy Voting Rules," Discussion paper, University of Michigan.

Brandenburger, A., and E. Dekel (1987): "Rationalizability and Correlated Equilibria," *Econometrica*, 55, 1391–1402.

Chen, Y., and S. Xiong (2010): "The Genericity of Belief-Determine-Preferences Models Revisited," *Journal of Economic Theory*, 146, 751–761.

——— (2011): "Genericity and Robustness of Full Surplus Extraction," Discussion paper, National University of Singapore and Rice University.

Choi, J., and T. Kim (1999): "A Nonparametric, Efficient Public Good Decision Mechanism: Undominated Bayesian Implementation," *Games and Economic Behavior*, 27, 64–85.

Chung, K.-S., and J. Ely (2001): "Efficient and Dominance Solvable Auctions with Interdependent Valuations," Discussion paper, Northwestern University.

Chung, K.-S., and J. Ely (2003): "Implementation with Near-Complete Information," *Econometrica*, 71, 857–871.

Chung, K.-S., and J. Ely (2007): "Foundations of Dominant Strategy Mechanisms," *Review of Economic Studies*, 74, 447–476.

Cremer, J., and R. McLean (1988): "Full Extraction of the Surplus in Bayesian and Dominant Strategy Auctions," *Econometrica*, 56, 1247–1258.

Dasgupta, P., P. Hammond, and E. Maskin (1979): "The Implementation of Social Choice Rules. Some General Results on Incentive Compatibility," *Review of Economic Studies*, 66, 185–216.

Dasgupta, P., and E. Maskin (2000): "Efficient Auctions," *Quarterly Journal of Economics*, 115, 341–388.

Dekel, E., D. Fudenberg, and S. Morris (2006): "Topologies on Types," *Theoretical Economics*, 1, 275–309.

Eso, P., and E. Maskin (2002): "Multi-Good Efficient Auctions with Multidimensional Information," Discussion paper, Northwestern University and Institute for Advanced Studies.

Fang, H., and S. Morris (2006): "Multidimensional Private Value Auctions," *Journal of Economic Theory*, 126, 1–30.

Gizatulina, A., and M. Hellwig (2010): "Informational Smallness and the Scope for Limiting Information Rents," *Journal of Economic Theory*, 145, 2260–2281.

——— (2011): "Beliefs, Payoffs, Information: On the Robustness of the BDP Property in Models with Endogenous Beliefs," Discussion paper, Max Planck Institute for Research on Collective Goods.

Harsanyi, J. (1967–1968): "Games with Incomplete Information Played by 'Bayesian' Players," *Management Science*, 14, 159–189, 320–334, 485–502.

HEIFETZ, A., AND Z. NEEMAN (2006): "On the Generic (Im)Possibility of Full Surplus Extraction in Mechanism Design," *Econometrica*, 74, 213–233.

HURWICZ, L. (1972): "On Informationally Decentralized Systems," in *Decisions and Organizations*, ed. by C. McGuire, and R. Radner, pp. 297–336. North-Holland, Amsterdam.

JACKSON, M. (1991): "Bayesian Implementation," *Econometrica*, 59, 461–477.

JEHIEL, P., M. MEYER-TER-VEHN, AND B. MOLDOVANU (2008): "*Ex-Post* Implementation and Preference Aggregation Via Potentials," *Economic Theory*, 37, 469–490.

——— (2010): "Locally Robust Implementation and its Limits," Discussion paper, Paris School of Economics.

JEHIEL, P., AND B. MOLDOVANU (2001): "Efficient Design with Interdependent Valuations," *Econometrica*, 69, 1237–1259.

JEHIEL, P., B. MOLDOVANU, M. MEYER-TER-VEHN, AND B. ZAME (2006): "The Limits of *Ex Post* Implementation," *Econometrica*, 74, 585–610.

KUNIMOTO, T., AND R. SERRANO (2010): "Evaluating the Conditions for Robust Mechanism Design," Discussion paper, McGill University and Brown University.

LAFFONT, J., AND D. MARTIMORT (2000): "Mechanism Design with Collusion and Correlation," *Econometrica*, 65, 309–342.

LEDYARD, J. (1979): "Dominant Strategy Mechanisms and Incomplete Information," in *Aggregation and Revelation of Preferences*, ed. by J.-J. Laffont, Chap. 17, pp. 309–319. North-Holland, Amsterdam.

LEDYARD, J. O. (1978): "Incentive Compatibility and Incomplete Information," *Journal of Economic Theory*, 18, 171–189.

LOPOMO, G., L. RIGOTTI, AND C. SHANNON (2009): "Uncertainty in Mechanism Design," Discussion paper, University of California at Berkeley and Duke University.

MASKIN, E. (1992): "Auctions and Privatization," in *Privatization: Symposium in Honor of Herbert Giersch*, ed. by H. Siebert, pp. 115–136. J.C.B. Mohr, Tuebingen.

——— (1999): "Nash Equilibrium and Welfare Optimality," *Review of Economic Studies*, 66, 23–38.

MATSUSHIMA, H. (1988): "A New Approach to the Implementation Problem," *Journal of Economic Theory*, 45, 128–144.

McAfee, P., and P. Reny (1992): "Correlated Information and Mechanism Design," *Econometrica*, 60, 395–421.

McLean, R., and A. Postlewaite (2002): "Informational Size and Incentive Compatibility," *Econometrica*, 70, 2421–2453.

Mertens, J., and S. Zamir (1985): "Formalization of Bayesian Analysis for Games with Incomplete Information," *International Journal of Game Theory*, 14, 1–29.

Milgrom, P., and J. Roberts (1990): "Rationalizability, Learning and Equilibrium in Games with Strategic Complementarities," *Econometrica*, 58, 1255–1277.

Moore, J., and R. Repullo (1988): "Subgame Perfect Implementation," *Econometrica*, 56, 1191–1220.

Mueller, C. (2009): "Robust Virtual Implementation under Common Strong Belief in Rationality," Discussion paper, University of Minnesota.

Myerson, R. (1981): "Optimal Auction Design," *Mathematics of Operations Research*, 6, 58–73.

Neeman, Z. (2004): "The Relevance of Private Information in Mechanism Design," *Journal of Economic Theory*, 117, 55–77.

Oury, M., and O. Tercieux (2011): "Continuous Implementation," Discussion paper, Paris School of Economics.

Penta, A. (2011): "Robust Dynamic Mechanism Design," Discussion paper, University of Wisconsin.

Perry, M., and P. Reny (2002): "An *Ex Post* Efficient Auction," *Econometrica*, 70, 1199–1212.

Peters, M. (2001): "Surplus Extraction and Competition," *Review of Economic Studies*, pp. 613–631.

Plum, M. (1992): "Characterization and Computation of Nash-Equilibria for Auctions with Incomplete Information," *International Journal of Game Theory*, 20, 393–418.

Postlewaite, A., and D. Schmeidler (1986): "Implementation in Differential Information Economies," *Journal of Economic Theory*, 39, 14–33.

Robert, J. (1991): "Continuity in Auction Design," *Journal of Economic Theory*, 55, 169–179.

Smith, D. (2010): "A Prior Free Efficiency Comparison of Mechanisms for the Public Good Problem," Discussion paper, University of Michigan.

Wilson, R. (1985): "Incentive Efficiency of Double Auctions," *Econometrica*, 53, 1101–16.

——— (1987): "Game-Theoretic Analyses of Trading Processes," in *Advances in Economic Theory: Fifth World Congress*, ed. by T. Bewley, pp. 33–70. Cambridge University Press, Cambridge, UK.

YAMASHITA, T. (2011): "Robust Welfare Guarantees in Bilateral Trading Mechanisms," Discussion paper, Stanford University.

YOKOO, M., Y. SAKURAI, AND S. MATSUBARA (2004): "The Effect of False-Name Bids in Combinatorial Auctions: New Fraud in Internet Auctions," *Games and Economic Behavior*, 46, 174–188.

CHAPTER 1

Robust Mechanism Design*

Dirk Bergemann and Stephen Morris

Abstract
The mechanism design literature assumes too much common knowledge of the environment among the players and planner. We relax this assumption by studying mechanism design on richer type spaces.

We ask when ex post implementation is equivalent to interim (or Bayesian) implementation for all possible type spaces. The equivalence holds in the case of *separable* environments; examples of separable environments arise (1) when the planner is implementing a social choice function (not correspondence) and (2) in a quasilinear environment with no restrictions on transfers. The equivalence fails in general, including in some quasilinear environments with budget balance.

In private value environments, ex post implementation is equivalent to dominant strategies implementation. The private value versions of our results offer new insights into the relationship between dominant strategy implementation and Bayesian implementation.

Keywords: Mechanism design, common knowledge, universal type space, interim equilibrium, ex post equilibrium, dominant strategies.

> "Game theory has a great advantage in explicitly analyzing the consequences of trading rules that presumably are really common knowledge; it is deficient to the extent it assumes other features to be common knowledge, such as one player's probability assessment about another's preferences or information."

*This research is supported by NSF Grant SES-0095321. We would like to thank the co-editor, three anonymous referees, and seminar participants at many institutions for helpful comments. We thank Bob Evans for pointing out errors in earlier examples and Sandeep Baliga, Matt Jackson, Jon Levin, Bart Lipman, Eric Maskin, Zvika Neeman, Andrew Postlewaite, Ilya Segal, and Tomas Sjöström for valuable discussions.

"I foresee the progress of game theory as depending on successive reductions in the base of common knowledge required to conduct useful analyses of practical problems. Only by repeated weakening of common knowledge assumptions will the theory approximate reality. Wilson (1987)."

1 Introduction

THE THEORY OF MECHANISM DESIGN helps us understand institutions ranging from simple trading rules to political constitutions. We can understand institutions as the solution to a well-defined planner's problem of achieving some objective or maximizing some utility function subject to incentive constraints. A common criticism of mechanism design theory is that the optimal mechanisms solving the well-defined planner's problem seem unreasonably complicated. Researchers have often therefore restricted attention to mechanisms that are "more robust" or less sensitive to the assumed structure of the environment.[1] However, if the optimal solution to the planner's problem is too complicated or sensitive to be used in practice, it is presumably because the original description of the planner's problem was itself flawed. We would like to see if improved modelling of the planner's problem endogenously generates the "robust" features of mechanisms that researchers have been tempted to assume.

As suggested by Robert Wilson in the above quote, the problem is that we make too many implicit common knowledge assumptions in our description of the planner's problem.[2] The modelling strategy must be to first make explicit the implicit common knowledge assumptions and then weaken them. The approach to modelling incomplete information introduced by Harsanyi (1967/1968) and formalized by Mertens and Zamir (1985) is

[1]Discussions of this issue are an old theme in the mechanism design literature. Hurwicz (1972) discussed the need for "nonparametric" mechanisms (independent of parameters of the model). Wilson (1985) states that a desirable property of a trading rule is that it "does not rely on features of the agents' common knowledge, such as their probability assessments." Dasgupta and Maskin (2000) "seek auction rules that are independent of the details — such as functional forms or distribution of signals — of any particular application and that work well in a broad range of circumstances."

[2]An important paper of Neeman (2004) shows how rich type spaces can be used to relax implicit common knowledge assumptions in a mechanism design context. For other approaches to formalizing robust mechanism design, see Chung and Ely (2003), Duggan and Roberts (1997), Eliaz (2002), Hagerty and Rogerson (1987), and Lopomo (1998, 2000).

ideally suited to this task. In fact, Harsanyi's work was intended to address the then prevailing criticism of game theory that the very description of a game embodied common knowledge assumptions that could never prevail in practice. Harsanyi argued that by allowing an agent's type to include his beliefs about the strategic environment, his beliefs about other agents' beliefs, and so on, any environment of incomplete information could be captured by a type space. With this sufficiently large type space (including all possible beliefs and higher order beliefs), it is true (tautologically) that there is common knowledge among the agents of each agent's set of possible types and each type's beliefs over the types of other agents. However, as a practical matter, applied economic analysis tends to assume much smaller type spaces than the universal type space, *and yet maintain the assumption that there is common knowledge among the agents of each agent's type space and each type's beliefs over the types of other agents.* In the small type space case, this is a very substantive restriction. There has been remarkably little work since Harsanyi to check whether analysis of incomplete information games in economics is robust to the implicit common knowledge assumptions built into small type spaces.[3] We will investigate the importance of these implicit common knowledge assumptions in the context of mechanism design.[4]

Formally, we fix a payoff environment, specifying a set of payoff types for each agent, a set of outcomes, utility functions for each agent, and a social choice correspondence (SCC) that maps payoff type profiles into sets of acceptable outcomes. The planner (*partially*) *implements*[5] the social choice correspondence if there exists a mechanism and an equilibrium strategy profile of that mechanism such that equilibrium outcomes for every payoff type profile are acceptable according to the SCC.[6] This is sometimes referred to as Bayesian implementation, but since we do not have a common prior, we will call it interim implementation.

While holding this environment fixed, we can construct many type spaces, where an agent's type specifies both his payoff type and his belief

[3]Battigalli and Siniscalchi (2003), Morris and Shin (2003).
[4]Neeman (2004) argued that small type space assumptions are especially important in the full surplus extraction results of Cremer and McLean (1985).
[5]"Partial implementation" is sometimes called "truthful implementation" or incentive compatible implementation. Since we look exclusively at partial implementation in this paper, we will write "implement" instead of "partially implement."
[6]In companion papers (Bergemann and Morris (2005a, 2005b)), we use the framework of this paper to look at full implementation, i.e., requiring that every equilibrium delivers an outcome consistent with the social choice correspondence.

about other agents' types. Crucially, there may be many types of an agent with the same payoff type. The larger the type space, the harder it will be to implement the social choice correspondence, and so the more "robust" the resulting mechanism will be. The smallest type space we can work with is the "payoff type space," where we set the possible types of each agent equal to the set of payoff types and assume a common knowledge prior over this type space. This is the usual exercise performed in the mechanism design literature. The largest type space we can work with is the union of all possible type spaces that could have arisen from the payoff environment. This is equivalent to working with a "universal type space," in the sense of Mertens and Zamir (1985). There are many type spaces in between the payoff type space and the universal type space that are also interesting to study. For example, we can look at all payoff type spaces (so that the agents have common knowledge of a prior over payoff types but the mechanism designer does not) and we can look at type spaces where the common prior assumption holds.

In the face of a planner who does not know about agents' beliefs about other players' types, a recent literature has looked at mechanisms that implement the SCC in *ex post equilibrium* (see references in footnote 10). This requires that in a payoff type direct mechanism, where each agent is asked to report his payoff type, each agent has an incentive to tell the truth if he expects others to tell the truth, whatever their types turn out to be. In the special case of private values, ex post implementation is equivalent to dominant strategies implementation. If an SCC is ex post implementable, then it is clearly interim implementable on *every* type space, since the payoff type direct mechanism can be used to implement the SCC.

The converse is not always true. In Examples 1 and 2, ex post implementation is impossible. Nonetheless, interim implementation is possible on every type space. The gap arises because the planner may have the equilibrium outcome depend on the agents' higher order belief types, as well as their realized pay-off type. The planner has no intrinsic interest in conditioning on non-payoff-relevant aspects of agents' types, but he is able to introduce slack in incentive constraints by doing so.

The main question we address in this paper is when the converse is true. A payoff environment is *separable* if the outcome space has a common component and a private value component for each agent. Each agent cares only about the common component and his own private component. The social choice correspondence picks a unique element from the common component and has a product structure over all components. In separable environments, interim implementation on all common prior payoff type spaces implies ex post implementation.[7] Whenever the social choice correspondence is a function, the environment has a separable representation (since we can make private value components degenerate). The other leading example of a separable environment is the problem of choosing an allocation when arbitrary transfers are allowed and agents have quasilinear utility. If the allocation choice is a function but the planner does not care about the level and distribution of transfers, then we have a separable environment.

This result provides a strong foundation for using ex post equilibrium as a solution concept in separable environments. Since ex post implementation implies interim implementation on all type spaces (with or without the common prior or the payoff type restrictions), we also have equivalence between ex post implementation and interim implementation on all type spaces. To the extent that the mechanisms required for ex post implementation are simpler than the mechanisms required for Bayesian implementation, our results contribute to the literature on detail-free implementation and the "Wilson doctrine."

For separable environments, the restriction to payoff type spaces is not important, but this is not true in general. In Example 3, we report a two agent quasilinear environment where we add the balanced budget requirement: transfers must add up to zero. In this example, ex post implementation and interim implementation on all type spaces are both impossible, but interim implementation on all payoff type spaces is possible. As a leading example of an important economic nonseparable environment, we look more generally at quasilinear environments with budget balance. With two agents, there is an equivalence between ex post implementation and interim implementation on all type spaces. With at most two payoff types for each agent, there is the stronger equivalence between ex post implementation and interim implementation on all payoff type spaces, but with three or more agents with three or more types, equivalence between ex post implementation and interim implementation on all type spaces breaks down.

In private values environments, ex post implementation is equivalent to dominant strategies implementation. Our positive and negative results

[7]This result extends to all common prior full support type spaces in the quasilinear case and when the environment is compact.

all have counterparts in private values environments. In particular, we (1) identify conditions when Bayesian implementation on all type spaces is equivalent to dominant strategies implementation, (2) give examples where the equivalence does not hold, and (3) show how and when the equivalence may depend on type spaces richer than the payoff type space. While related questions have long been discussed in the implementation literature (e.g., Ledyard (1978) and Dasgupta, Hammond, and Maskin (1979)) — we discuss the relationship in detail in the concluding Section 6 — our questions have not been addressed even under private values.

The paper is organized as follows. Section 2 provides the setup, introduces the type spaces, and provides the equilibrium notions. In Section 3 we present in some detail three examples that illustrate the role of type spaces in the implementation problem and point to the complex relationship between ex post implementation on the payoff type space and interim implementation on larger type spaces. In Section 4 we present equivalence results for separable social choice environments. The separable environment includes as special cases all social choice *functions* and the quasilinear environment without a balanced budget requirement. Section 5 investigates the quasilinear environment with a balanced budget requirement. We conclude with a discussion of further issues in Section 6.

2 Setup

2.1 *Payoff Environment*

We consider a finite set of agents $1, 2, \ldots, I$. Agent i's *payoff type* is $\theta_i \in \Theta_i$, where Θ_i is a finite set. We write $\theta \in \Theta = \Theta_1 \times \cdots \times \Theta_I$. There is a set of outcomes Y. Each agent has utility function u_i: $Y \times \Theta \to \mathbb{R}$. A social correspondence is a mapping F: $\Theta \to 2^Y \backslash \emptyset$. If the true payoff type profile is θ, the planner would like the outcome to be an element of $F(\theta)$.

An important special case — studied in some of our examples and results — is a *quasilinear environment* where the set of outcomes Y has the product structure $Y = Y_0 \times Y_1 \times \cdots \times Y_I$ where $Y_1 = Y_2 = \cdots = Y_I = \mathbb{R}$, and a utility function

$$u_i(y, \theta) = u_i(y_0, y_1, \ldots, y_I, \theta) \triangleq v_i(y_0, \theta) + y_i,$$

which is linear in y_i for every agent i. The planner is concerned only about choosing an "allocation" $y_0 \in Y_0$ and does not care about transfers. Thus there is a function $f_0\colon \Theta \to Y_0$ and

$$F(\theta) - \{(y_0, y_1, \ldots, y_I) \in Y : y_0 = f_0(\theta)\}.$$

Throughout the paper, this environment is fixed and informally understood to be common knowledge. We allow for interdependent types: one agent's payoff from a given outcome depends on other agents' payoff types. The payoff type profile is understood to contain all information that is relevant to whether the planner achieves his objective or not. For example, we do not allow the planner to trade off what happens in one state with what happens in another state. For the latter reason, this setup is somewhat restrictive. However, it incorporates many classic problems such as the efficient allocation of an object or the efficient provision of a public good.

2.2 *Type Spaces*

While maintaining that the above payoff environment is common knowledge, we want to allow for agents to have all possible beliefs and higher order beliefs about other agents' types. A flexible framework for modelling such beliefs and higher order beliefs is "type spaces."

A type space is a collection

$$\mathcal{T} = (T_i, \hat{\theta}_i, \hat{\pi}_i)_{i=1}^{I}.$$

Agent *i's type* is $t_i \in T_i$. A type of agent i must include a description of his payoff type. Thus there is a function

$$\hat{\theta}_i : T_i \to \Theta_i,$$

with $\hat{\theta}_i(t_i)$ being agent *i's payoff type* when his type is t_i. A type of agent i must also include a description of his beliefs about the types of the other agents. Write $\Delta(Z)$ for the space of probability measures on the Borel field of a measurable space Z. The belief of type t_i of agent i is a function

$$\hat{\pi}_i : T_i \to \Delta(T_i),$$

with $\hat{\pi}_i(t_i)$ being agent *i's beliefs* when his type is t_i. Thus $\hat{\pi}_i(t_i)[E]$ is the probability that type t_i of agent i assigns to other agents' types, t_{-i}, being an element of a measurable set $E \subseteq T_{-i}$. In the special case where each T_j is finite, we will abuse notation slightly by writing $\hat{\pi}_i(t_i)[t_{-i}]$ for the probability that type t_i of agent i assigns to other agents having types t_{-i}.

Our terminology is nonstandard relative to the mechanism design literature. In most of the mechanism design literature and indeed in much of the applied economics literature, it is common to fix a set of types for each agent, let agents' payoffs depend on their own and others' types, and then add on agents' beliefs (often through a common prior) as part of the description of the problem. Thus an agent's type implicitly defines his utility function but not his beliefs. By contrast, we are assuming that an agent's type implicitly contains a description of his beliefs and his payoffs. Our usage is in the tradition of Harsanyi (1967/1968) and Mertens and Zamir (1985), who originally introduced the idea of types into the economics and game theory literature, and also the literature on epistemic foundations of game theory since then. Our "payoff type" does correspond to the way types are often talked about in applied mechanism design. The foundations of this formalism are discussed in some detail in Section 2.5.

2.3 *Solution Concepts*

Fix a payoff environment and a type space $\mathcal{T}$. A mechanism specifies a message set for each agent and a mapping from message profiles to outcomes. Social choice correspondence F is interim implementable if there exists a mechanism and an interim (or Bayesian) equilibrium of that mechanism such that outcomes are consistent with F. However, by the revelation principle, we can restrict attention to truth-telling equilibria of direct mechanisms.[8] A direct mechanism is a function $f : T \to Y$.

Definition 1. A direct mechanism $f : T \to Y$ is interim incentive compatible on type space $\mathcal{T}$ if

$$\int_{t_{-i}\in T_{-i}} u_i(f(t_i,t_{-i}),\hat{\theta}(t_i,t_{-i}))d\hat{\pi}_i(t_i)$$
$$\geq \int_{t_{-i}\in T_{-i}} u_i(f(t_i',t_{-i}),\hat{\theta}(t_i,t_{-i}))d\hat{\pi}_i(t_i)$$

for all $i, t \in T$ and $t_i' \in T_i$.

The notion of interim incentive compatibility is often referred to as Bayesian incentive compatibility. We use the former terminology as there need not be a common prior on the type space.

[8]See Myerson (1991, Chapter 6).

Definition 2. A direct mechanism $f : T \to Y$ on $\mathcal{T}$ achieves F if

$$f(t) \in F(\hat{\theta}(t))$$

for all $t \in T$.

It should be emphasized that a direct mechanism f can prescribe varying allocations for a given payoff profile θ as different types, t and t', may have an identical payoff profile $\theta = \hat{\theta}(t) = \hat{\theta}(t')$.

Definition 3. A social choice correspondence F is interim implementable on $\mathcal{T}$ if there exists $f : T \to Y$ such that f is interim incentive compatible on $\mathcal{T}$ and f achieves F.

We will be interested in comparing interim implementation with the stronger solution concept of ex post implementation. Ex post implementation uses the stronger solution concept of ex post equilibrium for incomplete information games.[9] By the revelation principle, it is again enough to verify ex post incentive compatibility.

Definition 4. A direct mechanism $f : \Theta \to Y$ is ex post incentive compatible if, for all i and $\theta \in \Theta$,

$$u_i(f(\theta), \theta) \geq u_i(f(\theta_i', \theta_{-i}), \theta)$$

for all $\theta_i' \in \Theta_i$.

The notion of ex post incentive compatibility requires agent i to prefer truth-telling at θ if all the other agents also report truthfully. Ex post incentive compatibility is defined directly on the payoff type space, but observe that this is equivalent to requiring ex post incentive compatibility on any type space where all payoff types are possible (i.e., the range of each $\hat{\theta}_i$ is Θ_i).

In contrast, the notion of dominant strategy implementation requires agent i to prefer truth telling for all possible reports by the other agents, truth-telling or not.

[9] Ex post incentive compatibility was discussed as "uniform incentive compatibility" by Holmstrom and Myerson (1983). Ex post equilibrium is increasingly studied in game theory (see Kalai, 2004) and is often used in mechanism design as a more robust solution concept (Cremer and McLean, 1985). A recent literature on interdependent value environments has obtained positive and negative results using this solution concept: Dasgupta and Maskin (2000), Bergemann and Valimaki (2002), Perry and Reny (2002), and Jehiel *et al.* (2005).

Definition 5. A direct mechanism $f : \Theta \to Y$ is dominant strategies incentive compatible if, for all i and $\theta \in \Theta$,

$$u_i(f(\theta_i, \theta'_{-i}), \theta) \geq u_i(f(\theta'), \theta)$$

for all $\theta' \in \Theta$.

If there are private values (i.e., each $u_i(y, \theta)$ depends on θ only through θ_i), then ex post incentive compatibility is equivalent to dominant strategies incentive compatibility.

Definition 6. A social choice correspondence F is ex post implementable if there exists $f : \Theta \to Y$ such that f is ex post incentive compatible and $f(\theta) \in F(\theta)$ for all $\theta \in \Theta$.

2.4 *Questions*

Our main question is, when is F interim implementable on all type spaces? By requiring that F be interim implementable on all type spaces, we are asking for a mechanism that can implement F with no common knowledge assumptions beyond those in the specification of the payoff environment. In Sections 4 and 5, we provide sufficient conditions for ex post implementability to be equivalent to interim implementability on all type spaces, but Examples 1 and 2 in the next section show that it is possible to find social choice correspondences that are interim implementable on any type space but are not ex post implementable.

We also consider the implications of interim implementability on different type spaces. To describe these results, we must introduce some important properties of type spaces. A type space $\mathcal{T}$ is a *payoff type space* if each $T_i = \Theta_i$ and each $\hat{\theta}_i$ is the identity map. type space $\mathcal{T}$ is *finite* if each T_i is finite. Finite type space $\mathcal{T}$ has *full support* if $\hat{\pi}_i(t_i)[t_{-i}] > 0$ for all i and t. Finite type space $\mathcal{T}$ satisfies the *common prior assumption* (with prior p) if there exists $p \in \Delta(T)$ such that

$$\sum_{t_{-i} \in T_{-i}} p(t_i, t_{-i}) > 0 \quad \text{for all } i \text{ and } t_i$$

and

$$\hat{\pi}_i(t_i)[t_{-i}] = \frac{p(t_i, t_{-i})}{\sum_{t'_{-i} \in T_{-i}} p(t_i, t'_{-i})}.$$

The standard approach in the mechanism design literature is to restrict attention to a common prior payoff type space (perhaps with full support). Thus it is assumed that there is common knowledge among the agents of a common prior over the payoff types. A payoff type space can be thought of as the smallest type space embedding the payoff environment described above. Restricting attention to a full support, common prior, payoff type space is *with* loss of generality. We can relax the implicit common knowledge assumptions embodied in those restrictions by asking the following progressively tougher questions about interim implementability:

- Is F interim implementable on all full support common prior payoff type spaces?
- Is F interim implementable on all common prior payoff type spaces?
- Is F interim implementable on all common prior type spaces?
- Is F interim implementable on all type spaces?

We will see that relaxing common knowledge assumptions makes a difference. In particular, we will show that while the common prior assumption is not important and the full support assumption does not play a big role,[10] the payoff type space restriction *is* important. In Example 3 in the next section, it is possible to interim implement on any payoff type space (with or without the common prior) but not all type spaces. We are especially interested in the relationship between the ex post implementability of F and the interim implementability on all type spaces.

2.5 *Implicit versus Explicit Modelling of Higher Order Uncertainty and the universal type space*

Heifetz and Samet (1999) distinguish two ways of discussing higher order uncertainty about some state of nature. There is the *explicit* approach: an agent's possible higher order beliefs consist of his beliefs about nature, his beliefs about nature and other agents' beliefs about nature, and so on. Then there is the *implicit* approach, where there is a set of states of nature and a set of "types" of an agent, where each type corresponds to a belief over the state of nature and the types of the other agents. Each type encodes

[10]However, different type space assumptions will be important for different questions. The full support assumption is crucial when we look at full implementation (Bergemann and Morris, 2005b) and the common prior assumption is important when we look at revenue maximization.

implicitly the beliefs, and higher order beliefs about the state of nature. Harsanyi (1967/1968) argued that the implicit approach was sufficient to capture possible higher order beliefs, and Mertens and Zamir (1985) showed that the two approaches are — under some assumptions — equivalent.

We follow the implicit approach in this paper. The type spaces that we work with are thus "implicit type spaces" in the language of Heifetz and Samet (1999).[11] In this section, we briefly discuss what would happen if we had followed the explicit approach and what implications the explicit approach would have for our results.

We will describe a standard universal type space construction for our problem. The only nonstandard aspect is that we want to maintain the feature that each agent knows his payoff type.[12] Player i's zeroth level type is his payoff-relevant type $t_i^0 = \theta_i \in \Theta_i$. Let $T_i^0 \equiv \Theta_i$, be player i's set of zeroth level types. Player i's first level type must specify his payoff-relevant type and his belief about other players' zeroth level types. Thus $t_i^1 \in T_i^1 \equiv \Theta_i \times \Delta(T_{-i}^0)$. Player i's second level type must specify his payoff-relevant type and his belief about other players' first level types. Thus $t_i^2 \in T_i^2 \equiv \Theta_i \times \Delta(T_{-i}^1)$. Iterating this construction, we have $t_i^k \in T_i^k \equiv \Theta_i \times \Delta(T_{-i}^{k-1})$ and we obtain an infinite hierarchy of beliefs $(t_i^0, t_i^1, t_i^2, \ldots)$. We want to require that high level types, which intuitively contain more information than lower level types, are consistent with lower levels. Formally, an infinite hierarchy is *coherent* if all higher level types have the same payoff-relevant type as lower level types and if the projection of their beliefs over other players' types onto lower level type spaces is consistent with lower level types' beliefs. We can let player i's possible types, T_i, be the set of all coherent infinite hierarchies of beliefs. The universal type space literature[13] shows that — under some topological assumptions — the set of types, i.e., infinite hierarchies, can be identified with pairs of payoff-relevant types and beliefs, so that, for each i, there exists a

[11] As pointed out to us by a referee, they are therefore "Θ-based abstract belief spaces" in the language of Mertens and Zamir (1985).

[12] If we made a private values assumption — each agent's utility does not depend on others' payoff types — then the construction we describe is the same at the "private values universal type space" in Heifetz and Neeman (2004).

[13] Mertens and Zamir (1985), Brandenburger and Dekel (1993), Heifetz (1993), Mertens, Sorin, and Zamir (1994).

homeomorphism $f_i : T_i \to \Theta_i \times \Delta(T_{-i})$. Since each Θ_i is finite, such a construction is possible in our case. Now letting $\hat{\theta}_i$ be the projection of f_i onto Θ_i and letting $\hat{\pi}_i$ be the projection of f_i onto $\Delta(T_{-i})$, this canonical "known own payoff type" universal type space is an example of a type space $\mathcal{T} = (T_i, \hat{\theta}_i, \hat{\pi}_i)_{i=1}^{I}$, as described is Section 2.2, with the special property that for each $\theta_i \in \Theta_i$ and $\pi_i \in \Delta(T_{-i})$, there exists $t_i \in T_i$ such that $\hat{\theta}_i(t_i) = \theta_i$ and $\hat{\pi}_i(t_i) = \pi_i$.[14]

What is the connection between the explicit universal type space and the implicit type spaces we described above? An implicit type space has no "redundant types" if every pair of types differs at some level in their higher order belief types. Mertens and Zamir (1985, Property 5 and Proposition 2.16) show that any implicit type space that has no "redundant" types and satisfies some topological restrictions is a belief-closed subset of the universal type space (and the same result will be true in our setting). Thus modulo the redundancy and topological provisos, the union of all type spaces is the same as the universal type space.

How significant are the redundancy and topological restrictions required by Mertens and Zamir to show the equivalence of explicit and implicit type spaces? Heifetz and Samet (1999) show that — without topological restrictions — it is possible to find types that cannot be embedded in the universal type space.[15] In general, the no redundant types restriction is not innocuous either. To illustrate this point, consider the type space

$$T_1 = \{t_1, t_1'\},$$
$$T_2 = \{t_2, t_2'\},$$

with a single payoff-relevant type for each player,

$$\hat{\theta}_1(t_1) = \hat{\theta}_1(t_1') = \theta_1,$$
$$\hat{\theta}_2(t_2) = \hat{\theta}_2(t_2') = \theta_2,$$

[14]This "known own payoff type" universal type space has built in the feature that there is common knowledge that each agent i knows his payoff type θ_i. If instead we had allowed agents also to be uncertain about their own θ_i, we would be back to the standard universal type space concerning Θ, as constructed by Mertens and Zamir (1985).

[15]Heifetz and Samet (1998) provide a nonconstructive proof of the existence of a universal type space without topological restrictions.

and the associated belief types

$$\hat{\pi}_1(t_1)[t_2] = \frac{2}{3},$$
$$\hat{\pi}_1(t_1')[t_2] = \frac{1}{3},$$
$$\hat{\pi}_2(t_2)[t_1] = \frac{2}{3},$$
$$\hat{\pi}_2(t_2')[t_1] = \frac{1}{3}.$$

Since all types have the same payoff-relevant type, the infinite hierarchy of beliefs is degenerate: each type of player i is sure that he has payoff-relevant type θ_i, he is sure that his opponent j has payoff-relevant type θ_j, and so on. However, because of the opportunities for correlation, rational strategic behavior on this type space may be very different from the type space where each player has only a single possible type.[16]

So there are potential gaps between the explicit and implicit approaches. However, all the positive and negative results reported in this paper would be unchanged if we replaced "implementable on all type spaces" by "implementable on the (known own payoff type) universal type space." Since the universal type space is an example of a type space, implementability for all type spaces trivially implies implementability on the universal type space. On the other hand, when we show a failure of implementability for all type spaces, we do so by constructing a finite type space without redundancy where implementability is impossible, but those finite type spaces are isomorphic to belief closed subsets of the universal type space. In addition, if it is not possible to implement on a given type space, it is not possible to implement on any type space (such as the universal type space) that contains that given type space as a belief closed subset. Thus whenever implementability is impossible on those finite type spaces, it is also impossible on the universal type space.

3 Examples

This section presents three examples that illustrate the relationship between interim implementation on different type spaces and ex post implementation.

[16]This issue is important in Dekel, Fudenberg, and Morris (2005).

The first two examples exhibit social choice correspondences that are interim implementable on all type spaces, but are not ex post implementable. The first example is very simple, but relies on (i) a restriction to deterministic allocations, (ii) a social choice correspondence that depends on only one agent's payoff type, and (iii) interdependent types. In the second example, we show how to dispense with all three features. Since this second example has private values, we thus have an example where dominant strategies implementation is impossible, but interim implementation is possible on any type space.

The third example exhibits a social choice correspondence that is interim implementable on all payoff type spaces (with or without the common prior), but is not interim implementable on all type spaces. The social choice correspondence represents efficient allocations in a quasilinear environment with a balanced budget requirement. As such it also illustrates some of the results presented later in Section 5 on social choice problems with a balanced budget.

3.1 *F is Interim Implementable on all Type Spaces but not Ex Post Implementable*

Example 1. There are two agents. Each agent has two possible types: $\Theta_1 = \{\theta_1, \theta_1'\}$ and $\Theta_2 = \{\theta_2, \theta_2'\}$. There are three possible allocations: $Y = \{a, b, c\}$. The payoffs of the two agents are given by the following tables (each box describes agent 1's payoff, then agent 2's payoff):

a	θ_2	θ_2'
θ_1	$1, 0$	$-1, 2$
θ_1'	$0, 0$	$0, 0$

b	θ_2	θ_2'
θ_1	$-1, 2$	$1, 0$
θ_1'	$0, 0$	$0, 0$

c	θ_2	θ_2'
θ_1	$0, 0$	$0, 0$
θ_1'	$1, 1$	$1, 1$

The social choice correspondence is given by

F	θ_2	θ_2'
θ_1	$\{a, b\}$	$\{a, b\}$
θ_1'	$\{c\}$	$\{c\}$

These choices are maximizers of the sum of agents' utility. The key feature of this example is that the agents agree about the optimal choice when agent 1 is type θ_1'; when agent 1 is type θ_1, they agree that it is optimal to choose either a or b. However, each agent has strict and opposite preferences over outcomes a and b: 1 strictly prefers a when 2's type is θ_2, while 2 strictly prefers a when his type is θ_2'.

We now show — by contradiction — that this correspondence is not ex post implementable. If F were implementable, we would have to have c chosen at profiles (θ_1', θ_2) and (θ_1', θ_2') and either a or b chosen at profiles (θ_1, θ_2) and (θ_1, θ_2'). For type θ_1 to have an incentive to tell the truth when he is sure that agent 2 is type θ_2, we must have a chosen at profile (θ_1, θ_2); for type θ_1 to have an incentive to tell the truth when he is sure that agent 2 is type θ_2', we must have b chosen at profile (θ_1, θ_2'). However, if a is chosen at profile (θ_1, θ_2) and b is chosen at profile (θ_1, θ_2'), then both types of agent 2 will have an incentive to misreport their types when they are sure that agent 1 is type θ_1.

However, the correspondence is interim implementable on any type space using the very simple mechanism of letting agent 1 pick the outcome. There is always an equilibrium of this mechanism where agent 1 will pick outcome a if his type is θ_1 and he assigns probability at least $\frac{1}{2}$ to the other agent being type θ_2; and agent 1 will pick outcome b if his type is θ_1 and he assigns probability less than $\frac{1}{2}$ to the other agent being type θ_2; and agent 1 will pick outcome c if his type is θ_1'. By allowing the mechanism to depend on agent 1's beliefs about agent 2's type (something the planner does not care about intrinsically), the planner is able to relax incentive constraints that he cares about.

The failure of ex post implementation in this example relied on the assumption that only pure outcomes were chosen. This restriction can easily be dropped at the expense of adding a third payoff type for agent 1, so that the binding ex post incentive constraint for agent 1 is with a different type and outcome depending on 2's type. Example 1 also had the social choice correspondence depending only on agent 1's payoff type and had interdependent values. We can mechanically change these two assumptions by letting the planner want different outcomes depending on agent 2's type. Now instead of having agent 1's utility depend on agent 2's type, it can depend on the planner's refined choice.

Example 2. There are two agents. Agent 1 has three possible types, $\Theta_1 = \{\theta_1, \theta_1', \theta_1''\}$, and agent 2 has two possible types, $\Theta_2 = \{\theta_2, \theta_2'\}$. There are eight possible pure allocations, $\{a, b, c, d, a', b', c', d'\}$, and lotteries are allowed, so $Y = \Delta(\{a, b, c, d, a', b', c', d'\})$. The private value payoffs of agent 1 are given by the table

u_1	a	b	c	d	a'	b'	c'	d'
θ_1	1	−1	$\frac{1}{2}$	−1	−1	1	−1	$\frac{1}{2}$
θ_1'	0	0	1	0	0	0	1	0
θ_1''	0	0	0	1	0	0	0	1

The private value payoffs of agent 2 are given by the table

u_2	a	b	c	d	a'	b'	c'	d'
θ_2	0	1	0	0	0	1	−1	−1
θ_2'	1	0	−1	−1	1	0	0	0

The social choice correspondence F is described by the table[17]

	θ_2	θ_2'
θ_1	$\{a, b\}$	$\{a', b'\}$
θ_1'	$\{c\}$	$\{c'\}$
θ_1''	$\{d\}$	$\{d'\}$

We now show — by contradiction — that this correspondence is not ex post implementable. Let q be the probability that a is chosen at profile (θ_1, θ_2) and let q' be the probability that a' is chosen at profile (θ_1, θ_2'). For type θ_1 to have an incentive to tell the truth (and not report himself to be type θ_1') when he is sure that agent 2 is type θ_2, we must have

$$q - (1 - q) \geq \frac{1}{2} \Leftrightarrow q \geq \frac{3}{4}. \tag{1}$$

For type θ_1 to have an incentive to tell the truth (and not report himself to be type θ_l'') when he is sure that agent 2 is type θ_2, we must have

$$-q' + (1 - q') \geq \frac{1}{2} \Leftrightarrow q' \leq \frac{1}{4}. \tag{2}$$

[17]The SCC F in this example is not ex post Pareto efficient at (θ_1, θ_2) and (θ_1, θ_2'), as b' and a, respectively, Pareto dominate b and a', respectively. We choose this example for the simplicity of its payoffs, yet we have constructed examples with the same number of agents, states, and allocations such that the SCC F is ex ante Pareto efficient and interim implementable on all type spaces, but not ex post, and a fortiori, not dominant strategy implementable.

but for agent 2 to have an incentive to tell the truth when he is type θ_2 and he is sure that agent 1 is type θ_1, we must have

$$1 - q \geq 1 - q';$$

thus

$$q' \geq q. \tag{3}$$

However, (1), (2), and (3) generate a contradiction, so ex post implementation is not possible.

It is straightforward to implement on any interim type space. Consider the following indirect mechanism for any arbitrary type space where individual 1 chooses a message $m_1 \in \{m_1^1, m_1^2, m_1^3, m_1^4\}$, and individual 2 chooses a message $m_2 \in \{m_2^1, m_2^2\}$, and let outcomes be chosen as follows:

	m_2^1	m_2^2
m_1^1	a	a'
m_1^2	b	b'
m_1^3	c	c'
m_1^4	d	d'

There is always an equilibrium where type θ_1 of agent 1 sends message m_1^1 if he believes agent 2 is type θ_2 with probability at least $\frac{1}{2}$ and message m_1^2 if he believes agent 2 is type θ_2 with probability less than $\frac{1}{2}$; type θ_1' always sends message m_1^3 and type θ_1'' always sends message m_1^4. Type θ_2 of agent 2 sends message m_2^1 and type θ_2' sends message m_2^2, and this strategy is a dominant strategy for agent 2.

This private values example has the feature that dominant strategies implementation is impossible, but interim implementation is possible on any type space and seems to be the first example in the literature noting this possibility.[18]

As we will see in the next section, a necessary feature of the example is that we have a social choice correspondence (not function) that we are

[18]It is often noted that in public good problems with budget balance, dominant strategies implementation is impossible, whereas Bayesian implementation is possible. However, the positive Bayesian implementation results (d'Aspremont and Gerard-Varet, 1979 and d'Aspremont Cremer, and Gerard-Varet, 1995, 2004) hold only for "generic" priors on a fixed type space, not for all type spaces in our sense. They provide examples that show that Bayesian implementation fails for some type spaces.

trying to implement. In the example, it was further key that there were aspects of the allocation that the planner did not care about but the agents did. In the example, this may look a little contrived, but note that this is a natural feature of quasilinear environments where the planner wants to maximize the total welfare of agents. We will next present a quasilinear utility example that exploits this feature.

3.2 *F is Interim Implementable on All Payoff Type Spaces But not Interim Implementable on All Type Spaces*

Example 3. This example has two agents, denoted 1 and 2. Agent 1 has three possible payoff types, $\Theta_1 = \{\theta_1, \theta_1', \theta_1''\}$, and agent 2 has two possible payoff types, $\Theta_2 = \{\theta_2, \theta_2'\}$. The set of feasible "allocations" is given by

$$Y_0 = \{a, b, c, d\}.$$

The agents' gross utilities from the allocations, $v_1(y_0, \theta)$ and $v_2(y_0, \theta)$, respectively, are given by

a	θ_2	θ_2'
θ_1	$0, 2$	$0, 2$
θ_1'	$-4, 0$	$1, 0$
θ_1''	$-4, 0$	$-4, 0$

b	θ_2	θ_2'
θ_1	$0, 0$	$0, 0$
θ_1'	$0, 2$	$0, 0$
θ_1''	$-4, 0$	$0, 0$

c	θ_2	θ_2'
θ_1	$0, 0$	$-4, 0$
θ_1'	$0, 0$	$0, 2$
θ_1''	$0, 0$	$0, 0$

d	θ_2	θ_2'
θ_1	$-4, 0$	$-4, 0$
θ_1'	$1, 0$	$-4, 0$
θ_1''	$0, 2$	$0, 2$

The planner wants the allocation $y_0 \in Y_0$ to maximize the sum of the agents' utilities at every type profile θ; thus he wants the allocation to depend on type profile θ according to the function f_0 described in the table

f_0	θ_2	θ_2'
θ_1	a	a
θ_1'	b	c
θ_1''	d	d

(4)

In addition, balanced budget transfers are possible. Thus the planner must choose $(y_0, y_1, y_2) \in Y_0 \times \mathbb{R}^2$, with $y_1 + y_2 = 0$. Each agent has quasilinear utility, so agent i's utility from (y_0, y_1, y_2) in payoff profile θ is $v_i(y_0, \theta) + y_i$. The planner maximizes the sum of utilities and so does not care about transfers; thus

$$F(\theta) = \{(y_0, y_i, y_2) \in Y_0 \times \mathbb{R}^2 : y_0 = f_0(\theta) \quad \text{and} \quad y_2 = -y_1\}.$$

We first make a few observations regarding the ex post incentive constraints for truth-telling with *zero* transfers. Agent 1 always values the efficient alternatives at 0. The critical type for agent 1 is θ_1', where he values an inefficient alternative, either d or a (depending on the payoff type of agent 2 being θ_2 or θ_2'), at 1, and thus is higher than the efficient alternative at that type profile. The remaining negative entries, -4, for agent 1 simply ensure that no other incentive constraints become relevant. Agent 2 always values the efficient allocation at 2 and every inefficient allocation at 0.

It is straightforward to establish that ex post implementation with balanced transfers is not feasible. Writing $f_i(\theta)$ for the transfer received by i at payoff type profile θ, we have the following ex post incentive constraints for agent 1:

$$v_1(f_0(\theta_1, \theta_2), (\theta_1, \theta_2)) + f_1(\theta_1, \theta_2) \geq v_1(f_0(\theta_1', \theta_2), (\theta_1, \theta_2)) + f_1(\theta_1', \theta_2),$$
$$v_1(f_0(\theta_1', \theta_2), (\theta_1', \theta_2)) + f_1(\theta_1', \theta_2) \geq v_1(f_0(\theta_1'', \theta_2), (\theta_1', \theta_2)) + f_1(\theta_1'', \theta_2)$$

and

$$v_1(f_0(\theta_1'', \theta_2'), (\theta_1'', \theta_2')) + f_1(\theta_1'', \theta_2') \geq v_1(f_0(\theta_1', \theta_2'), (\theta_1'', \theta_2')) + f_1(\theta_1', \theta_2'),$$
$$v_1(f_0(\theta_1', \theta_2'), (\theta_1', \theta_2')) + f_1(\theta_1', \theta_2') \geq v_1(f_0(\theta_1, \theta_2'), (\theta_1', \theta_2')) + f_1(\theta_1, \theta_2')$$

Inserting the gross utilities $v_1(\cdot, \cdot)$, we can write the above set of inequalities as

$$f_1(\theta_1, \theta_2) \geq f_1(\theta_1', \theta_2) \geq f_1(\theta_1'', \theta_2) + 1 \tag{5}$$

and

$$f_1(\theta_1'', \theta_2') \geq f_1(\theta_1', \theta_2') \geq f_1(\theta_1, \theta_2') + 1. \tag{6}$$

Next we consider the ex post incentive constraints for agent 2 at θ_1 and θ_1'', respectively. Here the social choice mapping prescribes allocations constant in the reported type profile of agent 2 and ex post incentive compatibility hence requires constant transfers as well, or $f_2(\theta_1, \theta_2) = f_2(\theta_1, \theta_2')$

and $f_2(\theta_1'', \theta_2) = f_2(\theta_1'', \theta_2')$. Using the balanced budget requirement by writing $f_2(\theta) = -f_1(\theta)$, we thus obtain

$$f_1(\theta_1, \theta_2) = f_1(\theta_1, \theta_2')$$

and

$$f_1(\theta_1'', \theta_2) = f_1(\theta_1'', \theta_2'),$$

which lead to a contradiction with inequalities (5) and (6).

Despite the failure of ex post implementation, we now show that we can satisfy the interim incentive compatibility conditions for every prior on the payoff type space. The sole determinant of the appropriate transfers is the belief of agent 1 with payoff type θ_1'. If type θ_1' assigns probability at least $\frac{1}{2}$ to agent 2 being of payoff type θ_2, then the following transfers to agent 1 (and corresponding balanced budget transfers for agent 2) are interim incentive compatible:

$$\begin{aligned} f_1(\theta_1, \theta_2) &= 0, \quad f_1(\theta_1, \theta_2') = 0, \\ f_1(\theta_1', \theta_2) &= 0, \quad f_1(\theta_1', \theta_2') = -1, \\ f_1(\theta_1'', \theta_2) &= -1, \quad f_1(\theta_1'', \theta_2') = -1. \end{aligned} \tag{7}$$

Conversely, if type θ_1' assigns probability less than $\frac{1}{2}$ to the other agent being of payoff type θ_2, then the following transfers to agent 1 are interim incentive compatible:

$$\begin{aligned} f_1(\theta_1, \theta_2) &= -1, \quad f_1(\theta_1, \theta_2') = -1, \\ f_1(\theta_1', \theta_2) &= -1, \quad f_1(\theta_1', \theta_2') = 0, \\ f_1(\theta_1'', \theta_2) &= 0, \quad f_1(\theta_1'', \theta_2') = 0. \end{aligned} \tag{8}$$

By symmetry of the payoffs, it will suffice to verify the incentive compatibility conditions for the first case. We first observe that all the ex post incentive constraints hold except for agent 1 at type profile θ_1', θ_2', where he has a profitable deviation by misreporting himself to be of type θ_1. Suppose then that type θ_1' assigns probability p to the other agent being type θ_2. His expected payoff to truth-telling is

$$p(0 + 0) + (1 - p)(0 - 1) = -(1 - p),$$

while his expected payoff to misreporting type θ_1 is

$$p(-4 + 0) + (1 - p)(1 + 0) = 1 - 5p$$

and his expected payoff to misreporting type θ_1'' is given by

$$p(1-1)+(1-p)(-4-1)=-5(1-p).$$

Thus truth-telling is optimal as long as

$$-(1-p)\geq 1-5p \Leftrightarrow p\geq \frac{1}{3}. \tag{9}$$

The second set of transfers, described in (8), offers interim incentive compatibility for agent 1 provided that $p\leq \frac{2}{3}$. Whereas either of the above transfer schemes satisfies the ex post incentive constraints of agent 2, it follows for every belief p by type θ_1', we can find interim incentive compatible transfers and hence F is interim implementable for all payoff type spaces.

However, on richer type spaces than the payoff type space, there may be many types with payoff type θ_1', some of whom are sure that the other agent is type θ_2 while others are sure that he is type θ_2'. That is the idea behind the following example of a "complete information" type space where F cannot be interim implemented. We consider the following type space:

	t_2^1	t_2^2	t_2^3	t_2^4	t_2^5	t_2^6	
t_1^1	$\frac{1}{6}$	0	0	0	0	0	θ_1
t_1^2	0	$\frac{1}{6}$	0	0	0	0	θ_1'
t_1^3	0	0	$\frac{1}{6}$	0	0	0	θ_1''
t_1^4	0	0	0	$\frac{1}{6}$	0	0	θ_1''
t_1^5	0	0	0	0	$\frac{1}{6}$	0	θ_1'
t_1^6	0	0	0	0	0	$\frac{1}{6}$	θ_1
	θ_2	θ_2	θ_2	θ_2'	θ_2'	θ_2'	

Thus there are six types for each agent, t_1^k and t_2^l. The entries in the cell describe the probabilities of the common prior, which puts all probability mass on the diagonal. The payoff type that corresponds to each type appears at the end of the row/column corresponding to that type. Thus, for example, type t_1^3 of agent 1 has payoff type θ_1'' and believes that agent 2 has a payoff type θ_2 with probability 1. It is in this sense that we speak of complete information. We require that F is implemented even at "impossible" (zero probability) type profiles, but we could clearly adapt the example to have small probabilities off the diagonal.

Our impossibility argument will depend only on what happens at twelve critical type profiles: the diagonal profiles and the type profiles where agent 1 with type t_1^k claims to be one type higher, or t_1^{k+1}, and agent 2 with type t_2^l claims to be one type lower, or t_1^{l-1}. In the next table, we note which allocation must occur at these twelve profiles if F is to be implemented:

	t_2^1	t_2^2	t_2^3	t_2^4	t_2^5	t_2^6	
t_1^1	a					a	θ_1
t_1^2	b	b					θ_1'
t_1^3		d	d				θ_1''
t_1^4			d	d			θ_1''
t_1^5				c	c		θ_1'
t_1^6					a	a	θ_1
	θ_2	θ_2	θ_2	θ_2'	θ_2'	θ_2'	

We observe that the incentive constraints for agent 1 and agent 2 jointly form a cycle through the type space. We write y_{kl} for the transfer of agent 1 when the type profile is $t = (t_1^k, t_2^l)$. The incentive constraints that correspond to types t_1^k misreporting to be type t_1^{k+1} (modulo 6) imply (for $k = 1, 2, \ldots, 6$, respectively)

$$\begin{aligned} 0 + y_{11} &\geq 0 + y_{21}, \\ 0 + y_{22} &\geq 1 + y_{32}, \\ 0 + y_{33} &\geq 0 + y_{43}, \\ 0 + y_{44} &\geq 0 + y_{54}, \\ 0 + y_{55} &\geq 1 + y_{65}, \\ 0 + y_{66} &\geq 0 + y_{16}. \end{aligned} \tag{10}$$

The incentive constraints that correspond to types t_2^l misreporting to be type t_2^{l-1} imply, using the balanced budget to write the transfers to agent 2 as the negatives of agent 1 (for $l = 1, 2, \ldots, 6$, respectively),

$$\begin{aligned} 2 - y_{11} &\geq 2 - y_{16}, \\ 2 - y_{22} &\geq 2 - y_{21}, \\ 2 - y_{33} &\geq 2 - y_{32}, \\ 2 - y_{44} &\geq 2 - y_{43}, \\ 2 - y_{55} &\geq 2 - y_{54}, \\ 2 - y_{66} &\geq 2 - y_{65}. \end{aligned} \tag{11}$$

Inequalities (10) and (11) have a very simply structure. With very few exceptions, the payoffs that appear on the left- and right-hand sides of the inequalities are identical and only the transfers differ. These inequalities are generated either by true or misreported types, which induce only different transfer decisions but identical allocational decisions. The exceptions are the second and fifth inequality of agent 1, where a misreported type also leads to a different allocational decision. Rearranging the inequalities, we obtain

$$\begin{aligned} 0 &\geq y_{21} - y_{11}, & 0 &\geq y_{11} - y_{16}, \\ -1 &\geq y_{32} - y_{22}, & 0 &\geq y_{22} - y_{21}, \\ 0 &\geq y_{43} - y_{33}, & 0 &\geq y_{33} - y_{32}, \\ 0 &\geq y_{54} - y_{44}, & 0 &\geq y_{44} - y_{43}, \\ -1 &\geq y_{65} - y_{55}, & 0 &\geq y_{55} - y_{54}, \\ 0 &\geq y_{16} - y_{66}, & 0 &\geq y_{66} - y_{65}. \end{aligned}$$

When we sum these twelve constraints, the transfers on the right-hand side of the inequalities cancel out and we are left with the desired contradiction for any arbitrary choice of probabilities, namely $-2 \geq 0$. The transfers cancelled out because the set of incentive constraints for agent 1 and agent 2 jointly formed a cycle through the type space.

4 Separable Environments

We now present general results about the relationship between ex post implementability and interim implementability on different type spaces. The first result is an immediate implication from the definition of ex post equilibrium.

Proposition 1. *If F is ex post implementable, then F is interim implementable on any type space.*

Proof. If F is ex post implementable, then by hypothesis there exists $f^* : \Theta \to Y$ with $f^*(\theta) \in F(\theta)$ for all θ, such that for all i, all θ, and all θ_i',

$$u_i(f^*(\theta), \theta) \geq u_i(f^*(\theta_i', \theta_{-i}), \theta).$$

Consider then an arbitrary type space $\mathcal{T}$ and the direct mechanism $f : T \to Y$ with $f(t) = f^*(\hat{\theta}(t))$. Incentive compatibility now requires

$$\begin{aligned} t_i &\in \arg\max_{t'_i \in T_i} \int_{t_{-i} \in T_{-i}} u_i(f(t'_i, t_{-i}), (\hat{\theta}_i(t_i), \hat{\theta}_{-i}(t_{-i}))) d\hat{\pi}_i(t_i) \\ &= \arg\max_{t'_i \in T_i} \int_{t_{-i} \in T_{-i}} u_i(f^*(\hat{\theta}_i(t'_i), \hat{\theta}_{-i}(t_i)), (\hat{\theta}_i(t_i), \hat{\theta}_{-i}(t_{-i}))) d\hat{\pi}_i(t_i). \end{aligned}$$

This requires that

$$\begin{aligned} \hat{\theta}_i(t_i) &= \arg\max_{\theta_i \in \Theta_i} \int_{t_{-i} \in T_{-i}} u_i(f^*(\theta_i, \hat{\theta}_{-i}(t_{-i})), (\hat{\theta}_i(t_i), \hat{\theta}_{-i}(t_{-i}))) d\hat{\pi}_i(t_i) \\ &= \arg\max_{\theta_i \in \Theta_i} \sum_{\theta_{-i} \in \Theta_{-i}} \left(\int_{\{t_{-i} : \hat{\theta}_{-i}(t_{-i}) = \theta_{-i}\}} d\hat{\pi}_i(t_i) \right) \\ &\quad \times u_i(f^*(\theta_i, \theta_{-i}), (\hat{\theta}_i(t_i), \theta_{-i})), \end{aligned}$$

but by hypothesis of ex post implementability, truth-telling is a best response for every possible profile θ_{-i} and thus it remains a best response for arbitrary expectations over Θ_{-i}. Q.E.D.

The converse does not always hold, as shown by Examples 1 and 2 in the previous section, but we can identify important classes of problems for which the equivalence can be established.

4.1 *Separable Environments*

A social choice environment is *separable* if the outcome space has a common component and a private value component for each agent. Each agent cares only about the common component and his own private value component. The social choice correspondence picks a unique element from the common component and has a product structure over all components.

Thus the environment and SCC can be represented in the manner

$$Y = Y_0 \times Y_1 \times \cdots \times Y_I;$$

there exists $\tilde{u}_i : Y_0 \times Y_i \times \Theta \to \mathbb{R}$ such that

$$u_i((y_0, y_1, \ldots, y_I), \theta) = \tilde{u}_i(y_0, y_i, \theta)$$

for all $i, y \in Y$ and $\theta \in \Theta$; and there exists a function $f_0 : \Theta \to Y_0$ and, for each agent i, a nonempty valued correspondence $F_i : \Theta \to 2^{Y_i}/\emptyset$ such that

$$F(\theta) = f_0(\theta) \times F_1(\theta) \times \cdots \times F_I(\theta).$$

We observe that the private component for agent i, determined by $F_i(\theta)$, is allowed to depend on the payoff type profile θ of all agents. The common component is determined by a function, whereas the private components are allowed to be correspondences. The strength of the separability condition, represented by the product structure, is that the set of permissible private components for agent i does not depend on the choice of the private component for the remaining agents.

There are two subsets of separable environments in which we are particularly interested.[19] First, there is the case of the single-valued private component where $Y_i = \{\bar{y}_i\}$ is a single allocation for all i. In this case, there exists a representation of the utility function $\tilde{u}_i : Y_0 \times \Theta \to \mathbb{R}$ such that $\tilde{u}_i$ depends only on the common component y_0 and the payoff type profile θ. Thus any social choice *function* is separable. Second, there is the case of the classic quasilinear environment (described in Section 2). In this case, we set, for each agent i,

$$\begin{aligned} Y_i &= \mathbb{R}, \\ \tilde{u}_i(y_0, y_i, \theta) &= v_i(y_0, \theta) + y_i, \\ F_i(\theta) &= Y_i. \end{aligned}$$

In the quasilinear environment, the common component $f_0(\theta)$ will often represent the problem of implementing an efficient allocation, so that

$$f_0(\theta) = \arg\max_{y_0 \in Y_0} \sum_{i=1}^{I} v_i(y_0, \theta).$$

Whereas the designer is only interested in maximizing the social surplus and the utilities are quasilinear, there are no further restriction on the private components, here the monetary transfers, offered to the agents. In contrast, in the next section, we shall investigate the quasilinear environment *with* a balanced budget requirement as a canonical example of a nonseparable environment. By requiring a balanced budget, the SCC contains an element

[19]We would like to thank an anonymous referee for suggesting that we incorporate these two special cases in the unified language of a separable environment.

of interdependence in the choice of the private components as the transfers have to add up to zero.

Proposition 2. *In separable environments, if F is interim implementable on every common prior payoff type space $\mathcal{T}$, then F is ex post implementable.*

Proof. Suppose that F can be interim implemented on all type spaces. Then, in particular, it must be possible to interim implement F on the type space where agents other than i have type profile θ_{-i}. Thus for each i and $\theta_{-i} \in \Theta_{-i}$, there must exist $g^{i,\theta_{-i}} : \Theta_i \to Y$ such that i has an incentive to truthfully report his type,

$$\tilde{u}_i(g^{i,\theta_{-i}}(\theta_i), (\theta_i, \theta_{-i})) \geq \tilde{u}_i(g^{i,\theta_{-i}}(\theta_i'), (\theta_i, \theta_{-i})) \tag{12}$$

for all $\theta_i, \theta_i' \in \Theta_i$, and such that F is implemented, so that

$$g^{i,\theta_{-i}}(\theta_i) \in F(\theta). \tag{13}$$

If we have a separable environment, condition (13) can be rewritten as

$$\begin{aligned} g_0^{i,\theta_{-i}}(\theta_i) &= f_0(\theta_i, \theta_{-i}), \\ g_j^{i,\theta_{-i}}(\theta_i) &\in F_j(\theta_i, \theta_{-i}) \quad \text{for all } j = 1, \ldots, I; \end{aligned}$$

condition (12) can be rewritten as

$$\tilde{u}_i(f_0(\theta_i, \theta_{-i}), g_i^{i,\theta_{-i}}(\theta_i), (\theta_i, \theta_{-i})) \geq \tilde{u}_i(f_0(\theta_i', \theta_{-i}), g_i^{i,\theta_{-i}}(\theta_i'), (\theta_i, \theta_{-i})) \tag{14}$$

for all $\theta_i, \theta_i' \in \Theta_i$.

These conditions ensure ex post implementation by letting

$$f(\theta) = (f_0(\theta), g_1^{1,\theta_{-1}}(\theta_1), \ldots, g_i^{i,\theta_{-i}}(\theta_i), \ldots, g_I^{I,\theta_{-I}}(\theta_I)),$$

which completes the proof. Q.E.D.

Proposition 2 immediately implies the following strong equivalence result for a separable environment.

Corollary 1. *In separable environments, the following statements are equivalent:*

1. *F is interim implementable on all type spaces.*
2. *F is interim implementable on all common prior type spaces.*
3. *F is interim implementable on all payoff type spaces.*

4. *F is interim implementable on all common prior payoff type spaces.*
5. *F is ex post implementable.*

Proof. (1) $\Rightarrow$ (2), (3), and (4) follows by definition as we are asking for interim implementation on a smaller collection of type spaces. By Proposition 2, (4) $\Rightarrow$ (5). By Proposition 1, (5) $\Rightarrow$ (1). Q.E.D.

Given Proposition 1, whenever we can show that interim implementability on a class of type spaces implies ex post implementability, it follows that there is equivalence between ex post implementation and interim implementation on any collection of type spaces including that class. In the remainder of the paper, we do not report these immediate corollaries.

Our two leading examples of separable environments are (1) when the social choice correspondence is single-valued and (2) when the environment is quasilinear. Recent literature has established positive and negative results concerning ex post implementation in quasilinear environments (see footnote 10), motivating the ex post solution concept as reflective of the planner's ignorance about the true prior. Proposition 2 provides a foundation for the solution concept. In particular, it shows that the impossibility results in Jehiel *et al.* (2005) for ex post implementation with multidimensional signals extend to interim implementation.

Proposition 2 and Corollary 1 would be true even without the restriction to separable environments if attention were restricted to truth-telling payoff type direct mechanisms, where outcomes depend only on the reported *payoff* types. This would just be the interdependent value analogue of the classic private values observation that direct implementation for all priors implies dominant strategy implementation (Ledyard, 1978 and Dasgupta, Hammond, and Maskin, 1979). If the social choice correspondence is single-valued, then any implementing mechanism can only depend on payoff types, so the direct mechanism restriction is without loss of generality. However, the assumption is not usually without loss of generality, as Examples 1 and 2 showed.

The proof of Proposition 2 used the fact that the class of all common prior payoff type spaces contains as a special case priors where there is only uncertainty about the payoff profile of agent i, but no uncertainty about the payoff profile, $\theta_{-i} \in \Theta_{-i}$, of the remaining agents. Thus a necessary condition of implementation on all type spaces is that, for every i and every $\theta_{-i} \in \Theta_{-i}$, it is possible to solve the agent i single agent implementation problem when the payoff type profile of the remaining agents is known to be θ_{-i}. The separable condition is then enough to ensure that these necessary conditions spliced together replicate the ex post implementation problem for all agents (this is where the proof would break down in the cases of Examples 1 and 2). However, by construction, the priors used in this proof were not full support common priors. We will see in the next section the extent to which the equivalence result can be strengthened to full support common priors.

Proposition 2 used extreme looking type spaces to establish the necessity of ex post incentive compatibility. An interesting question is how rich the type space must be to make ex post incentive compatibility. In Bergemann and Morris (2004), we characterize interim incentive compatibility on arbitrary type spaces in quasilinear environments. These results can be used to construct less extreme looking type spaces where interim incentive compatibility implies ex post incentive compatibility.

4.2 *Full Support Conditions*

One obvious supplementary condition to the separable environment is to introduce compactness. Thus we say that the environment is compact if each $\tilde{u}_i(y_0, y_i, \theta)$ is continuous with respect to y_i and each $F_i(\theta)$ is a compact subset of Y_i. We observe that in the quasilinear environment, the private component is given by $F_i(\theta) = \mathbb{R}$ for all $\theta \in \Theta$ and hence $F(\theta)$ is not compact. For this reason, we will separately prove the equivalence result for the compact environment and the quasilinear environment.

Proposition 3. *In a compact separable environment, if F is interim implementable on every full support common prior payoff type space $\mathcal{T}$, then F is ex post implementable.*

Proof. Suppose that F is interim implementable on every common prior full support payoff type space. Then, for every $p \in \Delta_{++}(\Theta)$, there exists for each $i, g_i^p : \Theta \to Y_i$ such that $g_i^p(\theta) \in F_i(\theta)$ for all θ and

$$\sum_{\theta_{-i}} p(\theta_i, \theta_{-i}) \tilde{u}_i(f_0(\theta_i, \theta_{-i}), g_i^p(\theta_i, \theta_{-i}), (\theta_i, \theta_{-i}))$$
$$\geq \sum_{\theta_{-i}} p(\theta_i, \theta_{-i}) \tilde{u}_i(f_0(\theta_i', \theta_{-i}), g_i^p(\theta_i', \theta_{-i}), (\theta_i, \theta_{-i})) \qquad (15)$$

for all θ_i and θ_i'. Consider a sequence of priors with $p^n \to p^*$, where $p^*(\theta_{-i}) = 1$. By compactness of each $F_i(\cdot)$, we can choose a convergent subsequence of $g_i^{p^n}$. Writing $g_i^{\theta_{-i}}$ for the limit of that subsequence, we have

$$\tilde{u}_i(f_0(\theta_i, \theta_{-i}), g_i^{\theta_{-i}}(\theta_i, \theta_{-i}), (\theta_i, \theta_{-i}))$$
$$\geq \tilde{u}_i(f_0(\theta_i', \theta_{-i}), g_i^{\theta_{-i}}(\theta_i', \theta_{-i}), (\theta_i, \theta_{-i})) \quad (16)$$

for all i, θ, and θ_i', which ensures ex post incentive compatibility. Q.E.D.

Consider next the quasilinear environment in which the social choice correspondence is unbounded in the private component. With quasilinear utilities, it is useful to express the ex post incentive constraints as a set of linear constraints. The only data of the problem that interests us is the incentive of a payoff type θ_i to manipulate the choice of $y_0 \in Y_0$ by misreporting his payoff type. His ex post gain to reporting himself to be type θ_i' when he is type θ_i and he is sure that others have type profile θ_{-i}, is

$$\delta_i(\theta_i'|\theta_i, \theta_{-i}) \triangleq v_i(f_0(\theta_i', \theta_{-i}), \theta) - v_i(f_0(\theta_i, \theta_{-i}), \theta). \quad (17)$$

A set of transfer functions $f = (f_1, \ldots, f_I)$, each $f_i : \Theta \to \mathbb{R}$, then satisfy ex post incentive compatibility if

$$f_i(\theta_i, \theta_{-i}) - f_i(\theta_i', \theta_{-i}) \geq \delta_i(\theta_i'|\theta_i, \theta_{-i})$$

for all i, θ_i, θ_i', and θ_{-i}.

Proposition 4. *In a quasilinear environment, if F is interim implementable on every full support common prior payoff type space $\mathcal{T}$, then F is ex post implementable.*

Proof. We first show that a solution to the following maxmin problem exists for any fixed θ_{-i}:

$$\max_{f_i:\Theta_i \to \mathbb{R}} \left\{ \min_{(\theta_i, \theta_i') \in \Theta_i \times \Theta_i} \{f_i(\theta_i) - f_i(\theta_i') - \delta_i(\theta_i'|\theta_i, \theta_{-i})\} \right\}. \quad (18)$$

To show this, let M be the maximal gain or loss from misreporting of types,

$$M \triangleq \max_{(\theta_i, \theta_i') \in \Theta_i \times \Theta_i} |\delta_i(\theta_i'|\theta_i, \theta_{-i})|,$$

let $\bar{F}_i$ be the set of transfer rules bounded by $[-2M, 2M]$,

$$\bar{F}_i = \{f_i : \Theta \to [-2M, 2M]\},$$

and write $\Delta_i(f_i)$ for the lowest incentive to tell the truth under transfer rule f_i,

$$\Delta_i(f_i) \triangleq \min_{(\theta_i,\theta_i')\in\Theta_i\times\Theta_i} \{f_i(\theta_i) - f_i(\theta_i) - \delta_i(\theta_i'|\theta_i,\hat{\theta}_{-i})\}.$$

Now observe that for all $f_i \in F_i$, there exists $\bar{f}_i \in \bar{F}_i$ with $\Delta_i(f_i) \leq \Delta_i(\bar{f}_i)$. To see this, let $f_i^0(\theta_i) = 0$ for all θ_i; note that $f_i^0 \in \bar{F}_i$ and $\Delta_i(f_i^0) \geq -M$. If

$$\max_{(\theta_i,\theta_i')\in\Theta_i\times\Theta_i} |f_i(\theta_i) - f_i(\theta_i')| > 2M,$$

then $\Delta_i(f_i) < -M \leq \Delta_i(f_i^0)$. If

$$\max_{(\theta_i,\theta_i')\in\Theta_i\times\Theta_i} |f_i(\theta_i) - f_i(\theta_i')| \leq 2M,$$

fix any $\bar{\theta}_i$ and let $\tilde{f}_i(\theta_i) = f_i(\theta_i) - f_i(\bar{\theta}_i)$. Now $\tilde{f}_i \in \bar{F}_i$ and $\Delta_i(f_i) \leq \Delta_i(\tilde{f}_i)$, but now we have that the maximum in expression (18) is attained on a compact subset, so the maxmin exists.

Now suppose that ex post implementation is infeasible. Then there exist j and $\hat{\theta}_{-j}$ such that, for every $f_j : \Theta_j \to \mathbb{R}$,

$$f_j(\theta_j, \hat{\theta}_{-j}) - f_j(\theta_j', \hat{\theta}_{-j}) < \delta_j(\theta_j'|\theta_j, \hat{\theta}_{-j})$$

for some θ_j, θ_j'. Since we have shown that a solution to

$$\max_{f_j:\Theta_j\to\mathbb{R}} \left\{ \min_{(\theta_j,\theta_j')\in\Theta_j\times\Theta_j} \{f_j(\theta_j) - f_j(\theta_j') - \delta_j(\theta_j'|\theta_j,\hat{\theta}_{-j})\} \right\} \tag{19}$$

exists, there exists $\eta > 0$ such that, for every $f_j : \Theta_j \to \mathbb{R}$,

$$\min_{(\theta_j,\theta_j')\in\Theta_j\times\Theta_j} \{f_j(\theta_j) - f_j(\theta_j') - \delta_j(\theta_j'|\theta_j,\hat{\theta}_{-j})\} \leq -\eta. \tag{20}$$

Now suppose that F is interim equilibrium implementable on the payoff type space for all priors $p \in \Delta(\Theta)$. Consequently, for every p there must exist a set of transfer functions, $f_i^p : \Theta \to \mathbb{R}$, and associated interim payments,

$$f_i^p(\theta_i) \triangleq \sum_{\theta_{-i}\in\Theta_{-i}} f_i^p(\theta_i, \theta_{-i}) p(\theta_{-i}|\theta_i),$$

such that $\forall i$, $\forall \theta_i$, θ_i',

$$f_i^p(\theta_i) - f_i^p(\theta_i') \geq \sum_{\theta_{-i} \in \Theta_{-i}} \delta_i(\theta_i'|\theta_i, \theta_{-i}) p(\theta_{-i}|\theta_i). \tag{21}$$

Let

$$\xi(p) = \sup_{f_j : \theta_j \to \mathbb{R}} \left\{ \min_{(\theta_j, \theta_j') \in \Theta_j \times \Theta_j} \left\{ f_j^p(\theta_j) - f_j^p(\theta_j') \right.\right.$$
$$\left.\left. - \sum_{\theta_{-j} \in \Theta_{-j}} \delta_j(\theta_j'|\theta_j, \theta_{-j}) p(\theta_{-j}|\theta_j) \right\}\right\}.$$

For all full support p, we have

$$\xi(p) \leq -\eta + p(\hat{\theta}_{-j}|\theta_j) M$$

by (20) and

$$\xi(p) \geq 0$$

by (21). This yields a contradiction if we choose p with $p(\hat{\theta}_{-j}|\theta_j)$ sufficiently close to 1. Q.E.D.

The argument is straightforward, but distinct from the argument in Proposition 3. It proceeds by contrapositivity and relies on the linearity in monetary transfers f_i in two crucial steps. First, we can show that the problem of maximizing the minimal ex post benefits from truth-telling over all profiles and all agents is well-defined and admits a finite solution, even though the set of feasible transfers and utilities is unbounded. This allows us to conclude that if ex post implementation is infeasible, then the social choice function that maximizes the minimal benefits of ex post truth-telling (i.e., solves (18)) leads to a *strictly negative* solution. Second, we use the linearity to separate in the incentive constraints the contribution of the utility from the allocation $v_i(y_0, \theta)$ and the monetary transfer $f_i(\theta_i, \theta_{-i})$. The monetary transfer has the further property that the value of the transfer for agent i depends on neither the allocation y_0 nor on his own true payoff profile. This allows us to evaluate the value of transfers in expectations, thereby eliminating the payoff types of the other agents, exclusively on the basis of the reported type of agent i. However, then we are back at the ex post incentive constraints, from which we know from the first step that they have a strict gap and hence so do interim incentive constraints for distributions close by.

While a similar argument will apply under some weakenings of the quasilinear assumption, there is not a lot of slack. Suppose each agent's utility takes the form $u_i(y_0, \theta) + v_i(y_i, \theta_i)$, where each v_i is supermodular in (y_i, θ_i), strictly increasing in θ_i, and has range $\mathbb{R}_+$. Now each agent's benefit from his transfer is allowed to depend on his own type only. This seems like a minimal weakening of the quasilinear assumption, yet we have constructed a simple example where interim implementation on all full support payoff type spaces is possible, even though ex post implementation is impossible. We report this example in the Appendix (Bergemann and Morris, 2005c), along with an elaborate set of sufficient conditions that do extend the quasilinear result.

5 The Quasilinear Environment with Budget Balance

We now consider the quasilinear environment with budget balance as a canonical example of a nonseparable environment. There are three reasons for studying this case.

First, we are able to establish some more limited ex post equivalence results in this case. We show that if either there are only two agents or, for an arbitrary number of agents, the payoff space of each agent is binary, then the equivalence between ex post implementation and interim implementation on all type spaces holds.

Second, unlike in the case of separable environments in the previous section, we are able to identify an important class of economic environments when there is a gap between interim implementation on all type spaces and interim implementation on all payoff type spaces: in the two agent case, we show that ex post implementation is equivalent to the former but not to the latter. This confirms that our concern with the richness of the type space is not misplaced.

Finally, we know that our results are tight: once there are more than two agents and at least one agent has at least three types, we can show that there is no longer equivalence between ex post implementation and interim implementation on all type spaces. Thus within the budget balanced quasilinear environments of this section, we are able to establish the limits to ex post equivalence.

Formally, the budget balance requirement is introduced in the quasilinear environment by imposing budget balance on the private components. Thus we take the definition of a quasilinear environment in

Section 2.1 but let

$$Y = \left\{ (y_0, y_1, \ldots, y_I) \in Y_0 \times \mathbb{R}^I : \sum_{i=1}^{I} y_i = 0 \right\}.$$

Example 3 was an example of a quasilinear environment with budget balance.

We exploit a dual characterization of when ex post implementation is possible. The dual approach builds on the classic work of d'Aspremont and Gerard-Varet (1979) and the more recent works of d'Aspremont, Cremer, and Gerard-Varet (1995, 2004). In contrast to these works, we use the ex post rather than the interim dual. The dual variables of our characterization will be the multipliers of the budget balance constraints, ν, and the multipliers of the incentive constraints, λ.

Our first result concerns the two agent case. The critical type space in the argument will be the complete information type space. We used a subset of this type space earlier in Example 3 and describe it now more precisely. Let each $T_i = \Theta$ and hence a type of agent i will be written as $t_i = \theta^i \in \Theta$, where $\theta^i = (\theta_1^i, \ldots, \theta_I^i)$. We also write θ_{-i}^i for the vector θ^i excluding θ_i^i. We assume that $\hat{\theta}_i(\theta^i) = \theta_i^i$ and $\hat{\pi}_i$ satisfies

$$\hat{\pi}_i(\theta^i)[t_{-i}] = \begin{cases} 1, & \text{if } t_j = \theta^i \quad \text{for all } j \neq i, \\ 0, & \text{otherwise.} \end{cases}$$

Thus we require that for each θ, there is a type of agent i who has payoff type θ_i and assigns probability 1 to his opponents having types θ_{-i}. The complete information type space is $T = \text{X}_{i=1}^I T_i = [\text{X}_{i=1}^I \Theta_i]^I$.

Recall from (17) that we write $\delta_i(\theta_i'|\theta)$ for the ex post incentive of agent i to misreport himself to be type θ_i' when the true type profile is θ. With two agents, the ex post incentive constraints are given by

$$\begin{aligned} f_1(\theta) - f_1(r_1, \theta_2) &\geq \delta_1(r_1|\theta) \quad \forall\, r_1, \\ f_2(\theta) - f_2(\theta_1, r_2) &\geq \delta_2(r_2|\theta) \quad \forall\, r_2. \end{aligned} \tag{22}$$

We can use the budget balance condition $f_1(\theta)+f_2(\theta) = 0$ or $f_1(\theta) = -f_2(\theta)$ to combine the ex post incentive constraints (22) and observe that ex post implementation with budget balance exists if and only if there exists $f_1(\cdot)$ such that

$$f_1(\theta_1, r_2) - f_1(r_1, \theta_2) \geq \delta_1(r_1|\theta) + \delta_2(r_2|\theta) \quad \forall\, \theta,\, \forall\, r. \tag{23}$$

Proposition 5 (Equivalence with Budget Balance). $I = 2$: *If* $I = 2$ *and* F *is interim implementable on all complete information type spaces, then* F *is ex post implementable.*

Proof. We argue by contrapositivity and thus suppose that F is not ex post implementable. Then, by Farkas' lemma, there exists a nonnegative vector $(\lambda(\theta, r))_{(\theta,r)\in\Theta^2}$ such that for every $\theta \in \Theta$,

$$\sum_r \lambda((\theta_1, r_2), (r_1, \theta_2)) = \sum_r \lambda((r_1, \theta_2), (\theta_1, r_2)) \tag{24}$$

and

$$\sum_{\theta,r} \lambda(\theta, r)[\delta_1(r_1|\theta) + \delta_2(r_2|\theta)] > 0. \tag{25}$$

Let $\nu(\theta)$ denote the common value of the left- and right-hand side term in (24).

For $(\theta, r) \in \Theta^2$, we define $q(\theta, r)$ as

$$q(\theta, r) \triangleq \frac{\sum_{r'} \lambda((\theta_1, \theta_2), (r_1, r_2'))\lambda((r_1, r_2), (r_1', \theta_2))}{v(r_1, \theta_2)}. \tag{26}$$

Therefore, by (24),

$$\sum_{r_2} q((\theta_1, \theta_2), (r_1, r_2)) = \sum_{r_2} \lambda((\theta_1, \theta_2), (r_1, r_2)) \tag{27}$$

and

$$\sum_{r_1} q((r_1, r_2), (\theta_1, \theta_2)) = \sum_{r_1} \lambda((\theta_1, \theta_2), (r_1, r_2)). \tag{28}$$

so that

$$\sum_r q(\theta, r) - \sum_r q(r, \theta). \tag{29}$$

We now show that F is not implementable under the complete information common prior. In contradiction, suppose that $(f_1(\theta, \theta'), f_2(\theta, \theta'))_{(\theta,\theta')\in\Theta^2}$ is a budget balanced vector of transfers in the complete information setting and interim implements the social choice problem, i.e., for all $\theta \in \Theta$,

$$f_1(\theta, \theta) - f_1(r, \theta) \geq \delta_1(r_1|\theta) \quad \forall\, r \in \Theta \tag{30}$$

and

$$f_2(\theta,\theta) - f_2(\theta,r) \geq \delta_2(r_2|\theta) \quad \forall\, r \in \Theta. \tag{31}$$

It follows that with positive weights $q(\theta,r)$ and $q(r,\theta)$, as defined in (26), we can sum inequalities (30) and (31) to obtain

$$\sum_{\theta,r} q(\theta,r)[f_1(\theta,\theta) - f_1(r,\theta) - \delta_1(r_1|\theta)]$$
$$+\sum_{\theta,r} q(r,\theta)[f_2(\theta,\theta) - f_2(\theta,r) - \delta_2(r_2|\theta)] \geq 0.$$

Using the budget balance requirement, we can write the above inequality as

$$\sum_{\theta,r} q(\theta,r)[f_1(\theta,\theta) - f_1(r,\theta) - \delta_1(r_1|\theta)]$$
$$+\sum_{\theta,r} q(r,\theta)[f_1(\theta,r) - f_1(\theta,\theta) - \delta_2(r_2|\theta)] \geq 0. \tag{32}$$

Regarding the transfers, (29) implies that

$$\sum_{r} q(\theta,r) f_1(\theta,\theta) = \sum_{r} q(r,\theta) f_1(\theta,\theta) \quad \forall\, \theta,$$

and the remaining transfer terms cancel as well as

$$-\sum_{\theta,r} q(\theta,r) f_1(r,\theta) + \sum_{\theta,r} q(r,\theta) f_1(\theta,r) = 0.$$

The remaining terms in inequality (32) can be written as

$$\sum_{\theta,r_1} \delta_1(r_1|\theta_1,\theta_2) \sum_{r_2} q((\theta_1,\theta_2),(r_1,r_2))$$
$$+\sum_{\theta,r_2} \delta_2(r_2|\theta_1,\theta_2) \sum_{r_1} q((r_1,r_2),(\theta_1,\theta_2)). \tag{33}$$

Using (27) and (28), we can rewrite (33) as

$$\sum_{\theta,r_1} \delta_1(r_1|\theta_1,\theta_2) \sum_{r_2} \lambda((\theta_1,\theta_2),(r_1,r_2))$$
$$+\sum_{\theta,r_2} \delta_2(r_2|\theta_1,\theta_2) \sum_{r_1} \lambda((\theta_1,\theta_2),(r_1,r_2)). \tag{34}$$

Now (32) implies that expression (34) is less than or equal to zero, contradicting a property of the ex post dual solution (25). Q.E.D.

Since the equivalence holds for all complete information type spaces, it must also hold for *all type spaces*. Example 3 considered a balanced budget problem with two agents. It already indicated the crucial role of the type space for the interim implementation result. The main feature of the example was that ex post implementation and interim implementation on all type spaces was impossible, yet interim implementation on all payoff type spaces was possible. This illustrates that the equivalence result for $I = 2$ does not hold if all *complete information type spaces* are replaced with all *payoff type spaces*.

For $I = 2$ we directly used the budget balance to combine the ex post incentive constraints for agent 1 and agent 2 at a true payoff type profile θ against reports r_1 and r_2, respectively, into a single constraint for the true state θ and pair of misreports $r = (r_1, r_2)$. The resulting dual variable $\lambda(\theta, r)$ of the ex post constraint has the same dimension as the interim incentive constraints of the complete information type space with true type profile θ and report r. We then directly used the existence of $\lambda(\theta, r)$ to prove that interim implementation on the complete information type space is impossible.[20]

With more than two agents we have to consider the ex post incentive constraints of each agent separately and then link them through the additional budget balance constraints

$$f_i(\theta_i', \theta_{-i}) - f_i(\theta_i, \theta - i) + \delta_i(\theta_i'|\theta_i, \theta_{-i}) \leq 0 \quad \forall i,\ \forall \theta \tag{35}$$

and the balanced budget constraint

$$\sum_{i=1}^{I} f_i(\theta) = 0 \quad \forall \theta. \tag{36}$$

The dual problem to (35) and (36) with the multipliers $\lambda_i : \Theta_i \times \Theta_i \times \Theta_{-i} \to \mathbb{R}_+$ and $\nu : \Theta \to \mathbb{R}$ is given the ex post flow condition (EF)

$$v(\theta) = \sum_{\theta_i' \in \Theta_i} \lambda_i(\theta_i', \theta_i, \theta_{-i}) - \sum_{\theta_i' \in \Theta_i} \lambda_i(\theta_i, \theta_i', \theta_{-i}) \tag{37}$$

[20]We would like to thank an anonymous referee for suggesting the direct argument presented here.

for all $\theta \in \Theta$ and all i, and the ex post weighting condition (EW)

$$\sum_{i=1}^{I} \sum_{\theta \in \Theta} \sum_{\theta_i' \in \Theta_i} \lambda_i(\theta_i', \theta_i, \theta_{-i}) \delta_i(\theta_i' | \theta_i, \theta_{-i}) > 0. \tag{38}$$

Thus ex post implementation is impossible if and only if there exist (λ, ν) satisfying EF and EW. In the case where each agent has exactly two types, we can use this ex post dual characterization to show the impossibility of interim implementation on all payoff type spaces. In particular, if ex post implementation fails, we can construct a payoff type spaces where interim implementation fails: whenever

$$\sum_{\theta_{-i}} \lambda_i(\theta_i, \theta_i', \theta_{-i}) > 0$$

for some $\theta_i \neq \theta_i'$, let type θ_i assign probability

$$\frac{\lambda_i(\theta_i, \theta_i', \theta_{-i})}{\sum_{\theta_{-i}} \lambda_i(\theta_i, \theta_i', \theta_{-i}')} \tag{39}$$

to his opponents type profile θ_{-i} (this construction is well-defined exactly because there is only one possible $\theta_i' \neq \theta_i$). Now summing interim incentive compatibility constraints will give a contradiction.

We will show the stronger result that ex post implementation is equivalent to interim implementation on all *common prior* payoff type spaces. For this, it is necessary to establish properties of the ex post multipliers; we will show that any solution to EF and EW takes a simple form. Given a dual solution to the ex post program, we refer to $\lambda_i(\theta_i', \theta_i, \theta_{-i}) > 0$ as an *outflow* from (θ_i, θ_{-i}) and correspondingly as an inflow into (θ_i', θ_{-i}). Consistent with this language, we refer to the profile (θ_i, θ_{-i}) as a *source* if there are only outflows,

$$\sum_{\theta_i' \in \Theta_i} \lambda_i(\theta_i', \theta_i, \theta_{-i}) > 0 \quad \text{and} \quad \sum_{\theta_i' \in \Theta_i} \lambda_i(\theta_i, \theta_i', \theta_{-i}) = 0,$$

and refer to (θ_i, θ_{-i}) as a *sink* if there are only inflows,

$$\sum_{\theta_i' \in \Theta_i} \lambda_i(\theta_i', \theta_i, \theta_{-i}) = 0 \quad \text{and} \quad \sum_{\theta_i' \in \Theta_i} \lambda_i(\theta_i, \theta_i', \theta_{-i}) > 0.$$

In the simple solution, every payoff profile θ is either a sink or source, the ex post incentive multipliers $\lambda_i(\theta_i', \theta_i, \theta_{-i})$ are either 0 or 1, and the budget balance multipliers $\nu(\theta)$ are either -1 or $+1$. In graph-theoretic terms, the

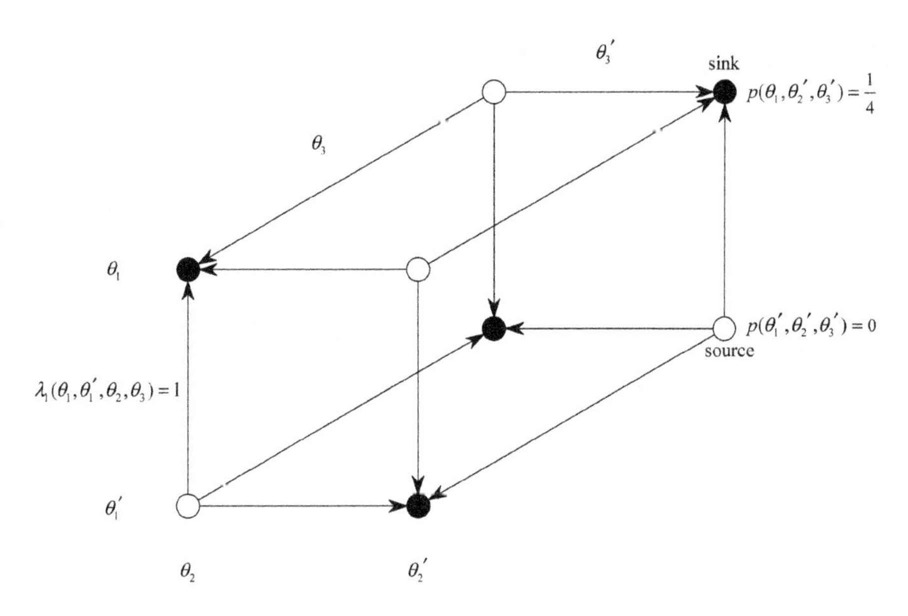

Fig. 1. Ex post dual solution.

multipliers (ν, λ) form the unique solution to the two-coloring problem, which we illustrate in Figure 1 for the case of $I = 3$.

Given this simple structure of the ex post dual, the flow equality ensures that the posteriors can be generated from a common prior. In fact, the resulting common prior $p(\cdot)$ puts uniform probability on all sources and zero probability on all the sinks as illustrated in Figure 1. The resulting type space is a common prior payoff type space with correlation. Finally, when we add up all the interim incentive constraints under these posteriors, due to the 0, 1 property of the posteriors and the balanced budget postulate, all the transfers cancel out and we are exactly left with the sum that appears in the ex post weighting inequality. By the hypothesis of the ex post dual, the sum is positive and hence the interim incentive constraints cannot be satisfied either.[21]

[21]The dual argument for the hypercube encompasses the cubical array lemma in Walker (1980) that establishes necessary and sufficient conditions for dominant strategy implementation with budget balance in a private value model. Walker considers dominant strategy implementation when the set of possible preferences is given by the class of all utility functions on a given set of allocations. This allows him to assert that the only dominant strategy incentive compatible transfer functions (without regard to budget balance) are the exact Groves schemes. Whereas the Groves schemes represent the marginal

Proposition 6 (Equivalence with Budget Balance). $I > 2$: *If* $\#\Theta_i \leq 2$ *for all* i *and* F *is interim implementable on all payoff type spaces, then* F *is ex post implementable.*

Proof. We first note that if any agent has only one type, then a well-known argument establishes that budget balance has no bite, since the single type can absorb the budget surpluses or deficit (see Mas-Collel, Whinston, and Green, 1995, p. 881). Thus suppose that $\#\Theta_i = 2$ for all i. The proof is by contradiction. Thus suppose F is not ex post implementable and hence there does not exist a solution to the ex post incentive constraints and budget balance constraints, (35) and (36). By Farkas' lemma with equality constraints, it then follows that there must exist a solution to the dual problem (37) and (38) that satisfies $\lambda_i(\theta_i', \theta_i, \theta_{-i}) \geq 0 \; \forall i, \; \forall \theta$.

Next we show that if a solution $\{v(\theta), \lambda_i(\theta_i', \theta_i, \theta_{-i})\}$ exists, then there also exists a solution such that for all i, θ and θ_i',

$$\nu(\theta) \in \{-1, 1\}, \quad \lambda_i(\theta_i', \theta_i, \theta_{-i}) \in \{0, 1\},$$
$$\lambda_i(\theta_i', \theta_i, \theta_{-i}) + \lambda_i(\theta_i, \theta_i', \theta_{-i}) = 1. \tag{40}$$

The binary payoff type space implies that for a given θ_i, payoff type $\theta_i' \neq \theta_i$ is uniquely determined. We first observe that a necessary condition for interim (and ex post) *incentive compatibility* on all payoff type spaces is that for all i and all θ_{-i},

$$f_i(\theta_i', \theta_{-i}) - f_i(\theta_i, \theta_{-i}) + \delta_i(\theta_i' | \theta_i, \theta_{-i}) \leq 0.$$

and

$$f_i(\theta_i, \theta_{-i}) - f_i(\theta_i', \theta_{-i}) + \delta_i(\theta_i | \theta_i', \theta_{-i}) \leq 0.$$

By summing the two inequalities, we obtain that for all i and all θ_{-i},

$$\delta_i(\theta_i' | \theta_i, \theta_{-i}) + \delta_i(\theta_i | \theta_i', \theta_{-i}) \leq 0. \tag{41}$$

contributions of each agent at each type profile, budget balance can be translated into an equality constraint on the sum of the differences of the social valuations at the true profiles on the hypercube. Generically, the social values will not satisfy the equality. In the current model, we are only considering a finite set of preferences for each agent, and hence the set of dominant strategy incentive compatible transfers (without regard to budget balance) is larger than the set of exact Groves schemes. For the hypercube, this implies that the sum of differences in Walker is always strictly larger than our weighting inequality, and thus if budget balance fails on the hypercube and the weighting inequality is positive, then the Groves schemes will necessarily fail as well.

Based on the given solution $\{\nu(\theta), \lambda_i(\theta_i', \theta_i, \theta_{-i})\}$, we then propose a new solution $\{\nu(\theta), \hat{\lambda}_i(\theta_i', \theta_i, \theta_{-i})\}$, which is defined by

$$\hat{\lambda}_i(\theta_i', \theta_i, \theta_{-i}) \triangleq \max\{\lambda_i(\theta_i', \theta_i, \theta_{-i}) - \lambda_i(\theta_i, \theta_i', \theta_{-i}), 0\}$$

and, correspondingly,

$$\hat{\lambda}_i(\theta_i, \theta_i', \theta_{-i}) \triangleq \max\{\lambda_i(\theta_i, \theta_i', \theta_{-i}) - \lambda_i(\theta_i', \theta_i, \theta_{-i}), 0\}.$$

By construction the new solution satisfies the equality constraints (37) under the original values $\nu(\theta)$ and by (41) weakly increases the right-hand side of the inequality constraint (38). Accordingly, equalities (37) simplify to either

$$\nu(\theta) = \hat{\lambda}_i(\theta_i', \theta_i, \theta_{-i}) \quad \forall_i \tag{42}$$

or

$$\nu(\theta) = -\hat{\lambda}_i(\theta_i, \theta_i', \theta_{-i}) \quad \forall_i. \tag{43}$$

Due to the binary property of the type space Θ_i and the fact that equalities (42) and (43) have to hold for all agents simultaneously, we obtain a bipartition of the type space Θ into subsets Θ' and Θ'' (in graph-theoretic terms they form the unique solution to the two-coloring problem) such that for all $\theta \in \Theta'$, $\nu(\theta) > 0$ and for all $\theta \in \Theta''$, $\nu(\theta) < 0$. We can finally normalize $\nu(\theta)$ and $\hat{\lambda}_i(\theta_i', \theta_i, \theta_{-i})$ by dividing through $|\nu(\theta)|$ to obtain a solution, denoted by $\{\nu^*(\theta), \lambda_i^*(\theta_i', \theta_i, \theta_{-i})\}$, with the desired properties described in (40). Inequality (38) now reads

$$\sum_{i=1}^{I} \sum_{\theta \in \Theta'} \delta_i(\theta_i' | \theta_i, \theta_{-i}) > 0. \tag{44}$$

We obtain a contradiction to (44) by considering the interim implementation for the payoff prior, which puts uniform probability on all $\theta \in \Theta'$ and zero probability on all $\theta \in \Theta''$. By the hypothesis of interim implementability, the interim incentive constraints for every i and every θ_i,

$$\sum_{(\theta_i, \theta_{-i}) \in \Theta'} [f_i(\theta_i', \theta_{-i}) - f_i(\theta_i, \theta_{-i}) + \delta_i(\theta_i' | \theta_i, \theta_{-i})] p(\theta_{-i} | \theta_i) \leq 0,$$

can be satisfied with a balanced budget transfer scheme. By summing the interim incentive constraints over all agents and omitting the constant (on Θ') probability $p(\theta_{-i}|\theta_i)$, we get

$$\sum_{i=1}^{I} \sum_{(\theta_i,\theta_{-i})\in\Theta'} [f_i(\theta_i', \theta_{-i}) - f_i(\theta_i, \theta_{-i}) + \delta_i(\theta_i'|\theta_i, \theta_{-i})] \leq 0,$$

and by the balanced budget stipulation, the transfers drop out and we are left with

$$\sum_{i=1}^{I} \sum_{(\theta_i,\theta_{-i})\in\Theta'} \delta_i(\theta_i'|\theta_i, \theta_{-i}) \leq 0,$$

which provides the desired contradiction to (44). Q.E.D.

In an earlier version of this paper (Bergemann and Morris, 2003), we demonstrated, by means of an example, the tightness of the ex post equivalence results obtained in Propositions 5 and 6. Example 4 consisted of three agents in which the first agent had three payoff types and the remaining two agents had binary payoff type spaces. With this minimal relaxation of either of the above sufficient conditions, we have an example where ex post implementation is impossible while interim implementation (using a single mechanism) is possible on all type spaces. We conjecture that ex post equivalence results may again be obtained in a general environment with $I > 2$ and $\#\Theta_i > 2$ only after imposing suitable restrictions on the environment such as single crossing or supermodularity conditions.

6 Discussion

6.1 *A Classical Debate*

An old debate in the Bayesian implementation literature went as follows. Some scholars pointed out that — as a practical matter — the planner was unlikely to know the true prior over the type space. Therefore, it would be desirable to have a mechanism that was going to work independently of the prior. For a private values environment, Dasgupta, Hammond, and Maskin (1979), Ledyard (1978, 1979), and Groves and Ledyard (1987) observed that if a direct mechanism was going to implement a social choice correspondence for every prior on the type space, then there must be dominant strategies implementation. Other scholars pointed out that if the planner did not know the prior (and the agents do), then we should not restrict attention to direct mechanisms; rather, we should allow the mechanism to elicit reports of the true prior from the agents (since this information is nonexclusive in the sense of Postlewaite and Schmeidler (1986), this elicitation will not lead to any incentive problems). A formal application of this folk argument appears in the recent work of Choi and Kim (1999). How do our results fit into this debate between the "practical designers" and the "implementation purists"?

Our results allow for interdependent values, but we believe they clarify this debate when restricted to private values (recall that ex post incentive compatibility implies dominant strategies incentive compatibility under private values).

- In some environments, even if the designer was allowed to elicit the true prior, implementation for every prior on the fixed type space implies dominant strategy implementation. In these environments, the practical designers' conclusion is immune to the purists' criticism. These environments include separable environments (Proposition 2) and quasilinear environments with budget balance but at most two types for each player (Proposition 6).[22]
- In some environments, the purists' criticism binds. That is, dominant strategy implementation is impossible, but Bayesian implementation is possible for every prior on a fixed type space. This was true of Examples 1 and 2, and we can also construct quasilinear environments with budget balance where it is true.
- A second practical criticism of the classical Bayesian implementation literature is that not only may the planner not know the true prior over the payoff types, but the agents may not know the true prior either. We have formalized this criticism by requiring implementation on type spaces larger than the payoff type space, and we have shown that in some environments, implementation on all type spaces implies dominant strategies implementation even when interim implementation for all priors on the payoff type space does not (Example 3 and Proposition 5).[23]

[22]Mookerjee and Reichelstein (1992) examine the relationship between Bayesian implementation and dominant strategy implementation in the private value environment. If dominant strategy implementation of an allocation rule is possible, then it is possible to do so in a way that generates any expected transfer levels achievable under Bayesian implementation of that allocation rule.

[23]Example 3 has interdependent values, but we could mechanically turn it into a private value example the same way we constructed Example 2 as a private value version of Example 1.

6.2 *Genericity*

If we restrict attention to "generic" priors on the payoff type space (or any fixed finite type space), it is possible to obtain very permissive implementation results. Thus arguments in d'Aspremont and Gerard-Varet (1979) and d'Aspremont, Cremer, and Gerard-Varet (1995, 2004) establish that it is possible to implement any allocation rule in a quasilinear environment with budget balance for a generic set of priors on all fixed type spaces. This contrasts with our results that show that implementation in some quasilinear environments with budget balance for all priors on (the fixed) payoff type space is equivalent to ex post implementation, which is known to be impossible under quite general conditions.

As emphasized by Neeman (2004), "generic" priors entail some counterintuitive properties, e.g., that a planner can infer an agent's valuation of an object from that agent's beliefs about other agents' types. In any case, the justification for fixing a set of types, "generically" picking a prior, and *then* assuming common knowledge of that prior is not clear. Some current work tries to identify more natural ways to think about genericity.[24] In this work, we have not discussed any results that rely on genericity notions.[25]

6.3 *Augmented Ex Post Equivalence*

In nonseparable environments, ex post implementability may be a strictly stronger requirement than interim implementability on all type spaces. Is there a natural weakening of ex post implementability that is necessary? Consider an augmented mechanism where each agent's report consists of his payoff type and a supplemental message. An agent's strategy is truthful if he always correctly reports his payoff type. A decision rule that maps message profiles into outcomes is augmented ex post incentive compatible if an agent who expects all other agents to report truthfully has a truthful best response. A social choice correspondence F is augmented ex post

[24] Morris (2002) and Dekel, Fudenberg, and Morris (2005) examine ways to define "strategic topologies" on types in the universal type space that might suggest useful topological notions of genericity. Heifetz and Neeman (2004) argue that among common prior belief closed subspaces of the universal type space, the type spaces often described as generic are not "prevalent" in the sense of Christiansen (1974), Hunt, Sauer, and Yorke (1992), and Anderson and Zame (2001).

[25] Genericity issues are discussed at greater length in the working paper version of this paper, Bergemann and Morris (2003).

implementable if there exists an augmented ex post incentive compatible decision rule that (under truthful strategies) always achieves outcomes in Γ. We showed in the working paper version of this paper (Bergemann and Morris (2003)) that (up to some technical restrictions) augmented ex post incentive compatibility is equivalent to interim implementability on all type spaces. Now an interesting way to characterize implementation problems is how many supplemental messages are needed. For separable environments, no extra messages are needed. In the worst case, the supplemental message might consist of the agent's belief over T_{-i}^{*} in the universal type space and we would be looking at a direct mechanism on the universal type space. An interesting problem for future research is the characterization of how many supplemental messages are required for different classes of problems.

References

ANDERSON, R., AND W. ZAME (2001): "Genericity with Infinitely Many Parameters," *Advances in Theoretical Economics*, 1, available at http://www.bepress.com/bejte/advances/vol1/iss1/art1

BATTIGALLI, P., AND M. SINISCALCHI (2003): "Rationalization and Incomplete Information," *Advances in Theoretical Economics*, 3, available at http://www.bepress.com/bejte/advances/vol3/iss1/art3

BERGEMANN, D., AND S. MORRIS (2003): "Robust Mechanism Design," Discussion paper 1421 Cowles Foundation, available at http://ssrn.com/abstract=412497.

——— (2004): "Bayesian Implementation of a Social Choice Rule in a Quasilinear Environment with Arbitrary Type Spaces."

——— (2005a): "Ex post Implementation," Discussion paper 1502, Cowles Foundation, available at http://ssrn.com/abstract=703042.

——— (2005b): "Robust Implementation: The Role of Large Type Spaces," Discussion paper 1519, Cowles Foundation, available at http://ssrn.com/abstract=748184.

——— (2005c): "Supplement to 'Robust Mechanism Design'," available at http:// www.econometric-society.org/ecta/supmat/4664appendix.pdf.

BERGEMANN, D., AND J. VALIMAKI (2002): "Information Acquisition and Mechanism Design," *Econometrica*, 70, 1007–1033.

BRANDENBURGER, A., AND E. DEKEL (1993): "Hierarchies of Beliefs and Common Knowledge," *Journal of Economic Theory*, 59, 189–198.

Choi, J., and T. Kim (1999): "A Nonparametric, Efficient Public Good Decision Mechanism: Undominated Bayesian Implementation," *Games and Economic Behavior*, 27, 64–85.

Christiansen, J. (1974): *Topology and Borel Structure*, Mathematical Studies, Vol. 10. Amsterdam, North-Holland.

Chung, K., and J. Ely (2003): "Implementation with Near-Complete Information," *Econometrica*, 71, 857–871.

Cremer, J., and R. McLean (1985): "Optimal Selling Strategies under Uncertainty for a Discriminating Monopolist When Demands are Interdependent," *Econometrica*, 53, 345–361.

Dasgupta, P., P. Hammond, and E. Maskin (1979): "The Implementation of Social Choice Rules: Some General Results on Incentive Compatibility," *Review of Economic Studies*, 46, 185–216.

Dasgupta, P., and E. Maskin (2000): "Efficient Auctions," *Quarterly Journal of Economics*, 115, 341–388.

D'Aspremont, C., J. Cremer, and L.-A. Gerard-Varet (1995): "Correlation, Independence and Bayesian Incentives," Working paper, University of Toulouse.

——— (2004): "Balanced Bayesian Mechanisms," *Journal of Economics Theory*, 115, 385–396.

D'Aspremont, C., and L.-A. Gerard-Varet (1979): "Incentives and Incomplete Information," *Journal of Public Economics*, 11, 25–45.

Dekel, E., D. Fudenberg, and S. Morris (2005): "Topologies on Types," available at http://ideas.repec.org/plcla/levrem/784828000000000061.html

Duggan, J., and J. Roberts (1997): "Robust Implementation," Working paper, University of Rochester.

Eliaz, K. (2002): "Fault Tolerant Implementation," *Review of Economic Studies*, 69, 589–610.

Groves, T., and J. Ledyard (1987): "Incentive Compatibility Since 1972," In: *Information, Incentives and Economic Mechanisms*, ed. by T. Groves, R. Radner, and S. Reiter, pp. 48–111. University of Minnesota Press, Minneapolis.

Hagerty, K., and W Rogerson (1987): "Robust Trading Mechanisms," *Journal of Economic Theory*, 42, 94–107.

Harsanyi, J. (1967/1968): "Games with Incomplete Information Played by Bayesian Agents," *Management Science*, 14, 159–182, 320–334, 486–502.

HEIFETZ, A. (1993): "The Bayesian Formulation of Incomplete information — The Non-Compact Case," *International Journal of Game Theory*, 21, 329–338.

HEIFETZ, A., AND Z. NEEMAN (2004): "On the Generic (Im)Possibility of Full Surplus Extraction in Mechanism Design," Discussion paper 349, Center for Rationality and Interactive Decision Theory, Hebrew University, Jerusalem.

HEIFETZ, A., AND D. SAMET (1998): "Topology-Free Typology of Beliefs," *Journal of Economic Theory*, 82, 324–341.

——— (1999): "Coherent Beliefs and Not Always Types," *Journal of Mathematical Economics*, 32, 475–488.

HOLMSTROM, B., AND R. MYERSON (1983): "Efficient and Durable Decision Rules with Incomplete Information," *Econometrica*, 51, 1799–1820.

HUNT, B., T. SAUER, AND J. YORKE (1992): "Prevalence: A Translation-Invariant 'Almost Every' on Infinite-Dimensional Spaces," *Bulletin of the American Mathematical Society (New Series)*, 27, 217–238.

HURWICZ, L. (1972): "On Informationally Decentralized Systems," In: *Decision and Organization*, ed. by R. Radner and C. McGuire, pp. 297–336. North-Holland, Amsterdam.

JEHIEL, P., AND B. MOLDOVANU (2001): "Efficient Design with Interdependent Valuations," *Econometrica*, 69, 1237–1259.

JEHIEL, P., M. MEYER-TER-VEHN, B. MOLDOVANU, AND W. R. ZAME (2005): "The Limits of Ex-Post Implementation," available at http://ideas.repec.org/plcla/lervem/66156000000000548.html

KALAI, E. (2004): "Large Robust Games," *Econometrica*, 72, 1631–1665.

LEDYARD, J. (1978): "Incentive Compatibility and Incomplete Information," *Journal of Economic Theory*, 18, 171–189.

——— (1979): "Dominant Strategy Mechanisms and Incomplete Information," In: *Aggregation and Revelation of Preferences*, ed. by J.-J. Laffont, pp. 309–319. North-Holland, Amsterdam.

LOPOMO, G. (1998): "The English Auction is Optimal among Simple Sequential Auctions," *Journal of Economic Theory*, 82, 144—166.

——— (2000): "Optimality and Robustness of the English Auction," *Games and Economic Behavior*, 36, 219–240.

MAS-COLELL, A., M. WHINSTON, AND J. GREEN (1995): *Microeconomic Theory*. Oxford University Press, Oxford, UK.

Maskin, E. (1992): "Auctions and Privatization," In: *Privatization: Symposium in Honor of Herbert Giersch*, ed. by H. Siebert. Tuebingen: J. C. B. Mohr, 115–136.

Mertens, J.-F., S. Sorin, and S. Zamir (1994): "Repeated Games. Part A: Background Material," CORE Discussion paper 9420, Universite Catholique de Louvain.

Mertens, J.-F., and S. Zamir (1985): "Formulation of Bayesian Analysis for Games of Incomplete Information," *International Journal of Game Theory*, 14, 1–29.

Mookherjee, D., and S. Reichelstein (1992): "Dominant Strategy Implementation of Bayesian Incentive Compatible Allocation Rules," *Journal of Economic Theory*, 56, 378–399.

Morris, S. (2002): "Typical Types," available at http://www.princeton.edu/~smorris/pdfs/typicaltypes.pdf

Morris, S., and H. S. Shin (2003): "Global Games: Theory and Applications," In: *Advances in Economics and Econometrics. Proceedings of the Eighth World Congress of the Econometric Society*, ed. by M. Dewatripont, L. Hansen, and S. Turnovsky, pp. 56–114. Cambridge University Press, Cambridge, UK.

Myerson, R. (1991): *Game Theory*. Harvard University Press, Cambridge, MA.

Neeman, Z. (2004): "The Relevance of Private Information in Mechanism Design," *Journal of Economic Theory*, 117, 55–77.

Perry, M., and P. Reny (2002): "An Ex Post Efficient Auction," *Econometrica*, 70, 1199–1212.

Postlewaite, A., and D. Schmeidler (1986): "Implementation in Differential Information Economies," *Journal of Economic Theory*, 39, 14–33.

Walker, M. (1980): "On the Nonexistence of a Dominant Strategy Mechanism for Making Optimal Public Decisions," *Econometrica*, 48, 1521–1540.

Wilson, R. (1985): "Incentive Efficiency in Double Auctions," *Econometrica*, 53, 1101–1115.

——— (1987): "Game-Theoretic Analyses of Trading Processes," In: *Advances in Economic Theory: Fifth World Congress*, ed. by T. Bewley, pp. 33–70. Cambridge University Press, Cambridge, UK.

CHAPTER 2

Ex Post Implementation*

Dirk Bergemann and Stephen Morris

Abstract
We analyze the problem of fully implementing a social choice set in ex post equilibrium. We identify an *ex post monotonicity* condition that is necessary and — in economic environments — sufficient for full implementation in ex post equilibrium. We also identify an ex post monotonicity no veto condition that is sufficient. Ex post monotonicity is satisfied in all single crossing environments with *strict* ex post incentive constraints.

We show by means of two classic examples that ex post monotonicity does not imply nor is it implied by Maskin monotonicity. The single unit auction with interdependent valuations is shown to satisfy ex post monotonicity but not Maskin monotonicity. We further describe a Pareto correspondence that fails ex post monotonicity but satisfies Maskin monotonicity.

Keywords: Ex post equilibrium, implementation, single crossing, interdependent values

1 Introduction

Recent research in auction theory, and mechanism design theory more generally, has led to a better understanding of models with interdependent rather than private values. Much of this work has used the solution concept of ex post rather than Bayesian equilibrium.[1] The analysis of ex post equilibrium is considerably more tractable, because incentive compatible

*This research has been supported by NSF Grants #SES-0095321 and #SES-0518929.

[1]See Dasgupta and Maskin (2000), Perry and Reny (2002) and Bergemann and Välimäki (2002) among many others.

transfers can frequently be derived with ease and single crossing conditions generating incentive compatibility are easy to identify. A conceptual advantage of ex post equilibrium is its robustness to the informational assumptions about the environment. In particular, it often seems unrealistic to allow the mechanism to depend on the designer's knowledge of the type space as Bayesian mechanisms do.[2]

Research on interdependent values has focussed almost exclusively on the incentive compatibility of the social choice problem in the direct mechanism. In contrast, this paper focuses on the problem of *full* rather than *partial* implementation. The task for the designer, who does not know the agents' types, is to choose a mechanism such that in *every* equilibrium of the mechanism, agents' play of the game results in the outcome specified by the social choice objective at every type profile. If the social choice problem is described by a social choice set, a set of social choice functions, then full implementation also requires that every selection from the set can be realized as an ex post equilibrium under the mechanism. This problem has been analyzed under the assumption of complete information, i.e., there is common knowledge among the agents of their types (see Maskin, 1999). It has also been analyzed under the assumption of incomplete information, with common knowledge among the agents of the prior (or the priors) according to which agents form their beliefs (see Postlewaite and Schmeidler, 1986; Palfrey and Srivastava, 1989a and Jackson, 1991). While complete information (or Nash) implementation and incomplete information (or Bayesian) implementation are well understood, the ex post implementation problem has not been analyzed. In this paper, we develop necessary and sufficient conditions for ex post implementation, both in general environments and also in settings of special interest for auction theory.

A strategy profile in an incomplete information game is an ex post equilibrium if each action profile is a Nash equilibrium at every type profile. Put differently, each player's incomplete information strategy mapping types to messages must remain a best response even if he knew the types of his opponents. We introduce an ex post monotonicity condition that — along

[2]In this paper, we take the solution concept of ex post equilibrium as given. At the end of the introduction, we discuss alternative justifications for focussing on ex post equilibrium in this context, including our earlier contribution Bergemann and Morris (2005c).

with ex post incentive compatibility — is necessary for ex post implementation. We show that a slight strengthening of ex post monotonicity — the ex post monotonicity no veto condition — is sufficient for implementation with at least three agents. The latter condition reduces to ex post monotonicity in economic environments. These results are the ex post analogues of the Bayesian implementation results of Jackson (1991), and we employ similar arguments to establish our results. But just as ex post incentive compatibility conditions are easier to verify and interpret, the ex post monotonicity condition is easier to verify and interpret than the Bayesian monotonicity condition, because it depends on complete information utilities and does not involve the prior or posterior distributions of the agents.

Because an ex post equilibrium is a Nash equilibrium at every type profile, there is a natural relationship between ex post and Nash implementation. When we compare the *complete* with the *incomplete* information settings, two important differences regarding the ability of the agents to sustain equilibrium behavior emerges. With complete information, the agents have the ability to coordinate their actions at every preference profile. Yet with complete information the designer can also detect individual deviations from the reports of the other agents. The ability of the agents to coordinate in complete information settings makes the task of implementing the social choice outcome more difficult for the designer, but it is made easier by the lack of individual incentive constraints. With incomplete information, the first problem becomes easier, but the second becomes harder. As these two effects are in conflict, we will show that ex post and Maskin monotonicity are not nested notions. In particular, either one of them can hold while the other fails. Interestingly, in the class of single crossing environments, ex post monotonicity is always guaranteed as is Maskin monotonicity. Even though ex post monotonicity has to include ex post incentive constraints absent in the complete information world, it turns out that the local property of single crossing indifference curves is sufficient to guarantee ex post monotonicity in the presence of strict rather than weak ex post incentive constraints.

The "augmented" mechanisms used to obtained our general positive ex post implementation results inherit some complex and unsatisfactory features from their complete information and Bayesian counterparts. The hope often expressed in the literature is that it should be possible to show in specific settings that less complex mechanisms are required. We are able to identify a number of important settings where ex post implementation is only possible when it is possible in the direct mechanism. This is true, for example, if the social choice function has a sufficiently wide range or if the environment is supermodular.

We also use the methods developed for the general case to show that the direct mechanism has a unique ex post equilibrium in the problem of efficiently allocating goods when bidders have interdependent values (see Dasgupta and Maskin, 2000 and Perry and Reny, 2002). And in this context, the interdependent value model delivers new and positive results. With at least three symmetric bidders, we show that the generalized Vickrey–Clark–Groves (VCG) allocation can be ex post implemented in the direct mechanism, even though Maskin monotonicity fails. This result is in stark contrast with the impossibility to Nash implement the single unit auction with private values. The positive result relies on interdependence. The latter intuition is also confirmed by contrasting our positive results with a recent result by Birulin (2003). He shows that with two bidders there are multiple and inefficient ex post equilibria in the single unit auction. With two agents, bidder i can use a non-truthful bidding strategy to exactly offset a non-truthful bidding strategy of bidder j. With more than two agents, the strategy of bidder i cannot incorporate anymore the bidding behavior by j and k and truthtelling becomes the unique ex post equilibrium strategy.

For twenty years from the mid-1970s to the mid-1990s, there was a large literature devoted to the problem of full implementation. While elegant characterizations of implementability were developed, the "augmented" mechanisms required to achieve positive results were complex and seemed particularly implausible. While the possibility of multiple equilibria does seem to be a relevant one in practical mechanism design problems, particularly in the form of collusion and shill bidding, the theoretical literature is not seen as having developed practical insights (with a few recent exceptions such as Ausubel and Milgrom, 2006 and Yokoo *et al.*, 2004). The gap between pure implementation theory and practical market design has appeared especially stark when thinking about full implementation. Following Wilson (1987), we hope that by relaxing unrealistic implicit common knowledge assumptions, we will deliver predictions that are more robust and practical. While the complete information implementation literature makes the assumption of common knowledge of preferences, the Bayesian implementation literature makes the assumption that there is common knowledge of a prior on a fixed set of types; this both seems unlikely to practical market designers and is a substantive constraint when viewed as a restriction on all possible beliefs and higher order beliefs.[3] Our hope is partially vindicated by the results of this paper: it does turn out that in many environments of interest, augmented mechanisms cannot deliver ex post implementation when direct mechanisms cannot. Of course, direct mechanisms also have their own robustness critiques, but a number of important papers have shown that, in the type of interdependent environments discussed in this paper, direct mechanisms can be replicated by more plausible auction mechanisms.[4]

This paper identifies the extra condition (ex post monotonicity) required for full implementation when the ex post incentive compatibility (EPIC) conditions already required for partial implementation are possible. Our results are only of interest if there are interesting environments where the EPIC conditions hold. A recent literature has identified sufficient conditions for EPIC in a widening variety of settings, when agents have one dimensional types (see the references in footnote 1). On the other hand, Jehiel *et al.* (2006) show in a recent paper that no non-trivial social choice functions are ex post implementable in a "generic" class of environments with multidimensional signals. The results in this paper are moot in such environments. But some economically natural problems with multidimensional signals will fail the genericity requirement.[5]

In this paper, we take the solution concept of ex post equilibrium as given and ask when full implementation in ex post equilibrium is possible. The interpretation of our results is sensitive to the justification for using ex post equilibrium as a solution concept. One justification would be a refinement argument: when ex post equilibria exist, they are more compelling as a solution concept for how the game might be played in a Bayesian setting, even when there are other Bayesian equilibria. Every ex post equilibrium has the important no regret property which means that no agent would have an incentive to change his report even if he were to be informed of the

[3]The common knowledge of a common prior assumption has particularly strong consequences in mechanism design, as argued by Neeman (2004) and Bergemann and Morris (2005c).

[4]See Dasgupta and Maskin (2000) and Perry and Reny (2002).

[5]Eso and Maskin (2000) obtain EPIC allocations in a multi-good setting: they assume that each agent has a distinct signal for each good; utility is additive across goods, but there may be constraints on what bundles of goods can be bought by a given agent; and a third buyer's signal has an identical effect on any pair of bidders' valuations. Bikhchandani (2006) shows that it is possible to construct EPIC allocations in a private good model with multidimensional signals, and argues that the genericity condition in Jehiel *et al.* (2006) requires externalities as well as interdependence.

true type profile of the other agents. In this sense, the refinement of an ex post equilibrium shares the spirit of the subgame perfection refinement with respect to the move by nature. A related justification is that only ex post equilibria have the property that agents do not have incentives to invest in finding out other agents' types; such incentives would lead to instability of the underlying mechanism; anticipating this, we might expect them to play the ex post equilibrium.[6] Under either of these justifications, our full implementation results have a natural positive interpretation: under weak additional assumptions, there are no further ex post equilibria.

In an earlier contribution, Bergemann and Morris (2005c), we formalized a different justification for focussing on ex post equilibrium in the context of truthful (or partial) implementation. We showed that in many environments a social choice problem can be truthfully Bayesian implemented for all beliefs and higher order beliefs about payoff types if and only if it can be truthfully ex post implemented. Thus for partial implementation, one may focus on ex post equilibrium. However, the fact that there is a unique ex post equilibrium does not imply that there are no other Bayesian equilibria on some type spaces. In companion papers, Bergemann and Morris (2005a, 2005b), we address the more demanding requirement of "robust implementation": all Bayesian equilibria on all type spaces must deliver the right outcome. While the incentive compatibility constraints for this problem are the same as for the ex post implementation problem,[7] the resulting "robust monotonicity" condition (equivalent to Bayesian monotonicity on all type spaces) is strictly stronger than ex post monotonicity (and Maskin monotonicity). The resulting robust monotonicity notions provide the full implementation counterparts to the robust mechanism design (i.e., partial implementation) questions pursued in Bergemann and Morris (2005b). Since robust implementation is a much stronger requirement than ex post implementation, our results in this paper can then be seen as highlighting the dangers of a naive adoption of ex post equilibrium as a "robust" solution concept.

The paper is organized as follows. Section 2 describes the formal environment and solution concepts. Section 3 introduces the notion of ex post monotonicity and compares it to Maskin monotonicity in a simple public

[6]This was a folk justification for focussing on dominant strategies equilibria (in private values settings) in the early days of the mechanism design literature.
[7]This follows from results in Bergemann and Morris (2005c).

good example. Section 4 shows that ex post monotonicity is necessary and, in economic environments, also sufficient for ex post implementation. We also provide a sufficient condition — ex post monotonicity no veto — for non-economic environments. Section 5 considers an important class of single crossing environments; we show that ex post monotonicity is satisfied in all single crossing environments if the social allocation problem satisfies strict rather than weak ex post incentive constraints. Section 6 provides sufficient conditions under which ex post implementation is possible in the direct mechanism. Section 7 considers the single unit auction environment. It is an important example as it fails Maskin monotonicity and has weak ex post incentive constraints almost everywhere; yet it satisfies ex post monotonicity and ex post implementation is possible in the direct mechanism. Section 8 presents a Pareto social choice set with the converse implementation properties: it satisfies Maskin monotonicity but fails ex post monotonicity. Section 9 extends the analysis to mixed strategy implementation and the use of stochastic mechanisms. Section 10 concludes.

2 Model

We fix a finite set of agents, $1, 2, \ldots, I$. Agent i's *type* is $\theta_i \in \Theta_i$. We write $\theta \in \Theta = \Theta_1 \times \cdots \times \Theta_I$. There is a set of outcomes Y. Each agent has utility function $u_i : Y \times \Theta \to \mathbb{R}$. Thus we are in the world of interdependent values, where an agent's utility may depend on other agents' types.[8] A *social choice function* is a function from states to allocations, or $f : \Theta \to Y$. The set of all social choice functions is $\mathcal{F} = \{f \mid f : \Theta \to Y\}$. A *social choice set* F is a subset of $\mathcal{F}$.[9]

[8]We represent the preferences of the agents by utility functions rather than by preference relations as much of the mechanism design literature with interdependent values uses utility functions rather than preferences. However, all our results (with the exception of the mixed strategy implementation results in Section 9) only rely on ordinal properties of the preferences and all results could be restated in terms of preferences rather than utility functions. This is made precise in Section 4.

[9]In the literature on complete information implementation, it is customary to use social choice correspondences (see Maskin, 1999) whereas in the literature on incomplete information implementation (see Postlewaite and Schmeidler, 1986 and Jackson, 1991) it is customary to use social choice sets. We shall discuss some of the issues regarding ex post implementation of functions, sets and correspondences in Section 8 in conjunction with the Pareto correspondence.

In the tradition of the implementation literature, we describe the implementation problem here for deterministic mechanisms and pure strategies. In Section 9 we extend the analysis to implementation with stochastic mechanisms and mixed strategies. We postpone the relevant modifications to accommodate mixed strategies until then.

A planner must choose a *game form* or *mechanism* for the agents to play in order to determine the social outcome. Let m_i be a message of agent i, M_i be the set of messages available to i and a message profile is denoted by $m = (m_1, m_2, \ldots, m_I) \in M = \times_{i=l}^{I} M_i$. Let $g: M \to Y$ be the outcome function and $g(m) = y$ be a specific outcome if message profile m is chosen. Thus a mechanism is a collection:

$$\mathcal{M} = (M_1, \ldots, M_I, g(\cdot)).$$

Unless otherwise stated, we make no additional assumptions on the structure of the type space Θ, the outcome space Y, or the message space M.[10]

For the given environment, we can combine the type space Θ with a mechanism $\mathcal{M}$ to get an incomplete information game. We wish to analyze the ex post equilibria of the incomplete information game (without a prior). A pure strategy in this game is a function $s_i: \Theta_i \to M_i$.

Definition 1 (*Ex post equilibrium*). A pure strategy profile $s^* = (s_1^*, \ldots, s_I^*)$ is an *ex post equilibrium* if

$$u_i(g(s^*(\theta)), \theta) \geq u_i(g(m_i, s_{-i}^*(\theta_{-i})), \theta),$$

for all i, θ and m_i.

An ex post equilibrium is a Nash equilibrium for every type profile θ. We observe that the notion of an ex post equilibrium does not refer to prior or posterior probability distributions of the types as the Bayesian Nash equilibrium does. The ex post equilibrium has an ex post no regret property in the incomplete information game, as no agent would like to change his message even if he were to know the true type profile of the remaining agents.[11]

---

[10]In Section 5, we assume that Y is compact subset of $\mathbb{R}^n$ with non-empty interior. In Section 9, we let Y be a lottery space.

[11]Ex post incentive compatibility was discussed as "uniform incentive compatibility" by Holmstrom and Myerson (1983). Ex post equilibrium is increasingly studied in game theory (see Kalai, 2004) and is often used in mechanism design as a more robust solution concept (see Cremer and McLean, 1985 and the references in footnote 1).

In an environment with private values, the notion of ex post equilibrium is equivalent to the notion of dominant strategy equilibrium. If in addition one could guarantee *strict* dominant strategy incentives, then full implementation can be achieved by fiat. The importance of the distinction between weak and strict incentive compatibility for implementation will be discussed in detail in the context of the single unit auction in Section 7. Results about the private value special case and dominant strategy incentive compatibility are collected in Appendix A.

Definition 2 (*Ex Post Implementation*). Social choice set F is ex post implementable (in pure strategies) if there exists a mechanism M such that:

(1) for every $f \in F$, there exists an ex post equilibrium s^* of the game that satisfies:

$$g(s^*(\theta)) = f(\theta), \quad \forall\, \theta \in \Theta;$$

(2) for every ex post equilibrium s^* of the game there exists $f \in F$ such that:

$$g(s^*(\theta)) = f(\theta), \quad \forall\, \theta \in \Theta.$$

Implementation then requires that the equilibria of the mechanism exactly coincide with the given social choice set. The notion of implementation defined above is sometimes referred to as "full" implementation (see Dasgupta *et al.*, 1979; Maskin, 1999 and Postlewaite and Schmeidler, 1986).

3 Monotonicity

3.1 *Ex Post Monotonicity*

Implementation is meant to address the problem that privately informed agents may consistently misrepresent their information and jointly establish equilibrium behavior which fails to realize the social choice objective of the planner. The notion of ex post monotonicity is easiest to grasp by considering the direct revelation game. If we were just interested in partially implementing F — i.e., constructing a mechanism with an ex post equilibrium achieving a selection $f \in F$ — then by the revelation principle we could restrict attention to the direct mechanism and a necessary and sufficient condition is the following ex post incentive compatibility condition.

Definition 3 (*Ex Post Incentive Compatibility*). F is ex post incentive compatible (*EPIC*) if for every $f \in F$:

$$u_i(f(\theta), \theta) \geq u_i(f(\theta_i', \theta_{-i}), \theta),$$

for all i, θ and θ_i'.

In the direct mechanism, a misrepresentation by an agent is a non-truthtelling strategy. As such it is an attempt by the agent to deceive the designer and we refer to such a misrepresentation as a deception α_i by agent i:

$$\alpha_i : \Theta_i \to \Theta_i.$$

The deception α_i represents i's reported type as a function of his true type. The entire profile of deceptions is denoted by:

$$\alpha(\theta) = (\alpha_1(\theta_1), \ldots, \alpha_I(\theta_I)).$$

In the direct mechanism, if agents report the deception $\alpha(\theta)$ rather than truthfully report θ, then the resulting social outcome is given by $f(\alpha(\theta))$ rather than $f(\theta)$. We write $f \circ \alpha(\theta) \triangleq f(\alpha(\theta))$. The notion of ex post monotonicity guarantees that there exists a whistle-blower (among the agents) who (i) will alert the designer of deceptive behavior α by receiving a reward for his alert; and (ii) will not falsely report a deception in a truth-telling equilibrium.

Definition 4 (*Ex Post Monotonicity*). Social choice set F satisfies ex post monotonicity (*EM*) if for every $f \in F$ and deception α with $f \circ \alpha \notin F$, there exists i, θ and $y \in Y$ such that

$$u_i(y, \theta) > u_i(f(\alpha(\theta)), \theta), \tag{1}$$

while

$$u_i(f(\theta_i', \alpha_{-i}(\theta_{-i})), (\theta_i', \alpha_{-i}(\theta_{-i}))) \geq u_i(y, (\theta_i', \alpha_{-i}(\theta_{-i}))), \quad \forall \theta_i' \in \Theta_i. \tag{2}$$

It is convenient to denote the set of allocations that make agent i worse off (relative to the social choice function f) at all of his types, $\theta_i' \in \Theta_i$, and a given type profile $\theta_{-i} \in \Theta_{-i}$ of the other agents by $Y_i^f(\theta_{-i})$:

$$Y_i^f(\theta_{-i}) \triangleq \{y : u_i(f(\theta_i', \theta_{-i}), (\theta_i', \theta_{-i})) \geq u_i(y, (\theta_i', \theta_{-i})), \quad \forall \theta_i' \in \Theta_i\}. \tag{3}$$

Thus (2) can be replaced with the requirement that $y \in Y_i^f(\alpha_{-i}(\theta_{-i}))$. The set $Y_i^f(\theta_{-i})$ depends on the selection $f \in F$ and is referred to as the *reward set*. It is the set of allocations which can be used to reward the whistle-blower without upsetting the truthtelling equilibrium realizing the social choice function f. If the social choice objective is a function rather than a set then we can omit the superscript on the reward set for notational ease. We refer to the subset of the reward set which also satisfies the reward inequality (1) as the *successful reward set* and denote it by $Y_i^*(\theta_{-i})$.

The definition of ex post monotonicity suggests a rather intuitive description as to why monotonicity is a necessary condition for implementation. Suppose that some selection $f \in F$ is ex post implementable. Then if the agents were to deceive the designer by misreporting $\alpha(\theta)$ rather than reporting truthfully θ and if the deception $\alpha(\cdot)$ would lead to an allocation outside of the social choice set, i.e., $f \circ \alpha \notin F$, then the designer should be able to fend off the deception. This requires that there is some agent i and profile θ such that the designer can offer agent i a reward y for denouncing the deception $\alpha(\theta)$ if the true type profile is θ. Yet, the designer has to be aware that the reward could be used in the wrong circumstances, namely when the true type profile of the remaining agents is $\alpha_{-i}(\theta_{-i})$ *and* truthfully reported to be $\alpha_{-i}(\theta_{-i})$. The strict inequality (1) then guarantees the existence of a whistle-blower, whereas the weak inequalities (2) guarantee ex post incentive compatible behavior by the whistle-blower.

3.2 *Maskin Monotonicity*

Maskin (1999) introduced a celebrated monotonicity notion which is a necessary and almost sufficient condition for complete information implementation. In the complete information environment, each agent i is assumed to know the entire type profile θ rather than just his private type θ_i. In consequence, the deception of each agent pertains to the entire type profile $\theta \in \Theta$, or:

$$\alpha_i : \Theta \to \Theta.$$

With complete information, it is easy to detect *individual* deceptions and hence it suffices to consider collective and coordinated deceptions in which all agents pursue a common deception strategy, $\alpha_i = \alpha$, for all i.

Definition 5 (*Maskin Monotonicity*). Social choice set F satisfies Maskin monotonicity (*MM*) if for every $f \in F, \alpha$ and θ with $f \circ \alpha(\theta) \notin \hat{f}(\theta)$ for all $\hat{f} \in F$, there exists i and $y \in Y$ such that

$$u_i(y, \theta) > u_i(f(\alpha(\theta)), \theta), \tag{4}$$

while

$$u_i(f(\alpha(\theta)), \alpha(\theta)) \geq u_i(y, \alpha(\theta)). \tag{5}$$

We state the notion of Maskin monotonicity in such a way as to facilitate a simple comparison with ex post monotonicity. Typically, Maskin monotonicity is defined for social choice correspondences rather than social choice sets. If we start with a social choice correspondence $\phi : \Theta \to 2^Y/\emptyset$, then we can define an associated social choice set $F = \{f \,|\, f : \Theta \to Y\}$ by including all social choice functions f which select at all profiles allocations in the image of the correspondence:

$$F = \{f \mid f(\theta) \in \phi(\theta), \ \forall \theta \in \Theta\}.$$

We compare the notions of social choice set and correspondence in more detail in Section 8.

Comparing ex post and Maskin monotonicity, it may initially appear that ex post monotonicity is a stronger requirement: the truthtelling constraint is required to hold at $(\theta_i', \alpha_{-i}(\theta_{-i}))$ for all $\theta_i' \in \Theta_i$ rather than just at $\alpha(\theta)$. Thus (2) is stronger requirement than (5) because the incomplete information reward set $Y_i^f(\theta_{-i})$ is (weakly) contained in the complete information counterpart

$$Y_i^f(\theta) = \{y : u_i(f(\theta), \theta) \geq u_i(y, \theta)\}.$$

The complete information reward set depends on the entire profile θ rather than the profile θ_{-i} of all agents but i. This difference in the reward sets arises from the informational assumption. With complete information, all individual deceptions can easily be detected and the designer only needs to worry about coordinated misrepresentations by all the agents. In the incomplete information environment, agent i has private information about θ_i' and hence incentive compatibility is required to hold for all types $\theta_i' \in \Theta_i$.

But for either ex post or Maskin monotonicity, we need a preference reversal relative to the allocation $f(\alpha(\theta))$. If the behavior of the incomplete information reward set $Y_i(\alpha_{-i}(\theta_{-i}))$ is locally similar to the complete information set $Y_i(\alpha(\theta))$, then the difference between them may not matter for implementation purposes. Indeed, we will show that in the important class of single crossing environments, ex post and Maskin monotonicity coincide. We will illustrate this coincidence in the public good example that follows. But this gap in reward sets may have important implications. While the Pareto correspondence is always Maskin monotonic, we will later give an example showing that it is not always ex post monotonic.

But outside the single crossing environment, ex post monotonicity is not necessarily a stronger notion than Maskin monotonicity. In the complete information environment, the agents are (implicitly) allowed to perfectly coordinate their misrepresentation for every societal type profile θ. In contrast, in the incomplete information world, agent i has to deceive, i.e., determine $\alpha_i : \Theta_i \rightarrow \Theta_i$, independently of the type profile of the other agents. For this reason, it is strictly more difficult to find a reward y for Maskin monotonicity than for ex post monotonicity. In other words, the independent choice of deception α_i leads to a strictly smaller number of feasible deceptions α in the incomplete information context. In the context of a single unit auction with interdependent valuations, this second difference will enable us to show that the single unit auction can be implemented in ex post equilibrium, yet fails to be implementable in complete information, and hence fails Maskin monotonicity.

3.3 *Public Good Example*

We will illustrate ex post monotonicity, and the relation to Maskin monotonicity, with the following public good example. The utility of each agent is given by:

$$u_i(\theta, x, t) = \left(\theta_i + \gamma \sum_{j \neq i} \theta_j\right) x + t_i, \tag{6}$$

where x is the level of public good provided and t_i is the monetary transfer to agent i. The utility of agent i depends on his own type $\theta_i \in [0, 1]$ and the type profile of other agents, with $\gamma \geq 0$. The cost of establishing the public good is given by $c(x) = \frac{1}{2}x^2$. The planner must choose $(x, t_1, \ldots, t_I) \in \mathbb{R}_+ \times \mathbb{R}^I$ to maximize social welfare, i.e., the sum of gross utilities minus the cost of the public good:

$$\left((1 + \gamma(I - 1)) \sum_{i=1}^{I} \theta_i\right) x - \frac{1}{2}x^2.$$

The socially optimal level of the public good is therefore equal to

$$x(\theta) = (1 + \gamma(I-1)) \sum_{i=1}^{I} \theta_i .$$

The social choice set F can then be described by:

$$F = \left\{ (x, t_1, \ldots, t_I) : \Theta \to \mathbb{R}^{I+1} \mid x(\theta) = (1 + \gamma(I-1)) \sum_{i=1}^{I} \theta_i \right\},$$

where the level of the public good is determined uniquely, but the designer is unrestricted in his choice of transfers. By standard arguments, ex post incentive compatibility pins down the levels of transfers[12]:

$$t_i(\theta) = h_i(\theta_{-i}) - (1 + \gamma(I-1)) \left(\gamma \left(\sum s_{j \neq i} \theta_j \right) \theta_i + \frac{1}{2} \theta_i^2 \right). \quad (7)$$

The complete information reward set $Y_i(\theta)$ is now characterized by an indifference curve in the (x, t_i) space. With the linear preferences here it is simply a straight line. The reward set is given by the set of allocations below the indifference curve. In contrast, the incomplete information reward set $Y_i(\theta_{-i})$ is characterized by the intersection of the reward sets for all $\theta_i' \in \Theta_i$, or:

$$Y_i(\theta_{-i}) = \bigcap_{\theta_i' \in \Theta_i} Y_i(\theta_i', \theta_{-i}).$$

The boundary of the set $Y_i(\theta_{-i})$ is the set of all truthtelling allocations

$$\{x(\theta_i', \theta_{-i}), t_i(\theta_i', \theta_{-i})\}_{\theta_i' \in \Theta_i}.$$

The respective reward sets are depicted for $I = 3, \gamma = \frac{1}{4}$, and $\theta_i = \frac{1}{4}$ for all i in Fig. 1 (setting $h_i(\theta_{-i}) = t_i(0, \theta_{-i}) = 0$ for all θ_{-i}).

The crucial observation is now that the slope of the boundary of the set $Y_i(\theta_{-i})$ at $\theta = (\theta_i, \theta_{-i})$ is identical to the slope of the boundary of the set $Y_i(\theta)$. In other words, locally, the slope of the boundary of $Y_i(\theta_{-i})$ is

[12]In this example, and in interdependent public good problems with more general functional forms, it is possible to find ex post incentive compatible transfers for all values of $\gamma \geq 0$. This can be established using conditions in Bergemann and Välimäki (2002). This contrasts with the case of allocating a private good with interdependent values, where ex post incentive compatibility puts an upper bound on the amount of interdependence (Dasgupta and Maskin, 2000). However, Fieseler *et al.* (2003) point out that negative interdependence, or $\gamma < 0$, relaxes the ex post incentive constraints in the private good problem.

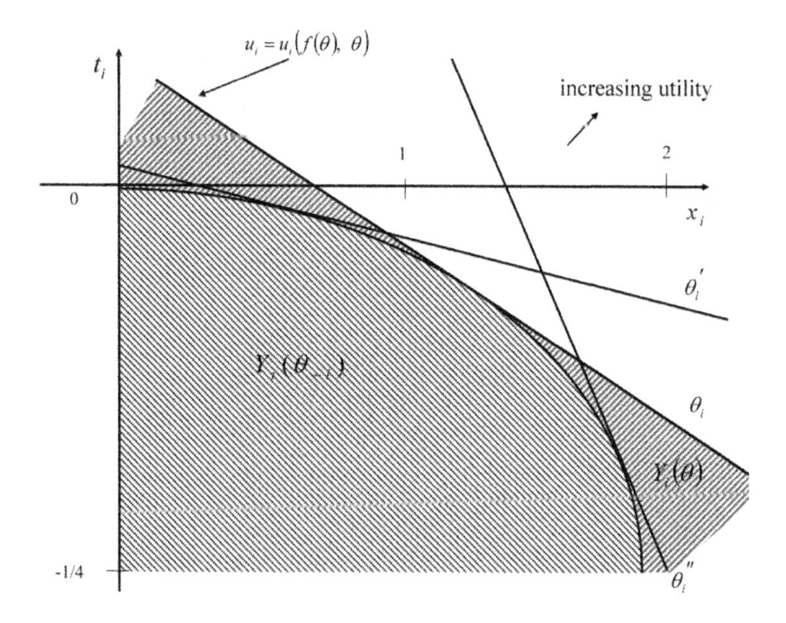

Fig. 1. Reward sets $Y_i(\theta), Y_i(\theta_{-i})$.

determined by the preferences of type θ. It then follows that if we can guarantee preference reversal at the allocation $f(\alpha(\theta))$, essentially the crossing of the indifference curves for θ and $\alpha(\theta)$, then the indifference curve of type θ will also cross with the boundary of the set $Y_i(\theta_{-i})$. This is illustrated with $\theta_i = \frac{3}{4}$ and $\alpha_i(\theta_i) = \frac{1}{4}$ for all i in Fig. 2.

This basic insight allows us later to conclude that despite the additional incentive constraints imposed by the ex post monotonicity condition, the single crossing environment by itself is strong enough to guarantee the ex post monotonicity condition. The only modification we need is to strengthen the necessary condition from weak to strict ex post incentive compatibility.

4 Ex Post Implementation

We present necessary and sufficient conditions for a social choice set F to be ex-post implementable. Our results extend the work of Maskin (1999) and Saijo (1988) for complete information implementation and of Postlewaite and Schmeidler (1986), Palfrey and Srivastava (1989a) and Jackson (1991) on Bayesian implementation to the notion of ex post equilibrium.

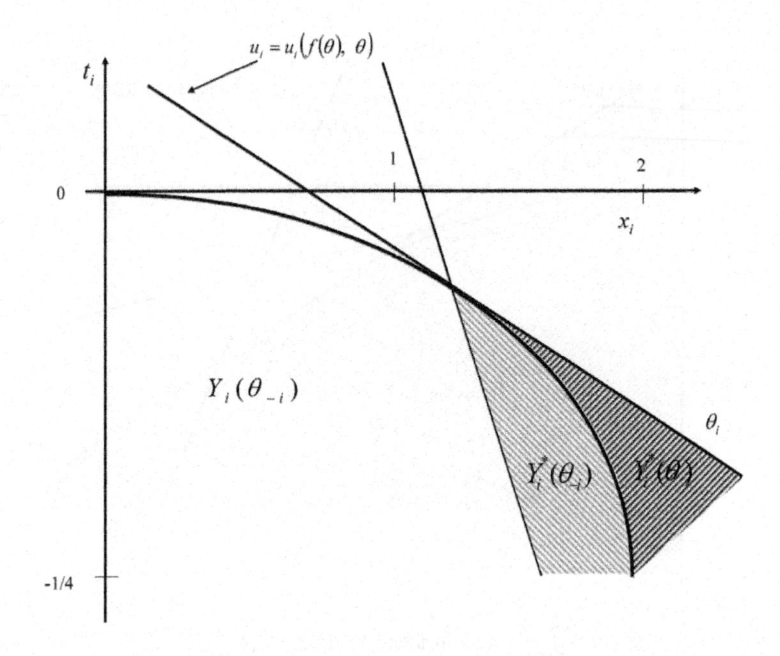

Fig. 2. Successful reward sets $Y_i^*(\theta), Y_i^*(\theta_{-i})$.

4.1 *Necessary Conditions*

Ex post incentive and monotonicity conditions are necessary conditions for ex post implementation.

Theorem 1 (*Necessity*). *If F is ex post implementable, then it satisfies (EPIC) and (EM).*

Proof. Let (M, g) implement F. Fix any $f \in F$. By the implementation hypothesis, there must exist an equilibrium $s = (s_1, \ldots, s_I)$, each $s_i : \Theta_i \to M_i$ such that $f = g \circ s$. Consider any $i, \theta_i' \in \Theta_i$. Since s is an equilibrium,

$$u_i(g(s(\theta)), \theta) \geq u_i(g(s_i(\theta_i'), s_{-i}(\theta_{-i})), \theta),$$

for all $\theta_i' \in \Theta_i$ and all $\theta \in \Theta$. Noting that $g(s(\theta)) = f(\theta)$ and $g(s_i(\theta_i'), s_{-i}(\theta_{-i})) = f(\theta_i', \theta_{-i})$ establishes (*EPIC*).

Now fix any deception α with $f \circ \alpha \notin F$. It must be that $s \circ \alpha$ is not an equilibrium at some $\theta \in \Theta$. Therefore there exists i and $m_i \in M_i$ such that we have

$$u_i(g(m_i, s_{-i}(\alpha_{-i}(\theta_{-i}))), \theta) > u_i(g(s(\alpha(\theta))), \theta).$$

Let $y \triangleq g(m_i, s_{-i}(\alpha_{-i}(\theta_{-i})))$. Then, from above,

$$u_i(y, \theta) > u_i(f(\alpha(\theta)), \theta).$$

But since s is an equilibrium and $f = g \circ s$, it follows that

$$\begin{aligned} &u_i(f(\theta_i', \alpha_{-i}(\theta_{-i})), (\theta_i', \alpha_{-i}(\theta_{-i}))) \\ &\quad = u_i(g(s(\theta_i', \alpha_{-i}(\theta_{-i}))), (\theta_i', \alpha_{-i}(\theta_{-i}))) \\ &\quad \geq u_i(g(m_i, s_{-i}(\alpha_{-i}(\theta_{-i}))), (\theta_i', \alpha_{-i}(\theta_{-i}))) \\ &\quad = u_i(y, (\theta_i', \alpha_{-i}(\theta_{-i}))), \quad \forall \theta_i' \in \Theta_i. \end{aligned}$$

This establishes the incentive compatibility of the whistle-blower, or $y \in Y_i^f(\alpha_{-i}(\theta_{-i}))$. □

We defined ex post monotonicity in terms of the type profiles and associated utility functions. As ex post monotonicity is the central condition in the subsequent analysis, we want to verify, as promised earlier, that ex post monotonicity is nonetheless an ordinal rather than a cardinal concept.

Definition 6 (*Ordinality*). The types θ_i and θ_i' are *ordinally equivalent* ($\theta_i \sim \theta_i'$) if for all $j, \theta_{-i} \in \Theta_{-i}, y$ an y',

$$u_j(y, (\theta_i, \theta_{-i})) \geq u_j(y', (\theta_i, \theta_{-i})) \Leftrightarrow u_j(y, (\theta_i', \theta_{-i})) \geq u_j(y', (\theta_i', \theta_{-i})).$$

In other words, any two types of agent i, θ_i and θ_i', are ordinally equivalent if the ranking of any pair of alternatives by any agent and for any profile of agents other than i remain unchanged. With interdependent values, it is important that the ranking remains unchanged not only for agent i but for all other agents as well.

Definition 7 (*Ordinal Social Choice Set*). Social choice set F is *ordinal* if $f \in F$ and $\alpha_j(\theta_j) \sim \theta_j$ for all j and θ_j imply $f \circ \alpha \in F$.

A social choice set is hence called ordinal if for any two profiles, θ and $\alpha(\theta)$, which only differ by ordinally equivalent types, the corresponding allocations remain in the social choice set.

Lemma 1. *If F satisfies ex post monotonicity, then F is ordinal.*

Proof. Suppose $f \in F$ and $\alpha_j(\theta_j) \sim \theta_j$ for all j and θ_j and that

$$u_i(f(\alpha(\theta)), \alpha(\theta)) \geq u_i(y, \alpha(\theta)).$$

By ordinality we have

$$u_i(f(\alpha(\theta)),\theta) \geq u_i(y,\theta).$$

But if ex post monotonicity holds, we must have $f \circ \alpha \in F$. □

We proceed by first showing that in a wide class of environments, to be referred to as economic environments, ex post incentive and monotonicity conditions are also sufficient conditions for ex post implementation. We then present weaker sufficiency conditions, in the spirit of the conditions used in Bayesian implementation, to obtain positive results outside of economic environments.

4.2 *Sufficient Conditions in Economic Environments*

The sufficiency arguments — for both the economic and the non-economic environment — will rely on the use of an augmented mechanism. The mechanism suggested here is similar to the one used to establish sufficiency in the complete information implementation literature (e.g., Saijo, 1988). Each agent sends a message of the form $m_i = (\theta_i, f_i, z_i, y_i)$, where θ_i is the reported type, f_i is the social choice function suggested by i, z_i is a positive integer from the set $\mathcal{I} = \{1, 2, \ldots, I\}$ and y_i is the reward claimed by i. The set of feasible messages for agent i is given by:

$$M_i = \Theta_i \times F \times \mathcal{I} \times Y.$$

The mechanism is described by three rules:

Rule 1. If $f_i = f$ for all i, then $g(m) = f(\theta)$.

Rule 2. If there exists j and f such that $f_i = f$ for all $i \neq j$ while $f_j \neq f$, then outcome y_j is chosen if $y_j \in Y_j^f(\theta_{-j})$; otherwise outcome $f(\theta)$ is chosen.

Rule 3. In all other cases, $y_{j(z)}$ is chosen, where $j(z)$ is the agent determined by the modulo game:

$$j(z) = \sum_{i=1}^{I} z_i \pmod{I}.$$

We refer to the mechanism described by Rules 1–3 as the *augmented mechanism.* A strategy profile in this game is a collection $s = (s_1, \ldots, s_I)$, with $s_i : \Theta_i \to M_i$ and we write:

$$s_i(\theta_i) = (s_i^1(\theta_i), s_i^2(\theta_i), s_i^3(\theta_i), s_i^4(\theta_i)) \in \Theta_i \times \mathcal{F} \times \mathcal{I} \times Y;$$

and $s^k(\theta) = (s_i^k(\theta))_{i=1}^I$. We observe that if Y and Θ are finite, then the above mechanism is finite.

Next we define the notion of an economic environment.

Definition 8 (*Economic Environment*). An environment is economic in state $\theta \in \Theta$ if, for every allocation $y \in Y$, there exist $i \neq j$ and allocations y_i and y_j, such that

$$u_i(y_i, \theta) > u_i(y, \theta),$$

and

$$u_j(y_j, \theta) > u_j(y, \theta).$$

An environment is economic if it is economic in every state $\theta \in \Theta$.

Theorem 2 (*Economic Environment*). *If $I \geq 3$, the environment is economic, and F satisfies (EPIC) and (EM), then F is ex post implementable.*

Proof. The proposition is proved in three steps, using the augmented mechanism.

Claim 1. *Fix any $f \in F$. There is an ex post equilibrium s with $g(s(\theta)) = f(\theta)$ for all θ.*

Any strategy profile s of the following form is an ex post equilibrium:

$$s_i(\theta_i) = (\theta_i, f, \cdot, \cdot).$$

Suppose agent i thinks that his opponents are types θ_{-i} and deviates to a message of the form

$$s_i(\theta_i) = (\theta_i', f_i, \cdot, \cdot);$$

if $y_i \notin Y_i^f(\theta_{-i})$, then the payoff gain is

$$u_i(f(\theta_i', \theta_{-i}), f(\theta_i, \theta_{-i})) - u_i(f(\theta_i, \theta_{-i}), f(\theta_i, \theta_{-i})),$$

which is non-positive by (*EPIC*); if $f_i \neq f$ and $y_i \in Y_i^f(\theta_{-i})$, then the payoff gain is

$$u_i(y_i, (\theta_i, \theta_{-i})) - u_i(f(\theta_i, \theta_{-i}), f(\theta_i, \theta_{-i})),$$

which is non-positive by the definition of $Y_i^f(\theta_{-i})$.

Claim 2. *In any ex post equilibrium, there exists $f \in F$ such that $s_i^2(\theta_i) = f$ for all i and θ_i.*

Suppose that for all $f \in F$, there exists i and θ_i such that $s_i^2(\theta_i) \neq f$. Then there exists θ such that Rule 1 does not apply in equilibrium.

First suppose that Rule 2 applies at θ, so that there exists j and f such that $f_i = f$ for all $i \neq j$. Then any agent $i \neq j$ of type θ_i who thought his opponents were types θ_{-i}, could send a message of the form $m_i = (\cdot, f_i, z_i, y_i)$, with $f_i \neq f$ and $i = \sum_{k=1}^{I} z_k$ and obtain utility $u_i(y_i, \theta)$. Thus we must have $u_i(g(s(\theta)), \theta) \geq u_i(y, \theta)$ for all y and all $i \neq j$. This contradicts the economic environment assumption.

Now suppose that Rule 3 applies at θ. Then every agent i of type θ_i who thought his opponents were types θ_{-i}, could send a message of the form $m_i = (\cdot, f_i, z_i, y_i)$, with $i = \sum_{k=1}^{I} z_k$ and obtain utility $u_i(y_i, \theta)$. Thus we must have $u_i(g(s(\theta)), \theta) \geq u_i(y, \theta)$ for all y and i. This again contradicts the economic environment assumption.

Claim 3. *For any $f \in F$ and in any ex post equilibrium with $s_i^2(\theta_i) = f$ for all i and θ_i, $f \circ s^1 \in F$.*

Suppose that $f \circ s^1 \notin F$. By (EM), there exists i, θ and $y \in Y_i^f(s_{-i}^1(\theta_{-i}))$ such that

$$u_i(y, \theta) > u_i(f(s^1(\theta)), \theta).$$

Now suppose that type θ_i of agent i believes that his opponents are of type θ_{-i} and sends message $m_i = (\cdot, f_i, \cdot, y)$, with $f_i \neq f$, while other agents send their equilibrium messages, then from the definition of $g(\cdot)$:

$$g(m_i, s_{-i}(\theta_{-i})) = y,$$

so that:

$$u_i(g(m_i, s_{-i}(\theta_{-i})), \theta) = u_i(y, \theta) > u_i(f(s^1(\theta)), \theta) = u_i(g(s(\theta)), \theta),$$

and this completes the proof of sufficiency. □

The economic environment condition was used to show that in equilibrium, the suggested social choice functions all have to agree: $f_i = f$ for all i. If not, then some agent j could profitably change his suggestion to $f_j \neq f$ and obtain a more desirable allocation than f. The economic environment assumption guaranteed the existence of an agent j with a preferred allocation.

4.3 *Sufficiency Conditions in Non-economic Environments*

We now establish sufficient conditions for ex post implementation outside of economic environments. For simplicity, we focus on the implementation of social choice functions, rather than social choice sets, in this section. The ex post sufficient conditions are the natural complements of the conditions obtained earlier for Bayesian implementation. But because only ex post utilities matter, they are more easily verified than their Bayesian analogues. We show that a joint strengthening of Maskin monotonicity and ex post monotonicity, together with a no veto condition, is sufficient for ex post implementation. In Section 9 we permit random mechanisms, which will allow us to strengthen the sufficient conditions presented here substantially.

Within the augmented mechanism there are essentially two ways in which the play of agents can lead to equilibrium behavior outside of the social choice objective. At any profile $\theta \in \Theta$, the agents can either misrepresent their true type and fail to alert the designer of the misrepresentation. Or, some agents alert the designer and thus lead him to choose an allocation different from $f(\theta)$. In the former case, Rule 1 of the augmented mechanism applies whereas in latter case, either Rule 2 or Rule 3 applies. With an economic environment, it was impossible that in any equilibrium Rule 2 or Rule 3 would apply. It followed that in every equilibrium Rule 1 would apply at all profiles $\theta \in \Theta$. The ex post monotonicity condition then guaranteed that the equilibrium conformed with the social choice set. As we abandon the assumption of an economic environment, we cannot anymore exclude the possibility that in equilibrium either Rule 2 or Rule 3 might apply. In consequence, the sufficient conditions have to account for these complications. There are now basically two ways to achieve this goal. Either behavior under Rule 2 or Rule 3 can be made to conform with the social choice, or a reward can be offered in the *subset* of profiles where Rule 1 applies. The sufficient condition will contain both elements: either an application of a no veto condition will make behavior under Rule 2 or Rule 3 consistent with the social choice objective or an ex post monotonicity condition on subsets of Θ guarantees that a reward can be offered.

The relevant no veto condition is simply the "no veto power" property of Maskin (1999).

Definition 9 (*No Veto Power*). Social choice function f satisfies no veto power (*NVP*) at θ if, for any j, if $u_i(b, \theta) \geq u_i(y, \theta)$ for all $y \in Y$ and $i \neq j$, then $f(\theta) = b$.

Note that no veto power is vacuously true at θ if the environment is economic at θ, since the latter implies that the premise in the definition is never satisfied.

Under either Rule 2 or Rule 3, (almost) every agent can change the outcome to his most preferred outcome. If the candidate allocation under either rule is part of an equilibrium, it follows that at least $I-1$ (if Rule 2 applies and I if Rule 3 applies) agents rank the candidate allocation y higher than any other allocation. The no veto power property guarantees that allocation y coincides with the social choice set. In other words, the possibility of *undesirable* equilibrium behavior is eliminated by no veto power. However, if the no veto power property fails, it might still be possible to generate a reward on the set of profiles where Rule 1 applies. For any strategy profile of the agents, the set of profiles at which Rule 1 applies always satisfies a product structure. Given a strategy profile of the agents, a subset Φ_i identifies the types of agent i at which Rule 1 applies. The product set Φ:

$$\Phi \triangleq \times_{i=1}^{I} \Phi_i,$$

is the set of profiles at which Rule 1 applies, and the complementary set $\Theta - \Phi$ at which either Rule 2 or Rule 3 applies. We state the sufficient condition combining ex post monotonicity on subsets and the no veto power property. We state the conditions for the case of a social choice function and the straightforward extension to general social choice sets is provided in Appendix A.

Definition 10 (*Ex Post Monotonicity No Veto (EMNV)*). Social choice function f satisfies ex post monotonicity no veto if, for any deception α and any product set $\Phi \subset \Theta$, the following holds: If the environment is non-economic at each $\theta \in \Theta - \Phi$, then

(1) f satisfies no veto power on $\Theta - \Phi$;
(2) if $f(\alpha(\theta)) \neq f(\theta)$ for some $\theta \in \Phi$ then there exists $i, \theta \in \Phi$ and y such that

$$u_i(y, \theta) > u_i(f(\alpha(\theta)), \theta),$$

while

$$u_i(f(\theta_i', \alpha_{-i}(\theta_{-i})), (\theta_i', \alpha_{-i}(\theta_{-i}))) \geq u_i(y, (\theta_i', \alpha_{-i}(\theta_{-i}))), \quad \forall \theta_i' \in \Theta_i.$$

The strategy profile of the agents could involve truthtelling on a subset Φ (and appealing to Rule 1) or involve whistle-blowing and misrepresentation on the complementary set $\Theta - \Phi$ (and appeal to Rule 2 or Rule 3). The sufficient condition guarantees that for all product sets Φ an appropriate reward can be found. The condition is weakened by the fact that we only need to consider those subsets Φ which guarantee that on the complementary subset $\Theta - \Phi$ *all* profiles are non-economic. It follows that the smaller the set Φ becomes (and hence restricting the ability of the designer to offer rewards), the more demanding is the requirement that *all* profiles are non-economic. This then conceivably puts a bound on the number of sets for which the ex post monotonicity part of the condition has to be verified.

Theorem 3 (*Sufficiency*). *For $I \geq 3$, if f satisfies (EPIC) and (EMNV), then f is ex post implementable.*

Proof. We use the same mechanism as before. The argument that there exists an ex post equilibrium s with $g(s(\theta)) = f(\theta)$ is the same as before. Now we establish three claims that hold for all equilibria. Let

$$\Phi_i = \{\theta_i : s_i(\theta_i) = (\cdot, f, \cdot, \cdot)\}.$$

Claim 1. *In any ex post equilibrium, for each $\theta \notin \Phi$, (a) there exists i such that $u_j(g(s(\theta)), \theta) \geq u_j(y, \theta)$ for all y and $j \neq i$; and thus (b) the environment is non-economic at θ.*

First, observe that for each $\theta \notin \Phi$, there exists i such that $s_i^2(\theta_i) \neq f$. Given the strategies of the other agents, any agent $j \neq i$ who thought his opponents were types θ_{-j} could send any message of the form

$$(\cdot, f_j, z_j, y_j),$$

and obtain utility $u_j(y_j, \theta)$. Thus we must have $u_j(g(s(\theta)), \theta) \geq u_j(y, \theta)$ for all y and $j \neq i$; thus the environment is non-economic for all $\theta \notin \Phi$.

Claim 2. *In any ex post equilibrium, for all $\theta \in \Phi$,*

$$u_i(f(s^1(\theta)), \theta) \geq u_i(y, \theta)$$

for all $y \in Y_i^f(s_{-i}^1(\theta_{-i}))$.

Suppose that $y \in Y_i^f(s_{-i}^1(\theta_{-i}))$ and that type θ_i of agent i believes that his opponents are of type θ_{-i} and sends message $m_i = (\cdot, f_i, z_i, y)$, while other agents send their equilibrium messages. Now

$$g(m_i, s_{-i}(\theta_{-i})) = y;$$

so ex post equilibrium requires that

$$\begin{aligned} u_i(g(s(\theta)), \theta) &= u_i(f(s^1(\theta)), \theta) \\ &\geq u_i(g(m_i, s_{-i}(\theta_{-i})), \theta) \\ &= u_i(y, \theta). \end{aligned}$$

Claim 3. *If EMNV is satisfied, then Claims 1 and 2 imply that $g(s(\theta)) = f(\theta)$ for all θ.*

Fix any equilibrium. Claim 1(b) establishes that the environment is non-economic at all $\theta \notin \Phi$. Suppose $g(s(\theta)) \neq f(\theta)$ for some $\theta \in \Phi$. Now EMNV implies that there exists $i, \theta \in \Phi$ and $y \in Y_i^f(s_{-i}^1(\theta_{-i}))$ such that $u_i(y, \theta) > u_i(f(s^1(\theta)), \theta)$, contradicting Claim 2. Suppose $g(s(\theta)) \neq f(\theta)$ for some $\theta \notin \Phi$. By Claim l(a), there exists i such that $u_j(g(s(\theta)), \theta) \geq u_j(y, \theta)$ for all y and $j \neq i$. This establishes that no veto power applies at θ. So again EMNV implies that $g(s(\theta)) = f(\theta)$. □

EMNV is almost equivalent to requiring ex post monotonicity and no veto power everywhere. More precisely, we have:

(1) If ex post monotonicity holds and no veto power holds at every type profile θ, then EMNV holds.
(2) If EMNV holds, then (1) ex post monotonicity holds and (2) if the environment is non-economic whenever $\theta_i = \theta_i^*$, then no veto power holds whenever $\theta_i = \theta_i^*$. To see (1), set $\Phi_i = \Theta_i$ for all i; to see (2), set α to be the truth-telling deception and, for some i, $\Phi_i = \Theta_i \backslash \{\theta_i^*\}$ and $\Phi_j = \Theta_j$ for all $j \neq i$.

In an economic environment, we only have to verify $\Phi = \Theta$. EMNV is then equivalent to ex post monotonicity as the no veto condition is vacuously satisfied. On the other hand, if the environment is non-economic at every profile $\theta \in \Theta$, then the EMNV condition simplifies considerably as it suffices to evaluate the hypothesis at the most restrictive sets, or $\Phi = \{\theta\}$ for every $\theta \in \Phi$. In particular, we can then state ex post monotonicity and no veto conditions separately.

Definition 11 (*Local Ex Post Monotonicity (LEM)*). f satisfies local ex post monotonicity if for all θ and all α such that $f(\alpha(\theta)) \neq f(\theta)$, there exists i and y with:

$$u_i(y, \theta) > u_i(f(\alpha(\theta)), \theta), \tag{8}$$

while

$$u_i(f(\theta_i', \alpha_{-i}(\theta_{-i})), (\theta_i', \alpha_{-i}(\theta_{-i}))) \geq u_i(y, (\theta_i', \alpha_{-i}(\theta_{-i}))), \quad \forall \theta_i' \in \Theta_i. \tag{9}$$

With LEM, the designer can offer a reward y at *every* type profile θ at which α leads to a different allocation, or $f(\alpha(\theta)) \neq f(\theta)$. In contrast, with ex post monotonicity it suffices to find *some* θ at which a reward y can be offered. The local version of ex post monotonicity is in fact identical to Maskin monotonicity with the additional ex post incentive constraints (see (9)).

Corollary 1 (*Sufficiency*). *For $I \geq 3$, if f satisfies (EPIC), (LEM) and (NVP), then f is ex post implementable.*

For a non-economic environment, the separate conditions of *LEM* and *NVP* are exactly identical to *EM*. If the environment is economic in some profiles but not all profiles, the joint conditions are more restrictive than *EMNV*.

5 Single Crossing Environment

In this section we consider ex post implementation in single crossing environments. To make full use of the crossing conditions, we restrict attention to social choice problems where Y is compact subset of $\mathbb{R}^n$ with non-empty interior and the social choice correspondence recommend allocations in the interior of Y. We show that under this mild restriction, single crossing preferences are essentially sufficient to guarantee ex post monotonicity.

Definition 12 (*Interior Social Choice Set*). Social choice set F is interior if for all $f \in F$ and for all $\theta \in \Theta, f(\theta) \in \text{int } Y$ for all $\theta \subset \Theta$.

The interior condition is essential to use the full strength of the single crossing environment. In this section we further assume that Y is a convex set and that $u_i(y, \theta)$ is continuous in y at all i and θ. The convexity and continuity assumptions appear in establishing that locally, around $f(\theta)$, it is only the single crossing condition with respect to the type profile θ, that matter for the monotonicity inequalities. We first give a general definition of preference reversal.

Definition 13 (*Preference Reversal*). The environment is an environment with preference reversal if for all θ, θ', every open set $\mathcal{O}$ contains allocations y, y' such that for some i:

$$u_i(y, \theta) > u_i(y', \theta), \quad \text{while} \quad u_i(y', \theta') > u_i(y, \theta').$$

The above definition is weak in the sense that the preference reversal is required to occur only for one rather than all agents. This weaker version is helpful as the type profiles of the agents may sometimes interact so as to precisely offset each other in their effect on the preferences of the agents. We simple require that at two distinct profiles of society, θ and θ', there is at least one agent with a preference reversal.[13,14]

As the above definition of single crossing applies to general allocation spaces, it is phrased as a preference reversal condition. In many applications of mechanism design, the allocation space for each agent is two-dimensional, say the level of private or public good and a monetary transfer. In this case, a sufficient condition for preference reversal is the well known intersection or single crossing condition:

Definition 14 (*Single Crossing*). The environment is a single crossing environment if for all $z \in \text{int } Y$, with $Y \subset \mathbb{R}^2$, the indifference curves for any two profiles θ and θ' generated by $u_i(z, \theta)$ and $u_i(z, \theta')$ intersect at z for some i.

With strictness and interiority of F, a local argument allows us to show that, even though relative to Maskin monotonicity, ex post monotonicity imposes additional ex post incentive constraints, these additional constraints do not bind. In consequence, the set of dominated allocations y is locally identical to the complete information set. For the local argument to

[13]With interdependent values the change in the type profile from θ to θ' may offset each other as can be easily verified within the earlier public good example (see Section 3.3). For example we can keep the utility of agent i constant as we move from θ to θ', where θ' is given by

$$\theta'_j = \begin{cases} \theta_j - \gamma\varepsilon, & \text{if } j = i, \\ \theta_k + \varepsilon, & \text{if } j = k, \\ \theta_j, & \text{if } j \neq i, k. \end{cases}$$

Now i has identical preferences at θ and θ', but it is easily verified that for all $j \neq i$, there is preference reversal.

[14]The condition can be weakened further: by requiring that each $f \in F$ satisfies $f(\theta) = f(\theta')$ if $u_i(\cdot, \theta) = u_i(\cdot, \theta')$ for all i, we can weaken the preference reversal condition to hold only at profile pairs, θ and θ', at which there exists i such that $u_i(\cdot, \theta) \neq u_i(\cdot, \theta')$.

go through, we need to strengthen the ex post incentive constraints in the direct mechanism to strict rather than weak inequalities.

Definition 15 (*Strict Ex Post Incentive Compatibility*). Social choice set F is strictly ex post incentive compatible if for all $f \in F$:

$$u_i(f(\theta), \theta) > u_i(f(\theta_i', \theta_{-i}), \theta),$$

for all i, θ and $\theta_i' \neq \theta_i$.

The public good example in Section 3.3 is an example where the (singleton) social choice set satisfies single crossing, strict EPIC and interiority.

Theorem 4 (*Single Crossing*). *In an environment with preference reversal, every strict ex post incentive compatible and interior F satisfies ex post monotonicity.*

Proof. We start with the contrapositive version of ex post monotonicity, which can be stated as follows. Fix a deception α. If, for all i and all $\hat{\theta} \in \Theta$, we have that

$$u_i(f(\theta_i', \alpha_{-i}(\hat{\theta}_{-i})), (\theta_i', \alpha_{-i}(\hat{\theta}_{-i}))) \geq u_i(y, (\theta_i', \alpha_{-i}(\hat{\theta}_{-i})))$$
$$\text{for all } \theta_i' \in \Theta_i \quad \text{and} \quad y \in Y,$$

implies that

$$u_i(f(\alpha(\hat{\theta})), \hat{\theta}) \geq u_i(y, \hat{\theta}),$$

then $f \circ \alpha \in F$. For a given $\hat{\theta} \in \Theta$ with $\alpha(\hat{\theta}) \neq \hat{\theta}$, let us define for notational ease $\theta \triangleq \alpha(\hat{\theta})$. Now consider the indifference curve for θ and $\hat{\theta}$ at $f(\theta)$. Since the environment has preference reversal, there is a sequence of allocations $\{y_n\}_{n=1}^{\infty}$ with $\lim_{n\to\infty} y_n = f(\theta)$ such that for all y_n:

$$u_i(f(\theta), \theta) > u_i(y_n, \theta)$$

and

$$u_i(f(\theta), \hat{\theta}) < u_i(y_n, \hat{\theta}).$$

We shall now argue that for every $\theta_i' \in \Theta_i$ there exists sequence $\{y_n\}_{n=1}^{\infty}$ such that:

$$u_i(f(\theta_i', \theta_{-i}), (\theta_i', \theta_{-i})) \geq u_i(y_n, (\theta_i', \theta_{-i})). \tag{10}$$

The proof is by contrapositive. Suppose that (10) did not hold, and that there exists θ_i' such that for all y_n:

$$u_i(f(\theta_i', \theta_{-i}), (\theta_i', \theta_{-i})) < u_i(y_n, (\theta_i', \theta_{-i})),$$

then it would follow from continuity of the utility function that:

$$u_i(f(\theta_i', \theta_{-i}), (\theta_i', \theta_{-i})) \leq u_i(f(\theta_i, \theta_{-i}), (\theta_i', \theta_{-i})).$$

But this violates the hypothesis of strict ex post incentive compatibility. We have thus shown that the hypothesis in the definition of ex post monotonicity is never satisfied and hence the implication is never required. It follows that ex post monotonicity is vacuously satisfied in the single crossing environment. □

The public good example of Section 3 is an example where the (singleton) social choice set satisfies single crossing, strict EPIC and interiority. The idea of the proof is that with strict ex post incentive compatibility, the set of allocations which are dominated by the social choice function is locally (around $f(\theta)$) determined by the preferences of the agents with type profile θ. The situation is represented in Fig. 3.

If the ex post incentive constraint only holds as an equality for some types, say θ_i and θ_i', then the set of allocations dominated by the social choice function is determined locally (around $f(\theta)$) by the preferences of both types. In this case, the hypothesis of Maskin monotonicity may be

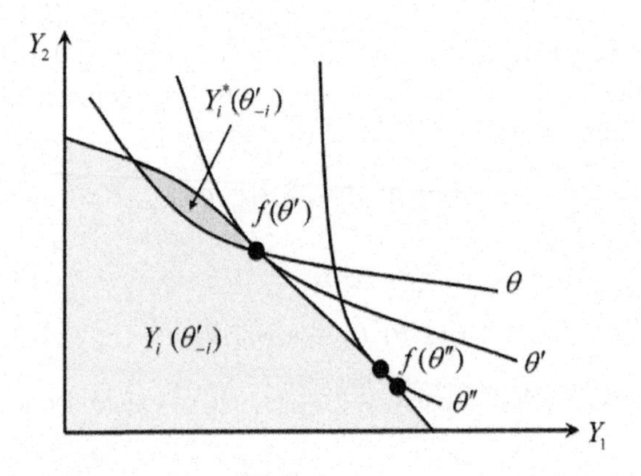

Fig. 3. Ex post monotonicity with strictness.

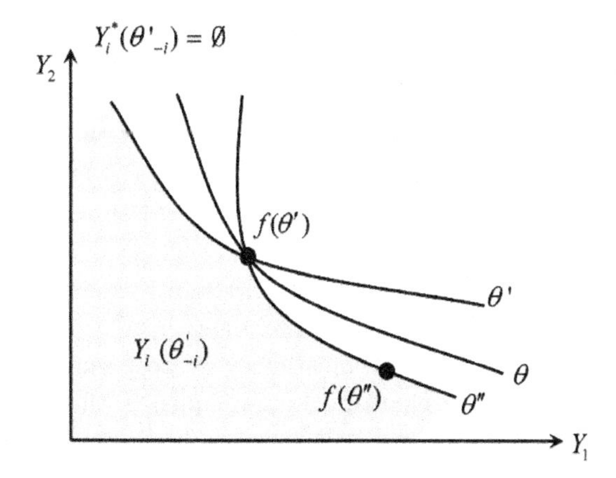

Fig. 4. Failure of strict ex post incentive compatibility.

satisfied and hence become a constraint. The second situation is represented in Fig. 4.

A recent contribution by Arya *et al.* (2000) explores in the private value environment the relationship between Maskin and Bayesian monotonicity and the single crossing condition. Interestingly, their sufficiency result regarding Bayesian monotonicity, also requires that the incentive compatibility conditions in the direct mechanism are satisfied as strict inequalities. They also present an example which shows that weak incentive compatibility and the single crossing condition alone do not guarantee Bayesian monotonicity. Incidentally, and for the same reason as in Arya *et al.* (2000) for the Bayesian incentive constraints, the strict ex post incentive compatibility condition actually allows a slightly stronger statement than actually stated in Proposition 4. The single crossing and strict ex post incentive constraints also imply a strict ex post monotonicity condition, where all the weak inequalities are replaced by strict inequalities.

The role of the interior assumption for the monotonicity condition has already been emphasized in work by Hurwicz *et al.* (1995). They presented an example of a Walrasian social choice correspondence where the Walrasian allocation for a given preference profile is on the boundary of the feasibility set. Naturally then, the indifference curves generated by a second and distinct set of preferences, intersect at the boundary. The crucial implication of the intersection at the boundary is that the set of allocations for which the Maskin monotonicity hypothesis fails is outside the feasible set, and hence Maskin monotonicity fails to hold.

6 Direct Mechanisms

In general, the ex post monotonicity conditions guarantee the existence of an incentive compatible reward. With the preference reversal environment and the strict ex post incentive constraints, we established that local changes in the report of the *types* are sufficient to establish the ex post monotonicity conditions. This in fact suggests that preference reversal and strictness of the ex post incentive constraints jointly guarantee implementation in the direct mechanism. We pursue this insight and show that in many economically important circumstances ex post implementation can be achieved in the direct mechanism. In consequence, the implementation does not have to rely on the augmentation on which much of the positive implementation results in the literature rest.

By definition, a direct mechanism cannot coordinate the selection of a particular social choice function f from a set F. Hence, we restrict our attention in this section to the implementation of a given social function f. The basic problem of implementation in the direct mechanism is that an agent must be able to claim the reward y by a report, possibly a misreport, of his type. A sufficient condition for direct implementation is therefore that for every allocation in the reward set $Y_i(\theta_{-i})$ of agent i, another allocation y' which is weakly preferred to y by agent i can be obtained by an appropriate report of agent i's type. In particular, if agent i can induce the choice of every y by the social choice function f through an appropriate reward, then the direct mechanism offers all the feasible rewards.

Definition 16 (*Full Range*). Social choice function f satisfies full range if for all i, all $\theta_{-i} \in \Theta_{-i}$ and $y \in Y$, there exists $\hat{\theta}_i$ such that $f(\hat{\theta}_i, \theta_{-i}) = y$.

Obviously, the full range condition is a very strong condition, but we shall now show that much weaker conditions will suffice in many environments. A common feature of many mechanism design models is that the allocation problem has two dimensions, the first is the assignment of the object and the second the monetary or quasi-monetary transfer. Within this two dimensional framework we can obtain positive results for ex post implementation in the direct mechanism. We thus suppose for the moment that the utility function of each agent permits the following representation:

$$u_i(y, \theta) = \hat{u}_i(y_0, y_i, \theta), \tag{11}$$

where $\hat{u}_i$ is strictly increasing in y_i, generalizing the monetary aspect in the quasilinear model. In the remainder of this section we hence investigate the implementation of a given social function f with $f = (f_0, f_1, f_2, \dots, f_I)$, where the pair (f_0, f_i) represents the relevant two dimensions of the allocation problem for agent i with $Y_i \subset \mathbb{R}$ for every i. We can now restrict the full range condition to the single dimension of every agent i.

Definition 17 (*One-Dimensional Full Range*). Social choice function f satisfies one-dimensional full range if for all i, $y_0 \in Y_0, \theta_{-i} \in \Theta_{-i}$, there exists $\hat{\theta}_i$ such that $f_0(\hat{\theta}_i, \theta_{-i}) = y_0$.

With the monotone utility u_i in y_i and the one dimensional full range condition we can indeed guarantee direct implementation.

Proposition 1 (*Direct Implementation with One-Dimensional Full Range*). *If f satisfies (EPIC), (EM) and one-dimensional full range, then direct ex post implementation is possible.*

Proof. We defined the reward set as:

$$Y_i(\theta_{-i}) = \{y : u_i(f((\theta_i', \theta_{-i})), (\theta_i', \theta_{-i})) \geq u_i(y, (\theta_i', \theta_{-i})) \text{ for all } \theta_i'\},$$

and we define the set of allocations attainable for i in the direct mechanism by:

$$Y_i^*(\theta_{-i}) = \{y : y = f(\theta_i', \theta_{-i}) \text{ for some } \theta_i'\}.$$

We now want to show that $y \in Y_i(\theta_{-i}) \Rightarrow \exists y' \in Y_i^*(\theta_{-i})$ such that $u_i(y', \theta) \geq u_i(y, \theta)$ for all θ. To do this, fix any $y \in Y_i(\theta_{-i})$. By one-dimensional full range, there exists θ_i' such that $f_0(\theta_i', \theta_{-i}) = y_0$. If $f_i(\theta_i', \theta_{-i}) < y_i$, then $u_i(y, (\theta_i, \theta_{-i})) > u_i(f(\theta_i', \theta_{-i}), (\theta_i', \theta_{-i}))$, contradicting $y \in Y_i(\theta_{-i})$. So $f_i(\theta_i', \theta_{-i}) \geq y_i$. So $u_i(f(\theta_i', \theta_{-i}), \hat{\theta}) \geq u_i(y, \hat{\theta})$ for all $\hat{\theta}$. □

The full range condition together with the monotonicity in the utility essentially guarantees that the agent can make a sufficiently large misreport to find an appropriate reward. The public good example on Section 3 satisfies the monotonicity condition and the one-dimensional full range condition if $\Theta_i = \mathbb{R}_+$. If we replace the monotonicity condition by the single crossing condition, then a local change in the report is sufficient to guarantee the reward to the whistle-blower. The type space Θ_i for every agent i now has to be an open set so that a local change in the report is always feasible.

Proposition 2 (*Direct Implementation with Single Crossing*). *In a single crossing environment, if Θ_i is an open set for every i, and f is interior,* *continuous in θ, and satisfies strict EPIC, then f can be ex post implemented in the direct mechanism.*

Proof. We first observe that with continuity, strict ex post incentive compatibility, and monotonicity in the second argument, it must be the case that for all θ_i, θ_i' and $\theta_i \neq \theta_i' f_0(\theta_i, \theta_{-i}) \neq f_0(\theta_i', \theta_{-i})$. Suppose not, then by strict ex post incentive compatibility the allocations have at least to differ in the second dimension, or, $f_i(\theta_i, \theta_{-i}) \neq f_i(\theta_i', \theta_{-i})$. But since $u_i(\cdot)$ is strictly increasing in y_i, it follows that this would violate the ex post incentive constraint for either θ_i or θ_i'. It now follows from continuity that $f_0(\theta_i, \theta_{-i})$ must be monotone in θ_i for every θ_{-i}.

We next show that the inequalities of ex post monotonicity can be satisfied for all θ, θ' and that a reward y can always be obtained by means of an allocation which is generated by the social choice function for some report $\hat{\theta}_i$ of agent i. Thus consider the indifference curve for θ and $\hat{\theta}$ at $f(\theta)$. Since the preferences are single crossing, there is a sequence of allocations $\{y_n\}_{n=1}^{\infty}$ with $\lim_{n\to\infty} y_n = f(\theta)$ such that for all y_n along the sequence

$$u_i(f(\theta), \theta) > u_i(y_n, \theta),$$

and

$$u_i(f(\theta), \hat{\theta}) < u_i(y_n, \hat{\theta}). \tag{12}$$

As in the earlier argument, we now argue that there exists an N such that for all $n \geq N$ and all y_n, we have

$$u_i(f(\theta_i', \theta_{-i}), (\theta_i', \theta_{-i})) \geq u_i(y_n, (\theta_i', \theta_{-i})). \tag{13}$$

The proof is by contrapositive. Suppose now that (13) were not to hold, and that there exists θ_i' such that for all N, we can find $n \geq N$ and y_n such that:

$$u_i(f(\theta_i', \theta_{-i}), (\theta_i', \theta_{-i})) < u_i(y_n, (\theta_i', \theta_{-i})),$$

then it would follow from continuity of the utility function that:

$$u_i(f(\theta_i', \theta_{-i}), (\theta_i', \theta_{-i})) \leq u_i(f(\theta_i', \theta_{-i}), (\theta_i', \theta_{-i})).$$

But this violates the hypothesis of strict ex post incentive compatibility. We have thus shown that for an appropriately chosen subsequence of $\{y_n\}_{n=1}^{\infty}$, converging to $y = f(\theta)$, all the elements satisfy (12) and (13).

By continuity and monotonicity of $f(\cdot)$, it follows that we can find a type $\hat{\theta}_i$ of agent i an element $\hat{y} \triangleq y_n$, such that $f_0(\hat{\theta}_i, \theta_{-i}) = \hat{y}_0$. By monotonicity of $u_i(\cdot)$ in y_i and the validity of (13), it follows that the corresponding component satisfies $f_i(\hat{\theta}_i, \theta_{-i}) \geq \hat{y}_i$. Again by monotonicity of $u_i(\cdot)$ in y_i, it now follows that $f(\hat{\theta}_i, \theta_{-i})$ is at least as desirable for agent i as $\hat{y}$. Thus agent i weakly prefers to claim $f(\hat{\theta}_i, \theta_{-i})$ to $\hat{y}$. But now it follows that agent i can claim the reward simply by reporting $\hat{\theta}_i$ in the direct mechanism without appealing to the augmented mechanism where he would claim $\hat{y}$. □

The public good example on Section 3 satisfies the condition of Proposition 2 provided the type space Θ_i is open for every i. The openness condition on the type space Θ_i simply guarantees that agent i can always "downward" and "upward" misreport and obtain a suitable reward y. It follows that we can easily relax the openness condition and obtain a quasi direct implementation by means of the following construction. For any given type space Θ_i, if we can find an open set $\bar{\Theta}_i$ such that $\Theta_i \subseteq \bar{\Theta}_i$ and the single crossing conditions extend to $\bar{\Theta}_i$, then we can directly apply the argument of Proposition 2 to the extended space $\bar{\Theta} = \times_{i=1}^{I} \bar{\Theta}_i$. Moreover, in equilibrium the agents will report only types $\theta \in \Theta$ belonging to the original type space.

7 Single Unit Auction

We consider the efficient social choice rule in the single unit auction with interdependent values as in Dasgupta and Maskin (2000). The auction model presents an interesting environment as it fails both the strict ex post incentive compatibility condition as well as the interiority condition. The assignment of the object among the agents changes only at pivotal types. As it stays constant for many reported types, it satisfies weak but not strict ex post incentive compatibility. The efficient assignment problem is also a canonical example of an exterior social choice function as for almost all preferences profiles, one agent receives the object with probability one and all other agents receive the object with probability zero. Despite the failure of the model to satisfy the conditions of Theorem 4, we will show that the local violations of strict ex post incentive compatibility and interior allocation can be overcome to establish implementation in ex post equilibrium, in particular by the direct mechanism. Incidentally, the local failure of these conditions leads to a failure of Maskin monotonicity when ex post monotonicity is still guaranteed.

7.1 *Model*

The utility function of agent i in the assignment problem is given by:

$$u_i(x_i, t_i, \theta) = x_i v_i(\theta) + t_i,$$

where x_i is the probability that agent i receives the object, t_i is his monetary payment and $v_i(\theta)$ is his interdependent valuation. We assume that $v_i(\theta)$ is continuously differentiable in θ_i and that:

$$\frac{\partial v_i(\theta)}{\partial \theta_i} > 0 \quad \text{and} \quad \frac{\partial v_i(\theta)}{\partial \theta_j} \neq 0. \tag{14}$$

The first condition simply says that a higher signal by i leads to a higher value of i and the second condition guarantees that we are in an interdependent rather than private value environment. The single crossing condition is, as in Dasgupta and Maskin (2000), that for all i, j and θ, if

$$v_i(\theta) = v_j(\theta) = \max_k \{v_k(\theta)\},$$

then

$$\frac{\partial v_i(\theta)}{\partial \theta_i} > \frac{\partial v_j(\theta)}{\partial \theta_i}. \tag{15}$$

We restrict our attention to a symmetric environment and a compact type space $\Theta_i = [0, 1]$. We consider the efficient allocation rule and in case of a tie at the top, we assign the object with equal probability among the agents with the highest valuation. The ex post incentive compatible transfer rule is of the form:

$$t_i(\theta) = -v_i(\underline{\theta}_i, \theta_{-i}), \tag{16}$$

where $\underline{\theta}_i$ is determined by:

$$\underline{\theta}_i = \min\{\theta_i' \in \Theta_i \mid v_i(\theta_i', \theta_{-i}) \geq v_j(\theta_i', \theta_{-i}), j \neq i\}.$$

The efficient direct mechanism satisfying (16) is the generalized Vickrey–Clark–Groves (VCG) mechanism.

7.2 *Monotonicity and the VCG Mechanism*

We present three results in this subsection. We first show that the generalized VCG mechanism satisfies the ex post monotonicity condition. In fact the positive result is strengthened to obtain ex post implementation in the direct mechanism. We then show that even though ex post monotonicity is satisfied, Maskin monotonicity fails.

Proposition 3 (*Ex Post Monotonicity*).

(1) *For $I \geq 3$, the generalized VCG mechanism satisfies ex post monotonicity.*
(2) *For $I \geq 3$, the direct generalized VCG mechanism has a unique pure ex post equilibrium.*

Proof. By Theorem 1, ex post monotonicity is a necessary condition for ex post implementation. It is therefore sufficient to show that the generalized VCG mechanism can be ex post implemented. We show that the unique equilibrium in the direct mechanism is the truthtelling equilibrium. Suppose not and there exists another ex post equilibrium. It follows that for some agent i and some type profile θ_i, we have that $s_i(\theta_i) \neq \theta_i$. We define the highest possible type across all agents for which we observe a report different from truthtelling:

$$\bar{\theta} = \max_{i \in \mathcal{I}} \sup\{\theta_i \in \Theta_i \mid s_i(\theta_i) \neq \theta_i\}. \tag{17}$$

We suppose initially that

$$\sup\{\theta_i \in \Theta_i \mid s_i(\theta_i) \neq \theta_i\} = \max\{\theta_i \in \Theta_i \mid s_i(\theta_i) \neq \theta_i\},$$

for all $i \in \mathcal{I}$, which we shall later relax.

Consider first $\bar{\theta} < 1$ and $s_i(\bar{\theta}) < \bar{\theta}$. We take agent i with $\theta = \bar{\theta}$ and $s_i(\bar{\theta})$ and consider for all other agents $j \neq i$, a type profile $\theta_j = \bar{\theta} + \varepsilon$ for some arbitrarily small $\varepsilon > 0$ so that $s_i(\bar{\theta}) < \bar{\theta} < \theta_j$. At θ_j, we know from (17), that $s_j(\theta_j) = \theta_j$. It follows that at the type profile (θ_i, θ_{-i}) and associated reports, all agents $j \neq i$ receive the object with the same probability but due to the misreport of agent i at a transfer $t_j(\cdot)$ (per unit of the object)

$$t_j(s_i(\bar{\theta}), \theta_j = \bar{\theta} + \varepsilon) > -v_j(\theta_i = \bar{\theta}, \theta_j = \bar{\theta} + \varepsilon), \tag{18}$$

which is strictly below the value of the object for agent j. It follows that every agent $j \neq i$ has a unilateral profitable deviation by reporting a higher type $\theta'_j > \theta_j$. With the higher report, he will still pay the same transfer (per unit) by the VCG mechanism, but agent j will then receive the object with probability one. By (18), the net utility of the transaction is strictly positive and hence agent j strictly increases his payoff with the deviation.

Consider next $\bar{\theta} < 1$ and $s_i(\bar{\theta}) > \bar{\theta}$. We now take agent i at $\theta_i = \bar{\theta}$ with $s_i(\theta)$ and consider for all other agents $j \neq i, \theta_j = \bar{\theta} + \varepsilon$ for some arbitrarily small $\varepsilon > 0$. At this profile $\theta_i = \bar{\theta}$ and $\theta_j = \bar{\theta} + \varepsilon$, agent i receives the object under the deception but at a transfer which is larger than the value of the object to him:

$$t_i(s_i(\bar{\theta}), \theta_j = \bar{\theta} + \varepsilon) < -v_i(\theta_i = \bar{\theta}, \theta_j = \bar{\theta} + \varepsilon). \tag{19}$$

It follows that by reporting a sufficiently low type profile so that he will not receive the object and receive a zero transfer, he can guarantee himself a zero net utility which is a strict improvement about his candidate negative net utility as displayed in (19).

Consider next $\bar{\theta} = 1$, then $s_i(\bar{\theta}) < \bar{\theta}$ has to hold. We observe first that at most one agent i can offer a downward biased report in equilibrium. If more than one agent would downward report at $\theta_j = 1$, then each one of the downward reporting agents would have a strict incentive to report the maximal type $\theta_j = 1$, as the monetary transfer (per unit) would be strictly less than the value of the object. It thus follows that all agents $j \neq i$ report truthfully at $\theta_j = 1$, for otherwise they would downward report and then agent i would always have a strict incentive to recover his bid and report truthfully.

By the same argument, it also follows that no agent other than i can ever downward report at any type profile $\theta_j \in \Theta_j$. Suppose to the contrary and that agent j downwards report at some θ'_j with $s_j(\theta'_j) < \theta_j$. Consider then the true profile $\theta = (\theta_j, 1, 1, \ldots, 1)$. With the downward report $s_j(\theta'_j) < \theta_j$, agent i would again have a strict incentive to report truthfully as the downward report by agent j leads to a transfer payment strictly less than the true value of the object. It hence follows that for all agents $j \neq i$ we only have to consider truthtelling or upward deceptions.

With agent i downward reporting at $\theta_i = 1$, we now argue that at least one agent $j \neq i$ has to consistently upward report for all $\theta_j \in (s_i(1), 1)$. For suppose not, then we can find some type profile θ_{-i} with $\theta_j \in (s_i(1), 1)$ for all $j \neq i$, such that all agents j report truthfully. But at such a type profile agent i would loose with his report of $s_i(1)$, when he values the object higher than everybody else and would have to pay a transfer strictly less than the value of the object to him. As this cannot be an equilibrium strategy profile, it follows that at least one agent most consistently misreport upwardly.

We finally argue that this cannot be an equilibrium strategy profile for agent j either. To see this consider the true profile $\theta_i = 0$ for agent

i. At $\theta_i = 0$, agent i cannot downward report, at least he has to report truthfully, and by the earlier argument all the other agents also report at least truthfully. But now consider agent j with an upward report at some θ'_j and $s_j(\theta'_j) > \theta'_j$. Consider now the type profile $\theta = (\theta_i = 0, \theta_j = \theta'_j, s_j(\theta'_j), \ldots, s_j(\theta'_j))$. In other words, at type profile θ all agents but i and j have a true type exactly equal to the reported profile of agent j. It follows that either agent j receives the object with positive probability at θ or that he looses out as some other agent, say k, different from i and j also misreports upwardly. But in either case, agent j or agent k will have to pay more than the object is worth to them. It follows that either the candidate strategy of agent j or k offers a profitable deviation by sufficiently lowering the report so that either j or k fails to get the object, receives a zero transfer and guarantees himself a zero net utility. We thus have obtained a contradiction to a candidate equilibrium involving a downward report at $\bar{\theta} = 1$.

It remains to consider the situation where

$$\bar{\theta} = \max_{i \in \mathcal{I}} \sup\{\theta_i \in \Theta_i \mid s_i(\theta_i) \neq \theta_i\} \neq \max_{i \in \mathcal{I}} \max\{\theta_i \in \Theta_i \mid s_i(\theta_i) \neq \theta_i\},$$

for some i. By definition of $\underline{\theta}$, for every $\varepsilon > 0$, we can then find an agent i and a type $\underline{\theta}_i$, such that $\bar{\theta} - \varepsilon < \theta_i < \bar{\theta}$, and $s_i(\theta_i) \neq \theta_i$. Now we can repeat the above arguments for $\bar{\theta} < 1$ with $s_i(\bar{\theta}) < \bar{\theta}$ or $s_i(\bar{\theta}) > \bar{\theta}$ at the type profile $\theta = (\theta_i, \bar{\theta}, \bar{\theta}, \ldots, \bar{\theta})$.

For $\bar{\theta} = 1$, we can still find for every $\varepsilon > 0$, an agent i and a type θ_i, such that $1 - \varepsilon < \theta_i < 1$, and $s_i(\theta_i) \neq \theta_i$. If for all such θ_i, we have $s_i(\theta_i) < \theta_i$, then the above argument goes through without modifications. The remaining possibility is that for all $\varepsilon > 0$, and θ_i satisfying $1 - \varepsilon < \theta_i < 1$, we have $s_i(\theta_i) > \theta_i$. But this cannot be an equilibrium either because at profile $\underline{\theta} = (\theta_i, 1, 1, \ldots, 1)$, all agents $j \neq i$ report by assumption truthfully, but with $s_i(\theta_i) > \theta_i$, every agent j would pay a transfer strictly exceeding the value of the object. Thus, we have excluded all other candidate ex post equilibria which involve misreporting for some agents at some type profiles. $\square$

The idea behind the proof is quite simply and relies essentially on interdependent rather than private valuations. Essentially we used two reward schemes. We either gave a currently winning agent the object with probability one at the current price or we released the current winner, i.e., did not give him the object and gave him a zero transfer. These rewards can be (implicitly) claimed in the direct mechanism. If an agent k finds it profitable to receive the object with probability one, then he can do so in the direct mechanism by slightly increasing his reported type. If an agent k would like not to receive the object, then he could always guarantee this by lowering his announced type profile.

This basic argument highlights the role of pivotal profiles at which an increase or decrease in the reported type leads to a change in the allocation. At type profiles at which the auctioneer is indifferent between assigning the object among two or more agents, the pivotal or competitive profiles, two important things happened. First, the allocation rule is now in the interior as the auctioneer awards the object to competitive bidders with the same probability. Second, even though at the competitive profile, the ex post incentive constraints are only weak inequalities, in any neighborhood of the competitive profile, we can find strict ex post incentive constraints. In light of the earlier results on the single crossing environment, notably Theorem 4, we then find that it is the existence of pivotal profiles which matters for ex post implementation rather than the everywhere strictness of the ex post incentive constraints.

In this respect, we should point out that the role of the symmetric valuations is precisely to facilitate the location of pivotal profiles θ. The proof of Proposition 3 would go through unchanged if we were to abandon symmetry everywhere except at the lowest and highest type profiles, or at $\theta = (0, \ldots, 0)$ and $\theta = (1, \ldots, 1)$. In fact, the only property of the symmetry at the bottom and the top we really need, is that for every type profile θ_i of agent i, there exists another agent j and type profile θ_{-i} such that j is competitive with respect to i at $\theta = (\theta_i, \theta_{-i})$. Hence a different sufficient condition for ex post implementation in the single unit auction model would be the full range condition introduced in Section 5.

Despite the positive ex post implementation results, the generalized VCG mechanism fails Maskin monotonicity. Indeed, the generalized VCG mechanism is an important example where the two monotonicity notions fail to coincide.

Proposition 4 (*Maskin Monotonicity*). *The generalized VCG mechanism fails Maskin monotonicity.*

Proof. Consider a profile θ such that

$$v_i(\theta) > v_j(\theta), \quad \forall j \neq i.$$

Then by the single crossing conditions (14) and (15) we can find θ' with

$$\theta_i' > \theta_i, \quad \theta_j' > \theta_j$$

such that

$$v_i(\theta') > v_i(\theta), \quad v_i(\theta') < v_j(\theta).$$

It then follows that

$$t_i(\theta) > t_i(\theta').$$

We now recall that a social choice function f is Maskin monotone, if for all $\theta, \theta' \in \Theta$:

$$u_i(f(\theta), \theta) \geq u_i(y, \theta) \Rightarrow u_i(f(\theta), \theta') \geq u_i(y, \theta') \tag{20}$$

for all i and y, then

$$f(\theta) = f(\theta'). \tag{21}$$

We can verify that the implication (20) holds for all i and y, but the conclusion to be drawn, $f(\theta) = f(\theta')$, obviously fails as the transfers offered to agent i have to be different to guarantee incentive compatibility. □

We should add that there are transfer rules which satisfy Maskin monotonicity, but necessarily fail ex post incentive compatibility. For example, a constant transfer rule satisfies Maskin monotonicity, but obviously is not ex post incentive compatible. Yet as Maskin monotonicity is concerned with complete information environments, this may be less of a concern for the notion of Maskin monotonicity.

The existence rather than ubiquitousness of the pivotal profiles also underlies the difference between ex post and Maskin monotonicity. With ex post monotonicity we can search for a competitive profile among all types of agent i given the type profile of the other agents, whereas with Maskin monotonicity and the inherent synchronicity of complete information, we cannot do that. In particular, the fact that for most type profiles we are at the exterior of the allocation space is a problem for Maskin, but not necessarily for ex post monotonicity. Provided that the social choice function is at least sometimes in the interior we can use the single crossing condition at the interior profiles.

7.3 *Private Versus Interdependent Values*

We can now make an interesting observation regarding the implementation of the single unit auction. Suppose we consider the Vickrey auction (for the private value model) and the generalized Vickrey–Clark–Groves mechanism (for the interdependent value model). With private values, we have a failure of Maskin and ex post monotonicity. However with interdependent values, this coincidence ceases to exist and while Maskin monotonicity continues to fail, ex post monotonicity can be reestablished. While it is well known that the second price auction cannot be Nash implemented (see Saijo *et al.*, 2004), the positive results here regarding interdependent valuations are new in the literature.

The intuition for the divergence between private and interdependent values goes as follows. In a private value model, if one agent submits the highest possible report and all other agents submit the lowest possible report, then the former receives the object at the lowest possible price. Any attempt to reward a whistle-blower will then inevitably violate the ex post incentive compatibility constraint. In contrast, consider the exact same reporting strategy with interdependent valuations. Now we can reward a loosing agent in all those instances where the loosing agents all report the lowest value but in fact all have a higher valuation. We can reward every loosing agent by giving him the object and asking him to pay only as much as the reported value would suggest. This satisfies ex post incentive compatibility, but at the same time provides the reward to break the undesired equilibrium. Yet, it is clear that the argument relies essentially on interdependent rather than private values. The distinction in the ex post implementation result between private and interdependent values continues to exist for any arbitrary small amount of interdependence.

The distinction between private and interdependent values also becomes apparent in the role of the three or more agent condition in Proposition 3. In a recent paper, Birulin (2003) shows that with two agents there is a continuum of inefficient undominated ex post equilibria in the single unit auction model with the generalized VCG mechanism. In contrast, Chung and Ely (2001) show that with two bidders, the efficient equilibrium is the only outcome which survives the process of iterative deletion of ex post weakly dominated strategies. Birulin (2003) shows there are other equilibria which involve monotone, but discontinuous, reporting strategies, which lead to inefficient equilibria. The idea behind the construction is that over an arbitrary interval of profiles, agent i overstates and agent j understates his type. The reporting strategies are constructed precisely on the basis of the indifference conditions of i and j at the lower and upper end of the interval where a misreport occurs. The deception strategy of agent i is basically determined by agent j's true valuation and vice versa. This construction is not feasible anymore when there are more than two agents. With more than two agents, the valuation of agent j will depend on the type of agent i, j and k. In consequence the derivation of the deception strategy of agent i will not only depend on agent j's critical type, but the *exact* type of agent k. But of course the strategy of agent i cannot depend simultaneously on the type of agent j and k. In fact, Proposition 3 showed that there is no deception strategy which will lead to an equilibrium different from the efficient equilibrium.

8 Social Choice Sets

In the initial discussion of ex post and Maskin monotonicity we argued that the notions diverge in two aspects: (i) the set of profiles at which rewards could be offered and (ii) the size of the reward set. The single unit auction demonstrated the relevance of the first aspect. The set of profiles at which the designer could offer a reward to the whistle-blower was larger with ex post monotonicity. In consequence, we could satisfy ex post monotonicity yet fail Maskin monotonicity. This section considers the reverse case, in which Maskin monotonicity is satisfied but ex post monotonicity fails. At the end of this section, we shall also discuss the relationship between functions, correspondences and sets in the context of ex post implementation.

8.1 *Pareto Correspondence*

Maskin (1999) observes that many prominent social choice correspondences, among them the Pareto, the Condorcet and the Walrasian correspondences, satisfy the complete information monotonicity notion. We now show with a specific Pareto correspondence that ex post monotonicity fails to share this property. Similar examples can be constructed for the Condorcet and the Walrasian correspondence. As we would expect, the divergence between the two notions arises from the difference in the respective reward sets.

The Pareto correspondence is generally defined by:

$$PO(\theta) = \{y \in Y \mid \forall z \in Y,\ \exists i \text{ s.t. } u_i(y, \theta) \geq u_i(z, \theta)\}.$$

We consider an example with three agents, $i = 1, 2, 3$ and each agent has two possible types: $\theta_i \in \Theta_i = \{0, 1\}$. A type profile is then given by $\theta = (\theta_1, \theta_2, \theta_3) \in \Theta = \times_{i=1}^{3} \Theta_i$. The set of allocations, Y, has the same cardinality as the type space, Θ. For simplicity, allocations and type profiles carry the same labels, but allocations are described as strings rather than vectors:

$$Y = \{000,\ 001,\ 010,\ 011,\ 100,\ 101,\ 110,\ 111\}.$$

The payoffs of the agents are described for every true type profile $\theta = (\theta_1, \theta_2, \theta_3)$ below. In each matrix, each cell identifies the utility from a specific allocation. As the type set coincides with the allocation set, the described utilities also represent the payoffs arising in the direct mechanism for any reported type profile by the agents:

$$\begin{array}{lccccccc}
 & \theta_3 = 0 & \theta_2 = 0 & \theta_2 = 1 & & \theta_3 = 1 & \theta_2 = 0 & \theta_2 = 1 \\
\theta = (0,0,0): & \theta_1 = 0 & 3,0,0 & 0,0,0 & & \theta_1 = 0 & 0,0,0 & 0,\varepsilon,0 \\
 & \theta_1 = 1 & 1,1,1 & 0,0,0 & & \theta_1 = 1 & 0,0,0 & \varepsilon,\varepsilon,\varepsilon
\end{array}$$

$$\begin{array}{lccccccc}
 & \theta_3 = 0 & \theta_2 = 0 & \theta_2 = 1 & & \theta_3 = 1 & \theta_2 = 0 & \theta_2 = 1 \\
\theta = (0,0,1): & \theta_1 = 0 & 0,0,0 & 0,0,0 & & \theta_1 = 0 & 0,3,0 & 1,1,1 \\
 & \theta_1 = 1 & 0,0,0 & \varepsilon,\varepsilon,\varepsilon & & \theta_1 = 1 & 0,0,0 & \varepsilon,0,0
\end{array}$$

$$\begin{array}{lccccccc}
 & \theta_3 = 0 & \theta_2 = 0 & \theta_2 = 1 & & \theta_3 = 1 & \theta_2 = 0 & \theta_2 = 1 \\
\theta = (0,1,0): & \theta_1 = 0 & 1,1,1 & 0,3,0 & & \theta_1 = 0 & 0,0,0 & 0,0,0 \\
 & \theta_1 = 1 & 0,0,\varepsilon & 0,0,\varepsilon & & \theta_1 = 1 & \varepsilon,\varepsilon,\varepsilon & 0,0,0
\end{array}$$

$$\begin{array}{lccccccc}
 & \theta_3 = 0 & \theta_2 = 0 & \theta_2 = 1 & & \theta_3 = 1 & \theta_2 = 0 & \theta_2 = 1 \\
\theta = (0,1,1): & \theta_1 = 0 & \varepsilon,0,0 & 1,1,1 & & \theta_1 = 0 & 0,0,0 & 0,0,3 \\
 & \theta_1 = 1 & \varepsilon,\varepsilon,\varepsilon & \varepsilon,\varepsilon,0 & & \theta_1 = 1 & 0,0,0 & 0,\varepsilon,0
\end{array}$$

$$\begin{array}{lccccccc}
 & \theta_3 = 0 & \theta_2 = 0 & \theta_2 = 1 & & \theta_3 = 1 & \theta_2 = 0 & \theta_2 = 1 \\
\theta = (1,0,0): & \theta_1 = 0 & 0,0,\varepsilon & \varepsilon,0,\varepsilon & & \theta_1 = 0 & 0,0,0 & \varepsilon,\varepsilon,\varepsilon \\
 & \theta_1 = 1 & 0,3,0 & 1,1,1 & & \theta_1 = 1 & 0,0,0 & \varepsilon,0,0
\end{array}$$

$$\begin{array}{lccccccc}
 & \theta_3 = 0 & \theta_2 = 0 & \theta_2 = 1 & & \theta_3 = 1 & \theta_2 = 0 & \theta_2 = 1 \\
\theta = (1,0,1): & \theta_1 = 0 & \varepsilon,0,0 & \varepsilon,\varepsilon,\varepsilon & & \theta_1 = 0 & 0,0,0 & 0,0,0 \\
 & \theta_1 = 1 & 1,1,1 & 0,0,0 & & \theta_1 = 1 & 0,0,3 & 0,0,0
\end{array}$$

$$\begin{array}{lccccccc}
 & \theta_3 = 0 & \theta_2 = 0 & \theta_2 = 1 & & \theta_3 = 1 & \theta_2 = 0 & \theta_2 = 1 \\
\theta = (1,1,0): & \theta_1 = 0 & 0,0,0 & 0,\varepsilon,0 & & \theta_1 = 0 & \varepsilon,\varepsilon,\varepsilon & \varepsilon,\varepsilon,0 \\
 & \theta_1 = 1 & \varepsilon,0,0 & 0,0,1 & & \theta_1 = 1 & 0,0,0 & 1,1,1
\end{array}$$

$$\begin{array}{lccccccc}
 & \theta_3 = 0 & \theta_2 = 0 & \theta_2 = 1 & & \theta_3 = 1 & \theta_2 = 0 & \theta_2 = 1 \\
\theta = (1,1,1): & \theta_1 = 0 & \varepsilon,\varepsilon,\varepsilon & 0,\varepsilon,0 & & \theta_1 = 0 & 0,0,0 & 1,1,1 \\
 & \theta_1 = 1 & 0,0,\varepsilon & 0,0,0 & & \theta_1 = 1 & 0,0,0 & 3,0,0
\end{array}$$

The above example has the property that in every state θ, there exist exactly two Pareto efficient allocations. The first Pareto allocation corresponds to the true state: $y = \theta$ and it favors one agent with payoff 3 and leaves the remaining two agents with payoff 0. The identity of the favored agent is determined by $1+\sum_i \theta_i$ (mod 3). The second Pareto allocation generates a uniform payoff of 1 across agents. The remaining allocations are all Pareto inferior. Besides the Pareto allocations, there is one more important allocation in this example, given by $y=\theta'$ with $\theta_i \neq \theta_i'$ for all i. It generates a payoff of ε for all agents in all states and is obviously Pareto dominated. Yet, we will show that it can be obtained as an ex post equilibrium under the deception $\alpha(\theta)=\theta'$. The remaining payoff vectors are combination of 0 and ε entries. In each one of the vectors, the 0 entries serve to support (always) truthtelling and (always) misreporting as ex post equilibria in the direct mechanism, whereas the ε entries serve to shrink the ex post reward set.

In this example, the Pareto correspondence is described by $PO\colon \Theta \to Y$:

$$\begin{array}{ccc} \theta_3=0 & \theta_2=0 & \theta_2=1 \\ \theta_1=0 & \{000,\ 100\} & \{010,\ 000\} \\ \theta_1=1 & \{100,\ 110\} & \{110,\ 111\} \end{array} \qquad \begin{array}{ccc} \theta_3=1 & \theta_2=0 & \theta_2=1 \\ \theta_1=0 & \{001,\ 011\} & \{011,\ 010\} \\ \theta_1=1 & \{101,\ 010\} & \{111,\ 011\} \end{array} \tag{22}$$

The matrices describe the set of Pareto efficient allocations as a function of the true type profiles of the agents. The corresponding social choice set F is the set of all functions which satisfy $f(\theta) \in PO(\theta)$ for all $\theta \in \Theta$.

Maskin Monotonicity. Maskin (1999) showed that the Pareto correspondence satisfies complete information monotonicity. For a better grasp of the difference between Maskin and ex post monotonicity, it may be instructive to see how these differences play out in this example. We therefore verify first that the Pareto selection $f(\theta)=\theta$ for all $\theta \in \Theta$, which assigns asymmetric utilities, is Maskin monotone. The only relevant deception is the complete deception with:

$$\alpha_i(\theta_i) \neq \theta_i, \quad \forall i, \forall \theta_i.$$

Without loss of generality we may consider $\theta=(0,0,0)$ and $\alpha(0,0,0)=(1,1,1)$. By setting $y=000$ and $i=1$, we clearly satisfy Maskin monotonicity as:

$$3 = u_1(y,(0,0,0)) > u_1(f(\alpha(0,0,0)),(0,0,0)) = \varepsilon,$$

and

$$3 = u_1(f(\alpha(0,0,0)),\alpha(0,0,0)) \geq u_1(y,\alpha(0,0,0)) = \varepsilon.$$

Ex Post Monotonicity. We continue with ex post monotonicity and again consider $\theta = (0,0,0)$ and $\alpha(0,0,0) = (1,1,1)$. We first observe that the candidate allocation $y = 000$, which we used in the complete information setting is not in the ex post reward set $Y_1(\theta_{-1})$ anymore. More precisely, if the true type of agent 1 is $\theta_1 = 0$, then agent 1 has an incentive to claim the reward $y = 000$ given the true type profile of the remaining two agents is $\theta_{-1} = (1,1)$, or:

$$0 = u_1(f(0, \alpha_{-1}(0,0)), (0, \alpha_{-1}(0,0))) < u_1(000, (0, a_{-1}(0,0))) = \varepsilon.$$

At the true profile $\theta = (0,0,0)$ and deception $\alpha(0,0,0) = (1,1,1)$, the designer could alternatively offer the reward $y = 100$, which is the second Pareto allocation. For every agent i, this allocation satisfies the reward equality:

$$1 = u_i(100, (0,0,0)) > u_i(f(\alpha(0,0,0)), (0,0,0)) = \varepsilon.$$

It also satisfies the ex post incentive compatibility at the true type profile $\alpha(0,0,0) = (1,1,1)$ for agent 1 and 2, but fails for agent 3, as:

$$0 = u_3(f(\alpha(0,0,0)), \alpha(0,0,0)) < u_3(100, \alpha(0,0,0)) = \varepsilon.$$

Moreover, for agent 1 and 2, it fails to be satisfied at $\theta_i' \neq \theta_i$ with:

$$0 = u_1(f(0, \alpha_{-1}(0,0)), (0, \alpha_{-1}(0,0))) < u_1(100, (0, \alpha_{-1}(0,0))) = \varepsilon,$$

and

$$0 = u_2(f(0, \alpha_{-2}(0,0)), (0, \alpha_{-2}(0,0))) < u_2(100, (0, \alpha_{-2}(0,0))) = 1.$$

By construction, the same argument goes through at every type profile and in consequence, ex post monotonicity fails in this example. The failure of ex post monotonicity comes as the reward set $Y_i(\theta_{-i})$ is strictly smaller than $Y_i(\theta)$ and in particular, eliminates all rewards which could satisfy the reward inequality.[15]

[15]The example is complicated as the Pareto correspondence is a set rather than a point everywhere. Yet, to display a difference between ex post and Maskin monotonicity, this appears to be necessary. First, observe, that if the payoffs were symmetric, then (generically) a Pareto efficient allocation would also constitute a strictly Pareto dominant allocation. The strictly dominant allocation clearly constitutes an ex post equilibrium and every deception α could be fended off by simply reestablishing the social choice allocation. The same argument continues to go through without symmetry if there is a unique Pareto allocation in every state. It follows that for a (generic) discrepancy between ex post and Maskin monotonicity, we need multiple Pareto efficient allocations,

8.2 *Functions, Sets and Correspondences*

The Pareto set is an example of a social choice set rather than a social choice function. This naturally leads us to inquire the relationship between functions and sets in the context of ex post implementation. In particular, we can ask whether implementation of a social choice set F is equivalent to the implementation of every social choice function $f \in F$ separately. The obvious direction is that if every $f \in F$ can be (ex post) implemented, then the social choice set F can be implemented. The more difficult direction is easy to understand with the following example. Let the social choice set F be the set of *all* ex post incentive compatible *plans*, i.e., mappings from profiles to allocations. By construction, it follows that F is ex post monotone and can be ex post implemented. But of course a single element $f \in F$ may not be ex post implementable by itself as there might be multiple equilibria under the social choice function f which do not correspond to f under some profiles. By the revelation principle, any such distinct equilibrium will also be an equilibrium under the corresponding direct mechanism. Hence a deception α which forms an equilibrium in f is an element in F and by contrast would not harm the implementability of F. It follows more generally that the implementation of a social choice set F does not imply that every social choice function $f \in F$ can be implemented separately.

The Pareto set was defined as a correspondence from the set of profiles to the set of allocations. On the basis of the Pareto correspondence, we naturally defined an associated social choice set. More generally, given a social choice correspondence $\phi : \Theta \to Y$, we can define an associated social choice set $F = \{f \mid f : \Theta \to Y\}$ by including all social choice functions f which select at all profiles allocations in the image of the correspondence:

$$F = \{f \mid f(\theta) \in \phi(\theta),\ \forall \theta \in \Theta\}.$$

Similarly, we can start with a social choice set F and define an associated social choice correspondence by including all allocations y at a profile θ which can be obtained by some selection f at θ from the social choice set F:

$$\phi(\theta) = \{y \mid y = f(\theta),\ f \in F\}.$$

---

which (generically) have to display some asymmetries in the way they affect the utilities of the agents.

With the above associations, we can then relate ex post implementation of sets and correspondences. For the purpose of this discussion, it might be useful to keep in mind the class of social choice problems in which a designer faces agents with quasilinear utility and wishes to implement the social efficient allocation without any balanced budget considerations. The social choice set of efficient and ex post incentive compatible allocations is then very large as the transfers to the agents are essentially only determined up to a constant. However if we consider the associated social choice correspondence, then we will typically loose the ex post incentive compatibility as arbitrary combinations of transfers across profiles will not satisfy ex post incentive compatibility. For this reason, incomplete information implementation typically considers sets rather than correspondences and even though we analyze ex post rather than Bayesian equilibrium, social choice correspondences typically still lack ex post incentive compatibility.

9 Mixed Strategy Implementation

Finally we extend the ex post implementation results to cover pure as well as mixed strategy equilibria. In the process we shall also propose significantly weaker sufficient conditions for ex post implementation. The proof strategy follows the argument for complete information implementation with mixed strategies presented in Maskin (1999) and refined in Maskin and Sjostrom (2004). The idea of their proof is to enlarge the strategy space of each agent by allowing him to make a contingent rather than deterministic reward proposal. In addition, they allow the agent to quote integers in the augmented mechanism to prevent the possibility of further equilibria arising in the augmented mechanism. We shall use the same basic idea but in addition use the integers to create lotteries in the augmented mechanism. The introduction of lotteries is natural in an environment which allows for mixed strategies. We then show that the use of lotteries has the additional advantage that we can dispense with the ex post monotonicity no veto hypothesis as a sufficient condition and replace it by a much weaker condition, called value distinction.[16]

[16]We would like to thank Andy Postlewaite and Phil Reny for comments during a Cowles Foundation Conference on "Robust Mechanism Design" which prompted us to pursue this argument.

The idea of using lotteries to weaken the sufficient condition also appears in a recent contribution by Benoît and Ok (2004). In the complete information environment, they show that by using simple lotteries the no veto condition can be replaced by a much weaker top-coincidence notion. As they focus on pure strategy implementation, their augmented mechanism makes use only of modulo rather than integer games. In sum, the use of random mechanism allows us to extend the earlier implementation result from pure to mixed strategy implementation and to substantially weaken the sufficiency conditions.

A *mixed* strategy for agent i is $\sigma_i : \Theta_i \to \Delta(M_i)$ and we denote the probability that type θ_i sends message m_i under strategy σ_i by $\sigma_i(m_i \mid \theta_i)$. The set of feasible allocations $\mathcal{Y}$ is now understood to be the set of all lotteries over a set of finite deterministic outcomes Y, or $\mathcal{Y} = \Delta(Y)$.

Definition 18 (*Ex Post Equilibrium in Mixed Strategies*). A mixed strategy profile $\sigma^* = (\sigma_1^*, \ldots, \sigma_I^*)$ is an *ex post equilibrium* if

$$\sum_{m \in M} u_i(g(m), \theta)\sigma^*(m \mid \theta) \geq \sum_{m_{-i} \in M_{-i}} u_i(g(m_i', m_{-i}), \theta)\sigma_{-i}^*(m_{-i} \mid \theta_{-i}),$$

for all i, θ and $m_i' \in M_i$.

The mixed strategy ex post equilibrium maintains all the features of the pure strategy ex post equilibrium. In particular, we observe that the no regret property is maintained conditional on the true type profile (but not conditional on every possible realization of messages). The notions of ex post monotonicity and ex post implementation remain unchanged with the understanding that the allocation y is possibly a lottery.

The necessary conditions for ex post implementation clearly remain necessary with the extension to mixed strategy equilibria and stochastic mechanisms. The focus of the remainder of this section is therefore on the sufficiency conditions. The no veto condition on the social choice set is replaced by a very weak condition on the preferences of the agents, referred to as value distinction.[17]

[17]The notion of value distinction is different from value distinguished types as defined by Palfrey and Srivastava (1989b). Their notion requires that for every pair, θ_i and θ_i', by agent i, there exists an allocation y which is valued differently by the two types, θ_i and θ_i'.

Definition 19 (*Value Distinction*). The environment satisfies value distinction if for all $\theta \in \Theta$ and all $y, y' \in Y$, there exists i such that $u_i(y, \theta) \neq u_i(y', \theta)$.

The basic novelty is the introduction of a contingent reward in the augmented mechanism. Each agent sends a message of the form $m_i = (\theta_i, f_i, z_i, y_i)$, where $\theta_i \in \Theta_i$, $f_i : \Theta \to \mathcal{Y}$ is the social choice function suggested by i, $z_i \in \mathbb{N} = \{1, 2, \ldots\}$, and $y_i \in \mathcal{Y}$. The set of feasible messages for agent i is given by

$$M_i = (\Theta_i \times \mathcal{F} \times \mathbb{N} \times \mathcal{Y}). \tag{23}$$

A strategy profile in this game is a collection $\sigma = (\sigma_1, \ldots, \sigma_I)$, with $\sigma_i : \Theta_i \to \Delta(M_i)$.

The proposal is required to satisfy:

$$f_i(\theta) \in \{y' \in \mathcal{Y} \mid u_i(f(\theta_i', \theta_{-i}), (\theta_i', \theta_{-i})) \geq u_i(y', (\theta_i', \theta_{-i})),\ \forall \theta_i' \in \Theta_i\}. \tag{24}$$

The outcome function $g : M \to \mathcal{Y}$ is defined by three rules:

Rule 1. If at m (and reported type profile θ), we have for all i, $f_i = f$ for some $f \in F$, then

$$g(m) = f(\theta). \tag{25}$$

Rule 2. If at m (and reported type profile θ), there exists $j \in N$ such that $f_i(\theta) = f_k(\theta), \forall i, k \neq j$ and $f_i(\theta) \neq f_j(\theta)$, then

$$g(y \mid m) = \begin{cases} 1 - \dfrac{1}{z_j + 1} & \text{if } y = f_j(\theta), \\ \dfrac{1}{z_j + 1} & \text{if } y = f_j(\theta), \\ 0 & \text{if otherwise.} \end{cases} \tag{26}$$

Rule 3. In all other cases, the agent j with the highest integer z_j is the winner (and, in the event of a tie, the agent with the highest label), and with probability $(1 - \frac{1}{z_j+1})$ pick y_j, and with probability $\frac{1}{\#Y} \frac{1}{z_j+1}$ pick

$y \in Y$, or

$$g(y \mid m) = \begin{cases} 1 - \dfrac{\#Y - 1}{\#Y} \dfrac{1}{z_j + 1} & \text{if } y = y_j \text{ and } z_j > z_i, \\ 1 - \dfrac{\#Y - 1}{\#Y} \dfrac{1}{z_j + 1} & \text{if } y = y_j \text{ and } z_j = \max_{i \neq j} z_i \text{ and } \forall\, i \\ & \quad \text{s.th. } z_i = z_j, j > i, \\ \dfrac{1}{z_j + 1} \dfrac{1}{\#Y} & \text{otherwise.} \end{cases} \tag{27}$$

The randomization in Rule 3 is simply a uniform randomization over the set of deterministic outcomes, and $\#Y$ is the cardinality of the set of deterministic outcomes.

We refer to the mechanism described by the message space $M = \times_{i=1}^{I} M_i$, described by (23), and the outcome function $g : M \to \mathcal{Y}$, described by (25)–(27), as the augmented mechanism. In contrast to the augmented mechanism presented in Section 4, the integer game is now defined on the natural numbers rather than a finite set of numbers.

Theorem 5 (*Mixed Strategy Implementation*). *For $I \geq 3$, if the environment satisfies value-distinction and F satisfies (EPIC) and (EM), then F is ex post implementable.*

Proof. We use the augmented mechanism described by (23)–(27).

Claim 1. *Every $f \in F$ can be realized as an ex post equilibrium with*

$$\sigma_i(m_i \mid \theta_i) > 0 \Rightarrow m_i = (\theta_i, f, \cdot, \cdot) \quad \textit{for all } \theta_i \in \Theta_i \textit{ and all } i.$$

Thus suppose that all agents but j pursue the above "truthtelling" strategy: $\sigma_i(m_i \mid \theta_i) > 0 \Rightarrow m_i = (\theta_i, f, \cdot, \cdot)$, for all $i \neq j$. By (*EPIC*), it follows that given $f_j = f$, truthtelling for agent j, is a best response. It remains to argue that it remains a best response if the restriction of $f_j = f$ is removed. But by definition of $f_j(\theta)$ and Rule 2, a deviation to $f_j(\theta) \neq f(\theta)$ does not increase the utility of agent j, and may even decrease the utility of agent j, and hence it follows that every $f \in F$ can be realized as an ex post equilibrium.

Claim 2. *In any ex post equilibrium with $f_i = f, \forall\, i, \forall\, \theta, g(m) = f(\theta)$ for some $f \in F$.*

Suppose not, then by ex post monotonicity, there exists an agent i, a type profile θ, and an allocation y which strictly improves the utility of agent i, and under Rule 2, he can obtain this improvement with arbitrarily large probability.

Claim 3. *In any ex post equilibrium,* $\sigma_i(m_i \mid \theta_i) > 0 \Rightarrow m_i = (\cdot, f, \cdot, \cdot)$.

Suppose not and hence there exists an ex post equilibrium with $\sigma_j(\cdot, f, \cdot, \cdot \mid \cdot) < 1$ for some j at some $\theta \in \Theta$ and $\sigma_i(\cdot, f, \cdot, \cdot \mid \cdot) = 1$ for all other i. Then with positive probability the allocation will be either $f_i(\theta)$ or $f_j(\theta)$. By value distinction, there exists an agent k who assigns different utilities to these two different allocations. If k happens to be j, then by proposing a sufficiently large integer z_j, he guarantees himself a higher payoff. If $k \neq j$, then k can make a proposal $f_k(\theta) \neq f(\theta)$ such that Rule 3 will be applied in which k can guarantee himself to be the winner with arbitrarily high probability, and thus lower the probability of receiving the low utility arbitrarily close to zero.

Consider finally an ex post equilibrium with i, j such that for $f_i, f_j \neq f$:

$$\sigma_i(\cdot, f_i, \cdot, \cdot \mid \cdot)\sigma_j(\cdot, f_j, \cdot, \cdot \mid \cdot) > 0.$$

Now the above argument applies again and it follows that there cannot exist an ex post equilibrium where Rule 3 is applied with positive probability.

□

Benoît and Ok (2004) show that by using "simple" stochastic mechanisms, the sufficient conditions for Nash implementation in pure strategies can be substantially weakened. In particular, the no veto condition of the social choice set can be replaced by much weaker conditions on (i) the social choice function, namely weak unanimity and (ii) the preferences, namely a top coincidence condition. Our Theorem 5 does not require weak unanimity and the value distinction condition is strictly weaker than their top coincidence condition. Their top coincidence condition says that if for every profile θ and every i there exists at most a single allocation z such that:

$$u_j(z, \theta) \geq u_j(y, \theta), \quad \forall j \neq i, \forall y \in Y. \tag{28}$$

To see why value distinction is (strictly) weaker, fix any arbitrary θ and i, and suppose that there does not exist an allocation z which is the preferred allocation for all agents $j \neq i$. Then there exist at least two agents, j and k, which differ in their most preferred allocation, call them y_j, y_k, respectively. In consequence, for at least one of the agents, say j, we have

$u_j(y_j, \theta) \neq u_j(y_k, \theta)$. The same argument goes through if there does exist a single allocation z satisfying (28). By implication, there must exist another allocation y which is not the most preferred allocation for all $j \neq i$. For the allocation y and a particular agent j, we then have value distinction again: $u_j(z, \theta) \neq u_j(y, \theta)$. It is further immediate that the two conditions, top coincidence and value distinction, do not coincide, which establishes the strict implication.

We can replace top coincidence by the weaker value distinction condition because we allow for more than "simple" stochastic allocations. In the augmented mechanism of Benoît and Ok (2004), the whistle-blower i claims the reward by choosing a lottery which selects the reward y and the social choice $f(\theta)$ with equal probability, the "simple" stochastic allocation. As $y \neq f(\theta)$, the top coincidence condition then implies that there will be another agent j who has a strict preference between y and $f(\theta)$, and can impose his choice by appealing to Rule 3. In our augmented mechanism, the whistle-blower i can always increase the probability of receiving the reward by increasing the integer z_i, and hence it suffices that for the pair of allocations, y and $f(\theta)$, there exists an agent who values the two allocations differently. The second condition in Benoît and Ok (2004), weak unanimity, is not necessary in our augmented mechanism either. The use of random allocations in Rule 3 excludes the possibility of equilibria with a strategy profile in which Rule 3 applies. Consequently, we can use the condition of value distinction as in Theorem 5 to obtain an even more permissive result for Nash implementation with mixed strategies, which we simply report here.

Corollary 2 (*Nash Implementation with Mixed Strategies*). *For $I \geq 3$, if the environment satisfies value distinction and if F satisfies Maskin monotonicity, then F is Nash implementable in mixed strategies.*

10 Conclusion

In this paper we reported a comprehensive set of results on the possibility of ex post implementation. The general necessary and sufficient conditions for ex post implementation have a similar structure as the well-known conditions for Nash and Bayesian Nash implementation. Ex post equilibrium requires that every strategy profile remains an equilibrium choice even if a given agent would know the true type profile of all the remaining agents. The informational assumptions underlying the notion of an ex post equilibrium are hence closer to complete information, i.e., Nash implementation.

In consequence, we pursued a close comparison of the conditions for Nash and ex post implementation. We showed that the respective necessary and sufficient conditions are not nested, and that neither Nash nor ex post notions imply the other.

However, in the important class of single crossing environments, we showed that ex post monotonicity is given virtually for free as it is known to be true for Maskin monotonicity. Yet in the single crossing environment, for Maskin monotonicity to hold, the social choice function must be everywhere in the interior of the set of feasible alternatives. Ex post monotonicity however can already be guaranteed even if the social choice function is rarely in the interior of the feasible set. As an important example we showed that the single unit auction with interdependent values can be ex post, but not Nash implemented. Interestingly, the positive implementation results relied on interdependent values and do not hold for private values. Additionally, we showed that in the single crossing environment ex post implementation is possible in the direct mechanism and does not have to rely on the augmented mechanisms which have been frequently criticized for various unrealistic features, such as modulo or integer games.

Acknowledgments

The first author gratefully acknowledges support through a DFG Mercator Research Professorship at the Center of Economic Studies at the University of Munich. We benefited from discussions with Jeff Ely, Matt Jackson, Andy Postlewaite, Phil Reny, Mike Riordan and Roberto Serrano. We would like to thank seminar audiences at California Institute of Technology, Columbia University, Cornell University, New York University, Nuffield College, Princeton University, University of Michigan, the Paris Roy seminar, the Cowles Foundation Conference on "Robust Mechanism Design" and the Second World Congress of the Game Theory Society for helpful comments. Parts of this paper were reported in early drafts of our work on Robust Mechanism Design (Bergemann and Morris, 2001).

Appendix A

A.1. *Private values*

The analysis of this paper applies to interdependent value environments. In the special case of private values, ex post incentive compatibility implies dominant strategy incentive compatibility, under which dominant strategy implementation (and thus ex post implementation) is trivially possible in the direct mechanism. In this section, we very briefly state without proof the simple connections between the properties described in this section in the special case of private values.

Definition 20 (*Private Values*). There are private values if

$$u_i(y, (\theta_i, \theta_{-i})) = u_i(y, (\theta_i, \theta'_{-i}))$$

for all $i, y, \theta_i, \theta_{-i}$ and θ'_{-i}.

Definition 21 (*Dominant Strategies Incentive Compatibility*). Social choice set F is dominant strategies incentive compatible if for every $f \in F$:

$$u_i(f(\theta), \theta) \geq u_i(f(\theta'), \theta),$$

for all i, θ, θ'.

Lemma 2. *Under private values, F is ex post incentive compatible if and only if F is dominant strategies incentive compatible.*

Definition 22 (*Strict Dominant Strategies Incentive Compatibility*). Social choice set F is strictly dominant strategies incentive compatible if for every $f \in F$:

$$u_i(f(\theta), \theta) > u_i(f(\theta'), \theta),$$

for all i, θ and θ' with $\theta'_i \neq \theta_i$.

Lemma 3. *Under private values, F is strictly ex post incentive compatible if and only if F is strictly dominant strategies incentive compatible.*

Lemma 4. *Under private values, if F satisfies strict dominant strategies incentive compatibility, then F satisfies ex post monotonicity.*

A.2. *Ex post monotonicity no veto for sets*

We now state the ex post monotonicity no veto condition *(EMNV)* for the case of a social choice set rather than social choice function. Given a social choice set F and a deception α, we define for each $f \in F$ and i a set $\Phi_i^f \subseteq \Theta_i$ and let $\Phi^f = \times_{i=1}^I \Phi_i^f \subseteq \Theta$. The set Φ^f represents the set of profiles at which the agents all agree to implement the selection $f \in F$.

The no veto power condition is now imposed on the complement set to the union of the sets Φ^f:

$$\Phi \triangleq \bigcup_{f \in F} \Phi^f.$$

Definition 23 (*Ex Post Monotonicity No Veto* (*For Sets*)). F satisfies ex post monotonicity no veto if for any deception α and any product set $\Phi \subseteq \Theta$ the following holds. If the environment is non-economic at each $\theta \in \Theta - \Phi$, then

(1) there exists $\hat{f} \in F$ which satisfies no veto power on $\Theta - \Phi$ and
(2) if $f(\alpha(\theta)) \neq \hat{f}(\theta)$ for some f and some $\theta \in \Phi^f$ then there exists $i, \theta \in \Phi^f$ and y such that

$$u_i(y, \theta) > u_i(f(\alpha(\theta)), \theta),$$

while

$$u_i(f(\theta_i', \alpha_{-i}(\theta_{-i})), (\theta_i', \alpha_{-i}(\theta_{-i}))) \geq u_i(y, (\theta_i', \alpha_{-i}(\theta_{-i}))), \quad \forall \theta_i' \in \Theta_i.$$

The proof of Theorem 3 now goes through simply by extending the argument from the a single set Φ^f to the union of sets $\cup_{f \in F} \Phi^f$. We simply record it is a corollary.

Corollary 3 (*Sufficiency for Social Choice Sets*). *For $I \geq 3$, if the social choice set F satisfies (EPIC) and (EMNV), then F is ex post implementable.*

References

ARYA, A., J. GLOVER, AND U. RAJAN (2000): "Implementation in principal-agent models of adverse selection," *J. Econ. Theory*, 93, 87–109.

AUSUBEL, L. M., AND P. MILGROM (2006): "The lovely but lonely Vickrey auction." In: *Combinatorial Auctions*, ed. by P. Cramton, R. Steinberg and Y. Shoham. MIT Press, Cambridge, pp. 17–40.

BENOÎT, J. P., AND E. OK (2008): "Nash implementation without no veto-power." *Games and Economic Behavior*, 64, 51–67.

BERGEMANN, D., AND S. MORRIS (2001): "Robust mechanism design." Technical report. Cowles Foundation, Yale University, available at http://www. princeton.edu/smorris/pdfs/robustmechanism2001.pdf.

Bergemann, D., and S. Morris (2005a): "Robust implementation: The case of direct mechanisms." Technical report. Cowles Foundation/ Discussion paper 1561. Yale University.

Bergemann, D., and S. Morris (2005b): "Robust implementation: The role of large type spaces." Technical report 1519. Cowles Foundation, Yale University.

Bergemann, D., and S. Morris (2005c): "Robust mechanism design," *Econometrica*, 73, 1771–1813.

Bergemann, D., and J. Välimäki (2002): "Information acquisition and efficient mechanism design," *Econometrica*, 70, 1007–1033.

Bikhchandani, S. (2006): "Ex post implementation in environments with private goods," *Theoret. Econ.*, 1, 369–393.

Birulin, O. (2003): "Inefficient ex post equilibria in efficient auctions," *Econ. Theory*, 22, 675–683.

Chung, K.-S., and J. C. Ely (2001): "Efficient and dominance solvable auctions with interdependent valuations," Technical report. Northwestern University.

Cremer, J., and R. McLean (1985): "Optimal selling strategies under uncertainty for a discriminating monopolist when demands are interdependent," *Econometrica*, 53, 345–361.

Dasgupta, P., and E. Maskin (2000): "Efficient auctions," *Quart. J. Econ.*, 115, 341–388.

Dasgupta, P., P. Hammond, and E. Maskin (1979): "The implementation of social choice rules. Some general results on incentive compatibility," *Rev. Econ. Stud.*, 66, 185–216.

Eso, P., and E. Maskin (2000): "Multi-good efficient auctions with multidimensional information," Technical report. Northwestern University and Institute for Advanced Studies.

Fieseler, K., T. Kittsteiner, and B. Moldovanu (2003): "Partnerships, lemons and efficient trade," *J. Econ. Theory*, 113, 223–234.

Holmstrom, B., and R. Myerson (1983): "Efficient and durable decision rules with incomplete information," *Econometrica*, 51, 1799–1819.

Hurwicz, L., E. Maskin, and A. Postlewaite (1995): "Feasible Nash implementation of social choice correspondences when the designer does not know endowments or productions sets." In: *The Economics of Informational Decentralization: Complexity, Efficiency and Stability*, ed. by J. Ledyard. Kluwer Academic, Dordrecht, pp. 367–133.

Jackson, M. O. (1991): "Bayesian implementation," *Econometrica*, 59, 461–177.

JEHIEL, P., B. MOLDOVANU, M. MEYER-TER-VEHN, AND B. ZAME (2006): "The limits of ex post implementation," *Econometrica*, 74, 585–610.

KALAI, E. (2004): "Large robust games," *Econometrica*, 72, 1631–1666.

MASKIN, E. (1999): "Nash equilibrium and welfare optimality," *Rev. Econ. Stud.*, 66, 23–38.

MASKIN, E., AND T. SJOSTROM (2004): "Implementation theory." In: *Handbook of Social Choice and Welfare*, ed. by K. Arrow, A. Sen and K. Suzumura. Vol. 1. North-Holland, Amsterdam.

NEEMAN, Z. (2004): "The relevance of private information in mechanism design," *J. Econ. Theory*, 117, 55–77.

PALFREY, T, AND S. SRIVASTAVA (1989a): "Implementation with incomplete information in exchange economies," *Econometrica*, 57, 115–134.

PALFREY, T. R., AND S. SRIVASTAVA (1989b): "Mechanism design with incomplete information: A solution to the implementation problem," *J. Polit. Economy*, 97, 668–691.

PERRY, M., AND P. RENY (2002): "An ex post efficient auction," *Econometrica*, 70, 1199–1212.

POSTLEWAITE, A., AND D. SCHMEIDLER (1986): "Implementation in differential information economies," *J. Econ. Theory*, 39, 14–33.

SAIJO, T. (1988): "Strategy space reduction in Maskin's theorem: Sufficient conditions for Nash implementation," *Econometrica*, 56, 693–700.

SAIJO, T., T. SJOSTROM, AND T. YAMATO (2004): "Secure implementation," Technical report. Osaka University, Pennsylvania State University and Tokyo Institute of Technology.

WILSON, R. (1987): "Game-Theoretic Analyses of Trading Processes. In: *Advances in Economic Theory: Fifth World Congress*, ed. by T. Bewley. Cambridge University Press, Cambridge, pp. 33–70.

YOKOO, M., Y. SAKURAI, AND S. MATSUBARA (2004): "The effect of false-name bids in combinatorial auctions: New fraud in Internet auctions," *Games Econ. Behav.*, 46, 174–188.

CHAPTER 3

Robust Implementation in Direct Mechanisms*

Dirk Bergemann and Stephen Morris

Abstract
A social choice function is robustly implementable if there is a mechanism under which the process of iteratively eliminating strictly dominated messages leads to outcomes that agree with the social choice for all beliefs at every type profile. In an interdependent value environment with single crossing preferences, we identify a contraction property on the preferences which together with strict ex post incentive compatibility is sufficient to guarantee robust implementation in the direct mechanism. Strict ex post incentive compatibility and the contraction property are also necessary for robust implementation in *any* mechanism, including indirect ones.

The contraction property requires that the interdependence is not too large. In a linear signal model, the contraction property is equivalent to an interdependence matrix having an eigenvalue less than one.

Keywords: Mechanism design, implementation, robustness, common knowledge, interim equilibrium, iterative deletion, direct mechanism

*This research is supported by NSF Grants #CNS-0428422 and #SES-0518929. Morris is grateful for financial support from the John Simon Guggenheim Foundation and the Center for Advanced Studies in the Behavioral Sciences. We thank the editor, Andrea Prat, and three anonymous referees for their instructive remarks. We thank Sandeep Baliga, Federico Echenique, Donald Goldfarb, Matt Jackson, Jim Jordan, Vijay Krishna, Tom Palfrey, Ilya Segal, Daniel Spielman, Richard van Weelden and Meixia Zhang for helpful discussions; seminar audiences at Northwestern, Ohio State, Penn State, Pittsburgh, Stanford, UBC, UCLA and UC San Diego for comments; and Tomasz Strzalecki for pointing out an error in an earlier draft. This paper supersedes and extends results we first reported in Bergemann and Morris (2005a).

1 Introduction

The mechanism design literature provides a powerful characterization of which social choice functions can be achieved when the designer has incomplete information about agents' types. If we assume a commonly known common prior over the possible types of agents, the revelation principle establishes that if the social choice function can arise as an equilibrium in some mechanism, then it will arise in a truth-telling equilibrium of the direct mechanism (where each agent truthfully reports his type and the designer chooses an outcome assuming they are telling the truth). Thus the Bayesian incentive compatibility constraints characterize whether a social choice function is implementable in this sense.

But even if a truth-telling equilibrium of the direct mechanism exists, there is no guarantee that there do not exist non truth-telling equilibria that deliver unacceptable outcomes. For this reason, the literature on *full* implementation has sought to show the existence of a mechanism all of whose equilibria deliver the social choice function. A classic literature on Bayesian implementation — Postlewaite and Schmeidler (1986), Palfrey and Srivastava (1989b) and Jackson (1991) — characterized when this is possible: a *Bayesian monotonicity*[1] condition is necessary for full implementation, in addition to the Bayesian incentive compatibility conditions. Bayesian monotonicity and Bayesian incentive compatibility are also "almost" sufficient for full implementation.

This important literature has had a limited impact on the more applied mechanism design literature, despite the fact that the problem of multiple equilibria is real. An important difficulty is that, in general, positive results rely on complicated indirect, or "augmented," mechanisms in which agents report more than their types. Such mechanisms appear impractical to many researchers. We believe that the difficulty arises because the standard formulation of the Bayesian implementation problem — assuming common knowledge of a common prior on agents' types and using equilibrium as solution concept — endows the planner with more information than would be available in practice. The implementing mechanism and equilibrium then rely on that information in an implausible way.

---

[1]The Bayesian monotonicity condition is an incomplete information analogue of the classic "Maskin monotonicity" condition shown to be necessary and almost sufficient for complete information implementation by Maskin (1999).

In this paper, we characterize when a social choice function can be *robustly* implemented. We fix a social choice environment including a description of the set of possible payoff types for each agent. We ask when does there exist a mechanism with the property that every outcome consistent with common knowledge of rationality agrees with the social choice function, making no assumptions about agents' beliefs and higher order beliefs about other agents' payoff types. This requirement gives rise to an iterative deletion procedure: fix a mechanism and iteratively delete messages for each payoff type that are strictly dominated by another message for each payoff type profile and message profile that has survived the procedure. Consequently, our notion of robust implementation requires that truthtelling is the unique rationalizable outcome in the incomplete information game defined by the mechanism. This notion of robust implementation is equivalent to requiring that every equilibrium on every type space corresponding to the social choice environment delivers the right outcome. An operational advantage of the iterative definition is that it is defined relative to the payoff type space rather than the much larger universal type space or union of all possible type spaces.

This paper identifies a class of environments where there are tight and easily understood characterizations of when robust implementation is possible. As always, there will be an incentive compatibility condition that is necessary: strict ex post incentive compatibility is necessary for robust implementation. We show that if, in addition, a *contraction* property — which we explain shortly — is satisfied, robust implementation is possible in the *direct mechanism*, where each agent reports only his payoff type. If strict ex post incentive compatibility or the contraction property fail, then robust implementation is not possible in *any* mechanism. Thus the *augmented mechanisms* used in the earlier complete information and Bayesian implementation literatures do not perform better than the simpler direct mechanisms. An intuition for this result is that the strong common knowledge assumptions used in the complete information and the classical Bayesian implementation literatures can be exploited via complex augmented mechanisms. Thus an attractive feature of our approach is that the robustness requirement reduces the usefulness of complexity in mechanism design (without any ad hoc restrictions on complexity).

In the case of private values, strict ex post incentive compatibility is equivalent to strict dominant-strategy incentive compatibility. Thus full implementation is obtained for free. It follows that the contraction property must have bite only if there are interdependent values. In fact, the contraction property requires exactly that there is *not too much* interdependence in agents' types. The contraction property can be nicely illustrated in a class of interdependent preferences in which the private types of the agents can be linearly aggregated. If θ_j is the type of agent j, then agent i's utility depends on $\theta_i + \gamma \sum_{j \neq i} \theta_j$. Thus if $\gamma \neq 0$, there are interdependent values — agent j's type will enter agent i's utility assessment — but each agent i cares differently about his own type than about other agents' types. In this example, the contraction property reduces to the requirement that $|\gamma| < 1/(I-1)$, where I is the number of agents. We provide characterizations of the contraction property — equivalent to the intuition that there is not too much interdependence — in more general environments.

An important paper of Chung and Ely (2001) analyzed auctions with interdependent valuations under iterated elimination of weakly dominated strategies. In a linear and symmetric setting, they reported sufficient conditions for direct implementation that coincide with the ones derived here. We show that in the environment with linear aggregation, under strict incentive compatibility, the basic insight extends from the single unit auction model to general allocations models, with elimination of strictly dominated actions only (thus Chung and Ely (2001) require deletion of weakly dominated strategies only because incentive constraints are weak). We also prove a converse result: if there is too much interdependence, then neither the direct nor any augmented mechanism can robustly implement the social choice function.[2]

The main results of this paper apply to environments where each agent's type profile can be aggregated into a one-dimensional sufficient statistic for each player, where preferences are single crossing with respect to that statistic. These restrictions incorporate many economic models with interdependence in the literature. In particular, these restrictions immediately hold in some well-known settings. They automatically hold in single or multi-unit auctions when each bidder demands at most one unit of the good. In this case, the aggregation function is the utility function itself. The restrictions also encompass a widely used statistical model with interdependent values. Since the seminal contributions of Wilson (1977) and Milgrom (1981), the canonical model of common values is one in which

[2]Bergemann and Morris (2008c) describe how to derive a strong converse to the original Chung and Ely (2001) result for iterated deletion of weakly dominated strategies.

each agent receives a conditionally independent and identical signal about a one-dimensional common value. In this case, the aggregator function is naturally given directly by Bayes rule. Subsequently, many allocations models with interdependent — but not necessarily common — values use the same conditionally independent and identically distributed information structure. Reny and Perry (2006) develop strategic foundations for the rational expectations equilibrium within a large double auction. Here the value of the object for each agent is determined by a (perhaps nonlinear) function of a private and common value. A private signal jointly informs the agent about the private value and the common value. In consequence, the value of the object is again given by a Bayesian estimate as the natural aggregator function. We shall illustrate informational aggregation via Bayes law and the relationship with the contraction property in Section 4.

We focus in this paper on economically important environments and well behaved mechanisms where we get clean and tight characterizations of the robust implementation problem with direct or augmented mechanisms. The ex post incentive constraints necessary for robust implementation are already strong (even without the contraction property): Jehiel, Moldovanu, Meyer-Ter-Vehn, and Zame (2006) have shown that in an environment with multidimensional signals, the ex post incentive constraints are "generically" impossible to satisfy. If ex post incentive compatibility fails, our positive results are moot. While this provides a natural limit for our analysis, there are many interesting applications for which ex post incentive compatibility holds and the contraction property binds.

First, there is the large and important literature on one-dimensional interdependent type models, including papers on auction environments (Dasgupta and Maskin, 2000; Perry and Reny, 2002; Bergemann and Välimäki, 2002), the bilateral trading model (Gresik, 1991, and Fieseler, Kittsteiner, and Moldovanu, 2003), and public and team decision problems without transferable utilities (Grüner and Kiel, 2004). We illustrate our results with a public good example with transfers, with a linear aggregator as described above; we also apply our results to the classic problem of allocating a single private good with quasilinear utility (i.e., a single unit auction with interdependent utility).

Second, even with multidimensional signals, there are many environments where economically natural "special" assumptions lead to a failure of the genericity conditions in Jehiel, Moldovanu, Meyer-Ter-Vehn, and Zame (2006). For example, in any environment with common interests and private information (e.g., Piketty, 1999) there will be ex post incentive compatibility (whatever the dimension of private information), yet our contraction property will imply the impossibility of robust implementation, as we discuss in Section 6 for the one-dimensional linear aggregator case. In Section 8.1, we use a multidimensional version of our public good example to illustrate an extension of our results to multidimensional signals without the aggregation property. In this case, symmetry allows the existence of ex post incentive compatible transfers. Further examples in the literature show that EPIC is satisfied in multidimensional signal models without allocative externalities (Bikhchandani, 2006) or with a separable structure (Eso and Maskin, 2002).

While we prove the necessity of the contraction property within the single crossing aggregator environments we study, the necessity argument extends to general environments. In Bergemann and Morris (2005a), we show that a robust monotonicity condition is necessary and almost sufficient for robust implementation in general environments with general mechanisms. The robust monotonicity condition is equivalent to assuming the Bayesian monotonicity necessary condition (from Postlewaite and Schmeidler, 1986; Palfrey and Srivastava, 1989b, and Jackson, 1991) on all possible type spaces. The contraction property is the simpler expression of this robust mono-tonicity condition in the single crossing aggregator environments of this paper. The definition of the contraction property depends on the aggregation of the preferences. But in Section 8.1, we discuss a generalized contraction property — again capturing the idea of moderate interdependence in preferences — that does not depend on an aggregation property or one-dimensional signals. While this generalized contraction property is rather strong, we show that a local version of it — holding only around the social choice function — is also sufficient and illustrate its use in a multidimensional public good example.

The remainder of the paper is organized as follows. Section 2 describes the formal environment and solution concepts. Section 3 considers a public good example that illustrates the main ideas and results of the paper. Section 4 establishes necessary conditions for robust implementation in the direct mechanism. Section 5 reports sufficient conditions for robust implementation. Section 6 considers the preference environment with a linear aggregation of the types and obtains sharp implementation results. Section 7 considers a single unit auction with interdependent values as a second example of robust implementation. Section 8 concludes.

2 Setup

Payoff Environment. We consider a finite set of agents, $1, 2, \ldots, I$. Agent i's *payoff type* is $\theta_i \in \Theta_i$, where Θ_i is a compact subset of the real line. We write $\theta = (\theta_1, \ldots, \theta_1) \in \Theta_1 \times \cdots \times \Theta_I = \Theta$. Let X be a compact set of deterministic outcomes and let $Y = \Delta(X)$ be the lottery space generated by the deterministic outcome space X. Each agent has a von Neumann Morgenstern expected utility function $u_i : Y \times \Theta \to \mathbb{R}$. Let agent i's utility if outcome y is chosen and agents' type profile is θ be $u_i(y, \theta)$. We emphasize that the utility function of agent i is allowed to depend on the type profile θ_{-i} of the other agents. A social choice function is a mapping $f : \Theta \to Y$.

Mechanisms. A planner must choose a *game form* or *mechanism* for the agents to play in order the determine the social outcome. Let M_i be a compact set of messages available to agent i. Let $g(m)$ be the outcome chosen if action profile m is chosen. Thus a mechanism is a collection:

$$\mathcal{M} = (M_1, \ldots, M_I, g(\cdot)),$$

where $g : M \to Y$. The *direct mechanism* has the property that $M_i = \Theta_i$ for all i and $g(\theta) = f(\theta)$.

Robust Implementation. In a fixed mechanism $\mathcal{M}$, we call a correspondence $S = (S_1, \ldots, S_I)$, with each $S_i : \Theta_i \to 2^{M_i}/\emptyset$, a *message profile* of the agents. We will refer to a message profile in the direct mechanism where truthtelling is always possible as a *report profile*. Thus a report profile $\beta = (\beta_1, \ldots, \beta_I)$ is described by

$$\beta_i : \Theta_i \to 2^{\Theta_i}/\emptyset, \quad \text{for all } i,$$

and $\theta_i \in \beta_i(\theta_i)$ for all i and θ_i. Let β^* be the truthful report, with $\beta_i^*(\theta_i) = \{\theta_i\}$ for all i and θ_i.

Next we define the process of iterative elimination of never best responses. We denote the belief of agent i over message and payoff type profiles of the remaining agents by a Borel measure λ_i:

$$\lambda_i \in \Delta(M_{-i} \times \Theta_{-i}).$$

Let $S_i^0(\theta_i) = M_i$ for all i and θ_i and define inductively:

$$S_i^{k+1}(\theta_i) = \left\{ m_i \in M_i \left| \exists \lambda_i \text{ s.th.: } \begin{array}{ll} (1) & \lambda_i[\{(m_{-i}, \theta_{-i}) | m_j \in S_j^k(\theta_j) \forall j \neq i\}] = 1 \\ (2) & \int u_i(g(m_i, m_{-i}), (\theta_i, \theta_{-i})) d\lambda_i \\ & \geq \int u_i(g(m_i', m_{-i}), (\theta_i, \theta_{-i})) d\lambda_i, \forall m_i' \in M_i. \end{array} \right. \right\}.$$

We observe that $S_i^k(\theta_i)$ is nonincreasing in k in the set inclusion order for each θ_i. We denote the limit set $S_i^{\mathcal{M}}(\theta_i)$ by:

$$S_i^{\mathcal{M}}(\theta_i) \triangleq \bigcap_{k \geq 0} S_i^k(\theta_i), \quad \text{for all } \theta_i \in \Theta_i \quad \text{and all } i.$$

We refer to the messages $m_i \in S_i^{\mathcal{M}}(\theta_i)$ as the *rationalizable messages* of type θ_i of agent i in mechanism $\mathcal{M}$. We call a social choice function f *robustly implementable* if there exists a mechanism $\mathcal{M}$ under which the social choice can be recovered through a process of iterative elimination of never best responses.

Definition 1 (Robust Implementation). Social choice function f is robustly implemented by mechanism $\mathcal{M}$ if $m \in S^{\mathcal{M}}(\theta) \Rightarrow g(m) = f(\theta)$.

We mentioned in the introduction that the above notion of robust implementation is equivalent to requiring that every equilibrium on every type space corresponding to the social choice environment delivers the right outcome. In other words, the set of rationalizable messages for mechanism $\mathcal{M}$ is equal to the set of messages that could be played in a Bayesian equilibrium of the game generated by the mechanism $\mathcal{M}$ and some type space. The basic logic of this equivalence result follows the well-known argument of Brandenburger and Dekel (1987) who establish, in complete information games, the equivalence of correlated rationalizable actions and the set of actions that could be played in a subjective correlated equilibrium. Battigalli and Siniscalchi (2003) describe a general incomplete information extension of this observation. We report a formal version of the equivalence result for our environment in Proposition 1 in Bergemann and Morris (2005a). Since all subsequent results work directly with the above iterative notion, we refer for the formal statements about the equivalence to Bergemann and Morris (2005a).

Monotone Aggregator. We now describe the structural assumptions that will be maintained throughout the rest of the paper. We assume the existence of a monotonic aggregator $h_i(\theta)$ for every i that allows us to rewrite the utility function of every agent i as:

$$u_i(y, \theta) \triangleq v_i(y, h_i(\theta)),$$

where $h_i : \Theta \to \mathbb{R}$ is assumed to be continuous, strictly increasing in θ_i and $v_i : Y \times \mathbb{R} \to \mathbb{R}$ is continuous in the aggregator (continuity with respect to lotteries follows from the vNM assumption). The content of the aggregation assumption comes from the continuity requirement and the following single-crossing condition.

Definition 2 (Strict Single Crossing). The utility function $v_i(y, \phi)$ satisfies strict single crossing (SSC) if for all $\phi < \phi' < \phi''$:

$$v_i(y, \phi) > v_i(y', \phi) \quad \text{and} \quad v_i(y, \phi') = v_i(y', \phi') \Rightarrow v_i(y, \phi'') < v_i(y', \phi'').$$

The single crossing property is defined relative to the aggregation $\phi = h_i(\theta)$ of all agents' types. The combination of monotonic aggregator representation of preferences and the strict single crossing condition will drive our results.

3 A Public Good Example

We precede the formal results with an example illustrating the main insights of the paper and reviewing some key ideas from the implementation literature. The example involves the provision of a public good with quasilinear utility. The utility of each agent is given by:

$$u_i(\theta, x) = \left(\theta_i + \gamma \sum_{j \neq i} \theta_j\right) x_0 + x_i,$$

where x_0 is the level of public good provided and x_i is the monetary transfer to agent i. The utility of agent i depends on his own type $\theta_i \in [0, 1]$ and the type profile of other agents, $\theta_{-i} \in [0, 1]^{I-1}$. The weight $\gamma \geq 0$ represents the strength of the interdependence in the preferences of agent i. The utility function of agent i has the aggregation property with

$$h_i(\theta) = \theta_i + \gamma \sum_{j \neq i} \theta_j,$$

but we notice that the aggregator function $h_i(\theta)$ depends on the identity of agent i. In particular, a given type profile θ leads to a different aggregation result for i and j, provided that $\theta_i \neq \theta_j$.

The cost of establishing the public good is given by $c(x_0) = \frac{1}{2}x_0^2$. The planner must choose $(x_0, x_1, \ldots, x_I) \in \mathbb{R}_+ \times \mathbb{R}^I$ to maximize social welfare, i.e., the sum of gross utilities minus the cost of the public good:

$$\left((1+\gamma(I-1)) \sum_{i=1}^{I} \theta_i \right) x_0 - \frac{1}{2} x_0^2.$$

The socially optimal level of the public good is therefore equal to

$$f_0(\theta) = (1+\gamma(I-1)) \sum_{i=1}^{I} \theta_i.$$

The generalized Vickrey-Clarke-Groves (VCG) transfers, unique up to a constant, that give rise to ex post incentive compatibility are:

$$f_i(\theta) = -(1+\gamma(I-1)) \left(\gamma\theta_i \sum_{j \neq i} \theta_j + \frac{1}{2}\theta_i^2 \right). \tag{1}$$

It is useful to observe that the generalized VCG transfers given by (1) guarantee ex post incentive compatibility for any $\gamma \in \mathbb{R}_+$. Hence, ex post incentive compatibility does not impose *any* constraint on the interdependence parameter γ. In contrast, the dominant-strategy property of the VCG mechanism only holds with private values, or $\gamma = 0$, and fails for all $\gamma > 0$.

Now we shall argue that if $\gamma < \frac{1}{I-1}$, the social choice function f is robustly implementable in the *direct mechanism* where each agent reports his payoff type θ_i and the planner chooses outcomes according to f on the assumption that agents are telling the truth. Consider an iterative deletion procedure. Let $\beta^0(\theta_i) = [0,1]$ and, for each $k = 1, 2, \ldots$, let $\beta^k(\theta_i)$ be the set of reports that agent i might send, for some conjecture over his opponents' types and reports, with the only restriction on his conjecture being that each type θ_j of agent j sends a message in $\beta^{k-1}(\theta_j)$.

Suppose that agent i has payoff type θ_i, but reports himself to be type θ_i' and has a point conjecture that other agents have type profile θ_{-i} and report their types to be θ_{-i}'. Then his expected payoff is a constant $(1+\gamma(I-1))$ times:

$$\left(\theta_i + \gamma \sum_{j \neq i} \theta_j \right) \left(\theta_i' + \sum_{j \neq i} \theta_j' \right) - \left(\gamma\theta_i' \sum_{j \neq i} \theta_j' + \frac{1}{2}(\theta_i')^2 \right).$$

The first order condition with respect to θ_i' is then

$$\theta_i + \gamma\left(\sum_{j\neq i}\theta_j\right) - \gamma\left(\sum_{j\neq i}\theta_j'\right) - \theta_i' = 0,$$

so he would wish to set

$$\theta_i' = \theta_i + \gamma\sum_{j\neq i}(\theta_j - \theta_j').$$

In other words, his best response to a misreport θ'_{-i} by the other agents is to report a type so that the aggregate type from his point of view is exactly identical to the aggregate type generated by the true type profile θ. Note that the above calculation also verifies the strict ex post incentive compatibility of f, since setting $\theta_i' = \theta_i$ is the unique best response if $\theta_j' = \theta_j$ for all $j \neq i$. The quadratic payoff / linear best response nature of this problem means that we can characterize $\beta^k(\theta_i)$ restricting attention to such point conjectures. In particular, we have

$$\beta^k(\theta_i) = [\underline{\beta}^k(\theta_i), \overline{\beta}^k(\theta_i)],$$

where

$$\begin{aligned}\overline{\beta}^k(\theta_i) &= \min\left\{1, \theta_i + \gamma \max_{\{(\theta'_{-i},\theta_{-i})|\theta_j'\in\beta^k(\theta_j)\text{ for all }j\neq i\}} \sum_{j\neq i}(\theta_j - \theta_j')\right\} \\ &= \min\left\{1, \theta_i + \gamma\max_{\theta_{-i}}\sum_{j\neq i}(\theta_j - \underline{\beta}^{k-1}(\theta_j))\right\}.\end{aligned}$$

Analogously,

$$\underline{\beta}^k(\theta_i) = \max\left\{0, \theta_i + \gamma\max_{\theta_{-i}}\sum_{j\neq i}(\overline{\beta}^{k-1}(\theta_j) - \theta_j)\right\}.$$

Thus

$$\overline{\beta}^k(\theta_i) = \min\{1, \theta_i + (\gamma(I-1))^k\},$$

and

$$\underline{\beta}^k(\theta_i) = \max\{0, \theta_i - (\gamma(I-1))^k\}.$$

Thus $\theta_i' \neq \theta_i \Rightarrow \theta_i' \notin \beta^k(\theta_i)$ for sufficiently large k, provided that $\gamma < \frac{1}{I-1}$.

Now consider what happens when this condition fails, i.e., $\gamma \geq \frac{1}{I-1}$. In this case, it is possible to exploit the large amount of interdependence to construct beliefs over the opponents' types such that all types are indistinguishable. In particular, suppose that every type $\theta_i \in [0,1]$ has a degenerate belief over the types of his opponents. In particular, type θ_i is convinced that each of his opponents is of type θ_j given by:

$$\theta_j[\theta_i] = \frac{1}{2} + \frac{1}{\gamma(I-1)}\left(\frac{1}{2} - \theta_i\right),$$

where the belief of i about j evidently depends on his type θ_i. In this case the aggregation of the types leads to:

$$\theta_i + \gamma \sum_{j \neq i} \theta_j[\theta_i] = \frac{1}{2}(1 + \gamma(I-1)),$$

independent of θ_i. Thus in *any* mechanism, for each type, we can construct beliefs so that there is no differences across types of agent i in terms of the actions which get deleted in the iterative process.

4 Robust Implementation

In our earlier work on robust mechanism design, Bergemann and Morris (2005b), we showed that ex post incentive compatibility is a necessary and sufficient for partial robust implementation (i.e., ensuring that there exists *an* equilibrium consistent with the social choice function).

Definition 3 (Ex Post Incentive Compatibility). Social choice function f satisfies ex post incentive compatibility (EPIC) if for all i, θ and θ_i' :

$$u_i(f(\theta_i, \theta_{-i}), (\theta_i, \theta_{-i})) \geq u_i(f(\theta_i', \theta_{-i}), (\theta_i, \theta_{-i})).$$

In the subsequent analysis we use the strict version of the incentive constraints as we require full implementation.

Definition 4 (Strict Ex Post Incentive Compatibility). Social choice function f satisfies strict ex post incentive compatibility (strict EPIC) if for all i, $\theta_i' \neq \theta_i$ and θ_{-i} :

$$u_i(f(\theta_i, \theta_{-i}), (\theta_i, \theta_{-i})) > u_i(f(\theta_i', \theta_{-i}), (\theta_i, \theta_{-i})).$$

The key property for our analysis is the following contraction property.

Definition 5 (Contraction Property). The aggregator functions $h = (h_i)_{i=1}^I$ satisfy the contraction property if, for all $\beta \neq \beta^*$, there exists i, θ_i and $\theta_i' \in \beta_i(\theta_i)$ with $\theta_i' \neq \theta_i$, such that

$$\text{sign}(\theta_i - \theta_i') = \text{sign}(h_i(\theta_i, \theta_{-i}) - h_i(\theta_i', \theta_{-i}')), \tag{2}$$

for all θ_{-i} and $\theta_{-i}' \in \beta_{-i}(\theta_{-i})$.

The contraction property essentially says that for some agent i the direct impact of his private signal θ_i on the aggregator $h_i(\theta)$ is always sufficiently strong such that the difference in the aggregated value between the true type profile and the reported type profile always has the same sign as the difference between the true and reported type of agent i by itself.

How strong is the aggregator restriction on the environment? It requires that the payoff types of the players can be aggregated into a variable that changes preferences in a monotonic way. To get some sense of the strength of this restriction, we next consider two examples. The first example involves a binary outcome space which automatically guarantees the aggregation property; the second example uses an informational foundation by means of Bayes' law to obtain the aggregation property.

Binary Allocation Model. Let each agent's utility from outcomes depend on the payoff type profile only via a binary partition of the deterministic outcome space X. Thus, for each i, there exists $f_i : X \to \{0,1\}, w_i^1 : \{0,1\} \times \Theta \to \mathbb{R}$ and $w_i^2 : X \to \mathbb{R}$ such that

$$u_i(x, \theta) = w_i^1(f_i(x), \theta) + w_i^2(x).$$

In this case, we can write the agent's utility over lotteries $y \in Y = \Delta(X)$ as

$$u_i(y, \theta) = \int_{\{x|f_i(x)=1\}} w_i^1(1, \theta)dy + \int_{\{x|f_i(x)=0\}} w_i^1(0, \theta)dy + \int_{x \in X} w_i^2(x)dy.$$

An equivalent representation of this agent's preferences is

$$\int_{\{x|f_i(x)=1\}} [w_i^1(1, \theta) - w_i^1(0, \theta)]dy + \int_{x \in X} w_i^2(x)dy = v_i(y, h_i(\theta)),$$

with

$$h_i(\theta) \triangleq w_i^1(1,\theta) - w_i^1(0,\theta),$$

and

$$v_i(y,\phi) \triangleq \int_{\{x: f_i(x)=1\}} dy \cdot h_i(\theta) + \int_{x \in X} w_i(x) dy.$$

Thus in such a binary allocation model, the aggregation property is satisfied automatically.

A natural example of a binary allocation model is an auction of many identical units of a good to agents with unit demand for the good and quasilinear preferences. In this case, an allocation is a pair $(Z, z) \in X = 2^{\{1,\ldots,I\}} \times \mathbb{R}^I$ where Z is the set of agents who are allocated a unit of the good and z_i is the payment of agent i. Now if agent i's utility from being allocated the good if the payoff type profile is θ is given by $v_i(\theta)$, then this fits the above framework since

$$u_i((Z,z),\theta) = \begin{cases} v_i(\theta) - z_i, & \text{if } i \in Z \\ -z_i, & \text{otherwise} \end{cases}.$$

Information Aggregation. A natural source of interdependence in preferences is informational, when an agent's payoff type corresponds to a signal which ends up being correlated with all agents' expected values of a state. In particular, suppose that each agent i's utility depends on the expected value of an additive random variable $\omega_0 + \omega_i$, where ω_0 is a common value component and ω_i is the private value component. We describe the additive model with two agents i and j (the generalization to many agents is immediate). The random variables $\omega_0, \omega_1, \omega_2$ are assumed to be independently and normally distributed with zero mean and variance σ_0^2, σ_1^2 and σ_2^2 respectively. Let each agent i observe one signal $\theta_i = \omega_0 + \omega_i + \varepsilon_i$, where each ε_i is independently normally distributed with mean 0 and variance τ_i^2. We are thus assuming that each agent observes only a one-dimensional signal, θ_i, of both the common and idiosyncratic component. Thus agent i is unable to distinguish with his noisy signal θ_i between the common and the private value components. But naturally his own signal is more informative about his valuation than the others' signals because it contains his own idiosyncratic shock.

Now standard properties of the normal distribution (see DeGroot, 1970) imply that agent i's expected value of $\omega_0 + \omega_i$, given the vector of signals

(θ_i, θ_j) is a constant

$$\frac{\sigma_0^2\tau_i^2 + \sigma_0^2\tau_j^2 + \sigma_0^2\sigma_i^2 + \sigma_0^2\sigma_j^2 + \tau_i^2\tau_j^2 + \tau_i^2\sigma_j^2 + \tau_j^2\sigma_i^2 + \sigma_i^2\sigma_j^2}{\sigma_0^2\tau_i^2 + \sigma_0^2\sigma_i^2 + \sigma_0^2\sigma_j^2 + \tau_j^2\sigma_i^2 + \sigma_i^2\sigma_j^2}$$

times

$$h_i(\theta) = \theta_i + \frac{\sigma_0^2\tau_i^2}{\sigma_0^2\tau_j^2 + \sigma_0^2\sigma_i^2 + \sigma_0^2\sigma_j^2 + \tau_j^2\sigma_i^2 + \sigma_i^2\sigma_j^2}\theta_j, \tag{3}$$

where we write j for the other agent $3 - i$. The calculations are reported in the appendix. Now if we assume each agent i's preferences conditional on $h_i(\theta)$ satisfy strict single crossing with respect to $h_i(\theta)$, then we have an informational microfoundation for the strict single crossing environment of the paper. Moreover, in this example the aggregator takes the linear form $h_i(\theta) = \theta_i + \gamma_{ij}\theta_j$, with

$$\gamma_{ij} = \frac{\sigma_0^2\tau_i^2}{\sigma_0^2\tau_j^2 + \sigma_0^2\sigma_i^2 + \sigma_0^2\sigma_j^2 + \tau_j^2\sigma_i^2 + \sigma_i^2\sigma_j^2}.$$

This conclusion is quite intuitive. If the variance of the common component (σ_0^2) is small or if the noise in one's own signal (τ_i^2) is small, then the interdependence goes away. But a reduction in variance of one's own idiosyncratic component (σ_i^2), in one's opponent's idiosyncratic component (σ_j^2) or in one's opponent's noise (τ_j^2) all tend to increase the interdependence.[3] With this interpretation the single crossing property with respect to the aggregator reduces to assuming that there is a one-dimensional parameter whose expected value effects the preferences and that there is a sufficient statistic for the vector of signals that agents observe.

We now state our first positive result.

Theorem 1 (Robust Implementation). *If a social choice function f satisfies strict EPIC and the aggregator functions satisfy the contraction property, then f can be robustly implemented in the direct mechanism.*

Proof. We argue by contradiction. Let $\beta = S^{\infty}$ and suppose that $\beta \neq \beta^*$. Continuity of each u_i with respect to θ implies that each $\beta_i(\theta_i)$ will be a

[3]The additive model with a private and a common component appears in Hong and Shum (2003). Interestingly, they prove the existence and uniqueness of an increasing bidding strategy by appealing to a dominant diagonal condition, which is implied by the contraction property. The example of a normal distribution fails the compact type space assumption of our model, but we use the normal distribution here merely for its transparent updating properties.

compact set. By the contraction property, there exists i and $\theta'_i \in \beta_i(\theta_i)$ with $\theta'_i \neq \theta_i$ such that

$$\text{sign } (\theta_i - \theta'_i) = \text{sign } (h_i(\theta_i, \theta_{-i}) - h_i(\theta'_i, \theta'_{-i})).$$

for all θ_{-i} and $\theta'_{-i} \in \beta_{-i}(\theta_{-i})$. Thus by compactness

$$\delta \triangleq \min_{\theta_{-i}\in\Theta_{-i} \text{ and } \theta'_{-i}\in\beta_{-i}(\theta_{-i})} |h_i(\theta_i, \theta_{-i}) - h_i(\theta'_i, \theta'_{-i})|,$$

is well defined and strictly positive. Suppose (without loss of generality) that $\theta_i > \theta'_i$. Let

$$\xi(\varepsilon) \triangleq \max_{\theta'_{-i}}\{h_i(\theta'_i + \varepsilon, \theta'_{-i}) - h_i(\theta'_i, \theta'_{-i})\}.$$

As $h_i(\cdot)$ is strictly increasing in θ_i, we know that $\xi(\varepsilon)$ is increasing in ε and by continuity of h_i in θ_i, $\xi(\varepsilon) \to 0$ as $\varepsilon \to 0$.

Thus we have

$$h_i(\theta_i, \theta_{-i}) - h_i(\theta'_i, \theta'_{-i}) \geq \delta, \tag{4}$$

for all θ_{-i} and $\theta'_{-i} \in \beta_i(0_{-i})$; and

$$h_i(\theta'_i, \theta'_{-i}) \geq h_i(\theta'_i + \varepsilon, \theta'_{-i}) - \xi(\varepsilon), \tag{5}$$

for all θ'_{-i}. By strict EPIC,

$$v_i(f(\theta'_i, \theta'_{-i}), h_i(\theta'_i, \theta'_{-i})) > v_i(f(\theta'_i + \varepsilon, \theta'_{-i}), h_i(\theta'_i, \theta'_{-i})),$$

for all $\varepsilon > 0$ and

$$v_i(f(\theta'_i + \varepsilon, \theta'_{-i}), h_i(\theta'_i + \varepsilon, \theta'_{-i})) > v_i(f(\theta'_i, \theta'_{-i}), h_i(\theta'_i, + \varepsilon, \theta'_{-i})),$$

for all $\varepsilon > 0$. Now continuity of u_i with respect to θ implies that for each $\varepsilon > 0$ and θ'_{-i}, there exists $\phi^*(\varepsilon, \theta'_{-i}) \in \mathbb{R}$ such that:

$$h_i(\theta'_i, \theta'_{-i}) < \phi^*(\varepsilon, \theta'_{-i}) < h_i(\theta'_i + \varepsilon, \theta'_{-i}), \tag{6}$$

and

$$v_i(f(\theta'_i, \theta'_{-i}), \phi^*(\varepsilon, \theta'_{-i})) = v_i(f(\theta'_i + \varepsilon, \theta'_{-i}), \phi^*(\varepsilon, \theta'_{-i}));$$

and SSC implies that:

$$v_i(f(\theta'_i, \theta'_{-i}), \phi) < v_i(f(\theta'_i + \varepsilon, \theta'_{-i}), \phi), \tag{7}$$

for all $\phi > \phi^*(\varepsilon, \theta'_{-i})$. Now fix any ε with

$$\xi(\varepsilon) < \delta. \tag{8}$$

Now for all $\theta'_{-i} \in \beta_{-i}(\theta_{-i})$,

$$\begin{aligned} h_i(\theta_i, \theta_{-i}) &\geq h_i(\theta'_i, \theta'_{-i}) + \delta, \text{ by (4)} \\ &\geq h_i(\theta'_i + \varepsilon, \theta'_{-i}) - \xi(\varepsilon) + \delta, \text{ by (5)} \\ &> h_i(\theta'_i + \varepsilon, \theta'_{-i}), \text{ by (8)} \\ &\geq \phi^*(\varepsilon, \theta'_{-i}), \text{ by (6)}. \end{aligned}$$

So

$$v_i(f(\theta'_i + \varepsilon, \theta'_{-i}), h_i(\theta_i, \theta_{-i})) > v_i(f(\theta'_i, \theta'_{-i}), h_i(\theta_i, \theta_{-i})),$$

for every θ_{-i} and $\theta'_{-i} \in \beta_{-i}(\theta_{-i})$ by (7). This contradicts our assumption that $\beta = S^{\mathcal{M}}$. □

A surprising element in this result is that we do not need to impose any conditions on how the social choice function varies with the type profile. It does not have to respond to the reported profile θ in a manner similar to the response of any of the aggregators h_i. The strict single crossing condition is sufficient to make full use of the contraction property.

The argument is based on a true type profile $\theta = (\theta_i, \theta_{-i})$ and a reported profile $\theta' = (\theta'_i, \theta'_{-i})$, with $\theta_i > \theta'_i$ without loss of generality. We use the contraction property to establish a positive lower bound on the difference $h_i(\theta_i, \theta_{-i}) - h_i(\theta'_i, \theta'_{-i})$ for all θ_{-i} and $\theta'_{-i} \in \beta_{-i}(\theta_{-i})$. With this positive lower bound, we then show that agent i is strictly better off to move his misreport θ'_i marginally upwards in the direction of θ_i, in other words to report $\theta'_i + \varepsilon$. This is achieved by showing that there is an intermediate value ϕ^* for the aggregator, with $h_i(\theta'_i, \theta'_{-i}) < \phi^* < h_i(\theta'_i + \varepsilon, \theta'_{-i})$, such that agent i with the utility profile corresponding to the aggregator value ϕ^* would be indifferent between the social allocations $f(\theta'_i, \theta'_{-i})$ and $f(\theta'_i + \varepsilon, \theta'_{-i})$. By choosing ε sufficiently small, we know that $h(\theta_i, \theta_{-i}) > \phi^*$ and strict single crossing then allows us to assert that an agent with a true preference profile $\theta = (\theta_i, \theta_{-i})$ would also prefer to obtain $f(\theta'_i + \varepsilon, \theta'_{-i})$ rather than $f(\theta'_i, \theta'_{-i})$. But this yields the contradiction to $\theta'_i \in \beta_i(\theta_i)$ being part of the fixed point of the iterative elimination. Consequently we show that the misreport θ'_i, which established the same sign on the difference between private type profiles and aggregated public profiles can be eliminated as a best response to the set of misreports of the remaining agents.

In the present environment with single crossing and aggregation, the contraction property is equivalent to a notion of "robust monotonicity" in Bergemann and Morris (2005a). Social choice function f satisfies *robust monotonicity* if for report profile $\beta \neq \beta^*$, there exist i, θ_i, $\theta_i' \in \beta_i(\theta_i)$ such that, for all $\theta_{-i}' \in \Theta_{-i}$, there exists y such that

$$u_i(y, (\theta_i, \theta_{-i})) > u_i(f(\theta_i', \theta_{-i}'), (\theta_i, \theta_{-i})) \tag{9}$$

for all θ_{-i} such that $\theta_{-i}' \in \beta_{-i}(\theta_{-i})$; and

$$u_i(f(\theta_i'', \theta_{-i}'), (\theta_i'', \theta_{-i}')) \geq u_i(y, (\theta_i'', \theta_{-i}')) \tag{10}$$

for all $\theta_i'' \in \Theta_i$.

It is now easy to see that the contraction property guarantees the validity of (9) and (10). Fix θ_i and θ_i' and without loss of generality assume $\theta_i > \theta_i'$. By the contraction property it follows that for every θ_{-i}', we have $h_i(\theta_i, \theta_{-i}) > h_i(\theta_i', \theta_{-i}')$. Hence we can find an $\varepsilon > 0$ such that

$$h_i(\theta_i, \theta_{-i}) > h_i(\theta_i' + \varepsilon, \theta_{-i}') > h_i(\theta_i', \theta_{-i}'). \tag{11}$$

But now we can choose the allocation y to be $y = f(\theta_i' + \varepsilon, \theta_{-i}')$. Now (9) follows from (11) and single crossing, and (10) follows from strict EPIC.

Bergemann and Morris (2005a) show that the above robust monotonicity condition is a necessary and almost sufficient condition for robust implementation, by following the classical implementation literature in allowing the use of complicated — perhaps unbounded — augmented mechanisms. In this paper, we show that the contraction property — equivalent to the robust monotonicity condition — is sufficient for implementation in the *direct* mechanism in single crossing aggregator environments.

5 Necessity of Contraction Property

We now show that the contraction property is necessary for robust implementation. We impose the following mild restriction on the social choice function for the necessity argument.

Definition 6 (Responsive Social Choice Function). Social choice function f is responsive if for all $\theta_i \neq \theta_i'$, there exists θ_{-i} such that

$$f(\theta_i, \theta_{-i}) \neq f(\theta_i', \theta_{-i}').$$

Responsiveness requires that a change in agent i's report changes the social allocation for some report of the other agents. The idea behind the necessity argument is to show that the hypothesis of robust implementation leads inevitably to a conflict with a report profile β which fails to satisfy the contraction property.

Theorem 2 (Necessity). *If f is robustly implementable and is responsive, then f satisfies strict EPIC and the aggregator functions satisfy the contraction property.*

Proof. Suppose that f is responsive and robustly implemented by mechanism $\mathcal{M}$. The restriction to compact mechanisms ensures that $S^{\mathcal{M}}$ is non-empty. Let $m_i^*(\theta_i)$ be any element of $S_i^{\mathcal{M}}(\theta_i)$. Because mechanism $\mathcal{M}$ robustly implements f, $g(m^*(\theta)) = f(\theta)$, for all $\theta \in \Theta$.

We first establish strict EPIC. Suppose strict EPIC fails. Then there exists i, θ and $\theta_i' \neq \theta_i$ such that

$$u_i(f(\theta_i', \theta_{-i}'), \theta) \geq u_i(f(\theta), \theta).$$

Now $m^*(\theta) = (m_i^*(\theta_i), m_{-i}^*(\theta_{-i})) \in S^{\mathcal{M}}(\theta)$ implies that

$$\begin{aligned}\max_{m_i'}\{u_i(g(m_i', m_{-i}^*(\theta_{-i})), (\theta_i, \theta_{-i}))\} &= u_i(g(m_i^*(\theta_i), m_{-i}^*(\theta_{-i})), (\theta_i, \theta_{-i})) \\ &= u_i(f(\theta), \theta).\end{aligned}$$

But

$$u_i(g(m_i^*(\theta_i'), m_{-i}^*(\theta_{-i})), (\theta_i, \theta_{-i})) = u_i(f(\theta_i', \theta_{-i}), \theta) \geq u_i(f(\theta), \theta).$$

So

$$m_i^*(\theta_i') \in \arg\max_{m_i'}\{u_i(g(m_i', m_{-i}^*(\theta_{-i})), (\theta_i, \theta_{-i}))\}$$

which implies that $m_i^*(\theta_i') \in S_i^{\mathcal{M}}(\theta_i)$. This in turn implies that

$$f(\theta_i', \theta_{-i}) = g(m_i^*(\theta_i'), m_{-i}^*(\theta_{-i})) = f(\theta_i, \theta_{-i}), \quad \text{for all } \theta_{-i},$$

contradicting our assumption that f is responsive and robustly implemented by mechanism $\mathcal{M}$.

Now we establish the contraction property. First, suppose that $m_i \in M_i$, $\theta_i' \in \Theta_i$, $\theta_{-i}' \in \Theta_{-i}$, $\hat{m}_{-i} \in S_{-i}^{\mathcal{M}}(\theta_{-i}')$ and

$$u_i(g(m_i, \hat{m}_{-i}), (\theta_i', \theta_{-i}')) > u_i(f(\theta_i', \theta_{-i}'), (\theta_i', \theta_{-i}')). \tag{12}$$

Then, we have

$$m_i^*(\theta_i') \notin \arg\max_{m_i'}\{u_i(g(m_i', \hat{m}_{-i}), (\theta_i, \theta_{-i}'))\},$$

since

$$u_i(g(m_i^*(\theta_i'), \hat{m}_{-i}), (\theta_i', \theta_{-i}'))$$
$$= u_i(f(\theta_i', \theta_{-i}'), (\theta_i', \theta_{-i}')) < u_i(g(m_i, \hat{m}_{-i}), (\theta_i', \theta_{-i}')).$$

Thus $m_i \in M_i$, $\theta_i' \in \Theta_i$, $\theta_{-i}' \in \Theta_{-i}$ and $\hat{m}_{-i} \in S_{-i}^{\mathcal{M}}(\theta_{-i}')$ imply

$$u_i(g(m_i, \hat{m}_{-i}), (\theta_i', \theta_{-i}')) \leq u_i(f(\theta_i', \theta_{-i}'), (\theta_i', \theta_{-i}')). \tag{13}$$

Now consider an arbitrary report profile $\beta \neq \beta^*$. Let $\hat{k}$ be the largest k such that for every i, θ_i and $\theta_i' \in \beta_i(\theta_i)$:

$$S_i^{\mathcal{M}}(\theta_i') \subseteq S_i^k(\theta_i).$$

We know that such a $\hat{k}$ exists because $S_i^0(\theta_i) = M_i$, and, since $\mathcal{M}$ robustly implements f, responsiveness implies $S_i^{\mathcal{M}}(\theta_i) \cap S_i^{\mathcal{M}}(\theta_i') = \varnothing$.

Now we know that there exists i and $\theta_i' \in \beta_i(\theta_i)$ such that

$$S_i^{\mathcal{M}}(\theta_i') \subsetneqq S_i^{\hat{k}+1}(\theta_i).$$

Thus there exists $\hat{m}_i \in M_i$ such that $\hat{m}_i \in S_i^{\hat{k}}(\theta_i) \cap S_i^{\mathcal{M}}(\theta_i')$ and $\hat{m}_i \notin S_i^{\hat{k}+1}(\theta_i) \cap S_i^{\mathcal{M}}(\theta_i')$. Since message $\hat{m}_i$ gets deleted for θ_i at round $\hat{k}+1$, we know that for every $\lambda_i \in \Delta(M_{-i} \times \Theta_{-i})$ such that

$$\lambda_i(m_{-i}, \theta_{-i}) > 0 \Rightarrow m_j \in S_j^{\hat{k}}(\theta_j) \quad \text{for all } j \neq i,$$

there exists m_i^* such that

$$\sum_{m_{-i},\theta_{-i}} \lambda_i(m_{-i}, \theta_{-i}) u_i(g(m_i^*, m_{-i}), (\theta_i, \theta_{-i}))$$
$$> \sum_{m_{-i},\theta_{-i}} \lambda_i(m_{-i}, \theta_{-i}) u_i(g(\hat{m}_i, m_{-i}), (\theta_i, \theta_{-i})).$$

Fix any $\theta_{-i}' \in \Theta_{-i}$ and any $\hat{m}_j \in S_j^{\mathcal{M}}(\theta_j')$, for each $j \neq i$. Now the above claim remains true if we restrict attention to distributions λ_i putting probability 1 on $\hat{m}_{-i}$. Thus for every $\psi_i \in \Delta(\Theta_{-i})$ such that

$$\psi(\theta_{-i}) > 0 \Rightarrow \hat{m}_j \in S_j^{\hat{k}}(\theta_j) \quad \text{for all } j \neq i,$$

there exists m_i^* such that

$$\sum_{\theta_{-i}} \psi_i(\theta_{-i}) u_i(g(m_i^*, \hat{m}_{-i}), (\theta_i, \theta_{-i})) > \sum_{\theta_{-i}} \psi_i(\theta_{-i}) u_i(g(\hat{m}_i, \hat{m}_{-i}), (\theta_i, \theta_{-i})).$$

Since $\hat{m}_i$ is never a best response, there must exist a mixed strategy $\mu_i \in \Delta(M_i)$ such that

$$\sum_{m_i} \mu_i(m_i) u_i(g(m_i, \hat{m}_{-i}), (\theta_i, \theta_{-i})) > u_i(g(\hat{m}_i, \hat{m}_{-i}), (\theta_i, \theta_{-i}))$$

for all θ_{-i} such that $\hat{m}_{-i} \in S_{-i}^{\hat{k}}(\theta_{-i})$ (by the equivalence of "strictly dominated" and "never a best response" (see Lemma 3 in Pearce, 1984).

But $\hat{m} \in S^{\mathcal{M}}(\theta')$, so (since $\mathcal{M}$ robustly implements f), $g(\hat{m}_i, \hat{m}_{-i}) = f(\theta')$. Also observe that if $\theta'_{-i} \in \beta_{-i}(\theta_{-i})$, then $\hat{m}_{-i} \in S_{-i}^{\hat{k}}(\theta_{-i})$. Thus

$$\sum_{m_i} \mu_i(m_i) u_i(g(m_i, \hat{m}_{-i}), (\theta_i, \theta_{-i})) > u_i(f(\theta'), (\theta_i, \theta_{-i})) \tag{14}$$

for all θ_{-i} such that $\theta'_{-i} \in \beta_{-i}(\theta_{-i})$. Now let y be the lottery outcome generated by selecting outcome $g(m_i, \hat{m}_{-i})$ with distribution μ_i on m_i. Now we have established that for any $\beta \neq \beta^*$, there exist i, θ_i and $\theta'_i \in \beta_i(\theta_i)$ with $\theta'_i \neq \theta_i$ such that, for any θ_{-i} and $\theta'_{-i} \in \beta_{-i}(\theta_{-i})$,

$$u_i(y, (\theta'_i, \theta'_{-i})) \leq u_i(f(\theta'_i, \theta'_{-i}), (\theta'_i, \theta'_{-i})),$$

by (13);

$$u_i(y, (\theta_i, \theta_{-i})) > u_i(f(\theta'), (\theta_i, \theta_{-i})),$$

by (14); and

$$u_i(y, (\theta_i, \theta'_{-i})) > u_i(f(\theta'), (\theta_i, \theta'_{-i})), \tag{15}$$

which also follows from (14), since $\theta'_{-i} \in \beta_{-i}(\theta'_{-i})$).

Thus using the aggregator representation $u_i(y, \theta) \triangleq v_i(y, h_i(\theta))$, we have

$$v_i(y, h_i(\theta'_i, \theta'_{-i})) \leq v_i(f(\theta'_i, \theta'_{-i}), h_i(\theta'_i, \theta'_{-i})), \tag{16}$$

and

$$v_i(y, h_i(\theta_i, \theta_{-i})) > v_i(f(\theta'), h_i(\theta_i, \theta_{-i})), \tag{17}$$

and

$$v_i(y, h_i(\theta_i, \theta'_{-i})) > v_i(f(\theta'), h_i(\theta_i, \theta'_{-i})), \tag{18}$$

Now strict monotonicity of h_i with respect to θ_i implies

$$\text{sign }(\theta_i - \theta_i') = \text{sign }(h_i(\theta_i, \theta_{-i}') - h_i(\theta_i', \theta_{-i}'));$$

combining this with the preference rankings (16)–(18) and single crossing property, we have

$$\text{sign }(\theta_i - \theta_i') = \text{sign }(h_i(\theta_i, \theta_{-i}) - h_i(\theta_i', \theta_{-i}')).$$

But now we have just stated the contraction property. □

The proof of the necessity of the contraction property (Theorem 2), but not of the sufficiency (Theorem 1), uses the fact that the outcome space includes lotteries. We do not know if the contraction property would be necessary for robust implementation with a deterministic domain. However, if the deterministic social choice function is continuous in θ, we can prove the weaker result that the contraction property is necessary for robust implementation in the direct mechanism.

Restricting attention to responsive social choice functions simplifies the statement of the necessity result. The result could be re-stated to allow for non-responsive social choice functions, with appropriate weakenings of the strict EPIC and contraction property conditions. The weakened strict EPIC condition would require only that $f(\theta_i, \theta_{-i}) \neq f(\theta_i', \theta_{-i})$ for some θ_{-i} implies $u_i(f(\theta_i, \theta_{-i}), (\theta_i, \theta_{-i})) > u_i(f(\theta_i', \theta_{-i}), (\theta_i, \theta_{-i}))$ for all θ_{-i}. The weakened contraction property would require only *unacceptable* report profiles β to satisfy the properties required for all $\beta \neq \beta^*$ in definition 5, where β is unacceptable only if there exists $\theta_i' \in \beta_i(\theta_i)$ with $f(\theta_i, \theta_{-i}) \neq f(\theta_i, \theta_{-i})$ for some θ_{-i}. The weakened strict EPIC and contraction properties are equivalent to the original strict EPIC and contraction properties if f is responsive, and are both automatically satisfied if f is constant. The weakened contraction property is a joint property of the aggregator functions and the social choice function.

We briefly sketch the idea of the proof. We establish the contraction property directly from the robust implementation of the social choice function. We fix an arbitrary report profile $\beta \neq \beta^*$ and consider the iterative process of deleting strictly dominated messages. We identify a step $\hat{k}$ in the process as follows: let $\hat{k}$ be the earliest step at which for some agent i a rationalizable action $\hat{m}_i$ for some type θ_i' fails to be rationalizable *at step* $\hat{k}+1$ for some other type θ_i of agent i given that $\theta_i' \in \beta_i(\theta_i)$. As message $\hat{m}_i$ is deleted for type θ_i, it is never a best response for any message and type profile by the remaining agents. It follows that the message $\hat{m}_i$ is strictly dominated for type θ_i of agent i by a possibly mixed strategy $\mu_i(m_i)$ of agent i. For every given message profile $\hat{m}_{-i}$ of the other agents, the mixed strategy $\mu_i(m_i)$ generates a lottery y over deterministic outcomes. We can now establish the preference ranking of agent i with respect to the allocations y and $f(\theta'_i, \theta'_{-i})$ for any $\hat{\theta}_{-i}$ such that $\hat{m}_{-i}$ is a rationalizable action for types θ'_{-i} of the remaining agents. In turn, the contraction property follows immediately from these rankings and the single crossing property.

6 The Linear Model

In this section, we consider the special case in which the preference aggregator $h_i(\theta)$ is linear for each i and given by:

$$h_i(\theta) = \sum_{j=1}^{I} \gamma_{ij}\theta_j,$$

with $\gamma_{ij} \in \mathbb{R}$ for all i, j and $\gamma_{ii} > 0$ for all i. Without loss of generality, we set $\gamma_{ii} = 1$ for all i:

$$h_i(\theta) = \theta_i + \sum_{j \neq i} \gamma_{ij}\theta_j.$$

The parameters γ_{ij} represent the influence of the signal of agent j on the value of agent i. With the exception of $\gamma_{ii} > 0$ for all i, we do not impose any further a priori sign restrictions on γ_{ij}. We denote the square matrix generated by the absolute values of γ_{ij}, namely $|\gamma_{ij}|$, for all i, j with $i \neq j$ and zero entries on the diagonal by Γ:

$$\Gamma \triangleq \begin{bmatrix} 0 & |\gamma_{12}| & \cdots & |\gamma_{1I}| \\ |\gamma_{21}| & 0 & & \\ \vdots & & \ddots & \\ |\gamma_{I1}| & & & 0 \end{bmatrix}.$$

We refer to the matrix Γ as the *interdependence matrix.* The matrix $\Gamma = 0$ then constitutes the case of pure private values. We shall give necessary and sufficient conditions for the matrix Γ to satisfy the contraction property. We then use duality theory to give a characterization of the contraction property in terms of the eigenvalue of the matrix Γ. The proofs of all auxiliary results are in the appendix.

Lemma 1 (Linear Aggregator). *The linear aggregator functions* $\{h_i(\theta)\}_{i=1}^I$ *satisfy the contraction property if and only if, for all* $\mathbf{c} \in \mathbb{R}_+^I$ *with* $\mathbf{c} \neq \mathbf{0}$, *there exists* i *such that*

$$c_i > \sum_{j \neq i} |\gamma_{ij}| c_j. \tag{19}$$

The absolute values in the matrix Γ are required to guarantee that the linear inequality (19) implies the contraction property. Condition (19) is required to hold only for a single agent i.

The proof of the contraction property is constructive. We identify for each player i an initial report of the form $\beta_i(\theta_i) = [\theta_i - c_i\varepsilon, \theta_i + c_i\varepsilon]$ for some $\varepsilon > 0$, common across all agents. The size of c_i is therefore proportional to the size of the set of candidate misreports by agent i. It can be thought of as the set of rationalizable strategies at an arbitrary stage k. The inequality of the contraction property then says that for any arbitrary set of reports, characterized by the vector $\mathbf{c}$, there is always an agent i whose set of reports is too large (in the sense of being rationalizable) relative to the set of reports by the remaining agents. It then follows that the set of reports for this agent can be chosen smaller than c_i, allowing us to reduce the set of possible reports for a given agent i with a given type θ_i. The inequality (19) asserts that for any given set of reports, there is always at least one agent i whose report β_i represents a set too large to be rationalizable. Moreover, if the set of reports by i is too large, then there is an "overhang" which can be "nipped and tucked". Now a dual interpretation of the condition (19) leads us from the idea of the overhang directly to the contraction property.

With the dual interpretation, we obtain the following simple test of the contraction property:

Proposition 1 (Contraction Property and Eigenvalue). *The matrix* Γ *has the contraction property if and only if its largest eigenvalue* $\lambda < 1$.

The matrix algebra underlying the above characterization of the contraction property arises in many economic problems which depend on the stability and uniqueness of solutions to a system of linear equations, e.g., the uniqueness of equilibrium and rationalizable outcomes in complete information games with linear best responses (see Luenberger, 1978; Gabay and Moulin, 1980; Bernheim, 1984, and Weinstein and Yildiz, 2007).

The linear model has the obvious advantage that the local conditions for contraction agree with the global conditions for contraction as the derivatives of the mapping $h_i(\theta)$ are constant and independent of θ.

We can naturally extend the idea behind the linear aggregator function to a general nonlinear and differentiable aggregator function $h_i(\theta)$, but with a gap between necessary and sufficient conditions. We report the results in the appendix of a working paper version of this paper (Bergemann and Morris, 2007b).

By linking the contraction property to the eigenvalue of the matrix Γ, we can immediately obtain necessary and sufficient conditions for robust implementation for different classes of linear aggregators.

Symmetric Preferences. In the symmetric model, the parameters for interdependent values are given by

$$\gamma_{ij} = \begin{cases} 1, & \text{if } i = i, \\ \gamma, & \text{if } j \neq i. \end{cases}$$

The eigenvalue λ of the resulting matrix satisfies: $1 + \lambda = 1 + \gamma(I - 1)$, and hence from Theorem 1, we immediately obtain the necessary and sufficient condition $\gamma < 1/(I - 1)$.

Common Interest Preferences. An important class of symmetric preferences is the case of common interest preferences with $\gamma = 1$. In this case, for every $I \geq 2$, the contraction property will fail. This result tells us that if there are exact common interests, then robust implementation will be impossible. This observation will hold for more general common interest models, beyond the linear aggregator model of this section. The common interest model is an interesting example in which the ex post incentive constraints will automatically be satisfied by any efficient social choice function, yet the contraction property imposes a constraint leading to the impossibility of robust implementation in common interest environments.

Cyclic Preferences. A weaker form of symmetry is incorporated in the following model of cyclic preferences. Here, the interdependence matrix is determined by the distance between i and j (modulo I), or $\gamma_{ij} = \gamma_{(i-j)_{\text{mod } I}}$. In this case, the positive eigenvalue is given by:

$$1 + \lambda = 1 + \sum_{j \neq i} \gamma_{(i-j)},$$

and consequently a necessary and sufficient condition for robust implementation is given by:

$$\sum_{j \neq i} \gamma_{(i-j)} < 1.$$

7 Single Unit Auction

We conclude our analysis with a second economic example, namely a single unit auction with symmetric bidders. The model has I agents and agent i's payoff type is $\theta_i \in [0,1]$. If the type profile is θ, agent i's valuation of the object is

$$\theta_i + \gamma \sum_{j \neq i} \theta_j,$$

where $0 \leq \gamma \leq 1$.

An allocation rule in this context is a function $y : \Theta \to [0,1]^I$, where $y_i(\theta)$ is the probability that agent i gets the object and so $\sum_i y_i(\theta) \leq 1$. The symmetric efficient allocation rule is given by:

$$y_i^*(\theta) = \begin{cases} \dfrac{1}{\#\{j|\theta_j \geq \theta_k \text{ for all } k\}}, & \text{if } \theta_i \geq \theta_k \text{ for all } k, \\ 0, & \text{if otherwise.} \end{cases}$$

Maskin (1992) and Dasgupta and Maskin (2000) have shown that the efficient allocation can be truthfully implemented in the generalized VCG mechanism, according to which the monetary transfer of the winning agent i is given by

$$x_i(\theta) = \max_{j \neq i} \theta_j + \gamma \sum_{j \neq i} \theta_j.$$

The winning probability $y_i^*(\theta)$ and the monetary transfer are piecewise constant. The generalized VCG mechanism therefore does not satisfy the strict EPIC conditions which we assumed as part of our analysis. We therefore modify the generalized VCG mechanism to a symmetric ε-efficient allocation rule given by:

$$y_i^{**}(\theta) = \varepsilon \frac{\theta_i}{I} + (1-\varepsilon) y_i^*(\theta).$$

Under this allocation rule, the object is not allocated with positive probability of order ε.[4] We show that the symmetric ε-efficient allocation rule can be robustly implemented if $\gamma < \frac{1}{I-1}$. Alternatively, we can say that

[4]If the realized payoff type profile is θ, the object will not be allocated to any agent with probability $\varepsilon \sum_i (1-\theta_i)/I$. At the cost of some additional algebra, we could modify the allocation rule so that it allocates the object with probability 1 by defining $y_i^{**}(\theta) = \varepsilon \theta_i \sum_j \theta_j + (1-\varepsilon) y_i^*(\theta)$.

the generalized VCG mechanism itself is robustly virtually implementable if $\gamma < \frac{1}{I-1}$.

It is easy to verify that the resulting generalized VCG transfers satisfy strict EPIC and show that this ε-efficient allocation is robustly implementable. The unique (up to a constant) ex post transfer rule is:

$$x_i(\theta) = \frac{\varepsilon}{2I}(\theta_i)^2 + \frac{\varepsilon\gamma}{I}\left(\sum_{j\neq i}\theta_j\right)\theta_i + (1-\varepsilon)\left(\max_{j\neq i}\left\{\theta_j + \gamma\sum_{j\neq i}\theta_j\right\}\right)y_i^*(\theta).$$

The first two components of the transfers guarantee incentive compatibility with the respect to the linear probability assignment and the third component with respect to the efficient allocation rule. The best response of agent i for misreports θ'_{-i} of the remaining agents at a true type profile θ is given as in the public good example by:

$$\theta'_i = \theta_i + \gamma\sum_{j\neq i}(\theta_j - \theta'_j).$$

We can therefore exactly repeat our earlier argument in the context of the public good and get robust implementation in the direct mechanism if $\gamma < 1/(I-1)$.

8 Discussion

8.1 *Dimensionality and Aggregation*

We assumed that agents have one-dimensional payoff types, and each agent's utility depends on the profile of agents' types via an aggregating function. If the agents had multidimensional payoff types, and there still existed an aggregator for the multidimensional types, then our results would still go through, as the single crossing condition is defined with respect to the aggregator rather than the types.

More interesting is what happens if the aggregator property fails, with one or many dimensional payoff types. The contraction property formalized the idea of moderate interdependence under which the process of iterative elimination of strictly dominated strategies continued until truthtelling remained the only surviving message for each type. Our definition of the contraction property explicitly used the existence of an aggregator. But the notion of moderate interdependence is meaningful in the absence of an aggregator. Informally, the idea of moderate interdependence is that for every possible set of type profiles of all agents, there exists an agent i and a type pair θ_i and θ_i' such that the ranking of any pair of alternatives, y and y', is determined by the payoff type of agent i, irrespective of the type profile of the other agents.

More formally, let each Θ_i be a compact subset of $\mathbb{R}^n$ (instead of compact subset of $\mathbb{R}$) and fix a report profile $\beta = (\beta_1, \ldots, \beta_I)$ of the agents and consider an agent i with type θ_i and $\theta_i' \in \beta_i(\theta_i)$. We consider the preference ranking of agent i at a type profile θ' with $\theta' \in \beta(\theta)$. A change in the preference profile of agent i from θ_i' in the direction of θ_i is represented by a convex combination $\varepsilon\theta_i + (1-\varepsilon)\theta_i'$. As we change the preference profile of agent i for small $\varepsilon > 0$ in this way, there may be preference reversals such that for some y and y', we observe

$$u_i(y', (\theta_i', \theta_{-i}')) > u_i(y, (\theta_i', \theta_{-i}')) \tag{20}$$

and

$$u_i(y', (\varepsilon\theta_i + (1-\varepsilon)\theta_i', \theta_{-i}')) < u_i(y, (\varepsilon\theta_i + (1-\varepsilon)\theta_i', \theta_{-i}')). \tag{21}$$

We then say that the preferences of agent i display the contraction property if the direction of the preference reversal at θ_i' is predicated on θ_i in the sense that (20) and (21) imply

$$u_i(y', (\theta_i, \theta_{-i})) < u_i(y, (\theta_i, \theta_{-i})), \tag{22}$$

for all $\theta_{-i}' \in \beta_{-i}(\theta_{-i})$. This new version of the contraction property is defined independently of an aggregation property or a dimensionality condition on the type of agent i. But it is supposed to hold for all allocations y, $y' \in Y$, even those far away from any realization of the social choice function. This makes it a very strong condition. We therefore propose a weaker, but still sufficient, condition for robust implementation in the direct mechanism by specializing the contraction property to the set of relevant allocations. The two prominent allocations to consider at the type profiles $(\theta_i', \theta_{-i}')$ and $(\varepsilon\theta_i + (1-\varepsilon)\theta_i', \theta_{-i}')$ are

$$y' \triangleq f(\theta_i', \theta_{-i}'), \tag{23}$$

and

$$y \triangleq f(\varepsilon\theta_i + (1-\varepsilon)\theta_i', \theta_{-i}'). \tag{24}$$

Now, if the social choice function f satisfies the ex post incentive compatibility, then y and y' as given by (23) and (24), satisfy the hypothesis (20) and (21) by force of the ex post incentive compatibility condition. A local contraction property can now be defined for the social choice function f.

Definition 7 (Local Contraction Property). Social choice function f satisfies the local contraction property if, for all $\beta \neq \beta^*$, there exists i, θ_i, $\theta_i' \in \beta_i(\theta_i)$ and $\bar{\varepsilon} > 0$, such that for all θ_{-i}, $\theta_{-i}' \in \beta_{-i}(\theta_{-i})$ and all $\varepsilon \in (0, \bar{\varepsilon}]$:

$$u_i(f(\theta_i', \theta_{-i}'), (\theta_i, \theta_{-i})) < u_i(f(\varepsilon\theta_i + (1-\varepsilon)\theta_i', \theta_{-i}'), (\theta_i, \theta_{-i})). \tag{25}$$

We can now that the local contraction property of the social choice function is a sufficient condition for robust implementation in the direct mechanism. It follows a fortiori that the above contraction property for the environment is also sufficient to establish the robust implementation result in the direct mechanism.

Theorem 3 (Sufficiency). *If social choice function f satisfies strict EPIC and the local contraction property, then f can be robustly implemented in the direct mechanism.*

It may be useful to illustrate the contraction property of the social choice function with a multidimensional generalization of the earlier public good example. We consider an environment with two agents, $i = 1, 2$, where the payoff type of each agent i is a two-dimensional vector $\theta_i = (\theta_i^a, \theta_i^b) \in [0, 1]^2$. Each agent i has a quasilinear utility from a two-dimensional public good $y = (y^a, y^b) \in \mathbb{R}_+^2$:

$$\begin{aligned} u_i(y, \theta) = {} & (\alpha_1\theta_i^a + \alpha_2\theta_i^b + \gamma_1\theta_j^a + \gamma_2\theta_j^b)y^a \\ & + (\alpha_1\theta_i^b + \alpha_2\theta_i^a + \gamma_1\theta_j^b + \gamma_2\theta_j^a)y^b. \end{aligned} \tag{26}$$

We assume that $\alpha_1, \alpha_2, \gamma_1, \gamma_2 > 0$. The cost function of providing the public good is given by

$$c(y) = \frac{1}{2}(y^a)^2 + \frac{1}{2}(y^b)^2.$$

The (efficient) social choice function $f(\theta) = (f^a(\theta),\ f^b(\theta))$ determines the level of the public good along each dimension l:

$$f^l(\theta) \triangleq (\alpha_1 + \gamma_1)(\theta_1^l + \theta_2^l) + (\alpha_2 + \gamma_2)(\theta_1^m + \theta_2^m), \tag{27}$$

for $l = a, b$ and $l \neq m$. The social choice function $f(\theta)$ can be implemented with the generalized VCG transfers:

$$\begin{aligned} t_{(\theta)} = {} & (\theta_i^a \theta_j^a + \theta_i^b \theta_j^b) \left(\sum_k (\alpha_k + \gamma_k)\gamma_k \right) \\ & + (\theta_i^a \theta_j^b + \theta_i^b \theta_j^a) \left(\sum_k (\alpha_k + \gamma_k)\gamma_{k+1} \right) \\ & + \frac{1}{2}((\theta_i^a)^2 + (\theta_i^b)^2) \left(\sum_k (\alpha_k + \gamma_k)\alpha_k \right) \\ & + \theta_i^a \theta_i^b \left(\sum_k (\alpha_k + \gamma_k)\alpha_{k+1} \right), \end{aligned} \tag{28}$$

where the sums over k are modulo 2. The generalized Vickrey–Clarke–Groves mechanism described by $(f(\theta), t_1(\theta), t_2(\theta))$ satisfies the strict ex post incentive constraints in the direct mechanism. The restriction to two agents and a two-dimensional type space is for expositional ease only and the results generalize easily to I agents and a K dimensional type and allocation space.

Given the two-dimensional type space for each agent and given the two dimensional allocation space, it is clear that in general, we cannot find an aggregator function to represent the preferences of the agents. Similarly, the preferences will not satisfy a single crossing condition due to the multidimensionality of the allocation space. But because of symmetry across agents and across allocations, the social choice function is strictly ex post incentive compatible.[5]

Proposition 2 (Multidimensional Public Goods). *The multidimensional social choice function (27) satisfies the local contraction property iff*

$$\alpha_1 + \alpha_2 > \gamma_1 + \gamma_2. \tag{29}$$

[5]The existence of an efficient and ex post incentive compatible mechanism in the public good model does not conflict with the generic impossibility of multidimensional ex post incentive compatible mechanism established by Jehiel, Moldovanu, Meyer-Ter-Vehn, and Zame (2006). The existence of ex post incentive compatible transfers here is due to the symmetry across agents and allocations. We emphasize that the objective of the current example is to show how the idea of moderate interdependence extends naturally beyond environments with the aggregation property.

8.2 *Relation to Partial and Ex Post Implementation*

The results in this paper concern full implementation. An earlier paper of ours, Bergemann and Morris (2005b), addresses the analogous questions of robustness to rich type spaces, but looking at the question of *truthtelling* in the *direct* mechanism. In the literature, this is frequently referred to as *partial* implementation. The notion of partial implementation asks whether there exists a mechanism such that *some* equilibrium under that mechanism implements the social choice function. By the revelation principle, it is then sufficient to look at truthtelling in the direct mechanism. In Bergemann and Morris (2005b), we showed that a social choice function robustly satisfies the interim incentive constraints, i.e., satisfies the interim incentive constraints for any type space, if and only if the ex post incentive constraints are satisfied.

It is important to note, however, that robust implementation is not equivalent to full ex post implementation, i.e., the requirement that every ex post equilibrium delivers the right outcome. Often ex post implementation will be possible — because there are no undesirable ex post equilibria — even though there exist type spaces and interim equilibria that deliver undesirable outcomes. In Bergemann and Morris (2008a), we identify the ex post monotonicity condition that is necessary and sufficient for full ex post implementation. It is much weaker than the contraction property (and its equivalent robust monotonicity condition).

8.3 *Robust and Virtual Implementation in General Environments*

The existing Bayesian implementation literature — Postlewaite and Schmeidler (1986), Palfrey and Srivastava (1989b) and Jackson (1991) — has shown that on a fixed type space with a common knowledge common prior, Bayesian incentive compatibility and a Bayesian monotonicity condition are necessary and almost sufficient for full implementation. The proof of the sufficiency part of the result relies on complex augmented mechanisms.

In Bergemann and Morris (2005a), we developed the results in this paper in the context of a general approach to robust implementation which allows for complex augmented mechanism. The results reported in this subsection appear in that working paper.

Our robust implementation notion is equivalent to requiring Bayesian implementation on all type spaces. Ex post *incentive compatibility* is equivalent to Bayesian *incentive compatibility* on all type spaces. It is possible to define a notion of robust monotonicity which is equivalent to Bayesian monotonicity on all type spaces. Ex post incentive compatibility and robust monotonicity are thus necessary and almost sufficient for full implementation. However, this result relies on allowing complex augmented mechanisms including integer games. If we restrict attention to well-behaved mechanisms — with the compact message space assumption of this paper — then strict EPIC is also necessary.

The contraction property is an implication of robust monotonicity in the environment studied in this paper. The robust monotonicity condition requires the existence of allocations that can be used to reward individuals for reporting deviations from desirable equilibria. In the environment of this paper, we are able to show that we can always use rewards from misreports in the direct mechanism.

In the single good auction example, we used an ε-efficient allocation rule to obtain strict EPIC. An alternative interpretation of the ε-efficient allocation rule is that it virtually implements the efficient social choice function.[6] This naturally leads to the question of how much could be achieved with a robust version of virtual implementation. We pursue this question in Bergemann and Morris (2008c) and provide a characterization of robust virtual implementation in general environments. Our general characterization requires complicated augmented mechanisms, building on those in Abreu and Matsushima (1992a) and (1992b). The reliance of positive virtual implementation results on such complex mechanisms has often been criticized (Glazer and Rosenthal, 1992). But in the single crossing monotonic aggregator environment studied in this paper, we show that ex post incentive compatibility and the contraction property are necessary for virtual robust implementation (as well as for exact implementation) and also sufficient for virtual robust implementation in the direct mechanism. Thus, in the current environment, the only implication of going from full to virtual implementation is a relaxation from strict EPIC to EPIC. Thus Bergemann and Morris (2008c) show that, by requiring robustness to beliefs and higher-order beliefs, the apparent permissiveness of virtual

[6]Abreu and Matsushima (1992a) obtain permissive results about virtual implementation under complete information, using iterated deletion of strictly dominated strategies as a solution concept. Abreu and Matsushima (1992b) obtain incomplete information analogues of those results.

implementation is greatly reduced relative to exact implementation (the notion studied in the current paper).[7]

8.4 *Social Choice Correspondences and Sets*

We considered necessary and sufficient conditions for the robust implementation of a social choice function. We briefly discuss the relevance of our results for the robust implementation of a social choice *correspondence* or a social choice *set*. A social choice correspondence is a set-valued mapping $F : \Theta \to 2^Y$. A social choice set is a set of social choice functions $\mathcal{F} = \{f | f : \Theta \to Y\}$. The concept of a social correspondence, prevalent in the complete information literature (see Maskin, 1999), differs from the concept of a social choice set, prevalent in the incomplete information literature (see Palfrey and Srivastava, 1989a, and Jackson, 1991).

Our robust implementation results for social choice functions can be applied directly to the notion of a social choice set. In other words a social choice set is robustly implementable if and only if the every element of the social choice set is robustly implementable. As the conditions for robust implementation were defined in terms of the utility environment and the social choice function, robust implementation for a social choice set simply requires that every element of the social choice set can be robustly implemented.

In contrast, it is substantially more difficult to obtain necessary or sufficient conditions for the robust implementation of social choice correspondences. In Bergemann and Morris (2005b) we showed that the difficulties with social choice correspondences already arise with respect to the incentive constraints. In Example 1 of Bergemann and Morris (2005b) we showed in a setting with two agents and two payoff types for each agent, that a social choice correspondence can satisfy the relevant interim incentive compatibility conditions for all type spaces, yet fail to satisfy the ex post incentive compatibility conditions. A comprehensive analysis of the robust implementation of social choice correspondences then first requires

[7]Abreu and Matsushima (1992b) show that an apparently weak "measurability" condition is the only requirement — beyond incentive compatibility — for virtual implementation under incomplete information. Bergemann and Morris (2008c) show that a "robust measurability" condition is the robust analogue of the measurability condition in Abreu and Matsushima (1992b) when beliefs and higher order beliefs are not known. This robust measurability condition is shown to be equivalent — in the single crossing monotonic aggregator environment of this paper — to the contraction property.

additional insights into the nature of incentive constraints in social choice correspondences.

8.5 *The Common Prior Assumption and Strategic Substitutes/Complements*

The definition of robust implementation in this paper is equivalent to requiring that every equilibrium on every type space delivers outcomes consistent with the social choice function. An interesting question is what happens when we look at an intermediate notion of robustness: allowing all possible common prior type spaces. In Bergemann and Morris (2008b) we pursue this question in the context of our leading example of Section 3.

Consider the case of negative interdependence in valuations, i.e., $\gamma < 0$, in the public good example. We recall the ex post best response function in that example: if type θ_i is sure that his opponents have type profile θ_{-i} and is sure that they will report themselves to be type profile θ'_{-i}, then his best response is to report himself to be type θ'_i with

$$\theta'_i = \theta_i + \gamma \sum_{j \neq i} (\theta_j - \theta'_j).$$

We see that there are strategic complements in misreporting strategies (if others misreport upwards, i has an incentive to misreport upwards). This means that when we carry out the iterated deletion procedure, the profile of largest and smallest misreports that survive must constitute an ex post equilibrium of the game (Milgrom and Roberts (1990)). Thus a failure of robust implementation also implies that there exists a bad equilibrium on any common prior type space.

In the case with positive interdependence, i.e., $\gamma > 0$, there is strategic substitutability in mis-reports and the above argument does not go through. In fact, Bergemann and Morris (2008b) show that if $\gamma \in (0, 1)$, even when the contraction property fails (i.e., $\gamma > \frac{1}{I-1}$), every equilibrium on any common prior type space delivers the right outcome. When we contrast the robust implementation condition for all types spaces with those for all common prior type spaces, it appears that the common prior leads to synchronized beliefs among the agents. This restricts the set of possible best responses and allows for positive implementation results. In the absence of a common prior, a sequential revelation of information may replace the role of the common prior by generating synchronized beliefs among the agents. This theme is developed in the context of an ascending auction in Bergemann and Morris (2007a).

8.6 *Informational Foundation of Interdependence*

In the discussion of the single crossing condition in Section 4 we presented a statistical model of noisy signals which naturally lead to the aggregation property of private signals by means of Bayes law. There is a possible criticism of using an informational justification for interdependent preferences like this one at the same time as insisting on a stringent robust implementation criterion.[8] This informational microfoundation for the environment depends on the common knowledge of the distribution of signals about the environment — among the agents and the planner. Thus there is common knowledge of a true distribution over the vectors of signals θ. However, we can show that if we allowed that each agent i might receive additional, conditionally independent information — not necessarily consistent with a common prior — about others' signals θ_{-i}, so that the information did not change his expectation of $\omega_0 + \omega_i$, conditional on the vector θ, then our robust implementation results would remain unchanged. Thus there is an admittedly stark story that reconciles the robust implementation environment with an informational justification of the reduced form representation of interdependent preferences.

9 Appendix

The appendix contains the arguments and proofs missing in the main text.

Informational Foundation for Interdependence. The vector $(\omega_0 + \omega_1, \theta_1, \theta_2)$ of random variables is normally distributed with mean zero and variance matrix:

$$\begin{pmatrix} \sigma_0^2 + \sigma_1^2 & \sigma_0^2 + \sigma_1^2 & \sigma_0^2 \\ \sigma_0^2 + \sigma_1^2 & \sigma_0^2 + \sigma_1^2 + \tau_1^2 & \sigma_0^2 \\ \sigma_0^2 & \sigma_0^2 & \sigma^2 + \sigma_2^2 + \tau_2^2 \end{pmatrix}.$$

By a standard property of the multivariate normal distribution this implies that the expectation of $\omega_0 + \omega_1$ conditional on θ_1 and θ_2 is given by:

$$(\sigma_0^2 + \sigma_1^2\sigma_0^2) \begin{pmatrix} \sigma_0^2 + \tau_1^2 + \sigma_1^2 & \sigma_0^2 \\ \sigma_0^2 & \sigma_0^2 + \sigma_2^2 + \tau_2^2 \end{pmatrix}^{-1} \begin{pmatrix} \theta_1 \\ \theta_2 \end{pmatrix},$$

[8]We thank Ilya Segal for prompting us to think about this in the context of robust implementation.

which equals

$$\frac{(\sigma_0^2\tau_2^2 + \sigma_0^2\sigma_1^2 + \sigma_0^2\sigma_2^2 + \tau_2^2\sigma_1^2 + \sigma_1^2\sigma_2^2)\theta_1 + \sigma^2\tau_1^2\theta_2}{\sigma_0^2\tau_1^2 + \sigma_0^2\tau_2^2 + \sigma_0^2\sigma_1^2 + \sigma_0^2\sigma_2^2 + \tau_1^2\sigma_2^2 + \tau_1^2\sigma_2^2 + \tau_2^2\sigma_1^2 + \sigma_1^2\sigma_2^2}.$$

If we multiply the above expression by the constant

$$\frac{\sigma_0^2\tau_1^2 + \sigma_0^2\tau_2^2 + \sigma_0^2\sigma_1^2 + \sigma_0^2\sigma_2^2 + \tau_1^2\tau_2^2 + \tau_1^2\sigma_2^2 + \tau_2^2\sigma_1^2 + \sigma_1^2\sigma_2^2}{\sigma_0^2\tau_2^2 + \sigma_0^2\sigma_1^2 + \sigma_0^2\sigma_2^2 + \tau_2^2\sigma_1^2 + \sigma_1^2\sigma_2^2},$$

we obtain, as reported in (3):

$$\theta_1 + \frac{\sigma^2\tau_1^2}{\sigma_0^2\tau_2^2 + \sigma_0^2\sigma_1^2 + \sigma_0^2\sigma_2^2 + \tau_2^2\sigma_1^2 + \sigma_1^2\sigma_2^2}\theta_2.$$

Proof of Lemma 1. We prove the contrapositive. Thus suppose there exists $\mathbf{c} \in \mathbb{R}_+^I$ with $\mathbf{c} \neq \mathbf{0}$, such that for all i:

$$c_i \leq \sum_{j \neq i} |\gamma_{ij}| c_j.$$

We now show that this implies that the contraction property fails. Choose $\varepsilon > 0$ such that $2c_i\varepsilon < \overline{\theta}_i - \underline{\theta}_i$ for all i. Now consider reports of the form:

$$\beta_i(\theta_i) = [\theta_i - \varepsilon c_i, \theta_i + \varepsilon c_i] \cap \Theta_i, \tag{30}$$

for all i. Then for all i and all $j \neq i$, let $\theta_j = \frac{1}{2}(\underline{\theta}_j + \bar{\theta}_j)$ and let $\theta_j' = \theta_j - \varepsilon c_i$ if $\gamma_{ij} \geq 0$ and $\theta_j' = \theta_j + \varepsilon c_i$ if $\gamma_{ij} < 0$. By (30), we have $\theta_j' \in \beta_j(\theta_j)$ for each $j \neq i$. Also observe that $\gamma_{ij}(\theta_j - \theta_j') = \varepsilon|\gamma_{ij}|c_j$. Thus

$$\sum_{j \neq i} \gamma_{ij}(\theta_j - \theta_j') = \varepsilon \sum_{j \neq i} |\gamma_{ij}| c_j \geq \varepsilon c_i.$$

Now if $\theta_i' = \theta_i + \varepsilon c_i, \theta_i - \theta_i'$ is strictly negative but

$$\theta_i - \theta_i' + \sum_{j \neq i} \gamma_{ij}(\theta_j - \theta_j'),$$

is non-negative. A symmetric argument works if $\theta_i > \theta_i'$. So the contraction property, which says that for all $\beta \neq \beta^*$, there exists i and $\theta_i' \in \beta_i(\theta_i)$ with

$\theta_i' \neq \theta_i$, such that

$$\begin{aligned}\text{sign}(\theta_i - \theta_i') &= \text{sign}(h_i(\theta_i, \theta_{-i}) - h_i(\theta_i', \theta_{-i}')) \\ &= \text{sign}\left(\theta_i - \theta_i' + \sum_{j\neq i} \gamma_{ij}(\theta_i - \theta_j')\right), \qquad (31)\end{aligned}$$

for all θ_{-i} and $\theta_{-i}' \in \beta_{-i}(\theta_{-i})$ fails. This proves the necessity of condition (19).

($\Leftarrow$) To show sufficiency, suppose that condition (19) of the lemma holds. Fix any report $\beta \neq \beta^*$. For all j, let:

$$c_j = \max_{\theta_j' \in \beta_j(\theta_j)} |\theta_j' - \theta_j|.$$

By hypothesis, there exists i such that $c_i > \sum_{j\neq i} |\gamma_{ij}| c_j$. Let $|\theta_i - \theta_i'| = c_i$, and suppose without loss of generality that $\theta_i > \theta_i'$. Observe that for all θ_{-i} and $\theta_{-i}' \in \beta_{-i}(\theta_{-i})$: $\gamma_{ij}(\theta_i - \theta_j') \leq |\gamma_{ij}| c_j$. Thus

$$\sum_{j\neq i} \gamma_{ij}(\theta_j - \theta_j') \leq \sum_{j\neq i} |\gamma_{ij}| c_j;$$

and so from

$$\begin{aligned}(\theta_i - \theta_i') + \sum_{j\neq i} \gamma_{ij}(\theta_i - \theta_i') &= c_i + \sum_{j\neq i} \gamma_{ij}(\theta_j - \theta_j') \\ &\geq c_i - \sum_{j\neq i} |\gamma_{ij}| c_j > 0,\end{aligned}$$

it follows that the contraction property, and equivalently (31), is satisfied. □

The following lemma gives a dual representation of the contraction property for the linear case. In turn, it allows us to characterize the contraction property in terms of the eigenvalue of the interdependence matrix Γ.

Lemma 2 (Duality). *The following two properties of Γ are equivalent:*

1. *for all $c \in \mathbb{R}_+^I$ with $c \neq \mathbf{0}$ there exists i such that:*

$$c_i > \sum_{j\neq i} |\gamma_{ij}| c_j; \qquad (32)$$

2. there exists $d \in \mathbb{R}^I_+$ *such that for all* i *:*

$$d_i > \sum_{j \neq i} |\gamma_{ji}| d_j. \tag{33}$$

Proof. Consider the following contrapositive restatement of condition (32): there does not exist $c \in \mathbb{R}^I_+$ such that

$$\sum_{i=1}^{I} c_i > 0, \tag{a}$$

and for all i:

$$\sum_{j \neq i} |\gamma_{ij}| c_j - c_i \geq 0. \tag{b$_i$}$$

Writing μ for the multiplier of constraint (a) and d_i for the multiplier of constraint (b_i), Farkas' lemma states that such a c does not exist if and only if there exist $d \in \mathbb{R}^I_+$ and $\mu \in \mathbb{R}_+$ such that

$$\mu - d_i + \sum_{j \neq i} |\gamma_{ji}| d_j = 0 \quad \text{for all } i, \tag{a'}$$

and

$$\mu > 0. \tag{b'}$$

But this is true if and only if condition (33) of the lemma holds. □

Proof of Proposition 1. If we try to find a solution for the strict inequalities (33):

$$d_i > \sum_{j \neq i} |\gamma_{ji}| d_j, \quad \text{for all } i$$

with the assistance of a contraction constant $\lambda < 1$, or

$$d_i \lambda = \sum_{j \neq i} |\gamma_{ji}| d_j,$$

then by the Frobenius-Perron Theorem for nonnegative matrices (see Minc (1988), Theorem 1.4.2), there exist positive right and left eigenvectors with the same positive eigenvalue λ. We can use the above dual property to establish that a (λ, d) solution exists for:

$$\lambda d_i = \sum_{j \neq i} |\gamma_{ji}| d_j,$$

but from the duality relationship (33), we know that for every $d > 0$,

$$d_i > \sum_{j \neq i} |\gamma_{ji}| d_j,$$

so it follows that $\lambda < 1$. $\square$

Proof of Proposition 2. Suppose first the inequality (29) holds. Given a report profile $\beta \neq \beta^*$, we consider agent i such that

$$\max_{\theta_i, \theta_i' \in \beta_i(\theta_i)} |\theta_i^a - \theta_i^{a'} + \theta_i^b - \theta_i^{b'}| \geq \max_{\theta_j, \theta_j' \in \beta_j(\theta_j)} |\theta_j^a - \theta_j^{a'} + \theta_j^b - \theta_j^{b'}|. \quad (34)$$

Consider a pair of types $\theta_i, \theta_i' \in \beta_i(\theta_i)$ which achieves the maximum in (34) and we may assume without loss of generality that the maximum is positive: $\theta_i^a - \theta_i^{a'} + \theta_i^b - \theta_i^{b'} > 0$. The indirect utility from the mechanism is given by:

$$\begin{aligned} u_i(f(m), \theta) = {} & (\alpha_1 \theta_i^a + \alpha_2 \theta_i^b + \gamma_1 \theta_j^a + \gamma_2 \theta_j^b) f^a(m) \\ & + (\alpha_1 \theta_i^b + \alpha_2 \theta_i^a + \gamma_1 \theta_j^b + \gamma_2 \theta_j^a) f^b(m) - t_i(m). \end{aligned}$$

We want to show that the inequality (25) holds for some $\bar{\varepsilon} > 0$. As the indirect utility is a quadratic function in the report profile, it is sufficient to consider the derivative of the indirect utility $u_i(f((1-\varepsilon)\theta_i' + \varepsilon\theta_i, \theta_{-i}'), \theta)$ with respect to ε evaluated at $\varepsilon = 0$. After collecting terms, the derivative results in:

$$\begin{aligned} & ((\theta_i^a - \theta_i^{a'} + \theta_i^b - \theta_i^{b'})(\alpha_1 + \alpha_2) \\ & \quad + (\theta_j^a - \theta_j^{a'} + \theta_j^b - \theta_j^{b'})(\gamma_1 + \gamma_2))(\alpha_1 + \alpha_2 + \gamma_1 + \gamma_2) > 0, \end{aligned} \quad (35)$$

where the strict inequality follows from the hypothesis of $\alpha_1 + \alpha_2 > \gamma_1 + \gamma_2$, the inequality (34):

$$|\theta_i^a - \theta_i^{a'} + \theta_i^b - \theta_i^{b'}| \geq |\theta_j^a - \theta_j^{a'} + \theta_j^b - \theta_j^{b'}|,$$

and the positivity of $\theta_i^a - \theta_i^{a'} + \theta_i^b - \theta_i^{b'} > 0$. (If the later sum were negative, then the sign of the derivative (35) would change and the same argument would go through.)

We prove the converse by contrapositive and assume that the inequality (29) is reversed to $\alpha_1 + \alpha_2 \leq \gamma_1 + \gamma_2$. We now consider a report profile

$\beta \neq \beta^*$ such that

$$\max_{\theta_i,\theta_i'\in\beta_i(\theta_i)} |\theta_i^a - \theta_i^{a'} + \theta_i^b - \theta_i^{b'}| = \max_{\theta_j,\theta_j'\in\beta_j(\theta_j)} |\theta_j^a - \theta_j^{a'} + \theta_j^b - \theta_j^{b'}|$$

and

$$\theta_i^a - \theta_i^{a'} + \theta_i^b - \theta_i^{b'} > 0 > \theta_j^a - \theta_j^{a'} + \theta_j^b - \theta_j^{b'}.$$

It follows that the strict inequality (35) is reversed and this establishes the failure of the contraction property of the social choice function. □

References

ABREU, D., AND H. MATSUSHIMA (1992a): "Virtual Implementation in Iteratively Undominated Strategies: Complete Information," *Econometrica*, 60, 993–1008.

——— (1992b): "Virtual Implementation In Iteratively Undominated Strategies: Incomplete Information," Discussion paper, Princeton University and University of Tokyo.

BATTIGALLI, P., AND M. SINISCALCHI (2003): "Rationalization and Incomplete Information," *Advances in Theoretical Economics*, 3, Article 3.

BERGEMANN, D., AND S. MORRIS (2005a): "Robust Implementation: The Role of Large Type Spaces," Discussion paper 1519, Cowles Foundation, Yale University.

——— (2005b): "Robust Mechanism Design," *Econometrica*, 73, 1771–1813.

——— (2007a): "An Ascending Auction for Interdependent Values: Uniqueness and Robustness to Strategic Uncertainty," *American Economic Review Papers and Proceedings*, 97, 125–130.

——— (2007b): "Robust Implementation in Direct Mechanisms," Discussion paper 1561R, Cowles Foundation, Yale University.

——— (2008a): "Ex Post Implementation," *Games and Economic Behavior*, 63, 527–566.

——— (2008b): "The Role of the Common Prior in Robust Implementation," *Journal of the European Economic Association Papers and Proceedings*, 6, 551–559.

——— (2008c): "Strategic Distinguishability and Robust Virtual Implementation," Discussion paper 1609R, Cowles Foundation, Yale University.

BERGEMANN, D., AND J. VÄLIMÄKI (2002): "Information Acquisition and Efficient Mechanism Design," *Econometrica*, 70, 1007–1033.

BERNHEIM, D. (1984): "Rationalizable Strategic Behavior," *Econometrica*, 52, 1007–1028.

BIKHCHANDANI, S. (2006): "Ex Post Implementation in Environments with Private Goods," *Theoretical Economics*, 1, 369–393.

BRANDENBURGER, A., AND E. DEKEL (1987): "Rationalizability and Correlated Equilibria," *Econometrica*, 55, 1391–1402.

CHUNG, K.-S., AND J. ELY (2001): "Efficient and Dominance Solvable Auctions with Interdependent Valuations," Discussion paper, Northwestern University.

DASGUPTA, P., AND E. MASKIN (2000): "Efficient Auctions," *Quarterly Journal of Economics*, 115, 341–388.

DEGROOT, M. (1970): *Optimal Statistical Decisions*. McGraw Hill, New York.

ESO, P., AND E. MASKIN (2002): "Multi-Good Efficient Auctions with Multidimensional Information," Discussion paper, Northwestern University and Institute for Advanced Studies.

FIESELER, K., T. KITTSTEINER, AND B. MOLDOVANU (2003): "Partnerships, Lemons and Efficient Trade," *Journal of Economic Theory*, 113, 223–234.

GABAY, D., AND H. MOULIN (1980): "On the Uniqueness and Stability of Nash's Equilibrium in Non-Cooperative Games," in *Applied Stochastic Control in Econometrics and Management Science*, ed. by A. Bensoussan, P. Kleindorfer and C.S. Tapiero, pp. 271–293. North-Holland, Amsterdam.

GLAZER, J., AND R. ROSENTHAL (1992): "A Note on Abreu-Matsushima Mechanisms," *Econometrica*, 60, 1435–1438.

GRESIK, T. (1991): "Ex-Ante Incentive Efficient Trading Mechanisms Without the Private Value Restriction," *Journal of Economic Theory*, 55, 41–63.

GRÜNER, H., AND A. KIEL (2004): "Collective Decisisions with Interdependent Valuations," *European Economic Review*, 48, 1147–1168.

HONG, H., AND M. SHUM (2003): "Econometric Models of Asymmetric Ascending Auctions," *Journal of Econometrics*, 112, 327–358.

JACKSON, M. (1991): "Bayesian Implementation," *Econometrica*, 59, 461–477.

JEHIEL, P., B. MOLDOVANU, M. MEYER-TER-VEHN, AND B. ZAME (2006): "The Limits of Ex Post Implementation," *Econometrica*, 74, 585–610.

LUENBERGER, D. G. (1978): "Complete Stability of Noncooperative Games," *Journal of Optimization Theory and Applications*, 25, 485–505.

MASKIN, E. (1992): "Auctions and Privatization," in *Privatization: Symposium in Honor of Herbert Giersch*, ed. by H. Siebert. Tuebingen: J. C. B. Mohr.

——— (1999): "Nash Equilibrium and Wellare Optimality," *Review of Economic Studies*, 66, 23–38.

MILGROM, P. (1981): "Rational Expectations, Information Acquistion and Competitive Bidding," *Econometrica*, 49, 921–943.

MILGROM, P., AND J. ROBERTS (1990): "Rationalizability, Learning and Equilibrium in Games with Strategic Complementarities," *Econometrica*, 58, 1255–1277.

MINC, H. (1988): *Nonnegative Matrices*. Wiley and Sons, New York.

PALFREY, T., AND S. SRIVASTAVA (1989a): "Implementation with Incomplete Information in Exchange Economies," *Econometrica*, 57, 115–134.

PALFREY, T., AND S. SRIVASTAVA (1989b): "Mechanism Design with Incomplete Information: A Solution to the Implementation Problem," *Journal of Political Economy*, 97, 668–691.

PEARCE, D. (1984): "Rationalizable Strategic Behavior and the Problem of Perfection," *Econometrica*, 52, 1029–1050.

PERRY, M., AND P. RENY (2002): "An Ex Post Efficient Auction," *Econometrica*, 70, 1199–1212.

PIKETTY, T. (1999): "The Information-Aggregation Approach to Political Institutions," *European Economic Review*, 43, 791–800.

POSTLEWAITE, A., AND D. SCHMEIDLER (1986): "Implementation in Differential Information Economies," *Journal of Economic Theory*, 39, 14–33.

RENY, P., AND M. PERRY (2006): "Toward a Strategic Foundation for Rational Expectations Equilibrium," *Econometrica*, 74, 1231–1269.

WEINSTEIN, J., AND M. YILDIZ (2007): "Impact of Higher-Order Uncertainty," *Games and Economic Behavior*, 60, 200–212.

WILSON, R. (1977): "A Bidding Model of Perfect Competition," *Review of Economic Studies*, 44, 511–518.

CHAPTER 4

Robust Implementation in General Mechanisms*

Dirk Bergemann and Stephen Morris

Abstract
A social choice function is *robustly implemented* if every equilibrium on every type space achieves outcomes consistent with it. We identify a *robust monotonicity* condition that is necessary and (with mild extra assumptions) sufficient for robust implementation.

Robust monotonicity is strictly stronger than both Maskin monotonicity (necessary and almost sufficient for complete information implementation) and ex post monotonicity (necessary and almost sufficient for ex post implementation). It is equivalent to Bayesian monotonicity on all type spaces.

Keywords: Mechanism design, implementation, robustness, common knowledge, interim equilibrium, dominant strategies

1 Introduction

The objective of mechanism design is to construct mechanisms (or game forms) such that privately informed agents have an incentive to reveal their information to a principal who seeks to realize a social choice function. The revelation principle establishes that if any mechanism can induce the agents

*This research is supported by NSF Grants #CNS-0428422 and #SES-0518929. We thank Matt Jackson and Andy Postlewaite for helpful discussions. This paper supersedes and incorporates results reported earlier in Bergemann and Morris (2005a).

to report their information, then the agents will also have an incentive to report truthfully in the direct mechanism. Given the beliefs of the agents, the truthtelling constraints then reduce in the direct mechanism to the Bayesian incentive compatibility conditions.

There are two important limitations of Bayesian incentive compatibility analysis. First, the analysis typically assumes a commonly known common prior over the agents' types. This assumption may be too stringent in practise. In the spirit of the "Wilson doctrine" (Wilson (1987)), we would like implementation results that are *robust* to different assumptions about what agents do or do not know about other agents' types. Second, the revelation principle only establishes that the direct mechanism has *an* equilibrium that achieves the social choice function. In general, there may be other equilibria that deliver undesirable outcomes. We would like to achieve *full* implementation, i.e., show the existence of a mechanism all of whose equilibria deliver the social choice function. We studied the first "robustness" problem in an earlier work, Bergemann and Morris (2005b). The second "full implementation" problem has been the subject of a large literature. In the incomplete information context, key full implementation references are Postlewaite and Schmeidler (1986), Palfrey and Srivastava (1989) and Jackson (1991). In this paper, we study "robust implementation" where we require robustness and full implementation simultaneously.

Interim implementation on all type spaces is possible if and only if it is possible to implement the social choice function using an iterative deletion procedure. We refer to the resulting notion as *rationalizable implementation*. We fix a mechanism and iteratively delete messages for each payoff type that are strictly dominated by another message for each payoff type profile and message profile that has survived the procedure. This observation about iterative deletion illustrates a general point well-known from the literature on epistemic foundations of game theory (e.g., Brandenburger and Dekel (1987), Battigalli and Siniscalchi (2003)): equilibrium solution concepts only have bite if we make strong assumptions about type spaces, i.e., we assume small type spaces where the common prior assumption holds.

We exploit this equivalence between robust and rationalizable implementation to obtain necessary and sufficient conditions for robust implementation in general environments. Our necessity argument is conceptually novel, exploiting the iterative characterization. The necessary conditions for robust implementation are ex post incentive compatibility of the social choice function and a condition — *robust monotonicity* — that is equivalent to requiring Bayesian monotonicity on every type space. Suppose that we fix a "deception" specifying, for each payoff type θ_i of each agent, a set of types that he might misreport himself to be. We require that for some agent i and a type misreport of agent i under the deception, for every misreport θ'_{-i} that the other agents might make under the deception, there exists an outcome y which is strictly preferred by agent i to the outcome he would receive under the social choice function for *every* possible payoff type profile that might misreport θ'_{-i}; where this outcome y satisfies the extra restriction that no payoff type of agent i prefers outcome y to the social choice function if the other agents were really types θ'_{-i}. This condition — while a little convoluted — is easier to interpret than the interim (Bayesian) monotonicity conditions. It is very strong and implies Maskin monotonicity — necessary and almost sufficient for complete information implementation — but is strictly weaker than dominant strategies.

The sufficiency argument requires only a modest strengthening of the necessary condition by guaranteeing that the preference profile of each agent satisfies a (conditional) no total indifference property. Under this no total indifference property, we show that the necessary conditions are also sufficient for robust implementation. The sufficient conditions guarantee robust implementation in pure, but more generally also in mixed strategies. Our robust analysis thus removes the frequent gap between pure and mixed strategy implementation in the literature.

In this paper, we follow the classic implementation literature in allowing for arbitrary mechanisms, including modulo and integer games. By allowing for these mechanisms, we are able to make tight connections with the existing implementation literature. Allowing for these badly behaved mechanisms does complicate our analysis: for example, we must allow for transfinite iterated deletion of best responses in our definition of rationalizable implementation. Given the complications arising from infinite mechanisms, we report new necessary conditions for robust implementation in the context of finite mechanisms. We also report how our earlier research can be used to show that these necessary conditions are sufficient conditions for finite mechanisms either in well-behaved, but restricted, environments (Bergemann and Morris, 2009a) or under a virtual rather than exact implementation requirement (Bergemann and Morris, 2009b).

Our results extend the classic literature on Bayesian implementation due to Postlewaite and Schmeidler (1986), Palfrey and Srivastava (1989) and Jackson (1991). We focus in this paper on an indirect approach to extending these results. We first note the equivalence between robust implementation and rationalizable implementation. We then exploit the equivalence to report a direct argument showing that robust monotonicity is a necessary and almost sufficient condition for rationalizable implementation. But in the light of the classic literature, we know that a necessary and almost sufficient condition for robust implementation must be Bayesian monotonicity on all type spaces. We confirm and clarify our results by directly checking that robust monotonicity is equivalent to Bayesian (or interim) monotonicity on all type spaces. Figure 1 gives a stylized account of the connection between these alternative approaches.

In the implementation literature, it is a standard practice to obtain the sufficiency results with augmented mechanisms. By augmenting the direct mechanism with additional messages, the designer may elicit additional information about undesirable equilibrium play by the agents. Yet, in many applied economic settings, single crossing or supermodular preference assumptions allow direct implementation. In a companion paper, Bergemann and Morris (2009a), we provide necessary and sufficient conditions for robust implementation in the direct mechanism. The main results of this paper apply to environments where each agent's type profile can be aggregated into a one dimensional sufficient statistic for each player, where preferences are single crossing with respect to that statistic. These restrictions incorporate many economic models with interdependence in the literature. We show that besides an incentive compatibility condition, in this case the strict ex post incentive compatibility condition, a *contraction* property which requires that there is *not too much* interdependence in agents' types, together present necessary and sufficient conditions for robust implementation in the direct mechanism.

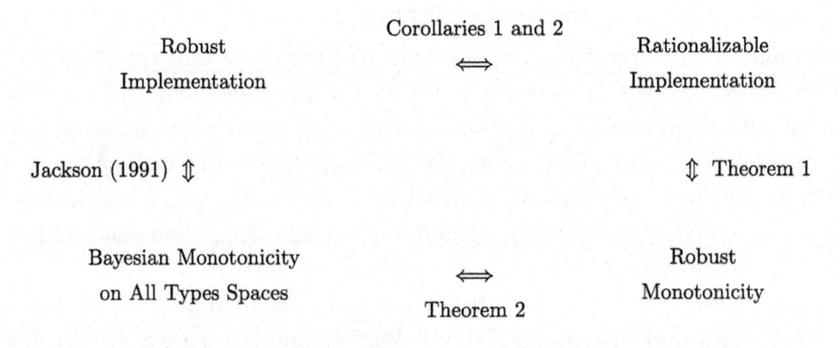

Fig. 1. Relationship between Bayesian and Robust Implementation/Monotonicity.

The robust monotonicity condition is stronger than both the Maskin and the Bayesian monotonicity conditions. In the context of robust implementation, it is then natural to ask whether a relaxation from the exact to the virtual implementation condition may lead to more permissive results. In Bergemann and Morris (2009b) we characterize the necessary and sufficient conditions for *robust virtual* implementation. There we show that a social choice function can be *robustly virtually implemented* if and only if the social choice function is ex post incentive compatible and robust measurable. In this contribution, we note that robust measurability remains a necessary condition for robust (exact) implementation, but it is not sufficient anymore.

The results in this paper concern full implementation. An earlier paper of ours, Bergemann and Morris (2005b), addresses the analogous questions of robustness to rich type spaces, but looking at the question of partial implementation, i.e., does there exist a mechanism such that *some* equilibrium implements the social choice function. We showed that ex post (partial) implementation of the social choice function is a necessary and sufficient condition for partial implementation on all type spaces.[1] This paper establishes that an analogous result does *not* hold for full implementation.

In a related paper, Bergemann and Morris (2008a), we therefore investigate the notion of ex post implementation. The necessary and sufficient conditions there straddle the implementation conditions for Nash and Bayesian-Nash respectively, as an ex post equilibrium is a Nash equilibrium at every incomplete information (Bayesian) type profile. However in contrast to the iterative argument pursued here, the basic reasoning in Bergemann and Morris (2008a) invokes more traditional equilibrium arguments. By comparing the conditions for ex post and robust implementation, it becomes apparent that robust implementation typically imposes additional constraints on the allocation problem. In Bergemann and Morris (2008a), we showed that in single crossing environments, the same single crossing conditions which guarantee incentive compatibility also guarantee full ex post implementation. In contrast, in the aggregation environment discussed above, we show that robust implementation imposes a strict bound on the interdependence of the preferences, which is not required by the truthtelling conditions. A contraction mapping behind the iterative argument directly points to the source of the restriction of the interaction term.

[1]This result does not extend to social choice correspondences.

The remainder of the paper is organized as follows. Section 2 describes the formal environment and solution concepts. Section 3 establishes necessary conditions for robust implementation in finite mechanisms. In addition, we present restrictions on the environment and weaker implementation notions under which the necessary conditions are also sufficient conditions. Section 4 establishes the relation between rationalizable and robust implementation in infinite mechanisms. Section 5 reports our main result on the necessary and sufficient conditions for robust implementation. Section 6 discusses extensions and variations of our implementation results, examining the role of lotteries and pure strategies and the relationship with Nash equilibrium and ex post equilibrium implementation. The appendix contains some additional examples.

2 Setup

2.1 *The Payoff Environment*

We consider a finite set of agents, $1, 2, \ldots, I$. Agent i's *payoff type* is $\theta_i \in \Theta_i$. We write $\theta \in \Theta = \Theta_1 \times \cdots \times \Theta_I$. There is a set of outcomes Z. We assume that each Θ_i and Z are countable.[2] Each individual has a von Neumann-Morgenstern utility function $u_i : Z \times \Theta \to \mathbb{R}$. Thus we are in the world of interdependent types, where an agent's utility depends on other agents' payoff types. We allow for lotteries over deterministic outcomes.[3] Let $Y \triangleq \Delta(Z)$ and extend u_i to the domain $Y \times \Theta$ in the usual way:

$$u_i(y, \theta) \triangleq \sum_{z \in Z} y(z) u_i(z, \theta).$$

A social choice function is a mapping $f : \Theta \to Y$. If the true payoff type profile is θ, the planner would like the outcome to be $f(\theta)$. In this paper, we restrict our analysis to the implementation of a social choice function rather than a social choice correspondence or social choice set.[4]

[2]The countable types restriction clarifies the relation to the existing literature. We postpone until Section 6.3 a discussion of what happens if we allow for uncountable payoff types, types and pure outcomes.

[3]The role of the lottery assumption and what happens when we drop it are discussed in Section 6.1.

[4]One reason why the extension to social choice correspondences is not straightforward is that, with social choice correspondences, the incentive compatibility conditions that arise from requiring partial implementation are typically weaker than ex post incentive compatibility, as shown by examples in Bergemann and Morris (2005b).

2.2 *Type Spaces*

We are interested in analyzing behavior in a variety of type spaces, many of them with a richer set of types than payoff types. For this purpose, we shall refer to agent i's *type* as $t_i \in T_i$, where T_i is a countable set. A type of agent i must include a description of his payoff type. Thus there is a function $\widehat{\theta}_i : T_i \to \Theta_i$ with $\widehat{\theta}_i(t_i)$ being agent i's payoff type when his type is t_i. A type of agent i must also include a description of his beliefs about the types of the other agents; thus there is a function $\widehat{\pi}_i : T_i \to \Delta(T_{-i})$ with $\widehat{\pi}_i(t_i)$ being agent i's *belief type* when his type is t_i. Thus $\widehat{\pi}_i(t_{-i})[t_i]$ is the probability that type t_i of agent i assigns to other agents having types t_{-i}. A *type space* is a collection:

$$\mathcal{T} = (T_i, \widehat{\theta}_i, \widehat{\pi}_i)_{i=1}^{I}.$$

2.3 *Mechanisms*

A planner must choose a *game form* or *mechanism* for the agents to play in order to determine the social outcome. Let M_i be the countably infinite set of messages available to agent i. We denote the generic message by $m_i \in M_i$ and let $m \in M = M_1 \times \cdots \times M_I$. Let $g(m)$ be the distribution over outcomes if action profile m is chosen. Thus a mechanism is a collection $\mathcal{M} = (M_1, \ldots, M_I, g(\cdot))$, where $g : M \to Y$.

2.4 *Solution Concepts*

Now holding fixed the payoff environment, we can combine a type space $\mathcal{T}$ with a mechanism $\mathcal{M}$ to get an incomplete information game $(\mathcal{T}, \mathcal{M})$. The payoff of agent i if message profile m is chosen and type profile t is realized is then given by

$$u_i(g(m), \widehat{\theta}(t)).$$

A pure strategy for agent i in the incomplete information game $(\mathcal{T}, \mathcal{M})$ is given by

$$s_i : T_i \to M_i.$$

A (behavioral) strategy is given by

$$\sigma_i : T_i \to \Delta(M_i).$$

The objective of this paper is to obtain implementation results for interim, or Bayesian Nash, equilibria on all possible types spaces.[5] The notion of interim equilibrium for a given type space $\mathcal{T}$ is defined in the usual way.

Definition 1 (Interim equilibrium). A strategy profile $\sigma = (\sigma_1, \ldots, \sigma_I)$ is an interim equilibrium of the game $(\mathcal{T}, \mathcal{M})$ if, for all i, t_i and m_i with $\sigma_i(m_i|t_i) > 0$,

$$\sum_{t_{-i}\in T_{-i}} \sum_{m_{-i}\in M_{-i}} \left(\prod_{j\neq i} \sigma_j(m_j|t_j)\right) u_i(g(m_i, m_{-i}), \widehat{\theta}(t))\widehat{\pi}_i(t_{-i})[t_i]$$

$$\geq \sum_{t_{-i}\in T_{-i}} \sum_{m_{-i}\in M_{-i}} \left(\prod_{j\neq i} \sigma_j(m_j|t_j)\right) u_i(g(m_i', m_{-i}), \widehat{\theta}(t))\widehat{\pi}_i(t_{-i})[t_i]$$

for all m_i'.

Requiring "robust" implementation, i.e., for "all type spaces", will push the solution concept in the direction of rationalizability. Consequently we define a message correspondence profile $S = (S_1, \ldots, S_I)$, where each

$$S_i : \Theta_i \to 2^{M_i}, \tag{1}$$

and we write $\mathcal{S}$ for the collection of message correspondence profiles. The collection $\mathcal{S}$ is a lattice with the natural ordering of set inclusion: $S \leq S'$ if $S_i(\theta_i) \subseteq S_i'(\theta_i)$ for all i and θ_i. The largest element is $\overline{S} = (\overline{S}_1, \ldots, \overline{S}_I)$, where $\overline{S}_i(\theta_i) = M_i$ for each i and θ_i. The smallest element is $\underline{S} = (\underline{S}_1, \ldots, \underline{S}_I)$, where $\underline{S}_i(\theta_i) = \emptyset$ for each i and θ_i.

We define an operator $b = (b_1, \ldots, b_I)$ to iteratively eliminate never best responses. To this end, we denote the belief of agent i over message and payoff type profiles of the remaining agents by

$$\lambda_i \in \Delta(M_{-i} \times \Theta_{-i}).$$

[5]We label these "interim" equilibria rather than "Bayesian" equilibria in light of the fact that our type space does not necessarily have a common prior.

The operator $b : \mathcal{S} \to \mathcal{S}$ is now defined as:

$$b_i(S)[\theta_i] = \left\{ m_i \in M_i \;\middle|\; \exists \lambda_i \text{ s.th.: } \begin{array}{ll} (1) & \lambda_i(m_{-i}, \theta_{-i}) > 0 \\ & \quad \Rightarrow m_j \in S_j(\theta_j),\ \forall\, j \neq i; \\ & \sum\limits_{m_{-i}, \theta_{-i}} \lambda_i(m_{-i}, \theta_{-i}) u_i(g(m_i, m_{-i}), \\ & \quad (\theta_i, \theta_{-i})) \geq \\ (2) & \sum\limits_{m_{-i}, \theta_{-i}} \lambda_i(m_{-i}, \theta_{-i}) u_i(g(m_i', m_{-i}), \\ & \quad (\theta_i, \theta_{-i})), \\ & \qquad \forall m_i' \in M_i; \end{array} \right\}. \tag{2}$$

We observe that b is increasing by definition: i.e., $S \leq S' \Rightarrow b(S) \leq b(S')$. By Tarski's fixed point theorem, there is a largest fixed point of b, which we label $S^{\mathcal{M}}$. Thus (i) $b(S^{\mathcal{M}}) = S^{\mathcal{M}}$ and (ii) $b(S) = S \Rightarrow S \leq S^{\mathcal{M}}$. We can also construct the fixed point $S^{\mathcal{M}}$ by starting with $\overline{S}$ — the largest element of the lattice — and iteratively applying the operator b. If the message sets and types are finite, we have

$$S_i^{\mathcal{M}}(\theta_i) \triangleq \bigcap_{n \geq 1} b_i(b^n(\overline{S}))[\theta_i].$$

But because the mechanism $\mathcal{M}$ may be infinite, transfinite induction may be necessary to reach the fixed point.[6] It is useful to define

$$S_i^{\mathcal{M},k}(\theta_i) \triangleq b_i(b^{k-1}(\overline{S}))[\theta_i],$$

again using transfinite induction if necessary. Thus $S_i^{\mathcal{M}}(\theta_i)$ are the set of messages surviving (transfinite) iterated deletion of never best responses. $S_i^{\mathcal{M}}(\theta_i)$ is the set of messages that type θ_i might send consistent with knowing that his payoff type is θ_i, common knowledge of rationality and the set of possible payoff types of the other players, but no restrictions on his beliefs and higher order beliefs about other types.

If message sets are finite (or compact), a well known duality argument implies that never best responses are equivalent to strictly dominated actions. However, the equivalence does not hold with infinite (non-compact)

[6]Lipman (1994) contains a formal description of the transfinite induction required (for the case of complete information, but nothing important changes with incomplete information). As he notes "we remove strategies which are never a best reply, taking limits where needed".

message sets.[7] In a compact message analysis, Chung and Ely (2001) consider a version of this solution concept in an incomplete information mechanism design context with dominated (not strictly dominated) messages deleted at each round. We observe that the solution concept defined through the iterative application of the operator b is weaker than the notion of interim rationalizability for a given type space T.[8] Under b, every agent i is allowed to hold arbitrary beliefs about Θ_{-i} and is not restricted to a particular posterior distribution over Θ_{-i}. On the other hand, if the type space $\mathcal{T}$ were the universal type space, then $S_i^{\mathcal{M}}(\theta_i)$ would be equal to the union of all interim rationalizable actions of agent i over all types $t_i \in T_i$ whose payoff type profile coincides with θ_i, or $\hat{\theta}_i(t_i) = \theta_i$. We refer to $S_i^{\mathcal{M}}(\theta_i)$ as the *rationalizable* messages of type θ_i of agent i in mechanism $\mathcal{M}$.

2.5 *Implementation*

We now define the notions of interim, robust and rationalizable implementation.

Definition 2 (Interim Implementation). Social choice function f is interim implemented on type space $\mathcal{T}$ by mechanism $\mathcal{M}$ if the game $(\mathcal{T}, \mathcal{M})$ has an equilibrium and every equilibrium σ of the game $(\mathcal{T}, \mathcal{M})$ satisfies

$$\sigma(m|t) > 0 \Rightarrow g(m) = f(\hat{\theta}(t)).$$

We note that a tradition in the implementation literature commonly restricts attention to pure strategy equilibria, but we allow mixed strategy equilibria.

Definition 3 (Robust Implementation). Social choice function f is robustly implemented by mechanism $\mathcal{M}$ if, for every $\mathcal{T}$, f is interim implemented on type space $\mathcal{T}$ by mechanism $\mathcal{M}$. Social choice function f is

---

[7]The following example, suggested to us by Andrew Postlewaite, illustrates the non-equivalence. Players 1 and 2 each choose a non-negative integer, k_1 and k_2 respectively. The payoff to player 1 from $k_1 = 0$ is 1. The payoff to player 1 from action $k_1 \geq 1$ is 2 if $k_1 > k_2$, 0 otherwise. For any belief that player 1 has about 2's actions, there is a (sufficiently high) action from player 1 that gives him a payoff greater than 1. Thus action 0 is never a best response for player 1. However, for any mixed strategy of player 1, there is a (sufficiently high) action of player 2 such that action 0 is a better response for player 1 than the mixed strategy. Thus action 0 is not strictly dominated.

[8]For the notion of interim rationalizability, see Battigalli and Siniscalchi (2003) and Dekel, Fudenberg, and Morris (2007).

robustly implementable if there exists a mechanism $\mathcal{M}$ such that f is robustly implemented by mechanism $\mathcal{M}$.

We observe that the notion of robust implementation requires that we can find a mechanism $\mathcal{M}$ which implements f for every type space $\mathcal{T}$. A weaker requirement would be to ask that for every type space $\mathcal{T}$ there exists a, possibly different, mechanism $\mathcal{M}$ such that f is implemented. This weaker notion would still lead to the same necessary condition as the stronger implementation version we pursue here, and we believe that it would not lead to a substantial change in the sufficiency conditions either.

The notion of robust implementation requires that a social choice function f can be interim implemented for all type spaces $\mathcal{T}$. As we look for necessary and sufficient conditions for robust implementation, conceptually there are (at least) two approaches to obtain the conditions.

One approach would be to simply look at the interim implementation conditions for every possible type space $\mathcal{T}$ and then try to characterize the intersection or union of these conditions for all type spaces. This is the approach we initially pursued, and it works in a brute force kind of way. In Section 6.1, we review what happens under this approach.

But we focus our analysis on a second, more elegant, approach. We first establish an equivalence between robust and "rationalizable implementation" and then derive the necessary conditions for robust implementation as an implication of rationalizable implementation. The advantage of the second approach is that after establishing the equivalence, we do not need to argue in terms of large type spaces, but rather derive the results from a novel argument using the iterative elimination process.

Definition 4 (Rationalizable Implementation). Social choice function f is implemented in rationalizable strategies by mechanism $\mathcal{M}$ if, for all θ, $S^{\mathcal{M}}(\theta) \neq \emptyset$ and if for all θ and m, $m \in S^{\mathcal{M}}(\theta) \Rightarrow g(m) = f(\theta)$.

We now report a formal epistemic argument that relates the rationalizable messages to the set of messages that might be played in any equilibrium on any type space.

Proposition 1 (Rationalizable Actions). *$m_i \in S_i^{\mathcal{M}}(\theta_i)$ if and only if there exists a type space $\mathcal{T}$, an interim equilibrium σ of the game $(\mathcal{T}, \mathcal{M})$ and a type $t_i \in T_i$ such that (i) $\sigma_i(m_i|t_i) > 0$ and (ii) $\widehat{\theta}_i(t_i) = \theta_i$.*

Proof. ($\Rightarrow$) Suppose $m_i^* \in S^{\mathcal{M}}(\theta_i^*)$. Now consider the following type space $\mathcal{T}$ defined through $T_i \triangleq \{(m_i, \theta_i)|m_i \in S_i^{\mathcal{M}}(\theta_i)\}$ and let $\widehat{\theta}_i(m_i, \theta_i) \triangleq \theta_i$.

By (2), we know that for each $m_i \in S_i^{\mathcal{M}}(\theta_i)$, there exists $\lambda_i^{m_i,\theta_i} \in \Delta(M_{-i} \times \Theta_{-i})$ such that:

$$\lambda_i^{m_i,\theta_i}(m_{-i},\theta_{-i}) > 0 \Rightarrow m_j \in S_j^{\mathcal{M}}(\theta_j) \quad \text{for each } j \neq i;$$

and

$$\sum_{m_{-i},\theta_{-i}} \lambda_i^{m_i,\theta_i}(m_{-i},\theta_{-i})[u_i(g(m_i,m_{-i}),(\theta_i,\theta_{-i})) - u_i(g(m_i',m_{-i}),(\theta_i,\theta_{-i}))] \geq 0, \quad \forall m_i' \in M_i.$$

Let $\bar{\pi}_i(m_{-i},\theta_{-i})[m_i,\theta_i] \triangleq \lambda_i^{m_i,\theta_{-i}}(m_{-i},\theta_{-i})$. Now by construction, there is a pure strategy equilibrium s with $s_i(m_i,\theta_i) = m_i$. But now $s_i(m_i^*,\theta_i^*) = m_i^*$ and $\widehat{\theta}(m_i^*,\theta_i^*) = \theta_i^*$.

($\Leftarrow$) Suppose there exists a type space $\mathcal{T}$, an equilibrium σ of $(\mathcal{T},\mathcal{M})$, and $m_i^* \in M_i$ and $t_i^* \in T_i$ such that $\sigma_i(m_i^*|t_i^*) > 0$ and $\widehat{\theta}_i(t_i^*) = \theta_i^*$. Let

$$S_i(\theta_i) = \{m_i : \sigma_i(m_i|t_i) > 0 \text{ and } \widehat{\theta}_i(t_i) = \theta_i \text{ for some } t_i \in T_i\}.$$

Now interim equilibrium conditions ensure that $b(S) \geq S$. Thus $S \leq S^{\mathcal{M}}$. Thus $m_i^* \in S_i^{\mathcal{M}}(\widehat{\theta}_i(t_i^*))$, which concludes the proof. □

Brandenburger and Dekel (1987) showed an equivalence for finite action complete information games between the set of actions surviving iterated deletion of strictly dominant actions and the set of actions that could be played in a subjective correlated equilibrium. Proposition 1 is a straightforward generalization of Brandenburger and Dekel (1987) to incomplete information and infinite actions. The infinite action extension (for complete information) was shown in Lipman (1994). The finite action incomplete information extension is reported in a paper of Battigalli and Siniscalchi (2003) (following an earlier analysis in Battigalli, 1999).

3 Finite Mechanisms

A complicating element in using the relationship between equilibrium strategies and rationalizable strategies in the implementation context is the fact that the augmented mechanisms often have infinite message spaces and that best responses may not exist. These complications are inherent to the entire implementation literature and we therefore have to carefully address these issues before we establish the implementation results. In this section we restrict attention to finite mechanisms, i.e., where each M_i is finite and we extend the argument to infinite mechanisms in the next section. All the results in this section will extend to compact or, more generally, "regular" mechanisms (e.g., mechanisms where best responses always exist as in Abreu and Matsushima, 1992b). With finite mechanisms, because $S_i^{\mathcal{M}}(\theta_i)$ is always non-empty, Proposition 1 immediately implies an equivalence between robust and rationalizable implementation.

Corollary 1 (Equivalence). *Social choice function f is robustly implemented by mechanism $\mathcal{M}$ if and only if it is rationalizably implemented by mechanism $\mathcal{M}$.*

We now establish necessary conditions for robust implementation which use the equivalence between robust and rationalizable implementation.

3.1 *Ex Post Incentive Compatibility*

The following ex post incentive compatibility condition is a necessary condition for robust truthful (or partial) implementation as established in Bergemann and Morris (2005b).

Definition 5 (EPIC). Social choice function f satisfies ex post incentive compatibility (EPIC) if

$$u_i(f(\theta_i, \theta_{-i}), (\theta_i, \theta_{-i})) \geq u_i(f(\theta_i', \theta_{-i}), (\theta_i, \theta_{-i})),$$

for all i, θ_i, θ_i' and θ_{-i}.

In the context of robust (full) implementation, we require a strict version of the ex post incentive compatibility conditions.

Definition 6 (Semi-Strict EPIC). Social choice function f satisfies semi-strict ex post incentive compatibility (semi-strict EPIC) if, for each i, θ_i, θ_i'

$$u_i(f(\theta_i, \theta_{-i}), (\theta_i, \theta_{-i})) > u_i(f(\theta_i', \theta_{-i}), (\theta_i, \theta_{-i})),$$

if there exists $\theta_{-i}' \in \Theta_{-i}$ such that $f(\theta_i, \theta_{-i}') \neq f(\theta_i', \theta_{-i}')$.

The necessity of the semi-strict version of ex post incentive compatibility now follows directly from the conditions imposed by rationalizable implementation.

Proposition 2 (Necessity of Semi-Strict EPIC). *If social choice function f is robustly implementable by a finite mechanism, then f satisfies semi-strict EPIC.*

Proof. If mechanism $\mathcal{M}$ robustly implements f, then, for each i, there exists $m_i^* : \Theta_i \to M_i$ such that $g(m^*(\theta)) = f(\theta)$ and $m^*(\theta) \in S^{\mathcal{M}}(\theta)$; we can simply let $m_i^*(\theta_i)$ be any element of $S_i^{\mathcal{M}}(\theta_i)$.

Suppose semi-strict EPIC fails. Then there exists i, θ and θ' such that

$$f(\theta') \neq f(\theta_i, \theta'_{-i}) \tag{3}$$

and

$$u_i(f(\theta'_i, \theta_{-i}), \theta) \geq u_i(f(\theta), \theta). \tag{4}$$

Now (4) implies that:

$$\begin{aligned} u_i(g(m_i^*(\theta'_i), m_{-i}^*(\theta_{-i})), (\theta_i, \theta_{-i})) &= u_i(f(\theta'_i, \theta_{-i}), \theta) \\ &\geq u_i(f(\theta), \theta) \\ &= u_i(g(m_i^*(\theta_i), m_{-i}^*(\theta_{-i})), (\theta_i, \theta_{-i})). \end{aligned}$$

Since $m_i^*(\theta_i) \in S_i^{\mathcal{M}}(\theta_i)$, this implies $m_i^*(\theta'_i) \in S_i^{\mathcal{M}}(\theta_i)$. But now

$$f(\theta_i, \theta'_{-i}) = g(m_i^*(\theta'_i), m_{-i}^*(\theta'_{-i})) = f(\theta'),$$

contradicting (3). □

Next we present two related, yet distinct, monotonicity conditions which are at the core of the robust implementation results.

3.2 *Robust Monotonicity*

To understand the robust monotonicity condition, it is useful to first think about agents playing the direct mechanism. In the direct mechanism, an agent i may or may not report truthfully. A *deception* is a set-valued profile $\beta = (\beta_1, \ldots, \beta_I)$, where

$$\beta_i : \Theta_i \to 2^{\Theta_i}/\emptyset,$$

with $\theta_i \in \beta_i(\theta_i)$ for all i and all θ_i. A deception of agent i with payoff type θ_i is a set of possible reports by agent i. By definition, a deception of payoff type θ_i includes, but is not restricted to, θ_i itself.

Definition 7 (Acceptable/Unacceptable Deception). A deception is acceptable if $\theta' \in \beta(\theta) \Rightarrow f(\theta') = f(\theta)$. A deception is unacceptable if it is not acceptable.

In this language, the "truthtelling" deception, defined by $\beta_i^*(\theta_i) \triangleq \theta_i$ for all θ_i is an acceptable deception. Other deceptions of agent i may also be acceptable if the social choice function does not vary with respect to some subset of reports of agent i for all type profiles of the other agents. The inverse mapping of a deception β_i represents the set of true type profiles θ_i which could lead to a report θ_i' and we write

$$\beta_i^{-1}(\theta_i') \triangleq \{\theta_i | \theta_i' \in \beta_i(\theta_i)\}.$$

A "robust monotonicity" condition is key to our main result. In the direct mechanism, where agents other than i report themselves to be types θ_{-i}, agent i can obtain outcomes $f(\theta_i', \theta_{-i})$ for any θ_i'. But once we allow augmented mechanisms, we could conceivably offer agent i a larger set of lotteries if he reports deviant behavior of his opponents. We need to identify, for any given report θ_{-i}, the set of lotteries with the property that whatever agent i's actual type, he would never prefer such an allocation to what he would obtain under the social choice function if other agents were reporting truthfully. Thus:

$$Y_i(\theta_{-i}) \triangleq \{y \in Y | u_i(y, (\theta_i', \theta_{-i})) \leq u_i(f(\theta_i', \theta_{-i}), (\theta_i', \theta_{-i})) \text{ for all } \theta_i' \in \Theta_i\}. \tag{5}$$

Henceforth, we refer to the set $Y_i(\theta_{-i})$ as the *reward set* (for agent i).

Suppose now that it was common knowledge that in the direct mechanism, type θ_i of agent i will send a report $\theta_i' \in \beta_i(\theta_i)$. If β is acceptable, we would know that f was being implemented. But if β is unacceptable, we must find a type of some agent who is prepared to report that other agents are misreporting. But for the "whistle-blower" who is going to report that we are in a bad equilibrium, we cannot know what he believes about the types of the other agents, nor can we know what message he expects to hear except that it is a message consistent with the deception. We thus have to allow for all possible beliefs ψ_i of agent i over payoff types $\theta_{-i} \in \Theta_{-i}$ consistent with a report θ_{-i}' from a given deception profile β, or

$$\psi_i \in \Delta\big(\beta_{-i}^{-1}(\theta_{-i}')\big).$$

Finally, the reward that he is offered must not mess up the truth-telling behavior in the good equilibrium. This gives the following condition:

Definition 8 (Dual Robust Monotonicity). Social choice function f satisfies dual robust monotonicity if, for every unacceptable deception β, there exist i, θ_i, $\theta_i' \in \beta_i(\theta_i)$ such that, for all $\theta_{-i}' \in \Theta_{-i}$ and

$\psi_i \in \Delta(\beta_{-i}^{-1}(\theta'_{-i}))$, there exists $y \in Y$ such that:

$$\sum_{\theta_{-i} \in \beta_{-i}^{-1}(\theta'_{-i})} \psi_i(\theta_{-i}) u_i(y, (\theta_i, \theta_{-i})) > \sum_{\theta_{-i} \in \beta_{-i}^{-1}(\theta'_{-i})} \psi_i(\theta_{-i}) u_i(f(\theta'_i, \theta'_{-i}), (\theta_i, \theta_{-i})) \quad (6)$$

and for all $\theta''_i \in \Theta_i$:

$$u_i(f(\theta''_i, \theta'_{-i}), (\theta''_i, \theta'_{-i})) \geq u_i(y, (\theta''_i, \theta'_{-i})). \quad (7)$$

Social choice function f satisfies dual strict robust monotonicity if for all θ''_i with $f(\theta''_i, \theta'_{-i}) \neq y$ the inequality (7) is strict.

We call this the "dual" version of robust monotonicity because, in the special case where the pure outcome space Z and payoff type spaces Θ_i are finite, dual robust monotonicity can be given a simpler expression.

Definition 9 (Robust Monotonicity). Social choice function f satisfies robust monotonicity if, for every unacceptable deception β, there exist i, θ_i, $\theta'_i \in \beta_i(\theta_i)$ such that, for all $\theta'_{-i} \in \Theta_{-i}$ there exists y such that for all $\theta_{-i} \in \beta_{-i}^{-1}(\theta'_{-i})$:

$$u_i(y, (\theta_i, \theta_{-i})) > u_i(f(\theta'_i, \theta'_{-i}), (\theta_i, \theta_{-i}))$$

and for all $\theta''_i \in \Theta_i$:

$$u_i(f(\theta''_i, \theta'_{-i}), (\theta''_i, \theta'_{-i})) \geq u_i(y, (\theta''_i, \theta'_{-i})). \quad (8)$$

Social choice function f satisfies strict robust monotonicity if for all θ''_i with $f(\theta''_i, \theta'_{-i}) \neq y$ the inequality (8) is strict.

The equivalence of robust monotonicity and dual robust monotonicity when the pure outcome space Z and payoff type spaces Θ_i are finite is established in Lemma 2 in the appendix.

Proposition 3 (Necessity of Dual Strict Robust Monotonicity). *If social choice function f is robustly implementable by a finite mechanism, then f satisfies dual strict robust monotonicity.*

Proof. Fix an unacceptable deception β. Let $\widehat{k}$ be the largest k satisfying $S_i^{\mathcal{M}}(\theta'_i) \subseteq S_i^{\mathcal{M},k}(\theta_i)$ or every i, θ_i and $\theta'_i \in \beta_i(\theta_i)$; we know that such a largest k exists since $S_i^{\mathcal{M}}(\theta'_i) \subseteq S_i^{\mathcal{M},0}(\theta_i) = M_i$ and, since $\mathcal{M}$ implements f, we have $S_i^{\mathcal{M}}(\theta_i) \cap S_i^{\mathcal{M}}(\theta'_i) = \emptyset$ and thus $S_i^{\mathcal{M}}(\theta'_i) \subsetneq S_i^{\mathcal{M},k}(\theta_i)$ for sufficiently large k. Thus there exists i and $\theta_i' \in \beta_i(\theta_i)$ such that $S_i^{\mathcal{M},\hat{k}}(\theta_i) \cap S_i^{\mathcal{M}}(\theta_i') = S_i^{\mathcal{M}}(\theta_i')$ and $S_i^{\mathcal{M},\hat{k}+1}(\theta_i) \cap S_i^{\mathcal{M}}(\theta_i') \neq S_i^{\mathcal{M}}(\theta_i')$. So there exists a message $\widehat{m}_i$ satisfying $\widehat{m}_i \in S_i^{\mathcal{M},\hat{k}}(\theta_i) \cap S_i^{\mathcal{M}}(\theta_i')$ and $\widehat{m}_i \notin S_i^{\mathcal{M},\hat{k}+1}(\theta_i) \cap S_i^{\mathcal{M}}(\theta_i')$. Since message $\widehat{m}_i$ gets deleted for θ_i at round $\widehat{k}+1$, we know that for every $\lambda_i \in \Delta(M_{-i} \times \Theta_{-i})$ such that

$$\lambda_i(m_{-i}, \theta_{-i}) > 0 \Rightarrow m_j \in S_i^{\mathcal{M},\hat{k}}(\theta_j) \quad \text{for all } j \neq i,$$

there exists m_i^* such that

$$\sum_{m_{-i},\theta_{-i}} \lambda_i(m_{-i}, \theta_{-i}) u_i(g(m_i^*, m_{-i}), (\theta_i, \theta_{-i}))$$
$$> \sum_{m_{-i},\theta_{-i}} \lambda_i(m_{-i}, \theta_{-i}) u_i(g(\widehat{m}_i, m_{-i}), (\theta_i, \theta_{-i})).$$

Let $\widehat{m}_j \in S_j^{\mathcal{M}}(\theta_j')$ for all $j \neq i$. Now the above claim remains true if we restrict attention to distributions λ_i putting probability 1 on $\widehat{m}_{-i}$. Thus for every $\psi_i \in \Delta(\Theta_{-i})$ such that

$$\psi_i(\theta_{-i}) > 0 \Rightarrow \widehat{m} \in S_j^{\mathcal{M},\hat{k}}(\theta_j) \text{ for all } j \neq i,$$

there exists m_i^* such that

$$\sum_{\theta_{-i}} \psi_i(\theta_{-i}) u_i(g(m_i^*, \widehat{m}_{-i}), (\theta_i, \theta_{-i})) > \sum_{\theta_{-i}} \psi_i(\theta_{-i}) u_i(g(\widehat{m}_i, \widehat{m}_{-i}), (\theta_i, \theta_{-i})).$$

But $\widehat{m} \in S^{\mathcal{M}}(\theta')$, so (since $\mathcal{M}$ robustly implements f), $g(\widehat{m}_i, \widehat{m}_{-i}) = f(\theta')$. Also observe that if $\theta'_{-i} \in \beta_{-i}(\theta_{-i})$, then $\widehat{m}_{-i} \in S_{-i}^{\mathcal{M},\hat{k}}(\theta_{-i})$. Thus for every $\psi_i \in \Delta(\beta_{-i}^{-1}(\theta'_{-i}))$, there exists m_i^* such that

$$\sum_{\theta_{-i}} \psi_i(\theta_{-i}) u_i(g(m_i^*, \widehat{m}_{-i}), (\theta_i, \theta_{-i})) > \sum_{\theta_{-i}} \psi_i(\theta_{-i}) u_i(f(\theta'), (\theta_i, \theta_{-i})),$$

which establishes the reward inequality, (6), of dual strict robust monotonicity.

Now suppose the incentive inequalities, (7), are not satisfied strictly, and hence

$$u_i(g(m_i^*, \widehat{m}_{-i}), (\widetilde{\theta}_i, \theta'_{-i})) \geq u_i(f(\widetilde{\theta}_i, \theta'_{-i}), (\widetilde{\theta}_i, \theta'_{-i}))$$

and $g(m_i^*, \widehat{m}_{-i}) \neq f(\widetilde{\theta}_i, \theta'_{-i})$. Now, for any

$$m_i \in \arg\max_{m_i'} u_i(g(m_i', \widehat{m}_{-i}), (\widetilde{\theta}_i, \theta'_{-i})), \tag{9}$$

since $\widehat{m}_{-i} \in S_i^{\mathcal{M}}(\theta'_{-i})$, we must have $m_i \in S_i^{\mathcal{M}}(\widetilde{\theta}_i)$ and thus $g(m_i, \widehat{m}_{-i}) = f(\theta_i, \theta'_{-i})$. Thus from (9) we also know that m_i^* achieves the maximum:

$$m_i^* \in \arg\max_{m_i'} u_i(g(m_i', \widehat{m}_{-i}), (\widetilde{\theta}_i, \theta'_{-i})),$$

and, for all $\widetilde{\theta}_i$, if

$$u_i(g(m_i^*, \widehat{m}_{-i}), (\widetilde{\theta}_i, \theta'_{-i})) \geq u_i(f(\widetilde{\theta}_i, \theta'_{-i}), (\widetilde{\theta}_i, \theta'_{-i})),$$

then $g(m_i^*, \widehat{m}_{-i}) = f(\widetilde{\theta}_i, \theta'_{-i})$.

Now setting $y \triangleq g(m_i^*, \widehat{m}_{-i})$, we have established that for each $\theta'_{-i} \in \beta_{-i}(\theta_{-i})$ and $\psi_i \in \Delta(\beta_{-i}^{-1}(\theta'_{-i}))$, there exists y such that $y \in Y_i(\theta'_{-i})$ and

$$\sum_{\theta_{-i}} \psi_i(\theta_{-i}) u_i(g(m_i^*, \widehat{m}_{-i}), (\theta_i, \theta_{-i})) > \sum_{\theta_{-i}} \psi_i(\theta_{-i}) u_i(f(\theta'), (\theta_i, \theta_{-i})),$$

which concludes the proof. □

3.3 *Robust Measurability*

We now report an additional necessary condition (from Bergemann and Morris (2009b)) for robust implementation in finite mechanisms. We will be interested in the set of preferences that an agent might have if his payoff type is θ_i and he knows that the type θ_j of each opponent j belongs to some subset Ψ_i of his payoff types Θ_j. Write $\mathcal{R}$ for the set of expected utility preference relations on lotteries Y. We will write $R_{\theta_i,\psi_i} \in \mathcal{R}$ for the preference relation of agent i if his payoff type is θ_i and he has belief $\psi_i \in \Delta(\Theta_{-i})$ about the types of others:

$$\begin{aligned} \forall\, y, y' \in Y:\quad y R_{\theta_i,\psi_i} y' \Leftrightarrow & \sum_{\theta_{-i}\in\Theta_{-i}} \psi_i(\theta_{-i}) u_i(y, (\theta_i, \theta_{-i})) \\ & \geq \sum_{\theta_{-i}\in\Theta_{-i}} \psi_i(\theta_{-i}) u_i(y', (\theta_i, \theta_{-i})); \end{aligned}$$

We write $\mathcal{R}_i(\theta_i, \Psi_{-i})$ for the set of preferences agent i might have if his payoff type is θ_i and he might have any beliefs over others' payoff types.

$$\mathcal{R}_i(\theta_i, \Psi_{-i}) = \{R \in \mathcal{R} | R = R_{\theta_i,\psi_i} \text{ for some } \psi_i \in \Delta(\Psi_{-i})\}.$$

Say that type set profile Ψ_{-i} *separates* Ψ_i if

$$\bigcap_{\theta_i \in \Psi_i} \mathcal{R}_i(\theta_i, \Psi_{-i}) = \emptyset.$$

Let $\Xi = (\Xi)_{i=1}^{I} \in \times_{i=1}^{I} 2^{\Theta_i}$ be a profile of type sets for each agent. Say that Ξ is *mutually inseparable* if, for each i and $\Psi_i \in \Xi_i$, there exists $\Psi_{-i} \in \Xi_{-i}$ such that Ψ_{-i} does not separate Ψ_i.

Definition 10 (Robust Measurability). Social choice function f satisfies robust measurability if Ξ mutually inseparable, $\Psi_i \in \Xi_i$ and $\{\theta_i', \theta_i''\} \subseteq \Psi_i \Rightarrow f(\theta_i', \theta_{-i}) = f(\theta_i'', \theta_{-i})$ for all θ_{-i}.

If payoff types are finite, one can give an alternative iterative definition of robust measurability: let $\Xi_i^0 = 2^{\Theta_i}$,

$$\Xi_i^{k+1} = \{\Psi_i \in \Xi_i^k | \Psi_{-i} \text{ does not separate } \Psi_i, \text{ for some } \Psi_{-i} \in \Xi_{-i}^k\}, \quad (10)$$

and

$$\Xi_i^* = \bigcap_{k \geq 0} \Xi_i^k. \quad (11)$$

Now we say that a social choice function f satisfies robust measurability if $\{\theta_i', \theta_i''\} \in \Xi_i^* \Rightarrow f(\theta_i', \theta_{-i}) = f(\theta_i'', \theta_{-i})$ for all θ_{-i}.[9]

Proposition 4 (Necessity of Robust Measurability). *If social choice function f is robustly implementable by a finite mechanism, then f satisfies robust measurability.*

Bergemann and Morris (2009b) showed that robust measurability was necessary for robust *virtual* implementation (as reported in Proposition 6 below); thus it must also be necessary of robust exact implementation. For completeness, we report a direct argument here.

Proof. Since f is robustly implementable, there exists a mechanism $\mathcal{M}$ such that

$$m \in S^{\mathcal{M}}(\theta) \Rightarrow g(m) = f(\theta).$$

Now suppose Ξ is mutually inseparable. We argue by induction that, for all i, $\Psi_i \in \Xi_i$ and k there exists a set of messages $\emptyset \neq M_i^k(\Psi_i) \subseteq S_i^{\mathcal{M},k}(\theta_i)$ for all $\theta_i \in \Psi_i$. This is true by definition for $k = 0$. Suppose that it is true

[9]See Lemma 3 of Bergemann and Morris (2009b).

for k. Now Ξ mutually inseparable implies that for any $\Psi_i \in \Xi_i$, there exists $\Psi_{-i} \in \Xi_{-i}$, R and, for each $\theta_i \in \Psi_i$, $\lambda_i^{\theta_i} \in \Delta(\Psi_{-i})$ such that $R_{\theta_i, \lambda_i^{\theta_i}} = R$. Now let $M_i^{k+1}(\Psi_i)$ be the optimal messages of agent i when he believes that his opponents will sent some message profile in $M_{-i}^k(\Psi_{-i})$ with probability 1 and has beliefs $\lambda_i^{\theta_i}$ about the type profile of his opponents, i.e.,

$$M_i^{k+1}(\Psi_i) = \bigcup_{m_{-i} \in M_{-i}^k(\Psi_{-i})} \arg\max_{m_i'} \sum_{\theta_{-i}} \lambda_i^{\theta_i}(\theta_{-i}) u_i(g(m_i', m_{-i}), (\theta_i, \theta_{-i})).$$

By construction, $\emptyset \neq M_i^{k+1}(\Psi_i) \subseteq S_i^{\mathcal{M},k}(\theta_i)$ for all $\theta_i \in \Psi_i$. Now for each $\Psi_i \in \Xi_i$, $M_i^k(\Psi_i)$ is a decreasing sequence under set inclusion. Since M_i is finite, there exists $\emptyset \neq M_i^*(\Psi_i) = \cap_{k \geq 0} M_i^k(\Psi_i)$. Thus $M_i^*(\Psi_i) \subseteq S_i^{\mathcal{M}}(\theta_i)$ for all $\theta_i \in \Psi_i$. Now if $\{\theta_i', \theta_i''\} \subseteq \Psi_i$, there exists $m_i \in M_i^*(\Psi_i) \subseteq S_i^{\mathcal{M}}(\theta_i')$ and $m_i \in M_i^*(\Psi_i) \subseteq S_i^{\mathcal{M}}(\theta_i'')$. Now fix any $m_{-i} \in S_{-i}^{\mathcal{M}}(\theta_{-i})$, and we have $(m_i, m_{-i}) \in S^{\mathcal{M}}(\theta_i', \theta_{-i}) \Rightarrow g(m_i, m_{-i}) = f(\theta_i', \theta_{-i})$ and $(m_i, m_{-i}) \in S^{\mathcal{M}}(\theta_i'', \theta_{-i}) \Rightarrow g(m_i, m_{-i}) = f(\theta_i'', \theta_{-1})$. Thus $f(\theta_i', \theta_{-i}) = f(\theta_i'', \theta_{-i})$. □

In the appendix of the working paper version, Bergemann and Morris (2008b), we show by means of two examples that robust monotonicity does not imply nor is it implied by robust measurability.

We have pursued two ways of deriving sufficient conditions in prior work. First, we identified natural restrictions on the environment that make these necessary conditions sufficient (Bergemann and Morris, 2009a). Second, we showed what happened if we weaken the implementation requirement to virtual implementation (Bergemann and Morris, 2009b). We briefly review these results below. If we neither put restrictions on the environment nor allow virtual implementation, then we do not know how to derive tight sufficient conditions for finite, or other well-behaved, mechanisms. However, as in the existing complete information and standard Bayesian implementation literature, it is possible to obtain tight conditions if we allow for badly behaved mechanisms. These results are reported in the remainder of the paper. We believe they improve our understanding about how the different elements in the incomplete information implementation literature fit together and highlight the role of infinite mechanisms.

3.4 *Single Crossing Aggregator Environments*

In Bergemann and Morris (2009a), we consider payoff environments in which each payoff type space Θ_i is completely ordered and where there exist for each i, an aggregator function $h_i : \Theta \to \mathbb{R}$ and a valuation function

$v_i : Y \times \mathbb{R} \to \mathbb{R}$ such that

$$v_i(y, h_i(\theta)) \triangleq u_i(y, \theta), \tag{12}$$

where h_i is continuous and strictly increasing in θ_i and $v_i : Y \times \mathbb{R} \to \mathbb{R}$ is continuous and satisfies the following strict single crossing property: for all $\phi < \phi' < \phi''$,

$$v_i(y, \phi) > v_i(y', \phi) \quad \text{and} \quad v_i(y, \phi') = v_i(y', \phi') \Rightarrow v_i(y, \phi'') < v_i(y', \phi''). \tag{13}$$

The aggregator functions $h = (h_i)_{i=1}^{I}$ are said to satisfy the *contraction property* if, for all deceptions $\beta \in \beta^*$, there exists i, θ_i and $\theta_i' \in \beta_i(\theta_i)$ with $\theta_i' \neq \theta_i$, such that

$$\operatorname{sign}(\theta_i - \theta_i') = \operatorname{sign}(h_i(\theta_i, \theta_{-i}) - h_i(\theta_i', \theta_{-i}')) \tag{14}$$

for all θ_{-i} and $\theta_{-i}' \in \beta_{-i}(\theta_{-i})$. In single crossing aggregator environments as described by (12) and (13), the contraction property is equivalent to both dual strict robust monotonicity and robust measurability.

We say that a social choice function f is *responsive* if for all $\theta_i \neq \theta_i'$, there exists θ_{-i} such that $f(\theta_i, \theta_{-i}) \neq f(\theta_i', \theta_{-i})$. If a social choice function is responsive, then semi-strict EPIC simplifies to *strict ex post incentive compatibility*, i.e., $u_i(f(\theta_i, \theta_{-i}), (\theta_i, \theta_{-i})) > u_i(f(\theta_i', \theta_{-i}), (\theta_i, \theta_{-i}))$, for all i, $\theta_i \neq \theta_i'$ and θ_{-i}.

Proposition 5 (Contraction Property). *In a single crossing aggregator environment, a responsive social choice function f is robustly implementable if and only if it satisfies strict ex post incentive compatibility and the contraction property.*

This result is reported in Theorem 1 and 2 of Bergemann and Morris (2009a). It follows that the necessary conditions of Propositions 2, 3 and 4 are also sufficient in these environments. Note that in the discrete type setting of this paper, the continuity properties are automatically satisfied if the payoff type spaces are finite. Bergemann and Morris (2009a) allowed for compact payoff type spaces and pure outcome spaces; they also showed that when robust implementation is possible, it is possible in a "direct" mechanism where agents report just their payoff types.

3.5 *Robust Virtual Implementation*

We can also derive sufficient conditions for robust implementation by relaxing the requirement from (robust) exact to (robust) virtual implementation. In Bergemann and Morris (2009b), we consider settings where the space of pure outcomes and payoff types are finite. By Corollary 1 we can therefore define robust virtual implementation directly with reference to the rationalizable messages in a given mechanism $\mathcal{M}$.

Definition 11 (Robust Virtual Implementation). Social choice function f is robustly virtually implementable if, for each $\varepsilon > 0$, there exists a mechanism $\mathcal{M}$ such that for all θ, $S^{\mathcal{M}}(\theta) \neq \emptyset$ and if for all θ and m, $m \in S^{\mathcal{M}}(\theta) \Rightarrow \|g(m) - f(\theta)\| \leq \varepsilon$.

We established in Theorem 1 and 2 of Bergemann and Morris (2009b) the following necessary and sufficient conditions for robust virtual implementation.

Proposition 6 (Robust Measurability). *A social choice function is robustly virtually implementable if and only if it satisfies ex post incentive compatibility and robust measurability.*

Thus strict robust monotonicity can be dropped and semi-strict EPIC can be weakened to EPIC if virtual implementation is enough. With these weakenings, the necessary conditions are sufficient.

3.6 *A Coordination Example*

We conclude this section with an example that demonstrates that while robust implementation is a strong requirement, it is weaker than dominant strategies. In the example there are two agents, $i = 1, 2$. Each agent i has two possible types, θ_i and θ_i'. There are six possible outcomes: $Z = \{a, b, c, d, z, z'\}$. The payoffs of the agents are a function of the allocation and the true payoff type profile, given by:

a	θ_2	θ_2'
θ_1	3, 3	0, 0
θ_1'	0, 0	1, 1

b	θ_2	θ_2'
θ_1	0, 0	3, 3
θ_1'	1, 1	0, 0

c	θ_2	θ_2'
θ_1	0, 0	1, 1
θ_1'	3, 3	0, 0

d	θ_2	θ_2'
θ_1	1, 1	0, 0
θ_1'	0, 0	3, 3

(15)

and

$\mathbf{z}$	θ_2	θ_2'
θ_1	$2,2$	$2,0$
θ_1'	$2,2$	$2,0$

$\mathbf{z}'$	θ_2	θ_2'
θ_1	$2,0$	$2,2$
θ_1'	$2,0$	$2,2$

.

The social choice function is given by the efficient outcome at each type profile:

f	θ_2	θ_2'
θ_1	a	b
θ_1'	c	d

. (16)

Clearly, the social choice function is strictly ex post incentive compatible. But in the "direct mechanism" where each agent simply reports his type, there will always be an equilibrium where each type of each agent misreports his type, and each agent gets a payoff of 1. This is also strictly ex post incentive compatible. The social choice function f which selects among $\{a, b, c, d\}$ embeds a coordination game. We further observe that the payoff for agent 1 from allocations z and z' are equal and constant for all type profiles. On the other hand, the payoff of agent 2 from z and z' depends on his type but not on the type of the other agent.

We now consider the following augmented, but finite, mechanism which responds to the messages of the agents as follows:

g	θ_2	θ_2'
θ_1	a	b
θ_1'	c	d
ζ	z	z'

The augmented mechanism enriches the message space of agent 1 by a single message ζ. The corresponding incomplete information game has the following payoffs:

	type	θ_2		θ_2'	
type	action	θ_2	θ_2'	θ_2	θ_2'
θ_1	θ_1	$3,3$	$0,0$	$0,0$	$3,3$
	θ_1'	$0,0$	$1,1$	$1,1$	$0,0$
	ζ	$2,2$	$2,0$	$2,0$	$2,2$
θ_1'	θ_1	$0,0$	$1,1$	$1,1$	$0,0$
	θ_1'	$3,3$	$0,0$	$0,0$	$3,3$
	ζ	$2,2$	$2,0$	$2,0$	$2,2$

Suppose we iteratively remove actions for each type that could never be a best response given the type action profiles remaining. Thus in the first round, we would observe that type θ_1 would never send message θ_1' and type θ_1' would never send message θ_1. Knowing this, we could conclude that type θ_2 would never send message θ_2' and type θ_2' would never send message θ_2. This in turn implies that neither type of agent 1 will ever send message ζ. Thus the only remaining message for each type of each agent is truth-telling. But now they must behave this way in any equilibrium on any type space.

4 Rationalizable and Robust Implementation in Infinite Mechanisms

In Section 3 we established the equivalence between rationalizable and robust implementation for finite mechanisms. A complicating element in the context of implementation is the fact that the augmented mechanisms often have infinite message spaces and that best responses may not exist. We now address these issues for infinite mechanisms and then establish the implementation results for general mechanisms.

4.1 *Best Response*

We observe that with infinite mechanism there is no a priori guarantee that $S^{\mathcal{M}}(\theta)$ is non-empty or that a game of incomplete information defined by $(T, \mathcal{M})$ has an interim equilibrium. The epistemic result of Proposition 1 which related the rationalizable messages with the equilibrium messages for some type space continues to hold vacuously in these cases. But for implementation results, we care about existence. We introduce the following two conditions that relate existence of equilibrium on all type spaces to the actions surviving iterated deletion. These conditions use the notion of message correspondence S defined in Section 2.4.

Definition 12 (Ex Post Best Response). Message correspondence S satisfies the ex post best response property if, for all i and $\theta_i \in \Theta_i$, there exists $m_i^* \in S_i(\theta_i)$ such that

$$m_i^* \in \arg\max_{m_i \in M_i} u_i(g(m_i, m_{-i}), (\theta_i, \theta_{-i})),$$

for all θ_{-i} and $m_{-i} \in S_{-i}(\theta_{-i})$.

We observe that for S to satisfy the ex post best response property, $S_i(\theta_i)$ must be non-empty for all i and all θ_i.

Definition 13 (Interim Best Response). Message correspondence S satisfies the interim best response property if, for all i and $\psi_i \in \Delta(\Theta_{-i})$, there exists $\lambda_i \in \Delta(M_{-i} \times \Theta_{-i})$ such that:

1. $\lambda_i(m_{-i}, \theta_{-i}) > 0 \Rightarrow m_j \in S_j(\theta_j)$ for each $j \neq i$;
2. for all $\theta_{-i} \in \Theta_{-i}$,

$$\sum_{m_{-i} \in M_{-i}} \lambda_i(m_{-i}, \theta_{-i}) = \psi_i(\theta_{-i});$$

3. for all $\theta_i \in \Theta_i$,

$$S_i(\theta_i) \cap \underset{m_{-i} \in M_i}{\arg\max} \sum_{m_{-i}, \theta_{-i}} \lambda_i(m_{-i}, \theta_{-i}) u_i(g(m_i, m_{-i}), (\theta_i, \theta_{-i})) \neq \emptyset.$$

The interim best response property only requires that for every conjecture over payoff type spaces, there exists some beliefs over messages consistent with the message correspondence S, such that a best response is in the message correspondence. In particular, it does not require that a best response exists for all possible beliefs over message profiles. Note that the ex post best response property is a stronger requirement than the interim best response property, but that the interim best response property also implies that $S_i^{\mathcal{M}}(\theta_i)$ is non-empty for all i and θ_i.

Proposition 1 linked every action profile in the set of rationalizable actions to an equilibrium action for some type space $\mathcal{T}$. Proposition 7 strengthens the relationship between rationalizable and equilibrium actions, after imposing some structure on the best response property of rationalizable and equilibrium actions, respectively.

Proposition 7 (Best Response Properties).

1. *If $S^{\mathcal{M}}$ has the ex post best response property, then $(\mathcal{T}, \mathcal{M})$ has an equilibrium for each $\mathcal{T}$.*
2. *If $(\mathcal{T}, \mathcal{M})$ has an equilibrium for each $\mathcal{T}$, then $S^{\mathcal{M}}$ satisfies the interim best response property.*

Proof. (1.) By the ex post best response property, there exists, for each i, $s_i^* : \Theta_i \to M_i$ such that

$$s_i^*(\theta_i) \in \underset{m_i \in M_i}{\arg\max}\ u_i(g(m_i, s_{-i}^*(\theta_{-i})), (\theta_i, \theta_{-i}))$$

for all θ_{-i}. Now fix any type space. The strategy profile s with $s_i(t_i) = s_i^*(\widehat{\theta}_i(t_i))$ is an equilibrium of the game $(\mathcal{T}, \mathcal{M})$.

(2.) Suppose $(\mathcal{T}, \mathcal{M})$ has an equilibrium for each $\mathcal{T}$. Fix any i and $\psi_i \in \Delta(\Theta_{-i})$. Fix any type space $\mathcal{T}$ with, for each $\theta_i \in \Theta_i$, a type $t_i^*(\theta_i)$ such that (a) $\widehat{\theta}_i(t_i^*(\theta_i)) = \theta_i$ for each θ_i, (b) there exists $\pi_i \in \Delta(T_{-i})$ such that $\widehat{\pi}_i(t_i^*(\theta_i)) = \pi_i$ for all θ_i and (c)

$$\sum_{\{t_{-i} | \widehat{\theta}_{-i}(t_{-i}) = \theta_{-i}\}} \pi_i(t_{-i}) = \psi_i(\theta_{-i}), \tag{17}$$

for all θ. The game has an equilibrium σ. Let m_i be any message with $\sigma_i(m_i | t_i^*(\theta_i)) > 0$. Let

$$\lambda_i(m_{-i}, \theta_{-i}) = \sum_{\{t_{-i} \in T_{-i} | \widehat{\theta}_{-i}(t_{-i}) = \theta_{-i}\}} \sigma_{-i}(m_{-i} | t_{-i}) \pi_i(t_{-i}).$$

Now $\sigma_i(m_i | t_i^*(\theta_i)) > 0$ implies

$$m_i(\theta_i) \in \arg\max_{m_i \in M_i} \sum_{m_{-i}, \theta_{-i}} \lambda_i(m_{-i}, \theta_{-i}) u_i(g(m_i, m_{-i}), (\theta_i, \theta_{-i})).$$

Proposition 1 implies that every message profile m_j which is played in equilibrium by type θ_j is part of the set $S^{\mathcal{M}}$, or that:

$$\lambda_i(m_{-i}, \theta_{-i}) > 0 \Rightarrow m_j \in S_j^{\mathcal{M}}(\theta_j) \quad \text{for each } j \neq i.$$

By construction of the type space $\mathcal{T}$, in particular property (c) as expressed by (17), this implies that

$$\sum_{m_{-i} \in M_{-i}} \lambda_i(m_{-i}, \theta_{-i}) = \psi_i(\theta_{-i}) \quad \text{for all } \theta_{-i} \in \Theta_{-i}.$$

Since these properties hold for arbitrary i and $\psi_i \in \Delta(\Theta_{-i})$, $S^{\mathcal{M}}$ satisfies the interim best response property, which concludes the proof. □

It is unfortunate that there is a gap between the necessary and sufficient conditions in the above proposition. However, an example in the appendix shows that it is possible to construct (admittedly silly) mechanisms where $(\mathcal{T}, \mathcal{M})$ has an equilibrium for each $\mathcal{T}$, but $S^{\mathcal{M}}$ fails the ex post best response property.

4.2 *Material Implementation*

We can maintain the relationship between rationalizable and robust implementation, despite the possibility of non-existence of an interim best response, by qualifying the implementation condition as being "material".

Definition 2M (Material Interim Implementation). Social choice function f is materially interim implemented on type space $\mathcal{T}$ by mechanism $\mathcal{M}$ if every equilibrium σ of the game $(\mathcal{T}, \mathcal{M})$ satisfies for all m and all t:

$$\sigma(m|t) > 0 \Rightarrow g(m) = f(\widehat{\theta}(t)).$$

In contrast to the earlier definition of interim implementation, given in Definition 2, we allow the premise of the definition to be vacuous. In other words, the mechanism $\mathcal{M}$ might have the property that on a given type space, there is no equilibrium. Our terminology mirrors the language of modal logic where proposition A materially implies B whenever A is false, as well as when both A and B are true, see Hughes and Creswell (1996). We similarly weaken the definition of robust and rationalizable implementation.

Definition 3M (Material Robust Implementation). Social choice function f is materially robustly implemented by mechanism $\mathcal{M}$ if, for every $\mathcal{T}$, f is materially interim implemented on type space $\mathcal{T}$ by mechanism $\mathcal{M}$.

Definition 4M (Material Rationalizable Implementation). Social choice function f is materially rationalizably implemented by mechanism $\mathcal{M}$ if for all θ and m, $m \in S^{\mathcal{M}}(\theta) \Rightarrow g(m) = f(\theta)$.

With these weaker notions of material implementation Proposition 1 now immediately implies an equivalence between material robust and material rationalizable implementation in the presence of infinite mechanisms.

Corollary 2 (Equivalence). *Social choice function f is materially rationalizably implemented by $\mathcal{M}$ if and only if f is materially robustly implemented by mechanism $\mathcal{M}$.*

Proposition 7 gave the slightly messier result relating equilibrium existence and properties of messages surviving iterated deletion. The following corollary gives the immediate implications for our implementation definitions:

Corollary 3 (Necessary Conditions).

1. *If social choice function f is materially rationalizably implemented by mechanism $\mathcal{M}$ and $S^{\mathcal{M}}$ satisfies the ex post best response property, then f is robustly implemented by $\mathcal{M}$.*
2. *If f is robustly implemented by $\mathcal{M}$, then f is materially rationalizably implemented by mechanism $\mathcal{M}$ and $S^{\mathcal{M}}$ satisfies the interim best response property.*

The 'material' qualification will only be used in the necessity part of Theorem 1 where we shall invoke the above Corollary 3.2. There we shall use the fixed-point property of $S^{\mathcal{M}}$, stated earlier in (2), to derive the robust monotonicity condition. In the sufficiency part of the proof, a non-empty set $S^{\mathcal{M}}$ is obtained in the augmented mechanism by virtue of ex post incentive compatibility. The following implication of rationalizable implementation will be used to establish robust monotonicity in Theorem 1.

Lemma 1 (Truthtelling as Best Response). *If f is materially rationalizably implemented by mechanism $\mathcal{M}$ and $S^{\mathcal{M}}$ satisfies the interim best response property, then for all i and $\theta_{-i} \in \Theta_{-i}$, there exists $\nu_i \in \Delta(S_{-i}^{\mathcal{M}}(\theta_{-i}))$ such that*

$$u_i(f(\theta_i, \theta_{-i}), (\theta_i, \theta_{-i})) \geq \sum_{m_{-i}} \nu_i(m_{-i}) u_i(g(m_i, m_{-i}), (\theta_i, \theta_{-i})), \tag{18}$$

for all $m_i \in M_i$ and $\theta_i \in \Theta_i$.

Proof. Applying the definition of the interim best response property for i and the degenerate distribution putting probability 1 on θ_{-i}, we have that there exists $\nu_i \in \Delta(S_{-i}^{\mathcal{M}}(\theta_{-i}))$ such that

$$\begin{aligned} \emptyset \neq \arg\max_{m_i} & \sum_{m_{-i}, \theta_{-i}} \nu_i(m_{-i}) u_i((m_i, m_{-i}), (\theta_i, \theta_{-i})) \\ & \subseteq S_i^{\mathcal{M}}(\theta_i) \text{ for all } \theta_i \in \Theta_i. \end{aligned}$$

But by material rationalizable implementability, $m \in S^{\mathcal{M}}(\theta) \Rightarrow g(m) = f(\theta)$. So

$$u_i(f(\theta_i, \theta_{-i}), (\theta_i, \theta_{-i})) \geq \sum_{m_{-i}} \nu_i(m_{-i}) u_i(g(m_i, m_{-i}), (\theta_i, \theta_{-i})),$$

for all $m_i \in M_i$ and $\theta_i \in \Theta_i$. □

Lemma 1 shows how small the gap between the ex post and interim best response property is. It establishes that truthtelling is a best response against some beliefs over messages m_{-i} for any given payoff type profile θ_{-i}.

5 Infinite Mechanisms

We will need a very weak economic condition to ensure that it is always possible to reward and punish each agent independently of the other agents.

Definition 14 (Conditional No Total Indifference). The conditional no total indifference (NTI) property is satisfied if, for all i, all θ and all $\psi_i \in \Delta(\Theta_{-i})$, there exists y, $y' \in Y_i(\theta_{-i})$ such that

$$\sum_{\theta'_{-i} \in \Theta_{-i}} \psi_i(\theta_{-i}) u_i(y, (\theta_i, \theta'_{-i})) > \sum_{\theta'_{-i} \in \Theta_{-i}} \psi_i(\theta_{-i}) u_i(y', (\theta_i, \theta'_{-i})).$$

The *conditional no total indifference property* imposes a very weak restriction on the preferences. For example, if there are a finite number of pure outcomes and an agent is never completely indifferent between all lotteries, then we can always find interior outcomes y and y' such that the conditional no total indifference condition is met. The conditional NTI property, together with the use of lotteries, allows us to dispense with any no veto property which typically appear in the sufficient conditions. In addition, we can omit the usual cardinality assumption of $I \geq 3$. A related no total indifference condition appears in the context of virtual implementation in Duggan (1997), who requires it to hold at every ex post profile θ and in Serrano and Vohra (2005), who require it at the interim level for a given belief $\psi_i(\theta_{-i})$ of player i.

Theorem 1 (Robust Implementation)

1. *If f is robustly implementable, then f satisfies EPIC and dual robust monotonicity;*
2. *If f satisfies EPIC, dual robust monotonicity and the conditional NTI property, then f is robustly implementable.*

Proof. (1.) We first prove that robust implementability implies EPIC and dual robust monotonicity. We do so by appealing to the necessary conditions for robust implementation in Corollary 3.

We first establish EPIC. By Lemma 1, for all $m_i \in M_i$ and $\theta_i \in \Theta_i$, there exists $\nu_i \in \Delta(S_{-i}^{\mathcal{M}}(\theta_{-i}))$,

$$u_i(f(\theta_i, \theta_{-i}), (\theta_i, \theta_{-i})) \geq \sum_{m_{-i}} (m_{-i}) u_i(g(m_i, m_{-i}), (\theta_i, \theta_{-i})).$$

If we choose $m_i \in S_i^{\mathcal{M}}(\theta_i')$, material rationalizable implementation implies that $g(m_i, m_{-i}) = f(\theta_i', \theta_{-i})$ for all $m_{-i} \in S_{-i}^{\mathcal{M}}(\theta_{-i})$. So for all $\theta_i' \in \Theta_i$:

$$u_i(f(\theta_i, \theta_{-i}), (\theta_i, \theta_{-i})) \geq u_i(f(\theta_i', \theta_{-i}), (\theta_i, \theta_{-i})).$$

We next establish dual robust monotonicity. Fix an unacceptable deception β and suppose that f is materially rationalizably implementable. There must exist a message correspondence profile S such that $b(S) \leq S$, and

$$S_i^{\mathcal{M}}(\theta_i') \subseteq S_i(\theta_i), \tag{19}$$

for all i, θ_i and $\theta_i' \in \beta_i(\theta_i)$; but

$$S_i^{\mathcal{M}}(\theta_i') \not\subseteq b_i(S)[\theta_i], \tag{20}$$

for all i, θ_i and $\theta_i' \in \beta_i(\theta_i)$. The existence of such an S can be established constructively. Clearly $\overline{S}$ satisfies (19). Iteratively apply the operator b. By rationalizable implementation, there exists k (perhaps transfinite) such that

$$S \triangleq b^k(\overline{S}) \tag{21}$$

satisfies (20). Thus there exists k such that $b^k(\overline{S})$ satisfies (19) and $b^{k+1}(\overline{S})$ satisfies (20).

By (20), we can pick

$$\widehat{m}_i \in S_i^{\mathcal{M}}(\theta_i') \subseteq S_i(\theta_i) \quad \text{with} \quad \widehat{m}_i \notin b_i(S)[\theta_i].$$

Since message $\widehat{m}_i \notin b_i(S)[\theta_i]$, we know that for every $\lambda_i \in \Delta(M_{-i} \times \Theta_{-i})$ such that

$$\lambda_i(m_{-i}, \theta_{-i}) > 0 \Rightarrow m_j \in S_j(\theta_j \quad \text{for all } j \neq i, \tag{22}$$

there exists m_i^* such that

$$\begin{aligned} \sum_{m_{-i}, \theta_{-i}} & \lambda_i(m_{-i}, \theta_{-i}) [u_i(g(m_i^*, m_{-i}), (\theta_i, \theta_{-i})) \\ & - u_i(g(\widehat{m}_i, m_{-i}), (\theta_i, \theta_{-i}))] > 0. \end{aligned} \tag{23}$$

Next we identify a particular belief $\lambda_i(m_{-i}, \theta_{-i})$ for which the inequality (23) holds. Fix $\theta'_{-i} \in \Theta_{-i}$. By (18) in Lemma 1, there exists $\nu_i \in \Delta(S_{-i}^{\mathcal{M}}(\theta'_{-i}))$ such that

$$\sum_{m_{-i}} \nu_i(m_{-i}) u_i(g(m_i, m_{-i}), (\theta''_i, \theta'_{-i})) \leq u_i(f(\theta''_i, \theta'_{-i}), (\theta''_i, \theta'_{-i})), \quad (24)$$

for all $m_i \in M_i$ and $\theta''_i \in \Theta_i$. Thus for any $\psi_i \in \Delta(S_{-i}^{\mathcal{M}}(\theta'_{-i}))$, we can set

$$\lambda_i(m_{-i}, \theta_{-i}) = \nu_i(m_{-i})\psi_i(\theta_{-i}).$$

Since this λ_i satisfies (22), we can apply the above claim (23) and there exists m_i^* such that:

$$\sum_{\theta_{-i}, m_{-i}} \psi_i(\theta_{-i})\nu_i(m_{-i})u_i(g(m_i^*, m_{-i}), (\theta_i, \theta_{-i}))$$
$$> \sum_{\theta_{-i}, m_{-i}} \psi_i(\theta_{-i})\nu_i(m_{-i})u_i(g(\widehat{m}_i, m_{-i}), (\theta_i, \theta_{-i})).$$

But $\nu_i(m_{-i}) > 0 \Rightarrow m_{-i} \in S_{-i}^{\mathcal{M}}(\theta'_{-i}) \Rightarrow (\widehat{m}_i, m_{-i}) \in S^{\mathcal{M}}(\theta')$, so by material rationalizable implementation:

$$g(\widehat{m}_i, m_{-i}) = f(\theta').$$

Thus for every $\psi_i \in \Delta(\beta_{-i}^{-1}(\theta'_{-i}))$, there exists m_i^* such that

$$\sum_{\theta_{-i}, m_{-i}} \psi_i(\theta_{-i})u_i(g(m_i^*, m_{-i}), (\theta_i, \theta_{-i}))$$
$$> \sum_{\theta_{-i}, m_{-i}} \psi_i(\theta_{-i})u_i(f(\theta'), (\theta_i, \theta_{-i})). \quad (25)$$

Now, the inequality (25) establishes the reward inequality for robust monotonicity. We can complete the argument by letting y be the lottery with

$$y \triangleq \sum_{m_{-i}} g(m_i^*, m_{-i})\nu_i(m_{-i}).$$

We now have established that for each $\theta'_{-i} \in \beta_{-i}(\theta_{-i})$ and $\psi_i \in \Delta(\beta_{-i}^{-1}(\theta'_{-i}))$, there exists y such that (by (24))

$$u_i(y, (\theta''_i, \theta'_{-i})) \leq u_i(f(\theta''_i, \theta'_{-i}), (\theta''_i, \theta'_{-i})),$$

for all $\theta_i'' \in \Theta_i$, and thus $y \in Y_i(\theta_{-i}')$.[10] And by (25) we then have:

$$\sum_{\theta_{-i}} \psi_i(\theta_{-i}) u_i(y, (\theta_i, \theta_{-i})) > \sum_{\theta_{-i}} \psi_i(\theta_{-i}) u_i(f(\theta'), (\theta_i, \theta_{-i})).$$

(2.) We now prove that EPIC, dual robust monotonicity and the conditional NTI property imply robust implementation. We do so by explicitly constructing the implementing mechanism. The mechanism will use "interior" lotteries over the deterministic outcome set Z and over the reward sets $Y_i(\theta_{-i})$. Given an arbitrary labelling of the outcome set $Z = \{z_0, z_1, \ldots, z_k, \ldots\}$, we define an "interior" lottery over the set Z by

$$\bar{y} = (\bar{y}_0, \bar{y}_1, \ldots, \bar{y}_k, \ldots), \tag{26}$$

where

$$\bar{y}_k \triangleq \Pr(z = z_k) = \frac{\delta^k}{1-\delta},$$

for some $\delta \in (0,1)$. For every given profile θ_{-i}, the reward set $Y_i(\theta_{-i})$ is by construction a convex set with at most a countable number of extreme points. We denote the set of extreme points of $Y_i(\theta_{-i})$ by $Y_i^*(\theta_{-i})$ and for some labelling of the points in the set we have $Y_i^*(\theta_{-i}) = \{y_{0,\theta_{-i}}, y_{1,\theta_{-i}}, \ldots, y_{l,\theta_{-i}}, \ldots\}$. An extreme point $y_{l,\theta_{-i}}$ in $Y_i^*(\theta_{-i})$ may be a deterministic or a random outcome and assigns probability $y_{l,\theta_{-i}}(z_k)$ to the pure outcome z_k. For every reward set $Y_{-i}(\theta_{-i})$, we define a "interior" lottery:

$$\bar{y}_{\theta_{-i}} = (\bar{y}_{0,\theta_{-i}}, \bar{y}_{1,\theta_{-i}}, \ldots, \bar{y}_{k,\theta_{-i}}) \tag{27}$$

with

$$\bar{y}_{k,\theta_{-i}} \triangleq \frac{1}{1-\delta} \sum_{l=0}^{\infty} \delta^l y_{l,\theta_{-i}}(z_k),$$

where the lottery $\bar{y}_{\theta_{-i}}$ is a compound lottery.

Each agent i sends a message $m_i = (m_i^1, m_i^2, m_i^3, m_i^4)$, where $m_i^1 \in \Theta_i$, $m_i^2 \in \mathbb{Z}_+$, $m_i^3 : \Theta_{-i} \to Y$ with $m_i^3(\theta_{-i}) \in Y_i(\theta_{-i})$, $m_i^4 \in Y$. The outcome

[10]Note that this step implies that even if we had restricted attention to mechanisms with deterministic outcomes, our robust monotonicity condition would only have established that there exists a lottery (not necessarily a deterministic outcome) sufficient to reward a whistle-blower.

$g(m)$ is determined by the following rules:

Rule 1: If $m_i^2 = 1$ for all i, pick $f(m^1)$.

Rule 2: If there exists $j \in I$ such that $m_i^2 = 1$ for all $i \neq j$ and $m_j^2 > 1$, then pick $m_j^3(m_{-j}^1)$ with probability $1 - \frac{1}{m_j^2+1}$ and $\bar{y}_{m_{-j}^1}$ (as defined in (27)) with probability $\frac{1}{m_j^2+1}$.

Rule 3: In all other cases, for each i, with probability $\frac{1}{I}(1 - \frac{1}{m_i^2+1})$ pick m_i^4, and with probability $\frac{1}{I}(\frac{1}{m_i^2+1})$ pick the interior lottery $\bar{y}$ (as defined in (26)).

We first show that it is never a best reply for type θ_i to send a message with $m_i^2 > 1$ (i.e., $m_i \in b_i(\overline{S}) \Rightarrow m_i^2 = 1$). Suppose that θ_i has conjecture $\lambda_i \in \Delta(M_{-i} \times \Theta_{-i})$. We can partition the messages of other agents as follows:

$$M_{-i}^*(\theta_{-i}) = \{m_{-i} : m_j^2 = 1 \text{ for all } j \neq i \text{ and } m_{-i}^1 = \theta_{-i}\},$$

and

$$\widehat{M}_{-i} = \{m_{-i} : m_j^2 > 1 \text{ for some } j \neq i\}.$$

By the conditional NTI property, we know that there exists $m_i^4 \in Y$ such that, if

$$\sum_{m_{-i} \in \widehat{M}_{-i}, \theta_{-i} \in \Theta_{-i}} \lambda_i(m_{-i}, \theta_{-i}) > 0,$$

then

$$\sum_{m_{-i} \in \widehat{M}_{-i}, \theta_{-i} \in \Theta_{-i}} \lambda_i(m_{-i}, \theta_{-i}) u_i(m_i^4, \theta) > \sum_{m_{-i} \in \widehat{M}_{-i}, \theta_{-i} \in \Theta_{-i}} \lambda_i(m_{-i}, \theta_{-i}) u_i(\bar{y}, \theta).$$

And we also know from the conditional NTI property that there exists m_i^3 such that, if

$$\sum_{m_{-i} \in M_{-i}^*(\theta'_{-i}), \theta_{-i} \in \Theta_{-i}} \lambda_i(m_{-i}, \theta_{-i}) > 0,$$

then

$$\sum_{m_{-i}\in M^*_{-i}(\theta'_{-i}),\theta_{-i}\in\Theta_{-i}} \lambda_i(m_{-i},\theta_{-i})u_i(m_i^3(\theta'_{-i}),\theta)$$
$$> \sum_{m_{-i}\in M^*_{-i}(\theta'_{-i}),\theta_{-i}\in\Theta_{-i}} \lambda_i(m_{-i},\theta_{-i})u_i(\bar{y}_{\theta_{-i}},\theta).$$

Thus if $(m_i^1, m_i^2, m_i^3, m_i^4)$ with $m_i^2 > 1$ were a best response, then $(m_i^1, m_i^2 + 1, m_i^3, m_i^4)$ would be an even better response, a contradiction.

Now fix any S with $m_i \in S_i(\theta_i) \Rightarrow m_i^2 = 1$. Let

$$\beta_i(\theta_i) = \{\theta'_i : (\theta'_i, 1, m_i^3, m_i^4) \in S_i(\theta_i) \text{ for some } (m_i^3, m_i^4)\}.$$

First observe that EPIC implies that $\theta_i \in \beta_i(\theta_i)$. We will argue that if β is not acceptable, then $b(S) \neq S$. By robust monotonicity, we know that there exists i, θ_i, $\theta'_i \in \beta_i(\theta_i)$ such that, for all $\theta'_{-i} \in \Theta_{-i}$ and $\psi_i \in \Delta(\beta_{-i}^{-1}(\theta'_{-i}))$, there exists $y \in Y_i(\theta'_{-i})$ such that

$$\sum_{\theta_{-i}\in\Theta_{-i}} \psi_i(\theta_{-i})u_i(y,(\theta_i,\theta_{-i})) > \sum_{\theta_{-i}\in\Theta_{-i}} \psi_i(\theta_{-i})u_i(f(\theta'_i,\theta'_{-i}),(\theta_i,\theta_{-i})).$$

But now for any conjecture $\lambda_i \in \Delta(\{(m_{-i},\theta_{-i}) : m_j^2 = 1 \text{ for all } j \neq i\})$, there exists m_i^3 (with $m_i^3(\theta_{-i}) \in Y_i(\theta_{-i})$) such that

$$\sum_{m_{-i},\theta_{-i}} \lambda_i(m_{-i},\theta_{-i})u_i(m_i^3(m_{-i}^1),\theta)$$
$$> \sum_{m_{-i},\theta_{-i}} \lambda_i(m_{-i},\theta_{-i})u_i(f(\theta'_i,m_{-i}^1),(\theta_i,\theta_{-i})).$$

Thus message $(\theta'_i, 1, m_i^3, m_i^4)$ is never a best response for type θ_i. We conclude that if

$$\beta_i(\theta_i) = \{\theta'_i : (\theta'_i, 1, m_i^3, m_i^4) \in S_i^{\mathcal{M}}(\theta_i) \text{ for some } (m_i^3, m_i^4)\},$$

then β is acceptable. Thus f is materially rationalizably implemented.

Finally observe that $S^{\mathcal{M}}$ must satisfy the ex post best response property, with type θ_i sending a message of the form $(\theta'_i, 1, m_i^3, m_i^4)$, so robust implementation is possible by Corollary 2. □

We deliberately allowed for very badly behaved infinite mechanisms in order to make a tight connection with the existing literature and to get tight results. Many authors have argued that "integer game" constructions, like that we use in Theorem 1, should be viewed critically (see, e.g., Abreu and Matsushima (1992a) and Jackson (1992)). In our analysis of finite mechanisms in Section 3, the best responses were always well defined. As we saw there, the relationship between rationalizable and robust implementation is much simpler with the restriction to "nice" mechanisms, where best responses exists for all conjectures.

Part 1 of the above theorem represents a slight weakening of the necessary conditions of Propositions 2 and 3: semi-strict EPIC is weakened to EPIC and strict dual robust monotonicity is weakened to dual robust monotonicity. These weaker conditions arise from allowing badly behaved mechanisms. Part 2 of the above theorem shows that they are also sufficient when combined with a no total indifference property.

The proof directly uses the link between rationalizable and robust implementation for the necessity as well as the sufficiency part. We briefly sketch the idea of the necessity part of the proof. If f is robustly implementable, then it is rationalizably implementable by Corollary 3. From rationalizable implementability, we then want to show that f satisfies strict robust monotonicity. We consider a given *and* unacceptable deception β. We start the process of iterative elimination and stop it at a specific round, denoted by k. This round k is the first round at which we can find an agent i, a true type profile θ_i and a report $\theta_i' \in \beta_i(\theta_i)$, such that a message, denoted by $\widehat{m}_i$, which will survive the process of iterated elimination for type θ_i', fails to survive the k-th round of elimination for type θ_i. We then show that the elimination of message $\widehat{m}_i$ at round k implies that the social choice function f satisfies strict robust monotonicity with respect to the deception β. Briefly, if $\widehat{m}_i$ survives the process of elimination for type θ_i', the message $\widehat{m}_i$ acts in the mechanism so as to report a payoff type θ_i'. If it is eliminated at round k for payoff type θ_i, then this means that for any belief agent i has over the remaining agents, there exists a message m_i^* which leads to an allocation through g which is strictly preferred by agent i when he has a payoff type θ_i. The significance of round k being the first round for which such an elimination relative to the deception β occurs, is that at this round, there do not yet exist any restrictions about message and payoff type profile regarding the other agents deception. The fact that $\widehat{m}_i$ can be eliminated allows us to use the full strength of the elimination argument to establish robust monotonicity.

6 Extensions, Variations and Discussion

6.1 *Lotteries, Pure Strategies and Bayesian Implementation*

In this section, we discuss how our main Theorem 1 is related to the classic literature on Bayesian implementation developed by Postlewaite and Schmeidler (1986), Palfrey and Srivastava (1989) and Jackson (1991). These authors asked whether it was possible to implement a social choice function in equilibrium on a fixed type space $\mathcal{T}$.[11] These authors analyzed the classic problem where attention was restricted to pure strategy equilibria and deterministic mechanisms. The assumption entails that the social choice function is a mapping $f : \Theta \to Z$ and the mechanism $g : M \to Z$. Note that in this classical approach it was not necessary to even define agent's preferences over lotteries and they certainly did not effect implementability.

Having fixed a type space, the natural notion of a pure strategy deception on the fixed type space is a collection $\alpha = (\alpha_1, \ldots, \alpha_I)$, with each $\alpha_i : T_i \to T_i$. Thus $\alpha : T \to T$ is defined by $\alpha(t) = (\alpha_i(t_i))_{i=1}^{I}$. The key monotonicity notion, translated into our language, is then the following:

Definition 15 (Bayesian Monotonicity). Social choice function f satisfies Bayesian monotonicity on type space $\mathcal{T}$ if, for every deception α with $f(\widehat{\theta}(t)) \neq f(\widehat{\theta}(\alpha(t)))$ for some t, there exists i, t_i and $k : T \to Z$ such that

$$\sum_{t_{-i}} u_i(k(\alpha(t)), \widehat{\theta}(t))\widehat{\pi}_i(t_{-i})[t_i] > \sum_{t_{-i}} u_i(f(\widehat{\theta}(\alpha(t))), \widehat{\theta}(t))\widehat{\pi}_i(t_{-i})[t_i],$$

and

$$\sum_{t_{-i}} u_i(f(\widehat{\theta}(t_i', t_{-i})), \widehat{\theta}(t_i', t_{-i}))\widehat{\pi}_i(t_{-i})[t_i']$$

$$\geq \sum_{t_{-i}} u_i(k(\alpha_i(t_i), t_{-i}), \widehat{\theta}(t_i', t_{-i}))\widehat{\pi}_i(t_{-i})[t_i'], \ \forall t_i'.$$

Jackson (1991) shows that this condition is necessary for Bayesian implementation, and that a slight strengthening, Bayesian monotonicity no veto, is sufficient. We can also show that our robust monotonicity condition is equivalent to the requirement that Bayesian monotonicity is satisfied on all type spaces.

[11] They allowed for more general social choice sets, but we restrict attention to functions for our comparison.

Theorem 2 (Equivalence). *Social choice function f satisfies Bayesian monotonicity on every type space if and only if it satisfies robust monotonicity.*

The equivalence is established by a constructive proof via a specific type space. The constructive element is the identification of a type space on which Bayesian monotonicity is guaranteed to fail if robust monotonicity fails. It is worthwhile to note that the specific type space is much smaller than the universal type space. The proof of this result is in the appendix of the working paper version, Bergemann and Morris (2008b).

In some sense, the notion of robustness is more subtle in the context of full rather than partial implementation. With partial implementation, i.e., truthtelling in the direct mechanism, the universal type space is by definition the most difficult type space to obtain truthtelling. In the universal type space, every agent has the maximal number of possible misreports and hence the designer faces the maximal number of incentive constraints. In the context of full implementation, the trade-off is ambiguous. As a larger type space contains by definition more types, it offers every agent more possibilities to misreport. But then, just as a larger type space made truthtelling more difficult to obtain, the other equilibria might also cease to exist after the introduction of additional types. This second part offers the possibility that larger type spaces facilitate rather than complicate the full implementation problem.

But note that this line of argument would establish the necessity of robust implementation if the planner is restricted to deterministic mechanisms (a disadvantage) but he can assume that agents follow pure strategies (an advantage). How do these assumptions matter?

First, observe that the advantage of restricting attention to pure strategies goes away completely when we require implementation on all type spaces: if there is a mixed strategy equilibrium that results in a socially sub-optimal outcome on some type space, we can immediately construct a larger type space (purifying the original equilibrium) where the socially sub-optimal outcome is played in a pure strategy equilibrium. Thus our robust analysis conveniently removes that unfortunate gap between pure and mixed strategy implementation that has plagued the implementation literature.

We use the extension to stochastic mechanisms in just two places. Ex post incentive compatibility and robust monotonicity would remain necessary conditions even if we restricted attention to deterministic mechanisms (the arguments would be unchanged). But, as we note in Footnote 10, even if lotteries were not used in the implementing mechanism, the implied robust monotonicity condition would involve lotteries (as rewards for whistle-blowers). But if lotteries were not allowed, our sufficiency argument would then require a slightly strengthened version of the robust monotonicity condition, with the lottery y replaced by a deterministic outcome. Our sufficiency argument also uses lotteries under Rules 1 and 2. As in recent papers by Benoit and Ok (2008) and Bochet (2007) on complete information implementation, we use lotteries to significantly weaken the sufficient conditions, so that we require only the conditional NTI property in addition to EPIC and robust monotonicity. If we did not allow lotteries in this part of the argument, we would require a much stronger economic condition in the spirit of Jackson's "Bayesian monotonicity no veto" condition. We have developed combined robust monotonicity and economic conditions (not reported here) sufficient for interim implementation on all full support types spaces. However, an additional complication is that, without lotteries in the implementing mechanism, we cannot establish sufficiency on type spaces where agents have disjoint supports.

It is possible to construct a simple example where EPIC and robust monotonicity are not sufficient for robust monotonicity without lotteries by taking the coordination example of Section 3.6 but removing the outcomes z and z'. As we show in the Appendix, robust implementation is then not possible in this example despite the fact that the social choice function selects a unique strictly Pareto-dominant outcome at every type profile.

6.2 *Ex Post and Robust Implementation*

In contrast to our earlier results in Bergemann and Morris (2005b), where we showed that robust partial implementation is equivalent to ex post incentive compatibility, robust implementation is in general a more demanding notion of implementation than ex post equilibrium implementation. The following simple example, introduced by Palfrey and Srivastava (1989), is useful to relate the different implementation notions and understand the role of interdependent types. In this example, there are three agents and each has two possible "payoff types", θ_a or θ_b. There are two possible choices for society, a or b. All agents have identical preferences. If a majority of agents (i.e., at least two) are of type θ_y, then every agent gets utility 1 from outcome y and utility 0 from the other outcome. The social choice function agrees with the common preferences of the agents. Thus $f : \{\theta_a, \theta_b\}^3 \to \{a, b\}$ satisfies $f(\theta) = y$ if and only if $\# \{i : \theta_i = \theta_y\} \geq 2$.

Clearly, ex post incentive compatibility is not a problem in this example. The problem is that in the "direct mechanism" — where all agents simply announce their types — there is the possibility that all agents will choose to always announce θ_a. Since no agent expects to be pivotal, he has no incentive to truthfully announce his type when he is in fact θ_b. What happens if we allow more complicated mechanisms?

If there were complete information about agents' preferences, then the social choice function is clearly implementable: the social planner could pick an agent, say agent 1, and simply follow that agent's recommendation.

But suppose instead that there is incomplete information about agents' preferences. In particular, suppose it is common knowledge that each agent's type is θ_b with independent probability q, with $q^2 > \frac{1}{2}$. This example fails the Bayesian monotonicity condition of Postlewaite and Schmeidler (1986) and Jackson (1991). Palfrey and Srivastava (1989) observe that it is also not possible to implement in undominated Bayesian Nash equilibrium in this example.

Bergemann and Morris (2008a) have analyzed the alternative "more robust" solution concept of ex post equilibrium in this context. It is easy to construct an augmented mechanism whose only ex post equilibrium delivers the social choice function. Let each agent send a message $m_i \in \{\theta_a, \theta_b\} \times \{\text{truth}, \text{lie}\}$, with the interpretation that an agent is announcing his own type and also sends the message "truth" if he thinks that others are telling the truth and sends the message "lie" if he thinks that someone is lying. Outcome y is implemented if a majority claim to be type θ_y and all agents announce "truth"; or if either 1 or 3 agents claim to be type θ_y and at least one agent reports lying.

There is a truthtelling ex post equilibrium where each agent truthfully announces his type and also announces "truth". Now suppose there exists an ex post equilibrium such that at some type profile, the desired outcome is not chosen. Note that whatever the announcements of the other agents, each agent always has the ability to determine the outcome y, by sending the message "lie" and — given the announcements of the other agents — choosing his message so that an odd number of agents have claimed to be type θ_y. So this is not consistent with ex post equilibrium.

Robust implementation is impossible in this example. Consider the type space where there is common knowledge that whenever an agent is type θ_y, he assigns probability $\frac{1}{2}$ to both of the other agents being type $y' \neq y$ and probability $\frac{1}{2}$ to one being type y and the other being y'. Thus every type of every agent thinks there is a 50% chance that outcome a is better and a 50% chance that b is better. Evidently, there is no way of designing a mechanism that ensures that agents do not fully pool. But if they fully pool, robust implementation is not possible.

6.3 *Extensions*

The previous sections examined the importance of our assumptions about lotteries over outcomes and restrictions on mechanisms. We also restricted attention in our main analysis to the case of discrete but infinite pure outcomes Z, payoff types Θ_i and types T_i. While most of our results would extend naturally to more general Z, Θ_i and T_i, the formal treatment of non-compact type spaces would raise technical issues that we have chosen to avoid.

7 Appendix

7.1 *Robust Monotonicity and Dual Robust Monotonicity*

Lemma 2

1. *If f satisfies (strict) robust monotonicity, then f satisfies dual (strict) robust monotonicity.*
2. *If deterministic outcomes and payoff types are finite, and f satisfies dual (strict) robust monotonicity, then f satisfies (strict) robust monotonicity.*

Proof. (1.) follows immediately from the definitions. To prove (2.), suppose that outcomes and payoff types are finite and that f satisfies dual (strict) robust monotonicity. Then for every unacceptable deception β, there exist i, θ_i, $\theta_i' \in \beta_i(\theta_i)$ such that, for all $\theta'_{-i} \in \Theta_{-i}$, there exists a compact set $\overline{Y} \subseteq Y$ such that $y \in \overline{Y}$ implies

$$u_i(f(\theta_i'', \theta_{-i}), (\theta_i', \theta'_{-i})) \geq (>) u_i(y, (\theta_i'', \theta'_{-i}))$$

for all θ_i'' with $f(\theta_i'', \theta_{-i}) \neq y$ and, for each $\psi_i \in \Delta(\beta_{-i}^{-1}(\theta'_{-i}))$, there exists $y \in \overline{Y}$ such that

$$\sum_{\theta_{-i} \in \beta_{-i}^{-1}(\theta'_{-i})} \psi_i(\theta_{-i}) u_i(y, (\theta_i, \theta_{-i}))$$
$$> \sum_{\theta_{-i} \in \beta_{-i}^{-1}(\theta'_{-i})} \psi_i(\theta_{-i}) u_i(f, (\theta_i', \theta'_{-i}), (\theta_i, \theta_{-i})).$$

By the equivalence between strict domination and never a best response (see Theorem 2.10 in Gale (1989)), there exists $y^* \in \overline{Y}$ with

$$u_i(y^*, (\theta_i, \theta_{-i})) > u_i(f(\theta_i', \theta_{-i}'), (\theta_i, \theta_{-i}))$$

for all $\theta_{-i} \in \beta_{-i}^{-1}(\theta_{-i}')$. This establishes (strict) robust monotonicity. □

7.2 *A Badly Behaved Mechanism*

The example illustrates the gap between the necessary and sufficient conditions in Proposition 7. Specifically, it shows that there can be an equilibrium for every type space $\mathcal{T}$ in a mechanism, yet $S^{\mathcal{M}}$ does not satisfy the ex post best response property.

In the example, there are two agents and there is complete information, so each agent has a unique type. There are a finite number of outcomes $Z = \{a, b, c\}$. The payoffs are given by the following table:

	a	b	c
agent 1	0	-1	$+1$
agent 2	0	0	0

The planner's choice (in the unique payoff state) is a. Thus it is trivial to robustly implement the social choice function. But suppose that the planner chooses the following (strange) mechanism: $M_1 = \{1, 2, 3, \ldots\}$, $M_2 = \{1, 2\}$ and

$$g(m_1, m_2) = \begin{cases} a, & \text{if } m_1 = 1; \\ b, & \text{if } m_1 > 1 \text{ and } m_2 = 1; \\ \left[\dfrac{1}{m_1}, b; \left(1 - \dfrac{1}{m_1}\right), c\right] & \text{if } m_1 > 1 \text{ and } m_2 = 2. \end{cases}$$

where $[\frac{1}{m_1}, b; (1 - \frac{1}{m_1}), c]$ is the lottery putting probability $\frac{1}{m_1}$ on b and probability $(1 - \frac{1}{m_1})$ on c. Thus $g(m_1, m_2)$ can be represented by the following table:

g	1	2
1	a	a
2	b	$\left[\frac{1}{2}, b; \frac{1}{2}, c\right]$
3	b	$\left[\frac{1}{3}, b; \frac{2}{3}, c\right]$
$\vdots$	$\vdots$	$\vdots$
k	b	$\left[\frac{1}{k}, b; 1 - \frac{1}{k}, c\right]$
$\vdots$	$\vdots$	$\vdots$

Thus the agents are playing the following complete information game:

m_1/m_2	1	2
1	$0,0$	$0,0$
2	$-1,0$	$0,0$
3	$-1,0$	$\frac{1}{3},0$
$\vdots$	$\vdots$	$\vdots$
k	$-1,0$	$1-\frac{2}{k},0$
$\vdots$	$\vdots$	$\vdots$

Now on any type space, there is always an equilibrium where agent 1 chooses action 1 and agent 2 chooses action 1, and outcome a is chosen. Moreover, on any type space, in any equilibrium, outcome a is always chosen: if agent 1 ever has a best response not to play 1 then he has no best response. So he always plays 1 in equilibrium. Thus the trivial social choice function is robustly implemented by this mechanism.

While only message 1 survives iterated deletion of never best responses for agent 1, both messages survive iterated deletion of never best responses for agent 2. Thus we have $S_1^{\mathcal{M}} = \{1\}$ and $S_2^{\mathcal{M}} = \{1, 2\}$. Note that $S^{\mathcal{M}}$ satisfies the interim best response property, see Definition 13, but not the ex post best response property, see Definition 12, as we observe that

$$u_1(g(1,2)) = u_1(a) = 0 < \frac{1}{2} = u_1(g(2,2)),$$

violating the ex post best response property.

The insight of the example is that the quantifier "for every type space $\mathcal{T}$" does not necessarily guarantee that all actions which will be chosen with positive probability in some equilibrium and for some type space, will also be chosen with probability one in some equilibrium for some type space. For this reason, the quantifier "for every type space $\mathcal{T}$" does not allow us to establish a local, i.e., ex post best response property of every action in $S^{\mathcal{M}}$.

7.3 *Coordination Example Continued*

The final example is the pure coordination game, which we first considered in Section 3.6, but without the additional allocations, z and z'. It illustrates the importance of lotteries for robust implementation. The example will satisfy EPIC and robust monotonicity, yet it cannot be robustly implemented without the use of lotteries. On the other hand the preferences clearly satisfy the conditional NTI property, and hence the sufiicient conditions for robust implementation would be satisfied with lotteries.

As in the example in Section 3.6, the payoffs of the player are given by (15) and the social choice function f is given by (16). The social choice function is strictly ex post incentive compatible but there is another equilibrium in the "direct mechanism" where each agent misreports his type, and each agent gets a payoff of 1.

Robust monotonicity is clearly satisfied even if the rewards $Y(\theta_{-i})$ are restricted to the deterministic allocations Z. We will show that robust implementation is not possible even in an infinite mechanism if we restrict attention to deterministic mechanisms. Fix a mechanism $\mathcal{M}$. Let

$$S_i^*(\theta_i) = \{m_i : g(m_i, m_j) = f(\theta_i, \theta_j) \text{ for some } m_j, \theta_j\},$$

be the set of messages for agent i which would select the allocation recommended by the social choice function for some m_j, θ_j. We now show by induction that, $S_i^*(\theta_i) \subseteq S_i^k(\theta_i)$ for all k using the structure of the payoffs. Suppose that this is true for k. Then for any $m_i \in S_i^*(\theta_i) \subseteq S_i^k(\theta_i)$, there exists $m_j \in S_j^*(\theta_j) \subseteq S_j^k(\theta_j)$ such that $g(m_i, m_j) = f(\theta_i, \theta_j)$. Thus there does not exist $\nu_i \in \Delta(M_i)$ such that

$$\sum_{m_i'} \nu_i(m_i') u_i(g(m_i', m_j), (\theta_i, \theta_j)) > u_i(g(m_i, m_j), (\theta_i, \theta_j)) = 3,$$

and so $m_i \in S_i^{k+1}(\theta_i)$.

Thus we must have that $(m_1, m_2) \in S_1^*(\theta_1) \times S_2^*(\theta_2)$ implies $g(m_1, m_2) = f(\theta_1, \theta_2)$. Let $m_i^*(\cdot)$ be any selection from $S_i^*(\cdot)$. Now let k^* be the lowest k such that, for some i,

$$m_i^*(\theta_i') \notin S_i^k(\theta_i).$$

Without loss of generality, let $i = 1$. Note $m_2^*(\theta_2') \in S_2^{k-1}$ by definition of k^*. If agent 1 was type θ_1 and was sure his opponent were type θ_2 and choosing action $m_2^*(\theta_2')$, we know that he could guarantee himself a payoff of 1 by choosing $m_1^*(\theta_1')$. Since $m_1^*(\theta_1')$ is deleted for type θ_1 at round k^*, we know that there exists $\nu_1 \in \Delta(M_1)$ such that

$$\sum_{m_1'} \nu_1(m_1') g_1(m_1', m_2^*(\theta_2')) > 1,$$

and thus there exists m_1' such that $g_1(m_1', m_2^*(\theta_2')) = f(\theta_1, \theta_2)$. This implies that $m_2^*(\theta_2') \in S_2^*(\theta_2)$, a contradiction.

The example uses the fact that the social choice function always selects an outcome that is strictly Pareto-optimal and — paradoxically — it is this feature which inhibits rationalizable implementation in the current example. Borgers (1995) proves the impossibility of complete information implementation of non-dictatorial social choice functions in iteratively undominated strategies when the set of feasible preference profiles includes such unanimous preference profiles and the argument here is reminiscent of Borgers' argument.

References

ABREU, D., AND H. MATSUSHIMA (1992a): "Virtual Implementation in Iteratively Undominated Strategies: Complete Information," *Econometrica*, 60, 993–1008.

——— (1992b): "Virtual Implementation In Iteratively Undominated Strategies: Incomplete Information," Discussion paper, Princeton University and University of Tokyo.

BATTIGALLI, P. (1999): "Rationalizability in Incomplete Information Games," Discussion paper ECO 99/17, European University Institute.

BATTIGALLI, P., AND M. SINISCALCHI (2003): "Rationalization and Incomplete Information," *Advances in Theoretical Economics*, 3, Article 3.

BENOIT, J.-P., AND E. OK (2008): "Nash Implementation Without No-Veto Power," *Games and Economic Behavior*, 64, 51–67.

BERGEMANN, D., AND S. MORRIS (2005a): "Robust Implementation: The Role of Large Type Spaces," Discussion paper 1519, Cowles Foundation, Yale University.

——— (2005b): "Robust Mechanism Design," *Econometrica*, 73, 1771–1813.

——— (2008a): "Ex Post Implementation," *Games and Economic Behavior*, 63, 527–566.

——— (2008b): "Robust Implementation in General Mechanisms," Discussion paper, Cowles Foundation, Yale University.

——— (2009a): "Robust Implementation in Direct Mechanisms," *Review of Economic Studies*.

——— (2009b): "Robust Virtual Implementation," Discussion paper 1609RR, Cowles Foundation, Yale University.

BOCHET, O. (2007): "Nash Implementation with Lottery Mechanisms," *Social Choice and Welfare*, 28, 111–125.

BORGERS, T. (1995): "A Note on Implementation and Strong Dominance," in *Social Choice, Welfare and Ethics*, ed. by W. Barnett, H. Moulin, M. Salles and N. Schofield. Cambridge University Press, Cambridge, UK.

BRANDENBURGER, A., AND E. DEKEL (1987): "Rationalizability and Correlated Equilibria," *Econometrica*, 55, 1391–1402.

CHUNG, K.-S., AND J. ELY (2001): "Efficient and Dominance Solvable Auctions with Interdependent Valuations," Discussion paper, Northwestern University.

DEKEL, E., D. FUDENBERG, AND S. MORRIS (2007): "Interim Correlated Rationalizability," *Theoretical Economics*, 2, 15–40.

DUGGAN, J. (1997): "Virtual Bayesian Implementation," *Econometrica*, 65, 1175–1199.

GALE, D. (1989): *The Theory of Linear Economic Models*. University of Chicago Press, Chicago.

HUGHES, G., AND M. CRESWELL (1996): *A New Introduction Into Modal Logic*. Routledge, London.

JACKSON, M. (1991): "Bayesian Implementation," *Econometrica*, 59, 461–477.

JACKSON, M. (1992): "Implementation in Undominated Strategies: A Look at Bounded Mechanisms," *Review of Economic Studies*, 59, 757–775.

LIPMAN, B. (1994): "A Note on the Implications of Common Knowledge of Rationality," *Games and Economic Behavior*, 6, 114–129.

PALFREY, T., AND S. SRIVASTAVA (1989): "Mechanism Design with Incomplete Information: A Solution to the Implementation Problem," *Journal of Political Economy*, 97, 668–691.

POSTLEWAITE, A., AND D. SCHMEIDLER (1986): "Implementation in Differential Information Economies," *Journal of Economic Theory*, 39, 14–33.

SERRANO, R., AND R. VOHRA (2005): "A Characterization of Virtual Bayesian Implementation," *Games and Economic Behavior*, 50, 312–331.

WILSON, R. (1987): "Game-Theoretic Analyses of Trading Processes," in *Advances in Economic Theory: Fifth World Congress*, ed. by T. Bewley. Cambridge University Press, Cambridge, UK, pp. 33–70.

CHAPTER 5

The Role of the Common Prior in Robust Implementation*

Dirk Bergemann and Stephen Morris

Abstract
We consider the role of the common prior for robust implementation in an environment with interdependent values. Specifically, we investigate a model of public good provision which allows for negative and positive informational externalities. In the corresponding direct mechanism, the agents' reporting strategies are strategic complements with negative informational externalities and strategic substitutes with positive informational externalities. We derive the necessary and sufficient conditions for robust implementation in common prior type spaces and contrast this with our earlier results without the common prior. In the case of strategic complements the necessary and sufficient conditions for robust implementation do not depend on the existence of a common prior. In contrast, with strategic substitutes, the implementation conditions are much weaker under the common prior assumption.

1 Introduction

We investigate the role of the common prior assumption in robust implementation. Robust implementation requires that every equilibrium on every type space delivers outcomes that are consistent with a social choice function. In using the term "every type space" we allow for multiple copies of the same payoff type with different beliefs over the types of others; we also allow for noncommon prior-type spaces. In this paper we want to look at

*This research is financially supported by NSF Grants SES 0518929 and CNS 0428422. We would like to thank an anonymous referee for his/her helpful comments.

an intermediate notion of robustness: allowing all possible common prior type spaces.

We develop our arguments in the context of a public good model with interdependent values. We have used this specific public good model as a leading example in our previous work on ex post implementation (Bergemann and Morris, forthcoming), robust implementation in direct mechanisms (Bergemann and Morris, 2007a), and robust virtual implementation (Bergemann and Morris, 2007b). The current objective is to analyze the importance of the common prior assumption for the possibility of robust implementation. In particular, we identify when the necessary and sufficient conditions for robust implementation depend on whether we allow for all type spaces (as in Bergemann and Morris, 2007a) or only for those type spaces with a common prior. The public good model allows for positive as well as negative informational externalities. When we consider the direct revelation mechanism, we show that the reporting strategies of the agents are strategic complements with negative informational externalities and strategic substitutes with positive informational externalities.

The analysis of the robust implementation with and without a common prior relies on epistemic results on games of incomplete information. Brandenburger and Dekel (1987) and Aumann (1987), respectively, reported formal epistemic arguments that — for complete information games — characterize the solution concepts of correlated equilibrium and rationalizability as the consequences of common knowledge of rationality with and without the common prior, respectively. In Bergemann and Morris (2007c), we report belief-free incomplete information generalizations of the solution concepts (incomplete information correlated equilibrium and incomplete information rationalizability) and their epistemic foundations; these solution concepts and results are variants or special cases of the work of Battigalli and Siniscalchi (2003) and Forges (1993), respectively. We apply these results to the direct mechanism design setting, where the strategy space is simply the payoff-type space itself. In this environment, a specific message of a payoff type is incomplete information rationalizable if and only if there exists a type space and an interim equilibrium such that the message is an equilibrium action for a type with a given payoff type in the type space. A similar result can be stated for the incomplete information correlated equilibrium — namely, a message of a specific payoff type is an element of an incomplete information correlated equilibrium if and only if there exists a type space with a common prior for which the specific message is an interim equilibrium action for a type with that payoff type. With these epistemic results in the background, we can rephrase the conditions for robust implementation with and without common prior as establishing conditions for a unique solution under incomplete information correlated equilibrium and incomplete information rationalizability, respectively. In the case of strategic complements, the necessary and sufficient conditions for robust implementation do not depend on the existence of a common prior. In other words, with strategic complements, if truthtelling is the unique incomplete information correlated equilibrium outcome, then it is also the unique incomplete information rationalizable outcome. This reflects the well-known property of supermodular environments that multiple rationalizable outcomes occur only when there are multiple equilibria (see, e.g., Milgrom and Roberts, 1990). In contrast, with strategic substitutes, the common prior assumption changes the implementation conditions. In particular, as the number of participating agents in the public good model increases, the conditions for a unique rationalizable outcome converge to requiring pure private values, whereas the conditions for robust implementation with a common prior are independent of the number of participating agents and accommodate moderate interdependence.

The public good example considered here has two notable features that facilitate the analysis. First, the willingness to pay of agent i for the public good is the weighted sum of the payoff types of all the agents. The valuation of agent i is therefore identified by an aggregator that summarizes the private information of all agents within a one-dimensional variable. Second, the cost function of the public good is quadratic, and the resulting ex post incentive compatible transfer of agent i is a quadratic function of the reports of the agents. The quadratic payoff environment leads to a linear best response property that allows us to analyze the reporting game in the direct mechanism as a potential game. The analysis of the incomplete information correlated equilibrium is then facilitated by using potential game arguments first developed by Neyman (1997) for complete information games.

2 Setup

There are I agents. Player i has a payoff type $\theta_i \in \Theta_i$, where each $\Theta_i = [0,1]$ is a compact interval of the real line. Each agent gets utility from a social choice $x \in X$, where X is a compact set, and transfers $t_i \in \mathbb{R}$; the agent's utility is given by $u_i(x, \theta) - t_i$. A direct mechanism specifies the social choice as a function of the profile of types, $f:\Theta \to X$, and a transfer rule for each agent, $t_i:\Theta \to \mathbb{R}$. Each agent sends an announcement of his type in the form of a message, $m_i \in \Theta_i$. This mechanism $(f,(t_i)_{i=1}^I)$ is ex post incentive compatible if, for all i, θ, m_i:

$$u_i(f(\theta_i,\theta_{-i}),(\theta_i,\theta_{-i})) - t_i(\theta_i,\theta_{-i}) \geq u_i(f(m_i,\theta_{-i}),(\theta_i,\theta_{-i})) - t_i(m_i,\theta_{-i}).$$

A number of papers have described single crossing characterizations under which ex post incentive compatible transfers exist (e.g., Dasgupta and Maskin, 2000; Bergemann and Valimaki, 2002). In this article we focus on the case where they exist. If the true type profile is θ and the announced profile is m, then agent i's payoff is

$$u_i^+(m,\theta) \triangleq u_i(f(m),\theta) - t_i(m).$$

By construction, truthtelling is an ex post equilibrium in the direct mechanism. We consider two solution concepts for this setting. The first notion is incomplete information rationalizability.

Definition 1. (Incomplete Information Rationalizability) The incomplete information rationalizable correspondences are $R = (R_i)_{i=1}^I$, with each $R_i:\Theta_i \to 2^{\Theta_i}/\varnothing$, being defined recursively. Let $R_i^0(\theta_i) = \Theta_i$; then

$$R_i^{k+1}(\theta_i) = \left\{ m_i \in R_i^k(\theta_i) \;\middle|\; \begin{array}{l} \exists\, \mu_i \in \Delta(\Theta_{-i} \times \Theta_{-i}) \text{ such that} \\ \mu_i[\{(m_{-i},\theta_{-i}) : m_j \in R_j^k(\theta_j) \forall j \neq i\}] = 1 \text{ and} \\ m_i \in \arg\max\limits_{m_i'} \displaystyle\int_{m_{-i},\theta_{-i}} u_i^+((m_i',m_{-i}),(\theta_i,\theta_{-i}))d\mu_i \end{array} \right\}$$

for each $k = 1, 2, \ldots$, and

$$R_i(\theta_i) = \bigcap_{k \geq 0} R_i^k(\theta_i).$$

The second notion is the incomplete information version of the correlated equilibrium.

Definition 2. (Incomplete Information Correlated Equilibrium) A probability distribution $\mu \in \Delta(\Theta \times \Theta)$ is an incomplete information correlated equilibrium (ICE) of the direct mechanism if, for each i and each measurable

$\phi_i : \Theta_i \times \Theta_i \to \Theta_i$,

$$\int_{m,\theta} u_i^+((m_i, m_{-i}), \theta) d\mu \geq \int_{m,\theta} u_i^+((\phi_i(m_i, \theta_i), m_{-i}), \theta) d\mu.$$

We define $C_i(\theta_i)$ as the set of messages that can be played by type θ_i in an incomplete information correlated equilibrium of the direct mechanism. We will say that $m_i^* \in C_i(\theta_i^*)$ if, for each $\varepsilon > 0$, there exists an ICE μ with

$$\mu[\{(m, \theta) | m_i \in [\theta_i^* - \varepsilon, \theta_i^* + \varepsilon] \quad \text{and} \quad \theta_i \in [\theta_i^* - \varepsilon, \theta_i^* + \varepsilon]\}] > 0.$$

In Bergemann and Morris (2007c) we report on these solution concepts in a general game theoretic environment. We observe that the solution concepts $R_i(\theta_i)$ and $C_i(\theta_i)$ are "belief-free" solution concepts in the sense that they do not refer to a specific common prior or to specific higher order beliefs of the agents. Rather, they represent the sets of actions that can be observed as rationalizable or correlated equilibrium actions for some beliefs of agent i given his payoff type θ_i. In Bergemann and Morris (2007c) we show that incomplete information rationalizability and correlated equilibrium share the same epistemic properties as their complete information equivalents, as outlined in the Introduction. With these belief-free notions in place, there is no further need to refer to beliefs and higher order beliefs of agent i. As a consequence, we shall hereafter refer to the payoff type θ_i simply as the type θ_i of agent i.

3 A Public Good Example

We discuss the role and the importance of the common prior for robust implementation in a public good example with interdependent values. In the final section, we discuss which special properties of the environment are used in establishing our results. We consider the provision of a public good $x \in \mathbb{R}_+$. The utility of each agent i for the public good is given by $u_i(x, \theta) = h_i(\theta) \cdot x$, where each

$$h_i(\theta) = \theta_i + \gamma \sum_{j \neq i} \theta_j$$

aggregates the agents' payoff types. Thus the utility of agent i depends on his own type with weight 1 and the types of the other agents with weight $\gamma \in \mathbb{R}$. The weight γ represents the preference interdependence among the agents. If $\gamma = 0$ we have a private values model, whereas $\gamma < 0$

represents negative informational externalities and $\gamma > 0$ represents positive informational externalities. The cost of establishing the public good is given by $c(x) = (1/2)x^2$. The planner must choose x to maximize social welfare-that is, the sum of gross utilities minus the cost of the public good. The socially optimal level of the public good is therefore equal to

$$f(\theta) = (1 + \gamma(I-1)) \sum_{i=1}^{I} \theta_i. \tag{1}$$

The generalized Vickrey–Clarke–Groves (VCG) transfers are given by

$$t_i(\theta) = (1 + \gamma(I-1)) \left(\frac{1}{2}\theta_i^2 + \gamma\theta_i \sum_{j \neq i} \theta_j \right). \tag{2}$$

The transfers $\{t_i(\theta)\}_{i=1}^{I}$ of the generalized VCG mechanism guarantee that truthtelling is an ex post incentive compatible strategy as long as $\gamma \geq -1/(I-1)$. If the (negative) externality γ falls below this threshold, then the single crossing condition ceases to hold.

Within the generalized VCG mechanism, we can define for every agent i an ex post best response as a mapping from the true payoff types of all agents and the reported types of all agents except i into a report of agent $i : b_i : \Theta \times \Theta_{-i} \to \Theta_i$. Given that agent i has type θ_i but reports himself to be of type m_i, and given that he has a point conjecture that other agents have type profile θ_{-i}, and report their types to be m_{-i}, it follows that the net payoff of agent i is a constant $(1 + \gamma(I-1))$ times:

$$\left(\theta_i + \gamma \sum_{j \neq i} \theta_j \right) \left(m_i + \sum_{j \neq i} m_j \right) - \left(\frac{1}{2} m_i^2 + \gamma m_i \sum_{j \neq i} m_j \right). \tag{3}$$

The ex post best response of agent i is simply the solution to the first-order condition of the payoff function (3) with respect to m_i:

$$b_i(\theta, m_{-i}) \triangleq \theta_i + \gamma \sum_{j \neq i} (\theta_j - m_j). \tag{4}$$

In other words, the best response by i to a (mis)report m_{-i} by the other agents is to report a type such that the aggregate type from his point of view is exactly identical to the true aggregate type (under the aggregation function $h_i(\theta)$ of agent i) generated by the true type profile θ. We note that the foregoing calculation also verifies the strict ex post incentive compatibility of f, because setting $m_i = \theta_i$ is the unique best response if $m_j = \theta_j$ for all $j \neq i$.

Whether there are strategic complements or substitutes plays an important role in determining the role of the common prior in implementation. We say that the strategies of i and j are *strategic complements* if $\partial b_i(\theta, m_j, m_{-ij})/\partial m_j > 0$, and they are *strategic substitutes* if $\partial b_i(\theta, m_j, m_{-ij})/\partial m_j < 0$, for all i, j, θ, m. Given the best response (4), it follows that the reports of the agents are strategic complements if there are negative informational externalities ($\gamma < 0$) but are strategic substitutes if there are positive informational externalities ($\gamma > 0$).

Bergemann and Morris (2007a) showed that if interdependence is small with $\gamma \in (-1/(I-1), +1/(I-1))$, then truthtelling is the unique rationalizable outcome and for all i and θ_i, $R_i(\theta_i) = \{\theta_i\}$, but if the interdependence is large with $\gamma \notin (-1/(I-1), +1/(I-1))$ then any message is rationalizable and for all i and θ_i, $R_i(\theta_i) = [0, 1]$. We refer the reader to Bergemann and Morris (2007a) for a formal statement and the derivation of this result. There we present necessary and sufficient conditions for robust implementation in environments where, for each agent i, the payoff types of all agents can be aggregated within a one-dimensional variable. The environment is general in the sense that neither the aggregator nor the utility function of each agent i must be linear as in the current example. We show that robust implementation is possible in any mechanism if and only if it is possible in the direct mechanism; and we show that robust implementation is possible if and only if the aggregator function satisfies a contraction property that reduces to the condition of a small γ with $\gamma \in (-1/(I-1), +1/(I-1))$.

In this article, we contrast the uniqueness result with incomplete information rationalizability with the incomplete information correlated equilibrium case. We use results regarding the uniqueness of the incomplete information correlated equilibrium in potential games derived in Bergemann and Morris (2007c). We say that a belief-free incomplete information game

$$\Gamma = \{I, \{A_i\}_{i=1}^I, \{\Theta_i\}_{i=1}^I, \{u_i(a, \theta)\}_{i=1}^I\}$$

has a *weighted potential* $v : A \times \Theta \to \mathbb{R}$ if there exists a $w \in \mathbb{R}_{++}^I$ such that

$$u_i((a_i, a_{-i}), \theta) - u_i((a_i', a_{-i}), \theta) = w_i[v((a_i, a_{-i}), \theta) - v((a_i', a_{-i}), \theta)]$$

for all i and all $a_i, a_i' \in A_i, a_{-i} \in A_{-i}$ and $\theta \in \Theta$. This is an incomplete information generalization of the definition of a weighted potential in Monderer and Shapley (1996); in particular, it is equivalent to requiring that each complete information game $(u_i(\cdot,\theta))_{i=1}^{I}$ be a weighted potential game in the sense of Monderer and Shapley, using the same weights for each $\theta \in \Theta$. We say that v is a *strictly concave potential* if $v(\cdot,\theta)$ is a strictly concave function of a for all $\theta \in \Theta$. In Bergemann and Morris (2007c) we show that, if Γ has a strictly concave smooth potential function *and* an ex post equilibrium s^*, then $s_i^*(\theta_i) = C_i(\theta_i)$ for all i and θ_i. In the direct mechanism, the set of actions is the set of types. We argued earlier that the direct mechanism has truthtelling as an ex post equilibrium provided that $\gamma \geq -1/(I-1)$. By the result in Bergemann and Morris (2007c), the sufficiency condition for a unique correlated equilibrium can then be established by verifying the existence of a potential of the direct mechanism that is strictly concave.

Proposition 1. (Incomplete Information Correlated Equilibrium) *The set of incomplete information correlated equilibrium actions for all i and θ_i is $C_i(\theta_i) = \{\theta_i\}$ if and only if $\gamma \in (-1/(I-1), 1)$.*

Proof. We first establish the sufficiency result. We consider the following function $v(m,\theta)$ as a potential function for the direct mechanism:

$$v(m,\theta) = -\sum_{i=1}^{I}(m_i-\theta_i)\left[(m_i-\theta_i)+\gamma\sum_{j\neq i}(m_j-\theta_j)\right].$$

The partial derivative of the function $v(m,\theta)$ with respect to m_i is

$$\frac{\partial v}{\partial m_i}(m,\theta) = -2(m_i-\theta_i) - 2\gamma\sum_{j\neq i}(m_j-\theta_j), \tag{5}$$

and the cross-derivatives are given by

$$\frac{\partial^2 v}{\partial m_i \partial m_j}(m,\theta) = \begin{cases} -\dfrac{1}{2} & \text{if } j = i, \\ -\dfrac{1}{2}\gamma & \text{if } j \neq i. \end{cases} \tag{6}$$

It follows from (6) that $v(m,\theta)$ is a potential function, and it follows from (5) and the ex post best response (4) that v is a potential for the belief free incomplete information game Γ with type- and person-independent weights $w_i = 1/2$. Finally, because $v(m,\theta)$ is a quadratic function, the (constant) Hessian H is given by definition (6). With elementary linear algebra, we can now verify that H is negative semi-definite if and only if $-1/(I-1) \leq \gamma \leq 1$ and that H is negative definite if $-1/(I-1) < \gamma < 1$. Now the potential function is strictly concave if and only if its Hessian H is negative definite, which establishes sufficiency.

The necessity result follows from the best-response function (4). It suffices to show that, for $\gamma \geq 1$, there exist complete information correlated equilibria that do not involve truthtelling. Consider a payoff type profile $\hat{\theta}$ given by $\hat{\theta} = (1/2, 1/2, \hat{\theta}_3, \ldots, \hat{\theta}_I$ for some $\hat{\theta}_{-12} = (\hat{\theta}_3, \ldots, \hat{\theta}_I)$. Consider the correlated equilibrium μ with $\mu((0, 1, \hat{\theta}_3, \ldots, \hat{\theta}_I), (1/2, 1/2, \hat{\theta}_3, \ldots, \hat{\theta}_I)) = 1$. It is easy to verify that the equilibrium conditions (4) will be satisfied at μ. We observe that, for $\gamma > 1$, the equilibrium condition (4) will be a corner solution and hence there will be strict inequalities for $i = 1, 2$. □

The results in Bergemann and Morris (2007a) together with Proposition 1 show how the common prior assumption has a considerable impact on the possibility of robust implementation with positive interdependence (and thus strategic substitutes) but no impact with negative interdependence (and thus strategic complementarities). The following corollary is then an immediate consequence of Bergemann and Morris (2007a) and Proposition 1.

Corollary 1. (Robust Implementation)

1. *If the reports are strategic complements, then robust implementation with common prior implies robust implementation without common prior.*
2. *If the reports are strategic substitutes, then robust implementation with common prior fails to imply robust implementation without common prior.*

The public good example shows how large the gap between robust implementation (with or without common prior) can be. In particular, as the number of agents increases, essentially only the private value model with $\gamma = 0$ can be robustly implemented without a common prior. In contrast, the interdependent model can be robustly implemented with a common prior for all $\gamma < 1$. This shows that the role of the common prior is critical in many mechanism design environments and for our understanding of robust implementation.

4 Discussion

Strategic complements and strategic substitutes. In the linear best-response environment of the public good problem, the notions of strategic complement and strategic substitute emerged directly from the best response. In general environments with differentiable mechanisms, the link between the information externality and the strategic properties of the reports have yet to be shown. For preciseness if we assume that the supermodularity conditions: $\partial f/\partial \theta_i > 0$ and $\partial^2 u_i/\partial x \partial \theta_i > 0$ support ex post incentive compatibility, then it follows that, locally at truthful reporting, the strategies of the agents are strategic substitutes if $\partial^2 u_i/\partial x \partial \theta_j > 0$ and strategic complements if $\partial^2 u_i/\partial x \partial \theta_j < 0$.

Potential and mechanism design. Our proof that there is a unique incomplete information correlated equilibrium action for each payoff type if $\gamma \in (-1/(I-1), 1)$ used the fact that we could construct a potential function for the direct mechanism. This turns out to be a very strong property. In a differentiable environment, it is straightforward to show that a sufficient condition for Bayesian potential games is that the cross-derivatives of agent i and j equalize at every true and reported type profile. Although the cross-derivative is equal to zero in the current linear quadratic environment (because f is linear), we lose that property if we replace the quadratic cost function with a general concave cost function. It remains an open question to identify a larger class of environments where the potential argument goes through.

Jehiel, Meyer-Ter-Vehn, and Moldovanu (forthcoming) have used potential arguments to characterize when ex post incentive compatible transfers exist. Their definition of the potential function implicitly assumes that the agents are telling the truth and so — as the authors note — this is a much weaker requirement than the one that the direct mechanism be a potential game.

References

Aumann, R. (1987): "Correlated Equilibrium as an Expression of Bayesian Rationality," *Econometrica*, 55, 1–18.

Battigalli, P., and M. Siniscalchi (2003): "Rationalization and Incomplete Information," *Advances in Theoretical Economics*, 3, Article 3.

Bergemann, D., and S. Morris (forthcoming). "Ex Post Implementation," *Games and Economic Behavior.*

Bergemann, D., and S. Morris (2007a): "Robust Implementation: The Case of Direct Mechanisms," Discussion paper No. 1561R, Cowles Foundation for Research in Economics, Yale University.

BERGEMANN, D., AND S. MORRIS (2007b): "Strategic Distinguishability with an Application to Robust Virtual Implementation," Discussion paper No. 1609, Cowles Foundation, Yale University.

BERGEMANN, D., AND S. MORRIS (2007c): "Belief Free Incomplete Information Games," Discussion paper no. 1629, Cowles Foundation for Research in Economics, Yale University.

BERGEMANN, D., AND J. VALIMAKI (2002): "Information Acquisition and Efficient Mechanism Design," *Econometrica*, 70, 1007–1033.

BRANDENBURGER, A., AND E. DEKEL (1987). "Rationalizability and Correlated Equilibria," *Econometrica*, 55, 1391–1402.

DASGUPTA, P., AND E. MASKIN (2000): "Efficient Auctions," *Quarterly Journal of Economics*, 115, 341–388.

FORGES, F. (1993): "Five Legitimate Definitions of Correlated Equilibrium in Games with Incomplete Information," *Theory and Decision*, 35, 277–310.

JEHIEL, P., M. MEYER-TER-VEHN, AND B. MOLDOVANU (forthcoming). "Ex-Post Implementation and Preference Aggregation Via Potentials," *Economic Theory*.

MILGROM, P., AND J. ROBERTS (1990): "Rationalizability, Learning and Equilibrium in Games with Strategic Complementarities," *Econometrica*, 58, 1255–1277.

MONDERER, D., AND L. SHAPLEY (1996): "Potential Games," *Games and Economic Behavior*, 14, 124–143.

NEYMAN, A. (1997): "Correlated Equilibrium and Potential Games," *International Journal of Game Theory*, 26, 223–227.

CHAPTER 6

An Ascending Auction for Interdependent Values: Uniqueness and Robustness to Strategic Uncertainty*

Dirk Bergemann and Stephen Morris

The important role of dynamic auctions, in particular ascending price auctions, for the revelation of private information has been recognized for a long time. The advantage of sequential procedures is the ability to reveal and communicate private information in the course of the mechanism. The revelation of private information can decrease the uncertainty faced by the bidders and ultimately improve the final allocation offered by the mechanism. In auctions, the source of the uncertainty can be payoff uncertainty (about others' payoff relevant information) or strategic uncertainty (about their bidding strategies).

The ability of dynamic auctions to reduce payoff uncertainty is well documented in the literature. In a setting with interdependent values, the seminal paper by Paul R. Milgrom and Robert J. Weber (1982) shows that the ascending price auction leads to larger expected revenues by weakening

* *Discussants:* Preston McAfee, California Institute of Technology; Eric Maskin, Institute for Advanced Studies, Princeton University; Paul Milgrom, Stanford University. This research is partially supported by National Science Foundation grants CNS 0428422 and SES-0518929. We thank our discussant Preston McAfee as well as Dilip Abreu, Tilman Borgers, Liran Einav, and Ilya Segal for comments. We thank Richard van Weelden for excellent research assistance.

the winner's curse problem. The ascending price auction leads to sequential revelation of good news for the active bidder. As the bidding proceeds and the price for the object increases, each active bidder revises upward his estimate of the private information of the remaining bidders. The continued presence of active bidders represents a flow of good news about the value of the object. As a consequence, each active bidder becomes less concerned about exposure to the winner's curse.

The objective of this paper is to argue that ascending auctions also offer benefits for the reduction of *strategic* uncertainty. We consider an environment with interdependent values and complete information. The complete information assumption removes payoff uncertainty and focuses our analysis on the role of strategic uncertainty. We show that, under ex post incentive compatible allocation rules, strategic uncertainty (i.e., multiple rationaliz-able outcomes) necessarily occurs in a static mechanism but does not occur in the ascending auction format.

We introduce strategic uncertainty by analyzing the rationalizable outcomes of static and dynamic versions of a generalized Vickrey–Clarke–Groves (VCG) mechanism. The relationship between rationalizability and strategic uncertainty has been established in Adam Brandenburger and Eddie Dekel (1987). In a complete information environment, they show that the set of rationalizable outcomes is equivalent to the set of outcomes of Nash equilibria in some type space. We appeal to this epistemic result and analyze the outcomes of static and dynamic auction formats under rationalizability.

An important difference emerges as we compare the set of rationalizable outcomes in the static and dynamic auction format. In the static auction, the efficient outcome is the unique rationalizable outcomes if and only if the interdependence in the valuation of the agents is moderate. In contrast, the efficient outcome will remain the unique rationalizable outcome in the dynamic auction as long as a much weaker single crossing condition prevails.

In the interdependent value environment, the reports of the bidders are strategic substitutes. If bidder i increases his bid for a given valuation, then bidder j has an incentive to lower his report. An increase in the report by bidder i makes the object more costly to obtain without changing its value. As a consequence, bidder j will lower his report to partially offset the increase in the payment for the object induced by bidder i. The element of strategic substitutes between the reports of bidder i and j is generated by the incentive compatible transfer scheme rather than the signal of the agents directly.

The discrepancy between the static and the dynamic version of the auction is due to the ability of the dynamic mechanism to partially synchronize the beliefs of the agents. In the static auction a low bid by agent i can be justified by high bids of the remaining bidders. But, in turn, a large bid by bidder j requires bidder j to believe in low bids by the remaining bidders. The beliefs of bidders i and j about the remaining bidders are thus widely divergent. In the dynamic auction, the current report of each bidder represents a lower bound on the beliefs of all the agents, and hence imposes a synchronization on the belief. In addition, in the dynamic auction, the bidders look ahead and consider only rationalizable future outcomes. This forces each bidder to have a belief about the future actions of the other bidders which is rationalizable.

1 Model

We consider an auction environment with interdependent values. There are I agents competing for a single object offered by a seller. The payoff type of agent i is given by a realization $\theta_i \in [0,1]$. The type profile is given by $\theta = (\theta_i, \theta_{-i})$, and agent i's valuation of the object at the type profile θ is given by $v_i(\theta_i, \theta_{-i}) = \theta_i + \gamma \sum_{j \neq i} \theta_j$, with $\gamma \in \mathbb{R}_+$. The net utility of agent i depends on his probability q_i of receiving the object and the monetary transfer y_i:

$$u_i(\theta, q_i, y_i) = \left(\theta_i + \gamma \sum_{j \neq i} \theta_j \right) q_i - y_i. \tag{1}$$

The socially efficient allocation rule is given by $\hat{q}_i(\theta) = 1/(\#\{j : \theta_j \geq \theta_k \text{ for all } k\})$, if $\theta_i \geq \theta_k$ for all $k, \hat{q}_i(\theta) = 0$, if otherwise.

Partha Dasgupta and Eric S. Maskin (2000) have shown that a generalized VCG auction leads to truthful revelation of private information in ex post equilibrium. In the generalized VCG auction, the monetary transfer of the winning agent i is given by

$$\hat{y}_i(\theta) = \max_{j \neq i} \theta_j + \gamma \sum_{k \neq i} \theta_k, \tag{2}$$

and the losing bidders all have a zero monetary transfer. The generalized VCG mechanism guarantees weak rather than strict ex post incentive compatibility conditions. We seek to analyze the strategic behavior in the auction in terms of rationalizable behavior. As rationalizabil-ity involves the iterative elimination of strictly dominated actions, we modify the generalized VCG mechanism to display strict ex post incentive constraints everywhere. We add to the VCG allocation rule, $\hat{q}_i$, an allocation rule that increases proportionally in the report of agent i:

$$q_i(\theta') = \frac{\theta'_i}{I} \quad \text{for all } i. \tag{3}$$

The modified VCG allocation rule is now denned for some $\varepsilon > 0$ by

$$q_i^*(\theta) = \varepsilon q_i(\theta) + (1-\varepsilon)\hat{q}_i(\theta). \tag{4}$$

The modified allocation rule is supported by an associated set of transfers:

$$\begin{aligned} y_i^*(\theta) = {} & \frac{\varepsilon}{2I}\theta_i^2 + \frac{\varepsilon\theta_i}{I}\left(\gamma\sum_{j\neq i}\theta_j\right) \\ & + (1-\varepsilon)\left(\max_{j\neq i}\theta_j + \gamma\sum_{k\neq i}\theta_k\right)\hat{q}_i(\theta). \end{aligned} \tag{5}$$

The transfer rule $y_i^*(\theta)$ leads to strict truth-telling incentives everywhere. The outcome function of the direct mechanism is denoted by $f^* = (q_i^*, y_i^*)_{i=i}^I$.

Truth-telling is a strict ex post equilibrium of the mechanism above. This means that whatever the agents' beliefs and higher-order beliefs about other agents' types, there exists a strict equilibrium where every agent tells the truth. This does not guarantee that there do not exist other, non-truth-telling equilibria, however. In the remainder of this paper, we fix this mechanism, which is designed to deal with incentive compatibility problems under general *incomplete information* structures, and examine the performance of static and dynamic versions of the mechanism under *complete information.*

2 Static Auction

First, we analyze the generalized VCG mechanism in a static environment. The purpose of this section is to provide a background for the analysis of the ascending auction. We then show that the ascending auction leads to a unique rationalizable outcome under very weak condition on the interaction parameter γ. More precisely, the set of rationalizable outcomes consists of a singleton for each bidder if $\gamma < 1$. This condition is weak, as $\gamma < 1$ is necessary and sufficient for the single crossing condition to hold. In contrast, the static version of a generalized VCG auction leads to the unique rationalizable outcome if and only if the interdependence is moderate, or $\gamma < 1/(I-1)$.

Proposition 1 is a special case of a general uniqueness result in environments with interdependent values and incomplete information in Bergemann and Morris (2006). The analysis in the present case is substantially simplified by the complete information assumption as well as the linear and symmetric valuation structure.

The net utility of agent i in the modified VCG mechanism depends on the true type profile θ and the reported profile θ':

$$u_i(\theta, f^*(\theta')) = \left(\theta_i + \gamma \sum_{j \neq i} \theta_j\right) q_i^*(\theta') - y_i^*(\theta').$$

We insert the outcome function f^* given by (4) and (5) to obtain the net utility of i:

$$\begin{aligned} u_i(\theta, f^*(\theta')) = {} & \left(\theta_i + \gamma \sum_{j \neq i} \theta_j\right) \left[\frac{\varepsilon}{I}\left(\theta_i' + \gamma \sum_{j \neq i} \theta_j'\right) + (1-\varepsilon)\hat{q}_i(\theta')\right] \\ & - \frac{\varepsilon}{2I}\theta_i'^2 - \frac{\varepsilon\gamma\theta_i'}{I}\sum_{j \neq i}\theta_j' - (1-\varepsilon)\left(\max_{j \neq i}\theta_j' + \gamma\sum_{k \neq i}\theta_k'\right)\hat{q}_i(\theta'). \end{aligned}$$

The net utility function is a linear combination of the efficient allocation rule and the proportional allocation rule. It is straightforward to compute the best response of each agent i given a point belief about the reports θ'_{-i} of the remaining agents. The best response is linear in the true valuation and the size of the downward or upward report of the other agents:

$$\theta_i' = \theta_i + \gamma \sum_{j \neq i} (\theta_j - \theta_j').$$

From here, it follows that the report of agent i and agent j are strategic substitutes. If agent j increases his report, then, in response, agent i optimally chooses to lower his report. The linear best response structure facilitates the analysis. In order to establish the largest possible report of agent i, it suffices to look at the lowest possible reports by all other agents. The process of elimination can therefore proceed based on specific point beliefs about minimal and maximal reports by the agents.

Proposition 1. (Static auction): *The rationalizable outcome is unique and coincides with truth-telling if and only if* $\gamma < \frac{1}{I-1}$.

3 Dynamic Auction

We now consider a dynamic version of the generalized VCG mechanism, namely an ascending auction in continuous time. The auction begins with the clock running and each bidder participating in the auction. Each bidder can choose to exit the auction at any point in time. The exit decision is irrevocable and presents a commitment. Similarly, a decision to stay in the game may be viewed as a partial commitment to bid at least as much as indicated by the current decision. The decision of each player is therefore to let the clock continue or to stop it. The time interval is the unit interval $t \in [0, 1]$, and the game ends at $T = 1$.

Agents will choose strategies that are rationalizable in every subgame. Given the perfect information, simultaneous move nature of the game, this will imply that we can characterize rationalizable outcomes essentially by backward induction in terms of recursive best response functions. Thus, we do not need to appeal to the forward induction logic built into David G. Pearce's (1984) notion of extensive form rationalizability.

In the ascending auction, the i-th bidder to exit the auction will choose a best response to the actions of the other bidders. Conditional on being the i-th bidder to leave, however, his best response will *distinguish* between the actions of the bidder who left before him and those who leave after him. As bidder i cannot influence the timing of bidders who already left the game, he will choose a best response to their actions. As for the actions of the bidders leaving the game after i, bidder i will have rational expectations as to how his timing will affect their future choices. Without loss of generality, we relabel the bidders so that we have ascending bidding times in the index $i : t_1 \leq t_2 \leq \cdots \leq t_1$. Given the stopping times of all other bidders, the stopping time t_I is simply the best response to the past stopping times. We denote the best response of bidder I by $\beta_I(t_1, t_2, \ldots, t_{I-1})$, and the best response of i to the stopping decisions of the preceding bidders is $\beta_i(t_1, \ldots, t_{i-1})$.

We can solve for the best response functions recursively. The stopping time of the last remaining bidder is his best response to the stopping times of the preceding bidders:

$$t_I = \theta_I + \gamma \sum_{i=1}^{I-1} (\theta_i - t_i). \tag{6}$$

The best response of the penultimate bidder $I-1$ is

$$\begin{aligned} t_{I-1} &= \theta_{I-1} + \gamma \sum_{i \neq I-1} (\theta_i - t_i) \\ &= \theta_{I-1} + \gamma \sum_{i=1}^{I-2} (\theta_i - t_i) + \gamma(\theta_I - t_I). \end{aligned}$$

Now, bidder $I-1$ anticipates the best response of bidder I to all previous stopping decisions. We can thus insert the best response of bidder I, (6), into the best response of bidder $I-1$. As a consequence, we obtain the best response of bidder $I-1$ to all preceding bidders:

$$t_{I-1} = \theta_{I-1} + \frac{\gamma}{1+\gamma} \sum_{i=1}^{I-2} (\theta_i - t_i).$$

We inductively obtain the best response for bidder $I-j$ for all $j = 0, 1, \ldots, I-1$ as:

$$\beta_i(t_1, t_2, \ldots, t_{i-1}) = \theta_i + \alpha_i \sum_{j=1}^{i-1} (\theta_j - t_j), \tag{7}$$

where the slope α_i of the best response depends on the exit position of bidder i:

$$\alpha_i = \frac{\gamma}{1 + (I-i)\gamma}. \tag{8}$$

While bidder i, with an early exit, responds more moderately to preceding bidders, at the same time an early exit by i gives bidder i the possibility to influence the decision of all succeeding bidders. In this sense, an early exit gives bidder i more strategic influence than a late exit would give bidder i. In order to induce truth-telling in the dynamic bidding game, we therefore have to account for the strategic influence in the monetary transfers. The strategic weight of each exit position suggests that the transfer functions should account for the difference across exit decisions. We therefore modify the monetary transfer (5) to account for the strategic weight:

$$y_i(t) = \frac{1}{2I} w_i t_i^2 + \frac{t_i}{I} \gamma \sum_{j \neq i} t_j, \tag{9}$$

where w_i is the strategic weight of bidder i. In the static mechanism, we implicitly assigned each agent the same strategic weight equal to one. In the dynamic game, the weights are given by the direct and indirect effects that bidder i has on the stopping times of all successive bidders. The weight w_i is simply the marginal effect that an increase in the stopping time of agent i has on the behavior of successive bidders, or

$$\begin{aligned} w_i &= \frac{1 + (I - i - 1)\gamma + (I - i)\gamma^2}{1 + (I - i - 1)\gamma} \\ &= 1 + \frac{(I - i)\gamma^2}{1 + (I - i - 1)\gamma} \end{aligned} \tag{10}$$

The dynamic game is solved recursively by means of the best response functions (7).

Proposition 2. (Dynamic Auction): *The rationalizable outcome is unique and coincides with truth-telling if* $\gamma < 1$.

The proof of Proposition 2 is in the Web Appendix available at www.e-aer.org/data/may07/P07044_app.pdf.[1]

The dynamic game introduces the possibility that a player can strategically commit in order to affect the behavior of the other agents. The analysis above suggests that the strategic value of the commitment does not interfere with our analysis. The absence of a strategic value of commitment is due, here, to the careful design of the monetary transfers that neutralize the strategic value of commitment.

4 Discussion

Incomplete Information. — In the current analysis, consider a game in which the bidding agents had complete information about their types. The focus on the complete information environment allowed us to interpret the

[1]The appendix is also available in the Cowles Foundation Discussion Paper 1600 available at http://ssrn.com/abstract=959444.

remaining uncertainty about the actions of the players as pure strategic uncertainty. We could then appeal to the epistemic analysis of Brandenburger and Dekel (1987) to interpret the set of rationalizable strategies as Nash equilibria in some type space. Naturally, it is of interest to ask how the results presented here would be affected by the introduction of incomplete information among the bidders. The nature of the argument in Proposition 2 suggests that the result may partially survive in an incomplete information environment. The best response of each bidder was largely a function of his own payoff type and his exit position in the auction. This information will still be available to him in the incomplete information game. Moreover, in the inductive process, the payoff type of agents exiting after i dropped out, and hence the information about future bidders, did not enter either.

Partial Commitment. — An ascending auction requires players to make partial commitments during the play of the game. By not dropping out at the current price, I commit to bid something strictly higher than the current price, but I do not commit to what it will be. There is a literature looking at how gradual partial commitment can help resolve multiplicity of equilibrium outcomes (e.g., Guillermo Caruana and Liran Einav, 2006). In our setting, the partial commitment reduces multiple static rationalizable outcomes to a unique dynamic rationalizable outcome. There is an important difference, however. For us, partial commitment reduces outcomes only by reducing strategic uncertainty, and we carefully adjust the transfers to ensure that players do not have an incentive to use their commitment to alter others' reports. Our dynamic selection is analogous to the classic observation that in a simple coordination game with multiple equilibria, a Pareto efficient equilibrium is selected if players choose sequentially (Douglas Gale, 1995).

Implementation in Refinements of Nash Equilibrium. — John Moore and Rafael Repullo (1988), Dilip Abreu and Arunava Sen (1990), and others have examined abstract settings where sequential rationality refinements of Nash equilibrium in dynamic mechanisms can be used to strengthen implementation results. Our example in this note belongs to this tradition, although our results seem to have a more direct intuition than the canonical mechanisms in that literature.

5 Conclusion

We analyzed strategic bidding behavior in a static and dynamic auction in an environment with interdependent values. We analyzed the static and dynamic version of the auction under rationalizability. We interpreted the results from rationalizability with a well-known epistemic point of view. The rationalizable behavior in the ascending auction was determined to be unique in a substantially larger class of environments than the static auction. The dynamic auction allowed the agents to update their beliefs about the behavior of their competitors. As the decision to stay or to exit is common knowledge among the bidders, thc ascending auction makes the strategic decision of the agent public. As a consequence, the ascending auction reduces the strategic uncertainty among the bidders and leads to tighter prediction of the behavior of the agents. This suggests a new and important advantage of ascending auctions over sealed bid auctions. By reducing the strategic uncertainty, the ascending auction severely limits the possibility of multiple equilibria to arise from the auction.

References

Abreu, D., and A. Sen (1990): "Subgame Perfect Implementation: A Necessary and Almost Sufficient Condition," *Journal of Economic Theory*, 50(2), 285–99.

Bergemann, D., and S. Morris (2006): "Robust Implementation: The Case of Direct Mechanisms," Yale University Cowles Foundation Discussion paper 1561.

Brandenburger, A., and E. Dekel (1987): "Rationalizability and Correlated Equilibria," *Econometrica*, 55(6), 1391–1402.

Caruana, G., and L. Einav (2006): "A Theory of Endogenous Commitment," Unpublished.

Dasgupta, P., and E. S. Maskin (2000): "Efficient Auctions," *Quarterly Journal of Economics*, 115(2), 341–88.

Gale, D. (1995): "Dynamic Coordination Games," *Economic Theory*, 5(1), 1–18.

Milgrom, P. R., and J. W. Robert (1982): "A Theory of Auctions and Competitive Bidding," *Econometrica*, 50(5), 1089–1122.

Moore, J., and R. Repullo (1988): "Sub-game Perfect Implementation," *Econometrica*, 56(5), 1191–1220.

Pearce, D. G. (1984): "Rationalizable Strategic Behavior and the Problem of Perfection," *Econometrica*, 52(4), 1029–50.

CHAPTER 7

Robust Virtual Implementation*

Dirk Bergemann and Stephen Morris

Abstract
In a general interdependent preference environment, we characterize when two payoff types can be distinguished by their rationalizable strategic choices without any prior knowledge of their beliefs and higher order beliefs. We show that two payoff types are *strategically distinguishable* if and only if they satisfy a separability condition. The separability condition for each agent essentially requires that there is not too much interdependence in preferences across agents.

A social choice function — mapping payoff type profiles to outcomes — can be *robustly virtually implemented* if there exists a mechanism such that every equilibrium on every type space achieves an outcome arbitrarily close to the social choice function. This definition is equivalent to requiring virtual implementation in iterated deletion of strategies that are strictly dominated for all beliefs. The social choice function is *robustly measurable* if strategically indistinguishable payoff types receive the same allocation. We show that ex post incentive compatibility and robust measurability are necessary and sufficient for robust virtual implementation.

Keywords: Mechanism design, virtual implementation, robust implementation, rationalizability, ex-post incentive compatibility

*This research is partially supported by NSF Grants #CNS 0428422 and #SES-0518929. We thank the co-editor, Jeffrey Ely, and two anonymous referees for their valuable comments and suggestions. We are grateful for discussions with Dilip Abreu, Faruk Gul, Matt Jackson, Eric Maskin, Wolfgang Pesendorfer, Phil Reny, Roberto Serrano and seminar audiences at Chicago, Georgetown, Hong Kong, IMPA, NUS, Penn, Princeton, Rutgers and UT Austin. This paper incorporates and replaces some preliminary results on robust virtual implementation appearing in Bergemann and Morris (2005b).

1 Introduction

Suppose that a social planner would like to design a mechanism that will induce self-interested agents to make strategic choices that will lead to the selection of socially desirable outcomes. A *social choice function* specifies the social desired outcomes as a function of unobserved payoff types of the agents. The planner would like to be sure that outcomes specified by the social choice function arise with probability arbitrarily close to 1: thus she requires *virtual* implementation; she would like *every* possible equilibrium to virtually implement the social choice function: thus she requires *full* implementation; and she would like every equilibrium to virtually implement the social choice function whatever the agents' beliefs and higher order beliefs about others' types; thus she requires *robust* implementation. In this paper, we provide a characterization of when *robust virtual implementation* is possible in a general interdependent preference environment.

One necessary condition for robust virtual implementation will be *ex post incentive compatibility*: under the social choice function, each agent must have an incentive to truthfully report his type if others report their types truthfully, whatever their types. Ex post incentive compatibility is sufficient to ensure the existence of desirable equilibria, but, as the existing incomplete information implementation literature has emphasized, further restrictions on the social choice function are required to rule out other, undesirable, equilibria. If a mechanism is to fully implement a social choice function, it must be that two types who are treated differently by the social choice function are guaranteed to behave differently in the implementing mechanism. The key result in this paper is a characterization of when two payoff types are *strategically distinguishable* in this sense that they can be guaranteed to behave differently. Now a second necessary condition for robust virtual implementation will be *robust measurability*: strategically indistinguishable types are treated the same by the social choice function. We show that ex post incentive compatibility and robust measurability are also sufficient for robust virtual implementation (under an economic assumption).

Thus the core of our contribution is an analysis of strategic distinguishability. Fix an interdependent preferences environment, with a finite set of agents, each with a finite set of possible *payoff types*, with expected utility preferences over lotteries depending on the whole profile of types. Two payoff types of an agent are *strategically distinguishable* if they have disjoint rationalizable strategic choices in some finite game for all possible beliefs and higher order beliefs about others' types. Thus a pair of payoff types are strategically *indistinguishable* if in every game, there exists some action which each type might rationally choose given some beliefs and higher order beliefs. We are able to provide an exact and insightful characterization of strategic distinguishability. If we have sets of types, Ψ_1 and Ψ_2, of agents 1 and 2, respectively, we say that Ψ_2 separates Ψ_1 if knowing agent l's preferences and knowing that agent 1 is sure that agent 2's type is in Ψ_2, we can rule out at least one type of agent 1. Now consider an iterative process where we start, for each agent, with all subsets of his type set and — at each stage — delete subsets of actions that are separated by every remaining subset of types of his opponents. A pair of types are said to be *pairwise inseparabley* if the set consisting of that pair of types survives this process. We show that two types are strategically indistinguishable if and only if they are pairwise inseparable.

If there are private values and every type is value distinguished, then every pair of types will be pairwise separable and thus strategically distinguishable. Thus strategic indistinguishability arises when the degree of interdependence in preferences is large. We can illustrate this with a simple example. Suppose that agent i's payoff type is $\theta_i \in [0, 1]$ and agent i's valuation of a private good is $\theta_i + \gamma \sum_{j \neq i} \theta_j$. Each agent has quasilinear utility, i.e., his utility from money is linear and additive. We show all distinct pairs of types are strategically distinguishable if $|\gamma| < \frac{1}{I-1}$ where I is the number of agents. All pairs of types are strategically indistinguishable if $|\gamma| \geq \frac{1}{I-1}$.

Our characterization result for strategic distinguishability (Theorem 1) comes in two parts. If two types of an agent are pairwise inseparable, then they belong to a set of types which are not separable by a profile of sets of types of that agent's opponents. The set of types of each opponent in that profile is then not separable by a profile of sets of types of that opponent's opponents. And there is a continuing chain of inseparable sets in the chain. We prove that pairwise inseparable types are strategically indistinguishable (Proposition 1) by induction, showing that in any mechanism at any stage in the iterated deletion of messages that are never best responses and for every set of types in the chain of inseparable type sets, there is a common action which is played. The inseparability property ensures that we can always construct beliefs for each type that make the same message a best response.

To show the converse result (Proposition 2), we construct a finite *maximally revealing mechanism* with the property that all pairwise separable types have disjoint sets of rationalizable actions. The construction exploits the linearity of expected utility preferences and duality theory. Whenever a set of types of one agent is separated by a profile of sets of types of other agents, we are able to construct a finite set of lotteries such that knowing the first agent's preference over those lotteries will always rule out at least one of his types. We can take the union over all such finite sets constructed for each profile of type sets where the separability property holds. We then construct a finite "test set" of lotteries such that knowing an agent's most preferred outcome in that test set implicitly reveals his ranking of outcomes in all the original sets. Finally, we consider a mechanism where each agent gets to pick a lottery with some positive probability, then guesses which lotteries others chose and gets to pick another lottery, with small probability, contingent on other agents making the choice he conjectured, and so on. With a large, but finite, number of stages this mechanism will eventually lead pairwise separable types to make distinct choices.

Our proof of the sufficiency of ex post incentive compatibility and robust measurability (corollary 1) for robust virtual implementation builds on an ingenious construction used by Abreu and Matsushinia (1992b) to establish an extremely permissive result for complete information virtual implementation; in Abreu and Matsushinia (1992c), they adapted the argument to a standard Bayesian virtual implementation problem; we in turn adapt the argument to our robust virtual implementation problem.

While our sufficiency argument for robust virtual implementation builds on Abreu and Matsushima (1992c), the interpretation of our results ends up being rather different. Abreu and Matsushima (1992c) characterized virtual implementation in a standard Bayesian environment, where there was common knowledge of a common prior over a fixed set of types, using the solution concept of iterated deletion of strictly dominated strategies and restricting attention to well-behaved (finite) mechanisms. Bayesian incentive compatibility of the social choice function is a necessary condition: a standard compactness argument shows that the weakening to *virtual* implementation does not weaken the incentive compatibility requirement. In addition, they showed that a *measurability* condition was necessary. Put each agent's types into equivalence classes that have the same preferences over outcomes, unconditional on other agents' types. Having distinguished some types by their unconditional preferences, we can then further refine agents' types, by distinguishing types with different preferences conditional on other agents' types in the first stage. We can continue this process of refining agents' types based on preferences conditional on other agents' types revealed so far. The social choice function is *Abreu–Matsushima measurable* if it is measurable with respect to the limit of this iterative refinement. This seems to be a weak restriction that is genericalfy satisfied.[1] Abreu and Matsushima (1992c) show that Bayesian incentive compatibility and Abreu–Matsushima nieasurabihty are sufficient as well as necessary for virtual implementation in iterated deletion of strictly dominated strategies.

Robust virtual implementation is equivalent to requiring that there is a single mechanism that implements a social choice function, for all possible type spaces that could be constructed for the environment with fixed payoff types and utility functions for the agents. It is instructive to see how to get from Abreu and Matsushima (1992c) to the robust virtual implementation results in this paper.

Observe that Abreu and Matsushima (1992c)'s solution concept naturally uses agents' given beliefs about others' types in their solution concept: when strategies are deleted, it is because they are strictly dominated conditional on their beliefs. We want implementation for all possible beliefs; we therefore establish our results under an incomplete information version of rationalizability that does not make use of any beliefs over others' types; it is equivalent to iteratively deleting strategies that are *ex post strictly dominated*, i.e., strictly dominated for all possible beliefs over others' types. We work with this solution concept throughout the paper. However, results from the epistemic foundations of game theory establish that an action is rationalizable in this sense for a payoff type if and only if it could be played in an equilibrium on some type space with beliefs and higher order beliefs, by a type with that payoff type (Brandenburger and Dekel, 1987, and Battigalli and Siniscalchi, 2003). Thus a bonus of our "robust" analysis is that the distinction between equilibrium and rationalizability (or iterated deletion of strictly dominated strategies) becomes moot.

Now ex post incentive compatibility is the robust analogue of Bayesian incentive compatibility and robust measurability is the robust analogue of measurability of Abreu and Matsushima (1992c). They could reasonably argue that — in a standard Bayesian setting — their measurability condition is a weak technical requirement.[2] As a result, the "bottom line" of

[1]Abreu and Matsushima (1992c) and Serrano and Vohra (2005) note that a simple sufficient condition for all social choice functions to be A–M measurable is *type diversity*: every type has distinct preferences over lotteries unconditional on others' types.

[2]Although Serrano and Vohra (2001) describe an economic example where all nontrivial individually rational and Bayesian incentive compatible social choice functions fail Abreu–Matsushima measurability because types have identical conditional preferences.

the virtual implementation literature has been that full implementation, i.e., getting rid of undesirable equilibria, does not impose any substantive constraints beyond incentive compatibility, i.e., the existence of desirable equilibria. By requiring the more demanding, but more plausible, robust formulation of incomplete information, we end up with a condition that is substantive (imposing significantly more structure in interdependent value environments than incentive compatibility) and easily interpretable.

This paper adds to a recent literature on robust mechanism design that provides one operationalization of the so-called "Wilson doctrine" that progress in practical mechanism design will come from relaxing the implicit common knowledge assumption in the formulation of mechanism design problems.[3] Neeman (2004) highlighted the fact that full surplus extraction with correlated type results (Myerson 1981, and Cremer and McLean, 1985) rely on the implicit assumption that there is common knowledge of a mapping from beliefs to payoff types of all agents (a "beliefs determine preferences" property). This (counterintuitive) assumption is implied by the "generic" choice of a common prior on a fixed type space where distinct types are assumed to have different preferences. The apparent weakness of the Abreu–Matsushima measurability condition (and the fact that it is satisfied for "generic" priors) relies on the same property. We believe that by relaxing this unnatural implicit assumption, we get a better insight into the nature of the extra requirement for full implementation over and above incentive compatibility conditions.

Our operationalization of the "Wilson doctrine" is rather strong: we put no restrictions on agents' beliefs and higher order beliefs. A recent paper of Artemov, Kunimoto, and Serrano (2008) examines what happens to the conditions for robust virtual implementation if the planner is given partial information about agents' beliefs, in particular, a subset of beliefs over others' payoffs types that can arise with each payoff type. We discuss this intermediate robustness approach in Section 6.3.

It is possible to interpret our result as rather negative: ex post incentive compatibility is already a very strong condition, as emphasized by the recent work of Jehiel, Moldovanu, Meyer-Ter-Vehn, and Zame (2006);[4]

[3]Neeman (2004), Bergemann and Morris (2005b), Heifetz and Neeman (2006), Chung and Ely (2007).

[4]Although we argue in Bergemann and Morris (2009) that ex post incentive compatibility is feasible in many economically important environments either because types are one

robust measurability adds the further substantive restriction that there not be too much interdependence of preferences; and, in any case, the mechanism that we use to robustly virtually implement social choice functions is complicated to describe and presumably hard to play. However, we can show that in one large and interesting class of economic environments with interdependent preferences, robust virtual implementation is not only possible but is possible in the direct mechanism where agents simply report their payoff types. Say that an environment has *aggregator single crossing* preferences if the profile of agents' types can be aggregated into a single number and preferences are single crossing with respect to that number. Efficient social choice functions satisfying ex post incentive compatibility often exist in such environments. Bergemann and Morris (2009) showed that in such an environment, exact robust implementation is possible if the social choice function satisfies strict ex post incentive compatibility and a contraction property. In this paper, we observe that the contraction property is equivalent to robust measurability, so that — under the weak condition that there exists some strictly ex post incentive compatible social choice function — whenever robust virtual implementation is possible, it is possible in the direct mechanism.

The remainder of the paper is organized as follows. Section 2 introduces the environment and the solution concept. Section 3 illustrates the notion of separability in the context of a single private good with interdependent preferences. Section 4 defines and characterizes strategic distinguishability, constructing the maximally revealing mechanism to show the equivalence between strategic distinguishability and pairwise separability. Section 5 reports our results on robust virtual implementation. Section 6 concludes with discussions of the formal relation between Abreu–Matsushima measurability and robust measurability, the role of moderate interdependence, intermediate notions of robustness, the epistemic foundations for the solution concept, weak rather than strict dominance, positive results in direct mechanisms and the relation to exact rather than virtual implementation.

dimensional or because natural economic features of the environment lead to a failure of the "generic" properties that lead to the non-existence of non-trivial ex post incentive compatible social choice functions in Jehiel, Moldovanu, Meyer-Ter-Vehn, and Zame (2006).

2 Setting

2.1 *Environment*

There is a finite set of agents $1, \ldots, I$ and each agent i has finite set of possible payoff types:

$$\Theta_i = \{\theta_i^1, \ldots, \theta_i^S, \ldots, \theta_i^S\}.$$

We assume without loss of generality that the cardinality of each set Θ_i is equal to S for all i. The finite set X of pure outcomes is given by

$$X = \{x_1, \ldots, x_n, \ldots, x_N\}.$$

The lottery space over the set of outcome is $Y = \Delta(X)$. A lottery y is an N dimensional vector $y = (y_1, \ldots, y_n, \ldots, y_N)$ with

$$y_n \geq 0, \quad \sum_{n=1}^{N} y_n = 1.$$

Each agent has a von Neumann-Morgenstern expected utility function $u_i : Y \times \Theta \to \mathbb{R}$ with

$$u_i(y, \theta) = \sum_{n=1}^{N} u_i(x_n, \theta) y_n.$$

We will abuse notation by writing x for the lottery putting probability 1 on outcome x and X for the set of degenerate lotteries.

It is often convenient to work with underlying preferences over lotteries rather than any of their representations. We write $\mathcal{R}$ for the collection of expected utility preference relations on Y. We will write $R_{\theta_i,\lambda_i} \in \mathcal{R}$ for the preference relation of agent i if his payoff type is θ_i and he has belief $\lambda_i \in \Delta(\Theta_{-i})$ about the types of others:

$$\forall y, y' \in Y : y R_{\theta_i,\lambda_i} y' \Leftrightarrow \sum_{\theta_{-i} \in \Theta_{-i}} \lambda_i(\theta_{-i}) u_i(y, (\theta_i, \theta_{-i}))$$
$$\geq \sum_{\theta_{-i} \in \Theta_{-i}} \lambda_i(\theta_{-i}) u_i(y', (\theta_i, \theta_{-i}));$$

and we write P_{θ_i,λ_i} for the strict preference relation corresponding to R_{θ_i,λ_i}.

We make a weak assumption on the preferences: every agent i, whatever his type $\theta_i \in \Theta_i$ and beliefs $\lambda_i \in \Delta(\Theta_{-i})$, has a strict preference over some pair of outcomes:

Assumption 1 (No Complete Indifference). For each $i, \theta_i \in \Theta_i$ and $\lambda_i \in \Delta(\Theta_{-i})$, there exist $x, x' \in X$ such that $x P_{\theta_i,\lambda_i} x'$.

We maintain this assumption throughout the paper.[5] An analogous condition appeared in Abreu and Matsushima (1992c) and Serrano and Vohra (2005) in the Bayesian setting for all types (and associated beliefs) of all agents. But in our robust context, it is a stronger assumption in the sense that it rules out the possibility that alternative payoff type profiles of others lead to a reversal in the preferences of agent i with respect to some x and x'.

We denote by $\bar{y}$ the *central lottery* which puts equal probability on each of the pure outcomes. Now no–complete-indifference implies that every agent i, whatever his type θ_i and beliefs $\lambda_i \in \Delta(\Theta_{-i})$, strictly prefers some pure outcome x to $\bar{y}$; and compactness implies that those strict preferences are uniformly strict:

Lemma 1. *There exists $c > 0$ such that, for each $i, \theta_i \in \Theta_i$ and $\lambda_i \in \Delta(\Theta_{-i})$, there exists $x \in X$ such that*

$$\sum_{\theta_{-i} \in \Theta_{-i}} \lambda_i(\theta_{-i}) u_i(x, (\theta_i, \theta_{-i})) > \sum_{\theta_{-i} \in \Theta_{-i}} \lambda_i(\theta_{-i}) u_i(\bar{y}, (\theta_i, \theta_{-i})) + c.$$

The lemma is proved in appendix and we will use c in our later constructions. We will also exploit the existence of an upper bound on payoff differences C which follows immediately from the finiteness of pure outcomes and states:

Lemma 2. *There exists $C > 0$ such that*

$$|u_i(y, \theta) - u_i(y', \theta)| \leq C,$$

for all i, y, y', θ.

2.2 *Mechanisms and Solution Concept*

A mechanism $\mathcal{M}$ is a collection $((M_i)_{i=1}^I, g)$ where each M_i is finite and $g : M \to Y$. We denote a belief of agent i over the product of payoff type and message spaces of the other agents by $\mu_i \in \Delta(\Theta_{-i} \times M_{-i})$. We consider

[5]Our results can be extended to allow for the presence of complete indifference as shown in the appendix of the working paper version, Bergemann and Morris (2007).

the process of iteratively eliminating never best responses, without making assumptions on agents' beliefs about others' payoff types. The set of messages surviving the k-th level of elimination for type θ_i in mechanism $\mathcal{M}$ are defined by

$$S_i^0[\mathcal{M}](\theta_i) \triangleq M_i,$$

and for each $k = 1, \ldots,$ by induction:

$$S_i^{k+1}[\mathcal{M}](\theta_i)$$
$$\triangleq \left\{ m_i \in S_i^k[\mathcal{M}](\theta_i) \;\middle|\; \begin{array}{l} \exists \mu_i \in \Delta(\Theta_{-i} \times M_{-i}) \text{ s.t. :} \\ (1)\ \mu_i(\theta_{-i}, m_{-i}) > 0 \Rightarrow m_{-i} \in S_{-i}^k[\mathcal{M}](\theta_{-i}) \\ (2)\ m_i \in \underset{m_i'}{\arg\max} \displaystyle\sum_{\theta_{-i}, m_{-i}} \mu_i(\theta_{-i}, m_{-i}) \\ \qquad \times u_i(g(m_i', m_{-i}), (\theta_i, \theta_{-i})) \end{array} \right\};$$

we let

$$S_i[\mathcal{M}](\theta_i) = \bigcap_{k \geq 0} S_i^k[\mathcal{M}](\theta_i).$$

We refer to $S_i[\mathcal{M}](\theta_i)$ as the *rationalizable* messages of type θ_i of agent i in mechanism $\mathcal{M}$. This incomplete information version of rationalizability was studied in Battigalli (1998) and Battigalli and Siniscalchi (2003). A standard and well known duality argument implies that this solution concept is equivalent to iterated deletion of ex post strictly dominated strategies.

$S_i[\mathcal{M}](\theta_i)$ is the set of messages that type θ_i might send consistent with knowing that his payoff type is θ_i, common knowledge of rationality and the set of possible payoff types of the other players, but no restrictions on his beliefs and higher order beliefs about other types. Equivalently, it is the set of messages that might be played in any equilibrium on any type space by a type of player i with payoff type θ_i and any possible beliefs and higher order beliefs about others' payoff types. In Section 6.4, we report a formal argument confirming this interpretation. In the body of the paper, we work directly with this solution concept.

2.3 *Separability*

We will be interested in the set of preferences that an agent might have if his payoff type is θ_i and he knows that the type θ_j of each opponent j belongs to some subset Ψ_j of his possible types Θ_j. Thus writing $\Psi_{-i} = \{\Psi_j\}_{j \neq i}$ for a profile of subsets of i's opponents, we define

$$\mathcal{R}_i(\theta_i, \Psi_{-i}) = \{R \in \mathcal{R} | R = R_{\theta_i, \lambda_i} \quad \text{for some } \lambda_i \in \Delta(\Psi_{-i})\}.$$

We adopt the convention that if for some $j \neq i$, $\Psi_j = \emptyset$, then $\mathcal{R}_i(\theta_i, \Psi_{-i}) = \emptyset$. Now suppose we observed i's preferences over lotteries and knew that i assigned probability 1 to his opponents' type profile θ_{-i} being an element of Ψ_{-i}, what would we be able to deduce about i's type? We will say that Ψ_{-i} *separates* Ψ_i if — whatever those realized preferences — we could rule out at least one possible type of i.

Definition 1 (Separation). Type set profile Ψ_{-i} separates Ψ_i if

$$\bigcap_{\theta_i \in \Psi_i} \mathcal{R}_i(\theta_i, \Psi_{-i}) = \emptyset.$$

We will be interested in a process by which we iteratively delete type sets of each agent that are separated by some type set profile of his opponents. Thus writing Ξ_i^k for the k-th level inseparable sets of player i, we have:

$$\Xi_i^0 = 2^{\Theta_i}, \tag{1}$$

and

$$\Xi_i^{k+1} = \{\Psi_i \in \Xi_i^k \mid \Psi_{-i} \text{ does not separate } \Psi_i, \text{for some } \Psi_{-i} \in \Xi_{-i}^k\}, \tag{2}$$

and a (finite) limit type set profile is defined by:

$$\Xi_i^* = \bigcap_{k \geq 0} \Xi_i^k. \tag{3}$$

Finally, we say that a pair of types are pairwise inseparable if they cannot be iteratively separated in this way:

Definition 2 (Pairwise Inseparability). Types θ_i and θ_i' are pairwise inseparable, written $\theta_i \sim \theta_i'$, if $\{\theta_i, \theta_i'\} \in \Xi_i^*$.

Note that the relation $\sim$ is reflexive and symmetric by construction, but it is not necessarily transitive. The following "fixed point" characterization of pairwise inseparability will be useful in the analysis that follows. Let $\Xi = (\Xi_i)_{i=1}^I \in \times_{i=1}^I 2^{\Theta_i}$ be a profile of type sets for each agent.

Definition 3 (Mutual Inseparability). Ξ is mutually inseparable if, for each i and $\Psi_i \in \Xi_i$, there exists $\Psi_{-i} \in \Xi_{-i}$ such that Ψ_{-i} does not separate Ψ_i.

Lemma 3. *Types* θ_i *and* θ_i' *are pairwise inseparable if and only if there exists mutually inseparable* $\Xi = (\Xi_i)_{i=1}^I$ *and* $\Psi_i \in \Xi_i$ *with* $\{\theta_i, \theta_i'\} \subseteq \Psi_i$.

Proof. (if) Suppose there exists $\widehat{\Xi} = (\widehat{\Xi}_i)_{i=1}^I$ and $\Psi_i \in \widehat{\Xi}_i$ with $\{\theta_i, \theta_i'\} \subseteq \Psi_i$. We claim that

$$\{\Psi_i \mid \Psi_i \subseteq \Psi_i' \quad \text{and} \quad \Psi_i' \in \widehat{\Xi}_i \quad \text{for some } \Psi_i'\} \subseteq \Xi_i^k$$

for each $k = 0, 1, \ldots$. The claim holds for $k = 0$ by definition. Suppose the claim holds for arbitrary k and suppose that $\Psi_i \subseteq \Psi_i'$ and $\Psi_i' \in \widehat{\Xi}_i$. Because $\widehat{\Xi}$ is mutually inseparable, there exists $\Psi_{-i} \in \widehat{\Xi}_{-i} \subseteq \Xi_i^k$ such that Ψ_{-i} does not separate Ψ_i'. By the definition of separation, since $\Psi_i \subseteq \Psi_i'$, Ψ_{-i} does not separate Ψ_i. So $\Psi_i \in \Xi_i^{k+1}$ and

$$\{\theta_i, \theta_i'\} \subseteq \Psi_i \in \Xi_i^* = \bigcap_{k \geq 0} \Xi_i^k.$$

(only if) Observe that $\Xi_i^{k+1} \subseteq \Xi_i^k$ for each $k = 0, 1, \ldots$ by construction. Thus $(\Xi_i^*)_{i=1}^I$ is mutually inseparable. Thus if $\theta_i \sim \theta_i'$, there exists mutually inseparable Ξ^* with $\{\theta_i, \theta_i'\} \in \Xi_i^*$. □

3 An Environment with Interdependent Values for a Single Good

We consider a quasi-linear environment with a single good with interdependent values to illustrate the notion of separability. There are I agents and agent i's payoff type is $\theta_i \in [0, 1]$. If the type profile is θ, agent i's valuation of an object is given by:

$$v_i(\theta_i, \theta_{-i}) = \theta_i + \gamma \sum_{j \neq i} \theta_j,$$

with $\gamma \in \mathbb{R}_+$. The parameter γ measures the amount of interdependence in valuations: the case of private values is given by $\gamma = 0$ and the case of pure common values is $\gamma = 1$. The net utility of agent i depends on his probability y_i of receiving the object and the monetary transfer t_i:

$$u_i(\theta, y_i, t_i) = \left(\theta_i + \gamma \sum_{j \neq i} \theta_j \right) y_i - t_i.$$

We determine the conditions for separability of types in this preference environment.[6]

Type set profile Ψ_{-i} separates Ψ_i if, knowing i's preferences and knowing that he is sure that others' type profile is Ψ_{-i}, we can always rule out some θ_i. In this example, because the utility function $u_i(\cdot)$ is linear in the monetary transfer for all types and all agents, separability must come from different valuations of the object. For given type set profile Ψ_{-i} of all but i, we can identify the set of possible (expected) valuations of agent i with type θ_i by writing:

$$\begin{aligned} V_i(\theta_i, \Psi_{-i}) & \\ &= \left\{ v_i \in \mathbb{R}_+ \,\middle|\, \exists \lambda_i \in \Delta(\Psi_{-i}) \text{ s.t. } v_i = \theta_i + \gamma \sum_{\theta_{-i} \in \Psi_{-i}} \lambda_i(\theta_{-i}) \sum_{j \neq i} \theta_j \right\} \\ &= \left[\theta_i + \gamma \sum_{j \neq i} \min \Psi_j,\ \theta_i + \gamma \sum_{j \neq i} \max \Psi_j \right]. \end{aligned} \tag{4}$$

Now Ψ_{-i} separates Ψ_i if and only if

$$\bigcap_{\theta_i \in \Psi_i} V_i(\theta_i, \Psi_{-i}) = \emptyset.$$

This is equivalent to requiring that

$$V_i(\max \Psi_i, \Psi_{-i}) \cap V_i(\min \Psi_i, \Psi_{-i}) = \emptyset.$$

By (4), this will hold if and only if

$$\max \Psi_i + \gamma \sum_{j \neq i} \min \Psi_j > \min \Psi_i + \gamma \sum_{j \neq i} \max \Psi_j.$$

We can rewrite the inequality as

$$\max \Psi_i - \min \Psi_i > \gamma \sum_{j \neq i} (\max \Psi_j - \min \Psi_j).$$

[6]This example has a continuum of types and a continuum of deterministic monetary allocations while the general model was defined for a finite number of types and pure outcomes. We could rewrite the example and the corresponding results without loss in the finite setting. With a finite model, integer problems would need to be taken into account; in particular, the exact value of the critical threshold for moderate interdependence would depend on the size of the grid. But as the grid becomes finer, the critical thresholds converge to the ones of the continuum example here.

Thus Ψ_{-i} separates Ψ_i if and only if the difference between the smallest and the largest element in the set Ψ_i is larger than the weighted sum of the differences of the smallest and the largest element in the remaining sets Ψ_j for all $j \neq i$. Conversely, Ψ_{-i} does not separate Ψ_i if the above inequality is reversed, i.e.,

$$\max \Psi_i - \min \Psi_i \leq \gamma \sum_{j \neq i} (\max \Psi_j - \min \Psi_j). \tag{5}$$

Now we can identify the k-th level inseparable sets, described in (1)–(3), for our example. We have

$$\Xi_i^0 = 2^{[0,1]}$$

and, by (5),

$$\Xi_i^k = \left\{ \Psi_i \in \Xi_i^k \,\middle|\, \max \Psi_i - \min \Psi_i \leq \gamma \sum_{j \neq i} \max_{\Psi_j \in \Xi_j^k} (\max \Psi_j - \min \Psi_j) \right\},$$

Now by induction, we have that

$$\Xi_i^{k+1} = \{\Psi_i | \max \Psi_i - \min \Psi_i \leq (\gamma(I-1))^k\}.$$

Thus if $\gamma(I-1) < 1, \Xi_i^*$ consists of singletons, $\Xi_i^* = (\{\theta_i\})_{\theta_i \in [0,1]}$, while if $\gamma(I-1) \geq 1, \Xi_i^*$ consists of all subsets, $\Xi_i^* = 2^{[0,1]}$.

Thus if $\gamma < \frac{1}{I-1}$, so that interdependence is not too large, every distinct pair of types are pairwise separable. If $\gamma > \frac{1}{I-1}$, every pair of types are pairwise inseparable. We note that the linear structure of the valuations $v_i(\cdot)$ leads to the strong converse result. But the example illustrates the general principle that pairwise separability corresponds to not too much interdependence. We shall state a more general result about the relationship between pairwise separability and not too much interdependence in Section 6.2. We also note that the argument surrounding the pairwise separability result relies on the boundedness of the payoff type space. In particular if $\Theta_i = \mathbb{R}$, then pairwise separability could only be achieved in the case of of pure private values, i.e. $\gamma = 0$.

Our later results will show that if $\gamma \geq \frac{1}{I-1}$, no social choice function (except for a constant one) is robustly virtually implementable; but if $\gamma < \frac{1}{I-1}$, any ex post incentive compatible allocation can be robustly virtually implemented. One can construct generalized VCG payments such that efficient allocation is ex post incentive compatible in this environment if $\gamma \leq 1$. Thus the efficient allocation is robustly virtually implementable if and only if $\gamma < \frac{1}{I-1}$.

Our result on robust virtual implementation in this environment will contrast with what would happen with standard Bayesian implementation. Suppose we assumed there was common knowledge of a common prior on the set of payoff types $[0,1]^I$. Suppose first that agents' types were drawn independently. Then each type would have different expected valuations of the object and could easily be separated. Even if priors were not independent, for a "typical" choice of prior, the measurability condition of Abreu and Matsushima 1992b, and Bayesian virtual implementation would be possible as long as incentive compatibility conditions were satisfied. Ex post incentive compatibility (and thus Bayesian incentive compatibility for any prior) is satisfied by the efficient allocation if $\gamma \leq 1$.

4 Strategic Distinguishability

4.1 *Main Result*

Two payoff types are strategically distinguishable if there exists a mechanism where the rationalizable actions of those payoff types are disjoint; thus they are strategically indistinguishable if they have a rationalizable action in common in every mechanism.

Definition 4 (Strategically Indistinguishable). Types θ_i and θ_i' are strategically indistinguishable if $S_i[\mathcal{M}](\theta_i) \cap S_i[\mathcal{M}](\theta_i') \neq \emptyset$ for every $\mathcal{M}$.

The notion of strategic indistinguishability is related to the idea of incentive compatibility in the context of information revelation in a mechanism. The difference between distinguishability and incentive compatibility arises from the two central features of strategic indistinguishability. First, we say that two payoff types can be strategically distinguished if there exists *some* mechanism and hence *some* outcome function for which the types have disjoint rationalizable actions. In contrast, the analysis of incentive compatibility is typically concerned with a specific mechanism and hence a specific outcome function. Second, strategic distinguishability requires that the two payoff types display disjoint rationalizable actions for *all possible* beliefs and higher order beliefs. In contrast, the analysis of incentive compatibility is typically concerned with a fixed and common prior belief of the agents.

The characterization of strategic indistinguishability is the key result in our characterization of robust virtual implementation.

Theorem 1 (Equivalence). *Types θ_i and θ_i' are strategically indistinguishable if and only if they are pairwise inseparable.*

This result will be proved in two parts. First, Proposition 1 shows that under any finite mechanism, if θ_i and θ_i' are pairwise inseparable, then the intersection of the set of rationalizable messages for θ_i and θ_i' will always be non-empty. This observation follows easily from our definitions.

Proposition 1. *If θ_i and θ_i' are pairwise inseparable ($\theta_i \sim \theta_i'$), then $S_i[\mathcal{M}](\theta_i) \cap S_i[\mathcal{M}](\theta_i') \neq \emptyset$ in any mechanism $\mathcal{M}$.*

Proof. By Lemma 3, if $\theta_i \sim \theta_i'$, there exists mutually inseparable Ξ with $\{\theta_i, \theta_i'\} \subseteq \Psi_i^* \in \Xi_i$.

Now fix any mechanism $\mathcal{M}$. We will show, by induction on k, that for each k, i and $\Psi_i \in \Xi_i$, there exists $m_i^k(\Psi_i) \in M_i$ such that $m_i^k(\Psi_i) \in S_i^k[\mathcal{M}](\tilde{\theta}_i)$ for each $\tilde{\theta}_i \in \Psi_i$. This is true by definition for $k = 0$. Suppose that it is true for k. Now fix any i and $\Psi_i \in \Xi_i$. Since Ξ is mutually inseparable, there exists $\Psi_{-i} \in \Xi_{-i}, R$ and, for each $\tilde{\theta}_i \in \Psi_i, \lambda_i^{\tilde{\theta}_i} \in \Delta(\Psi_{-i})$ such that $R_{\tilde{\theta}_i, \lambda_i^{\tilde{\theta}_i}} = R$. Now let $m_i^{k+1}(\Psi_i)$ be any optimal message of agent i when he believes that his opponents will sent message profile $m_{-i}^k(\Psi_{-i})$ with Probability 1 and has beliefs $\lambda_i^{\tilde{\theta}_i}$ about the type profile of his opponents, i.e.,

$$m_i^{k+1}(\Psi_i) \in \arg\max_{m_i'} \sum_{\theta_{-i}} \lambda_i^{\tilde{\theta}_i}(\theta_{-i}) u_i(g(m_i', m_{-i}^k(\Psi_{-i})), (\tilde{\theta}_i, \theta_{-i})).$$

By construction, $m_i^{k+1}(\Psi_i) \in S_i^{k+1}[\mathcal{M}](\tilde{\theta}_i)$ for all $\tilde{\theta}_i \in \Psi_i$.

By the finiteness of the mechanism, there exists K such that $S_i^k[\mathcal{M}](\tilde{\theta}_i) = S_i[\mathcal{M}](\tilde{\theta}_i)$ for all $i, \tilde{\theta}_i$ and $k \geq K$. Thus for each $\Psi_i \in \Xi_i$, there exists $m_i(\Psi_i) \in M_i$ such that $m_i(\Psi_i) \in S_i[\mathcal{M}](\tilde{\theta}_i)$ for each $\tilde{\theta}_i \in \Psi_i^*$. Thus there exists $m_i \in S_i[\mathcal{M}](\theta_i) \cap S_i[\mathcal{M}](\theta_i')$. □

The second part of the theorem's proof is the converse result.

Proposition 2 (Existence of Maximally Revealing Mechanism). *There exists $\mathcal{M}^*$ such that $\theta_i \nsim \theta_i' \Rightarrow S_i[\mathcal{M}^*](\theta_i) \cap S_i[\mathcal{M}^*](\theta_i') = \emptyset$.*

Propositions 1 and 2 immediately imply Theorem 1. Proposition 2 is proved by the explicit construction of a mechanism which will lead every pair of distinguishable types to choose different messages. We refer to the specific mechanism as the "maximally revealing mechanism", and spend the rest of this section describing its construction and finding its properties.

4.2 *The Maximally Revealing Mechanism*

We will construct a mechanism that will work for any environment. In the canonical mechanism, each agent is given K simultaneous opportunities to select a preferred allocation from a given "test set" of allocations. For each opportunity k to select a preferred allocation, with $k = 1, \ldots, K$, the agent is asked to report a profile of possible choices by the remaining agents in the opportunities preceding the k-th opportunity. If the report of the agent at opportunity k matches the choices of the other agents in the opportunities below k, then he will be given the right to choose a preferred allocation. On the other hand, if his report fails to replicate the choices of the other agents in the opportunities before k, then the designer will simply select the central lottery $\bar{y}$. While the mechanism is entirely static, it requires each agent to make a series of choices, each one contingent on the choices of the other agents. In particular, by asking the agent at opportunity k to match his report with the choices of the other agents at the opportunities before k, we introduce an inductive structure into the series of choices by each agent. We therefore refer to the k-th opportunity as the k-th stage or k-th step of the mechanism even though the mechanism itself is entirely static.

The central aspect of the inductive structure of the choice mechanism is that it allows us to analyze the behavior of the agent in the mechanism in terms of the iterative elimination of dominated strategies. The precise construction of the choice mechanism is based on two central concepts, the notion of a test set and the notion of an augmentation of a given mechanism. A test set will give each agent a finite set of choices and the choice behavior by the agent allows us to distinguish between different types of the agent. The construction of the set of test allocations relies on a few critical implications of our notion of separation. In turn, the notion of an augmentation permits us to show that we can always construct a more informative mechanism on the basis of a given mechanism.

4.2.1 *A class of maximally revealing mechanisms*

Fix a finite "test set" of lotteries Y^*. A maximally revealing mechanism offers each agent i a series of K opportunities to select a preferred allocation from Y^*. The set of messages for each agent in a maximally revealing mechanism is defined as follows. Let $M_i^0 = \{\bar{m}_i^0\}$ and inductively define

$$M_i^{k+1} = M_i^k \times M_{-i}^k \times Y^*.$$

Thus $M_i^0 = \{\bar{m}_i^0\}, M_i^1 = \{\bar{m}_i^0\} \times M_{-i}^0 \times Y^*$, $M_i^2 = \{\bar{m}_i^0\} \times M_{-i}^0 \times Y^* \times M_{-i}^1 \times Y^*$, and so on. The message m_i^{k+1} of agent i in stage $k+1$ thus reiterates his message from step k and announces a possible message profile of the remaining agents in step k. Due to the inductive structure of the messages, we can write a typical element $m_i^k \in M_i^k$ as a list of the form

$$m_i^k = \{m_i^0, r_i^1, y_i^1, r_i^2, y_i^2, \ldots, r_i^k, y_i^k\}, \tag{6}$$

with $m_i^0 = \bar{m}_i^0$ and each $r_i^k \in M_{-i}^{k-1}$ and each $y_i^k \in Y^*$. The entry r_i^k constitutes the report of agent i regarding the message of the other agents in the previous stage $k-1$. The message set of agent i is then given by M_i^K.

The outcome function in the revealing mechanism is defined as:

$$g^{K,\varepsilon}(m) \triangleq \bar{y} + \left(\frac{1-\varepsilon^k}{1-\varepsilon}\right)\frac{1}{I}\left(\sum_{k=1}^{K}\varepsilon^{k-1}\sum_{i=1}^{I}\mathbb{I}(r_i^k, m_{-i}^{k-1})(y_i^k - \bar{y})\right), \tag{7}$$

for some $\varepsilon > 0$ and where $\mathbb{I}$ is the indicator function:

$$\mathbb{I}(r_i^k, m_{-i}^{k-1}) \triangleq \begin{cases} 1, & \text{if } r_i^k = m_{-i}^{k-1}, \\ 0, & \text{otherwise.} \end{cases}$$

For a given $\varepsilon > 0$ and positive integer K, we refer to the associated revealing mechanism as

$$\mathcal{M}_\varepsilon^K \triangleq (M^K, g_\varepsilon^K).$$

In words, the mechanism has K stages. In each stage k, an agent is asked to announce a stage $k-1$ message profile of messages he thinks his opponents might have sent and — with positive probability — gets to pick a lottery from Y^*. Lotteries from early stages are much more likely to be chosen than lotteries from later stages. We can now analyze how the series of messages can iteratively and interactively identify the types of each agent.

4.2.2 *Characterizing rationalizable behavior for small* ε

For sufficiently small $\varepsilon > 0$, an agent's choice of a message at the k-th stage will be independent of what messages he thinks others will send at stage k and higher and thus also independent of K, the total number of stages of messages that will be sent. We first propose an inductive characterization of the set of types of player i who could possibly send k-th stage message

m_i^k and we denote this set by $\bar{\Theta}_i^k(m_i^k)$. We then verify with Lemmas 6 and 8 that our proposed inductive characterization of rationalizable messages is correct for sufficiently small ε.

Write $B_i^{Y^*}(\theta_i, \lambda_i)$ for agent i's most preferred lotteries in the set Y^* if he has payoff type θ_i and beliefs $\lambda_i \in \Delta(\Theta_{-i})$ and (with a minor abuse of notation) let $B_i^{Y^*}(\theta_i, \Psi_{-i})$ be agent i's possible most preferred lotteries if he has payoff type θ_i and assigns Probability 1 to his opponents having types in Ψ_{-i}, so that

$$B_i^{Y^*}(\theta_i, \lambda_i) \triangleq \{y \in Y^* | y R_{\theta_i, \lambda_i} y' \quad \text{for all } y' \in Y^*\},$$

and

$$B_i^{Y^*}(\theta_i, \Psi_{-i}) \triangleq \bigcup_{\lambda_i \in \Delta(\Psi_{-i})} B_i^{Y^*}(\theta_i, \lambda_i).$$

We adopt the convention that if $\Psi_j = \emptyset$ for some $j \neq i$, then $B_i^{Y^*}(\theta_i, \emptyset) = \emptyset$ as well.

Let $\bar{\Theta}_i^1(m_i^1)$ be the set of types of player i who could possibly send first stage message m_i^1. Since we will ignore later stages, this will be independent of ε and K. Taking these sets as given, we will then find the set $\bar{\Theta}_i^2(m_i^2)$ of types of player i who could possibly send second stage message m_i^2, and so on. We will end up with an inductive characterization of the set $\bar{\Theta}_i^k(m_i^k)$ types of player i who could possibly send k-th stage message m_i^k. Thus

$$\bar{\Theta}_i^0(\bar{m}_i^0) \triangleq \Theta_i,$$

and inductively define $\bar{\Theta}_i^{k+1}(m_i^{k+1})$, where we recall that by the inductive description of the message m_i^{k+1} in (6), we have $m_i^{k+1} = (m_i^k, r_i^{k+1}, y_i^{k+1})$:

$$\bar{\Theta}_i^{k+1}(m_i^{k+1}) \triangleq \left\{ \theta_i \in \Theta_i \;\middle|\; \begin{array}{l} \text{(i) } \theta_i \in \bar{\Theta}_i^k(m_i^k); \\ \text{(ii) } \bar{\Theta}_{-i}^k(r_i^{k+1}) \neq \emptyset; \text{ and} \\ \text{(iii) } y_i^{k+1} \in B_i^{Y^*}(\theta_i, \bar{\Theta}_{-i}^k(r_i^{k+1})). \end{array} \right\} \tag{8}$$

The set $\bar{\Theta}_i^k(m_i^k)$ is meant to approximate the set of types of agent i for whom a specific message m_i^k is rationalizable in stage k. In some sense, the set $\bar{\Theta}_i^k(m_i^k)$ is the dual to $S_i^k[\mathcal{M}](\theta_i)$, which describes the set of messages m_i which are rationalizable for a specific type θ_i in stage k. The role of the set $\bar{\Theta}_i^k(m_i^k)$ is to track the information that can be inferred from the choices of messages m_i about the type θ_i of agent i.

The analysis of the limit behavior of $\bar{\Theta}_i^{k+1}(m_i^{k+1})$ is heuristic in the sense that the inductive process assumes the properties (ii) and (iii) in (8). In particular, it is simply assumed that agent i in stage $k+1$ announces a past message profile of the remaining agents which could have been sent by some type profile of the other agents, and it is simply assumed that agent i will select an allocation which is a best response to some belief in stage $k+1$.

We will use two preliminary results to establish formally that these sets characterize limit behavior for small ε and large K. The routine proofs are reported in the Appendix. First, we note that for any fixed finite mechanism $\mathcal{M}$, when we iteratively delete messages that are not best responses, they are uniformly worse responses, i.e., there exists $\eta_{\mathcal{M}} > 0$ such that each of those deleted messages is not even an $\eta_{\mathcal{M}}$-best response.

Lemma 4 (Uniformly Worse Responses). *For any mechanism $\mathcal{M}$, there exists $\eta_{\mathcal{M}} > 0$ such that if $m_i \in S_i^k[\mathcal{M}](\theta_i)$, $m_i \notin S_i^{k+1}[\mathcal{M}](\theta_i)$ and $\mu_i \in \Delta(\Theta_{-i} \times M_{-1})$ satisfies*

$$\mu_i(\theta_{-i}, m_{-i}) > 0 \Rightarrow m_j \in S_j^k[\mathcal{M}](\theta_j) \quad \text{for each } j \neq i,$$

then there exists $\bar{m}_i$ such that

$$\sum_{\theta_{-i}, m_{-i}} \mu_i(\theta_{-i}, m_{-i}) u_i(g^*(\bar{m}_i, m_{-i}), (\theta_i, \theta_{-i}))$$
$$> \sum_{\theta_{-i}, m_{-i}} \mu_i(\theta_{-i}, m_{-i}) u_i(g^*(m_i, m_{-i}), (\theta_i, \theta_{-i})) + \eta_{\mathcal{M}}.$$

Second, we use the uniform lower bound in stating a key result about "augmenting" mechanisms. We use this "augmentation lemma" in the construction of both the maximally revealing mechanism (in this section) and the canonical mechanism for robust virtual implementation (in the next section). For each player i, fix finite message sets M_i^0 and M_i^1 and let $M_i = M_i^0 \times M_i^1$. Fix $g^0 : M^0 \to Y, g^1 : M^1 \to Y$ and $g^+ : M \to Y$. Fix $\pi^0, \pi^1, \pi^+ \geq 0$ with $\pi^0 + \pi^1 + \pi^+ = 1$ and let $g : M \to Y$ be defined by

$$g(m) \triangleq \pi^0 g^0(m^0) + \pi^1 g^1(m^1) + \pi^+ g^+(m).$$

We now consider the mechanism

$$\mathcal{M}^0 \triangleq ((M_i^0)_{i=1}^I, g^0),$$

and the augmented mechanism

$$\mathcal{M} \triangleq ((M_i)_{i=1}^I, g).$$

We recall that the constant $C > 0$ is a finite upper bound on the difference in payoffs across all agents and all pairs of lotteries y and y', which we established earlier in Lemma 2.

Lemma 5 (Augmentation). *If $\pi^+ C \leq \pi^0 \eta_{\mathcal{M}^0}$, then*

$$(m_i^0, m_i^1) \in S_i[\mathcal{M}](\theta_i) \Rightarrow m_i^0 \in S_i[\mathcal{M}^0](\theta_i).$$

The lemma states that if the weight π^0 put on the original payoff function g^0 in the augmented mechanism is much larger than the weight π^+ put on the other component of the mechanism at which m^0 effects the allocation, then any rationalizable message in the augmented mechanism must entail sending a message m_i^0 that was rationalizable in the original mechanism.

We now show that these choices are indeed the result of iteratively elimination of strictly dominated strategies. More precisely, we verify that $\bar{\Theta}_i^k(m_i^k)$ is an upper bound on the set of types who could send kth stage message m_i^k in any $\mathcal{M}_\varepsilon^k$ for sufficiently small ε.

Lemma 6 (Limit). *Suppose that $B_i^{Y^*}(\theta_i, \lambda_i) \neq Y^*$ for each i, θ_i and $\lambda_i \in \Delta(\Theta_{-i})$. Then, for each k, there exists $\bar{\varepsilon} > 0$ such that:*

$$\{\theta_i \in \Theta_i | m_i^k \in S[\mathcal{M}_\varepsilon^k](\theta_i)\} \subseteq \bar{\Theta}_i^k(m_i^k),$$

for all $\varepsilon < \bar{\varepsilon}$ and $m_i^k \in M_i^k$.

Proof. By induction. The claim of holds for $k = 0$, since

$$\{\theta_i \in \Theta_i | m_i^0 \in S[\mathcal{M}_\varepsilon^0](\theta_i)\} = \Theta_i = \bar{\Theta}_i^0(m_i^0).$$

Now suppose that the claim holds for k. Thus there exists $\bar{\varepsilon}_k > 0$, such that

$$\{\theta_i \in \Theta_i | m_i^k \subset S[\mathcal{M}_\varepsilon^k](\theta_i)\} \subseteq \bar{\Theta}_i^k(m_i^k) \quad \text{for all } \varepsilon \leq \bar{\varepsilon}_k \quad \text{and} \quad m_i^k \in M_i^k.$$

Now observe that $\mathcal{M}_\varepsilon^{k+1}$ is an augmentation of $\mathcal{M}_\varepsilon^k$ and thus — by Lemma 5 — there exists $\bar{\varepsilon}_{k+1} \in (0, \bar{\varepsilon}_k]$, such that for all $\varepsilon \leq \bar{\varepsilon}_{k+1}$,

$$m_i^{k+1} = (m_i^k, r_i^{k+1}, y_i^{k+1}) \in S[\mathcal{M}_\varepsilon^{k+1}](\theta_i) \Rightarrow m_i^k \in S[\mathcal{M}_\varepsilon^k](\theta_i). \quad (9)$$

Now by the inductive hypothesis, we also have

$$\theta_i \in \bar{\Theta}_i^k(m_i^k). \quad (10)$$

We further observe that $m_i^{k+1} \in S[\mathcal{M}_\varepsilon^{k+1}](\theta_i)$ also implies there must exist $\mu_i \in \Delta(\Theta_{-i} \times M_{-i}^{k+1})$ such that (1):

$$\mu_i(\theta_{-i}, m_{-i}^{k+1}) > 0 \Rightarrow m_j^{k+1} \in S[\mathcal{M}_\varepsilon^{k+1}](\theta_j) \quad \text{for each } j \neq i$$

and (2):

$$m_i^{k+1} \in \underset{\bar{m}_i^{k+1} \in M_i^{k+1}}{\arg\max} \sum_{\theta_{-i}, m_{-i}^{k+1}} \mu_i(\theta_{-i}, m_{-i}^{k+1}) \times [u_i(g^{k+1,\varepsilon}(\bar{m}_i^{k+1}, m_{-i}^{k+1}), (\theta_i, \theta_{-i}))].$$

But note that (r_i^{k+1}, y_i^{k+1}) — the last components of m_i^{k+1} — effect only one additively separable component of the above expression. In particular, (r_i^{k+1}, y_i^{k+1}) must maximize:

$$\sum_{\theta_{-i}, m_{-i}^{k+1}} \mu_i(\theta_{-i}, m_{-i}^{k+1}) \mathbb{I}(r_i^{k+1}, m_{-i}^k)(u_i(y_i^{k+1}, (\theta_i, \theta_{-i})) - u_i(\bar{y}, (\theta_i, \theta_{-i}))), \tag{11}$$

which we can rewrite as

$$\sum_{\theta_{-i}} \sum_{\{m_{-i}^{k+1} | m_{-i}^k = r_i^{k+1}\}} \mu_i(\theta_{-i}, m_{-i}^{k+1})(u_i(y_i^{k+1}, (\theta_i, \theta_{-i})) - u_i(\bar{y}, (\theta_i, \theta_{-i}))).$$

In particular, the later expression is zero if

$$\mu_i(r_i^{k+1}) \triangleq \sum_{\theta_{-i}} \sum_{\{m_{-i}^{k+1} | m_{-i}^k = r_i^{k+1}\}} \mu_i(\theta_{-i}, m_{-i}^{k+1}) = 0.$$

But if $\mu_i(r_i^{k+1}) > 0$ and $y_i^{k+1} \in B_i^{Y^*}(\theta_i, \lambda_i)$, where

$$\lambda_i(\theta_{-i}) = \frac{\sum_{\{m_{-i}^{k+1} | m_{-i}^k = r_i^{k+1}\}} \mu_i(\theta_{-i}, m_{-i}^{k+1})}{\sum_{\theta'_{-i}} \sum_{\{m_{-i}^{k+1} | m_{-i}^k = r_i^{k+1}\}} \mu_i(\theta'_{-i}, m_{-i}^{k+1})},$$

then (11) must be strictly positive, by the premise of the lemma. Thus we must have (r_i^{k+1}, y_i^{k+1}) chosen such that $\mu_i(r_i^{k+1}) > 0$ and $y_i^{k+1} \in B_i^{Y^*}$ (θ_i, λ_i). Now $\mu_i(r_i^{k+1}) > 0$, (9) and the inductive hypothesis imply that

$$\bar{\Theta}_{-i}^k(r_i^{k+1}) \neq \emptyset; \tag{12}$$

and

$$\lambda_i \in \Delta(\bar{\Theta}_{-i}^k(r_i^{k+1})) \quad \text{and} \quad y_i^{k+1} \in B_i^{Y^*}(\theta_i, \lambda_i). \tag{13}$$

To wit, by the construction of the revealing mechanism (see (7)), the lottery y_i^{k+1} specified in (13) only affects the (expected) payoff of agent i when $r_i^{k+1} = m_{-i}^k$. It follows that y_i^{k+1} should be a best reply to some belief conditioned on the event that $r_i^{k+1} = m_{-i}^k$.

Now (10), (12) and (13) together imply that any message $m_i^{k+1} \in S[\mathcal{M}_\varepsilon^{k+1}](\theta_i)$ satisfies the three requirements in the construction of $\bar{\Theta}_i^{k+1}(r_i^{k+1})$ in (8) and hence we have that for any $m_i^{k+1} \in S[M_\varepsilon^{k+1}(\theta_i), \theta_i \in \bar{\Theta}_i^{k+1}(m_i^{k+1})$. □

4.3 Constructing a Rich Enough Test Set

Finally, we show that we can choose the "test set" Y^* to be sufficiently large so that Lemma 6 will imply that — for sufficiently small $\varepsilon > 0$ and sufficiently large K — any pair of mutually separable types are sending distinct messages in the (K, ε) revealing mechanism.

Lemma 7 (Existence of Finite Test Set). *There exists a finite test set $Y^* \subseteq Y$ such that:*

1. *for each i, θ_i and $\lambda_i \in \Delta(\Theta_{-i}), B_i^{Y^*}(\theta_i, \lambda_i) \neq Y^*$;*
2. *for each i, Ψ_i and Ψ_{-i}, if Ψ_{-i} separates Ψ_i, then for each $\theta_i \in \Psi_i$ and $\lambda_i \in \Delta(\Psi_{-i})$, there exists $\theta_i' \in \Psi_i$ such that*

$$B_i^{Y^*}(\theta_i, \lambda_i) \cap B_i^{Y^*}(\theta_i', \Psi_{-i}) = \emptyset.$$

The proof of Lemma 7 is in the Appendix. Now the proof of Proposition 2 is completed by the following lemma, establishing that the sets $\bar{\Theta}_i^k$ are closely related to k-th level inseparable sets Ξ_i^k, as defined earlier in (1)–(3).

Lemma 8. *For all i, all k, and all $m_i^k \in M_i^k, \bar{\Theta}_i^k(m_i^k) \subseteq \Psi_i$ for some $\Psi_i \in \Xi_i^k$.*

Proof. By induction. The claim holds for $k = 0$ by definition. Suppose for all $m_{-i}^k \in M_{-i}^k$, we have $\bar{\Theta}_i^k(m_{-i}^k) \subseteq \Psi_i$ for some $\Psi_i \in \Xi_i^k$. Now fix any $m_i^{k+1} = (m_i^k, r_i^{k+1}, y_i^{k+1}) \in M_i^{k+1}$. If $\bar{\Theta}_i^{k+1}(m_i^{k+1}) = \emptyset$, then we are done as the empty set is included in every $\Psi_i \neq \emptyset$. If $\bar{\Theta}_i^{k+1}(m_i^{k+1}) \neq \emptyset$, then we let $\Psi_i = \bar{\Theta}_i^{k+1}(m_i^{k+1})$ and let $\Psi_{-i} = \bar{\Theta}_{-i}^k(r_i^{k+1})$. Lemma 7.1 ensures that for every θ_i and λ_i, there exist $y, y' \in \tilde{Y}$ such that $yP_{\theta_i,\lambda_i}y'$. Thus any best response will involve setting r_i^{k+1} equal to some m_{-i}^k that he assigns positive probability to and choosing a strictly preferred lottery. By our inductive assumption, $\Psi_{-i} \in \Xi_{-i}^k$. Now suppose Ψ_{-i} separates Ψ_i and fix

$\theta_i \in \Psi_i$. By Lemma 7.2, there exists $\theta_i' \in \Psi_i$ such that $y_i^{k+1} \notin B_i^{Y^*}(\theta_i', \Psi_{-i})$. Thus $\theta_i' \notin \bar{\Theta}_i^{k+1}(m_i^{k+1})$, a contradiction. We conclude that Ψ_{-i} does not separate Ψ_i. □

5 Robust Virtual Implementation

In this section, we use the notions of strategic distinguishability and the maximally revealing mechanism to establish necessary and sufficient conditions for robust virtual implementation. Virtual implementation of a social choice function requires a mechanism such that the desired outcomes are realized with probability arbitrarily close to 1 (see Abreu and Matsushima, 1992b and Abreu and Matsushima, 1992c). Robust implementation requires implementation of a social choice function depending on agents' "payoff types" independent of their beliefs and higher order beliefs about others' payoff types (see Bergemann and Morris, 2005a and Bergemann and Morris, 2009). Our definition of robust virtual implementation is the natural one incorporating both these notions.

5.1 *Definitions*

Write $\|y - y'\|$ for the rectilinear norm between a pair of lotteries y and y', i.e.,

$$\|y - y'\| \triangleq \sum_{x \in X} |y(x) - y'(x)|.$$

Definition 5 (Robust ε-Implementation). The mechanism $\mathcal{M}$ robustly ε-implements the social choice function f if

$$m \in S[\mathcal{M}](\theta) \Rightarrow \|g(m) - f(\theta)\| \leq \varepsilon;$$

f is robustly ε-implementable if there exists a mechanism $\mathcal{M}$ that robustly ε-implements f.

We can now define the notion of robust virtual implementation.

Definition 6 (Robust Virtual Implementation). Social choice function f is robustly virtually implementable if, for every $\varepsilon > 0, f$ is robustly ε-implementable.

The relevant incentive compatibility condition required for our robust problem is ex post incentive compatibility.

Definition 7 (EPIC). Social choice function f satisfies ex post incentive compatibility (EPIC) if, for all i, θ_i, θ_{-i} and θ_i':

$$u_i(f(\theta_i, \theta_{-i}), (\theta_i, \theta_{-i})) \geq u_i(f(\theta_i', \theta_{-i}), (\theta_i, \theta_{-i})).$$

Now "robust measurability" requires that if θ_i is pairwise inseparable from θ_i', then the social choice function must treat the two types the same. This condition is the robust analogue of the measurability condition in Abreu and Matsushima (1992c) as we formally establish in Section 6.1.

Definition 8 (Robust Measurability). Social choice function f is robust measurable if $\theta_i \sim \theta_i' \Rightarrow f(\theta_{-i}, \theta_{-i}) = f(\theta_i', \theta_{-i})$ for all θ_{-i}.

5.2 *Necessity*

It is well known from the literature on virtual Bayesian implementation (e.g., Abreu and Matsushima, 1992c) that the relaxation to virtual implementation does not relax incentive compatibility conditions by a standard compactness argument.[7]

Theorem 2 (Necessity). *If f is robustly virtually implementable, then f is ex post incentive compatible and robustly measurable.*

Proof. We first establish ex post incentive compatibility. Fix any mechanism $\mathcal{M}$ that robustly ε-implements f. Fix θ_{-i} and $m_{-i} \in S_{-i}^{\mathcal{M}}(\theta_{-i})$. For any $m_i' \in S_i[\mathcal{M}](\theta_i')$, virtual implementation requires

$$\|g(m_i', m_{-i}) - f(\theta_i', \theta_{-i})\| \leq \varepsilon. \tag{14}$$

Now suppose that player i is type θ_i and is convinced that his opponent is type θ_{-i} sending message m_{-i}. Let m_i be any message which is a best response to that belief. Then $m_i \in S_i[\mathcal{M}](\theta_i)$, implying that

$$\|g(m_i, m_{-i}) - f(\theta_i, \theta_{-i})\| \leq \varepsilon. \tag{15}$$

[7]Dasgupta, Hammond, and Maskin (1979) and Ledyard (1979) argued in a private value environment that dominant strategy incentive compatibility was implied by Bayesian incentive compatibility for all priors on a fixed type space. In the case of a social choice function, this argument — generalized to interdependent values — shows the necessity of ex post incentive compatibility (see Bergemann and Morris, 2005b).

In particular, by the best response property of m_i:

$$u_i(g(m_i, m_{-i}), (\theta_i, \theta_{-i})) \geq u_i(g(m_i', m_{-i}), (\theta_i, \theta_{-i})). \tag{16}$$

Now (14) and Lemma 2 imply

$$|u_i(g(m_i', m_{-i}), (\theta_i, \theta_{-i})) - u_i(f(\theta_i', \theta_{-i}), (\theta_i, \theta_{-i}))| \leq \frac{1}{2}\varepsilon C, \tag{17}$$

and (15) and Lemma 2 imply

$$|u_i(g(m_i, m_{-i}), (\theta_i, \theta_{-i})) - u_i(f(\theta_i, \theta_{-i}), (\theta_i, \theta_{-i}))| \leq \frac{1}{2}\varepsilon C. \tag{18}$$

Now combining (16), (17) and (18), we obtain

$$u_i(f(\theta_i, \theta_{-i}), (\theta_i, \theta_{-i})) \geq u_i(f(\theta_i', \theta_{-i}), (\theta_i, \theta_{-i})) - \varepsilon C.$$

But virtual implementation implies that this holds for all $\varepsilon > 0$, so we have

$$u_i(f(\theta_i, \theta_{-i}), (\theta_i, \theta_{-i})) \geq u_i(f(\theta_i', \theta_{-i}), (\theta_i, \theta_{-i})),$$

and this establishes EPIC as necessary condition.

Next we establish robust measurability. Suppose that f is robustly virtually implementable. Fix any $\varepsilon > 0$. Since f is robustly virtually implementable, there exists a mechanism $\mathcal{M}_\varepsilon$ such that

$$m \in S[\mathcal{M}_\varepsilon](\theta) \Rightarrow \|g(m) - f(\theta)\| \leq \varepsilon.$$

Now fix any θ_{-i} and $m_{-i}^\varepsilon \in S_{-i}[\mathcal{M}_\varepsilon](\theta_{-i})$. Also fix any $\theta_i \sim \theta_i'$, so by Proposition 1, there exists

$$m_i^\varepsilon \in S_i[\mathcal{M}_\varepsilon](\theta_i) \cap S_i[\mathcal{M}_\varepsilon](\theta_i').$$

Now $\|g(m_i^\varepsilon, m_{-i}^\varepsilon) - f(\theta_i, \theta_{-i})\| \leq \varepsilon$ and $\|g(m_i^\varepsilon, m_{-i}^\varepsilon) - f(\theta_i', \theta_{-i})\| \leq \varepsilon$. Thus $\|f(\theta_i, \theta_{-i}) - f(\theta_i', \theta_{-i})\| \leq 2\varepsilon$. This is true for each $\varepsilon > 0$, so $f(\theta_i, \theta_{-i}) = f(\theta_i', \theta_{-i})$. □

While we maintain the assumption the mechanism is finite, the same argument implies the necessity of EPIC and robust measurability if we allow "regular mechanisms" (Abreu and Matsushima, 1992c), i.e., mechanisms where best replies always exist of any conjecture over opponents' behavior.

5.3 *Sufficiency*

We first describe the construction of a *canonical mechanism* that will be used to establish sufficiency. Our construction follows the logic of Abreu and Matsushima (1992c), which in turn builds on Abreu and Matsushima (1992b). In the mechanism we construct, each agent simultaneously announces (i) a message in the maximally revealing mechanism described above; (ii) L announcements of his payoff type. With probability close to $\frac{1}{L}$, the outcome is chosen according the agents' lth announcement of their payoff types in part (ii) of their messages. But with small probability, the outcome is chosen according to the maximally revealing mechanism and their part (i) messages. The mechanism then checks to see which agents were the "first" to "lie", in the sense that his lth report of his type is not consistent with the message he sent in the maximally revealing mechanism and no other agent sent an inconsistent message in an "earlier" report. If an agent is not one of the first to lie, then the agent is rewarded. For this part of the mechanism, we need an economic property.

Definition 9 (Economic Property). The uniform economic property is satisfied if there exist a profile of lotteries, $(z_i)_{i=1}^{I}$, such that, for each i and $\theta, u_i(zi, \theta) > u_i(\bar{y}, \theta)$ and $u_j(\bar{y}, \theta) \geq u_j(z_i, \theta)$ for all $j \neq i$.

Under the uniform economic property, there will exist a constant c_0 such that

$$u_i(z_i, \theta) > u_i(\bar{y}, \theta) + c_0 \tag{19}$$

for all i and θ.

In the canonical mechanism, part (i) announcements for the maximally revealing mechanism are made as if the maximally revealing mechanism was being played as a stand alone mechanism (since the probability of rewards can be chosen sufficiently small). An agent will never allow himself to be one of the first to lie: sending a message that ensures that he is not the first to lie (given his beliefs about others' strategies) will always strictly improve on his expected payoff, since if others are telling the truth, truth-telling is a weak best response by ex post incentive compatibility, and if they are lying, for sufficiently large L, the reward will outweigh the cost of not lying in one stage of the mechanism.

We write $\mathcal{M}^* = ((M_i^*)_{i=1}^{I}, g^*)$ for the maximally revealing mechanism. We use three numbers in defining the canonical mechanism: c_0 is the uniform lower bound on an agent's utility gain from having his uniformly preferred lottery rather than the central lottery; recall from Lemma 2 that C is an upper bound on payoff differences in the environment; recall from Lemma 4 whenever a message is deleted in the iterated deletion process for the maximally revealing mechanism $\mathcal{M}^*$, it is not even an $\eta_{\mathcal{M}^*}$-best response to any conjecture. We will use these three numbers c_0, C and $\eta_{\mathcal{M}^*}$, together with the number of players I, to define two further numbers δ and L that will be used in the construction of the canonical mechanism. Choose $\delta > 0$ such that

$$\delta < \frac{\eta_{\mathcal{M}^*}}{C}, \tag{20}$$

and an integer L such that

$$L > \frac{IC}{\delta^2 c_0}. \tag{21}$$

Now the message space of the canonical mechanism is

$$M_i = M_i^* \times \overbrace{\Theta_i \times \cdots \times \Theta_i}^{L \text{ times}} = M_i^* \times \Theta_i^L.$$

Thus a typical message will be written as $m_i = (m_i^0, m_i^1, \ldots, m_i^L)$, with $m_i^0 \in M_i^*; m_i^l \in \Theta_i$ for each $l = 1, \ldots, L$. The idea is that an agent is "supposed" to truthfully report his payoff type in each stage $l = 1, \ldots, L$ and will receive a small punishment if he is one of the "first" to report a type that is not consistent with his 0-th message. The small individual rewards and punishments are provided by

$$r_i(m) = \begin{cases} \bar{y}, \text{ if} & \begin{array}{l} \exists k \in \{1, \ldots, L\} \quad \text{s.t. } m_i^0 \notin S_i[\mathcal{M}^*](m_i^k), \text{ and} \\ m_j^0 \in S_j[\mathcal{M}^*](m_j^l) \forall_j = 1, \ldots, I \text{ and } l = 1, \ldots, k-1; \end{array} \\ z_i, \text{ if} & \text{otherwise.} \end{cases}$$

(In slight abuse of notation, we use $r_i(m)$ here to denote rewards whereas we used r_i^k earlier in subsection 4.2.1). Now the outcome function of the canonical mechanism is:

$$g(m) = (1 - \delta - \delta^2)\frac{1}{L}\sum_{l=1}^{L} f(m^l) + \delta g^*(m^0) + \frac{\delta^2}{I}\sum_{i=1}^{I} r_i(m).$$

The mechanism $g(m)$ has three components. The first component, which carries the largest probability, is the social choice function f itself. The appropriate allocation $f(m^l)$ will be selected by L replicas, each one of which is chosen with a small Probability $1/L$. The second component is the maximally revealing mechanism outcome function g^* which receives a smaller weight of δ. The third and final component, $r_i(m)$, represents a small reward or punishment. It is designed to give each agent an incentive to replicate in stage l the report issued in the previous stage. It provides a small "punishment" ($\bar{y}$) if player i is the first to report in the message component, m_i^l, a type inconsistent with previous reports, otherwise $r_i(m)$ provides the small "reward" (z_i).

Theorem 3. *Under the uniform economic property, if f satisfies EPIC and robust measurability, then the canonical mechanism $\delta(1+\delta)$ robustly implements f.*

This immediately implies the sufficiency part of our characterization of robust virtual implementation, since we can choose δ arbitrarily close to 0 in the canonical mechanism.

Corollary 1 (Sufficiency). *Under the uniform economic property, if f satisfies EPIC and robust measurability, then f is robustly virtually implementable.*

Proof. To prove the theorem, it is enough to establish that, for each $i, m_i = (m_i^0, m_i^1, \ldots, m_i^L) \in S_i[\mathcal{M}](\theta_i)$ implies that (1) $m_i^0 S_i[\mathcal{M}^*(\theta_i)$ and (2) $m_i^0 \in S_i[\mathcal{M}^*](m_i^l)$ for each $l = 1, \ldots, L$. To see why, observe that $m_i^0 \in S_i[\mathcal{M}^*](\theta_i) \cap S_i[\mathcal{M}^*](m_i^l)$ implies θ_i is strategically indistinguishable from m_i^l, which implies, by robust measurability, that $f(m_i^l, m_{-i}^l) = f(\theta_i, m_{-i}^l)$. Since this holds for each i, we have $f(m^l) = f(\theta)$. Since this is true for each l, we have that the mechanism selects $f(\theta)$ with probability at least $1 - \delta - \delta^2$.

Claim (1) above — that $(m_i^0, m_i^1, \ldots, m_i^L) \in S_i[\mathcal{M}](\theta_i) \Rightarrow m_i^0 \in S_i[\mathcal{M}^*](\theta_i)$ — follows from Lemma 5 and inequality (20), since m^0 influences the outcome only through weight δ on $g^*(m^0)$ and weight δ^2 on $\frac{1}{I}\sum_{i=1}^{I} r_i(m)$.

We will now establish claim (2) above — that $(m_i^0, m_i^1, \ldots, m_i^L) \in S_i[\mathcal{M}](\theta_i) \Rightarrow m_i^0 \in S_i[\mathcal{M}^*](m_i^l)$ for all i and $l = 1, \ldots, L$.

Suppose this claim were false. Then there must exist a smallest l for which the claim fails. Thus there exists $l^* \in \{1, \ldots, L\}$ such that, for all $j, m_j \in S_j[\mathcal{M}^*](\theta_j) \Rightarrow m_j^0 \in S_j[\mathcal{M}^*](m_j^l$ for all $1 \leq l < l^*$; but there exists i and $m_i = (m_i^0, m_i^1, \ldots, m_i^L) \in S_i[\mathcal{M}^*](\theta_i)$ with $m_i^0 \notin S_i[\mathcal{M}^*](m_i^{l^*})$. Now fix any conjecture $\mu_i \in \Delta(\Theta_{-i} \times M_{-i})$ with $\mu_i(\theta_{-i}, m_{-i}) > 0 \Rightarrow m_j \in$

$S_j[\mathcal{M}^*](\theta_i)$ for all $j \neq i$. Consider two cases. First, suppose that

$$\mu_i(\theta_{-i}, m_{-i}) > 0 \Rightarrow m_j^0 \in S_j[\mathcal{M}^*](m_j^l) \quad \text{for all } j \neq i \text{ and } l = 1, \ldots, L. \tag{22}$$

In this case, sending the message

$$\bar{m}_i = (m_i^0, \overbrace{\theta_i, \theta_i, \ldots, \theta_i}^{L \text{ times}})$$

instead of m_i will strictly increase i's utility: since he is certain that each agent is reporting a type that is strategically indistinguishable in each of the L stages, EPIC and robust measurability ensure that his utility will not decrease from truthtelling in the L stages; his utility will be unchanged in the maximally revealing mechanism; and his utility will be strictly increased in the punishment component. Secondly, i's conjecture μ_i is such that (22) fails. In this case, we can define

$$\hat{l} = \min\{l \in \{1, \ldots, L\} : \exists(\theta_{-i}, m_{-i}) \quad \text{with } \mu_i(\theta_{-i}, m_{-i}) > 0 \text{ and} \\ m_j^0 \in S_j[\mathcal{M}^*](m_j^l) \quad \text{for some } j \neq i\}.$$

Note that $\hat{l} \geq l^*$. Now sending the message

$$\bar{m}_i = (m_i^0, \overbrace{\theta_i, \theta_i, \ldots, \theta_i}^{\hat{l} \text{ times}}, \; m_i^{\hat{l}+1}, \ldots, m_i^L)$$

instead of m_i will strictly increase i's utility: since he is certain that each agent is reporting a type that is strategically indistinguishable in each of the first $\hat{l} - 1$ stages, EPIC and robust measurability ensure that his utility will not decrease from truthtelling in the first $\hat{l} - 1$ stages; his utility will be unchanged in the maximally revealing mechanism; if it turns out that $m_j^0 \in S_j[\mathcal{M}^*](m_j^{\hat{l}})$ for some $j \neq i$, then i's utility will also not be reduced in the $\hat{l}$-th stage or in the punishment component; but if it turns out that $m_j^0 \notin S_j[\mathcal{M}^*](m_j^{\hat{l}})$ for all $j \neq i$, then i's utility will be reduced in the $\hat{l}$-th stage by at most $(1 - \delta - \delta^2)\frac{1}{L}C$ and will increase in his own punishment component $r_i(\cdot)$ by at least $\frac{\delta^2}{I}c_0$ (and by the economic property, will not decrease in his opponents' punishment components $r_{-i}(\cdot)$). The second term exceeds the first term by (21).

We conclude that for no conjecture is m_i a best response, contradicting our original assumption. This proves our second claim. □

While the basic construction of this proof follows Abreu and Matsushima (1992c), there are some complications that arise in our robust formulation. The messages sent in the maximally revealing mechanism do not partition an agent's types. Rather, for each set of types that survives the iterated deletion of sets that can always be separated, there is a message that may be sent by all types in that set. So we say that message m_i^l is consistent with m_i^0 if message m_i^0 is one that might be sent by $m_i^0 \in S_i[\mathcal{M}^*](m_i^l)$.

The economic property can be weakened along the lines of assumption 2 in Abreu and Matsushima (1992c). It would be enough to have the economic property hold for any type set profile Ψ in the inseparable type set Ξ^*, i.e., for each set profile $\Psi = (\Psi_i)_{i=1}^I \in \Xi^*$, there exists $(z_i)_{i=1}^I$, such that, for each i and $\theta \in X_{i=1}^I \Psi_j$, $u_i(z_i, \theta) > u_i(\bar{y}, \theta)$ and $u_j(\bar{y}, \theta) \geq u_j(z_i, \theta)$ for all $j \neq i$.

6 Discussion

6.1 *Abreu–Matsushima Measurability*

We established in the preceding section that robust measurability, jointly with ex post incentive compatibility, is a necessary and sufficient condition for robust virtual implementation. Ex post incentive compatibility is equivalent to Bayesian incentive compatibility on the union of all type spaces (Bergemann and Morris, 2005b). We now show that robust measurability is equivalent to requiring that the notion of measurability originally suggested by Abreu and Matsushima (1992c) holds on the union of all type spaces.[8] To spell out the details of this equivalence result, we need a formal language for epistemic type spaces in the sense of Harsanyi (1967–68) and Mertens and Zamir (1985).

A *type space* is defined by $\mathcal{T} \triangleq (T_i, \hat{\pi}_i, \hat{\theta}_i)_{i=1}^I$, where each T_i is a countable set of types, where the function $\hat{\pi}_i : T_i \to \Delta(T_{-i})$ defines the beliefs that agent i assigns to other agents having types t_{-i} and where the function $\hat{\theta}_i : T_i \to \Theta_i$ defines the agent i's payoff types. A type space is *finite* if each T_i is finite. We fix a type space $\mathcal{T}$ and write $\succeq_{t_i}^{\mathcal{T}}$ for the induced preferences of type t_i of agent i over type-contingent lotteries $\tilde{y}_i : T_{-i} \to Y$. Thus

$$\tilde{y}_i \succeq_{t_i} \tilde{y}_i'$$

[8] We would like to thank an anonymous referee who suggested to investigate the exact relationship between Abreu–Matsushima measurability and robust measurability.

if and only if

$$\sum_{t_{-i}\in T_{-i}} \hat{\pi}_i(t_{-i}|t_i)u_i(\tilde{y}_i(t_{-i}),t) \geq \sum_{t_{-i}\in T_{-i}} \hat{\pi}_i(t_{-i}|t_i)u_i(\tilde{y}'_i(t_{-i}),t).$$

Fix a partition profile $\mathcal{H} = (\mathcal{H}_i)_{i=1}^I$, where each $\mathcal{H}_i$ is a partition of T_i. A function $\tilde{y}_i : T_{-i} \to Y$ is $\mathcal{H}$-measurable if for all $j \neq i$:

$$\{t_j, t'_j\} \subseteq H_j \in \mathcal{H}_j \Rightarrow \tilde{y}(t_j, t_{-\{i,j\}}) = \tilde{y}(t'_j, t_{-\{i,j\}}).$$

Say that a pair of types t_i and t'_i are $(\mathcal{T},\mathcal{H})$-distinguishable if there exists $\mathcal{H}$-measurable $\tilde{y}_i : T_{-i} \to Y$, such that

$$\tilde{y}_i \succ_{t_i}^{\mathcal{T}} \bar{y} \quad \text{and} \quad \bar{y} \succ_{t_i}^{\mathcal{T}} \tilde{y}_i,$$

where we continue to denote by $\bar{y}$ the constant uniform lottery.

Now iteratively define a sequence of partitions $\mathcal{H}^k = (\mathcal{H}_i^k)_{i=1}^I$ be letting each $\mathcal{H}_i^0$ be the coarsest partition of the type set T_i, namely $\{T_i\}$ and let each $\mathcal{H}_i^{k+1}$ consist of sets of types of agent i that are $(\mathcal{T},\mathcal{H}^k)$-indistinguishable.

Let $\mathcal{H}^*$ be the limit of the sequence of partitions. We say that types t_i and t'_i are Abreu-Matsushima, or "AM", indistinguishable on type space $\mathcal{T}$, written $t_i \sim_{AM}^{\mathcal{T}} t'_i$, if t_i and t'_i are in the same element of the partition $\mathcal{H}_i^*$.

Proposition 3 (Equivalence).

1. *If θ_i and θ'_i are pairwise inseparable, then there exists a finite type space $\mathcal{T}$ and a pair of types $t_i, t'_i \in T_i$ such that (i) $\hat{\theta}_i(t_i) = \theta_i$; (ii) $\hat{\theta}_i(t'_i) = \theta'_i$; and (iii) $t_i \sim_{AM}^{\mathcal{T}} t'_i$.*
2. *Conversely, if there exists a type space $\mathcal{T}$ (perhaps infinite but countable) and a pair of types $t_i, t'_i \in T_i$ such that (i) $\hat{\theta}_i(t_i) = \theta_i$; (ii) $\hat{\theta}_i(t'_i) = \theta'_i$; and (iii) $t_i \sim_{AM}^{\mathcal{T}} t'_i$, then θ_i and θ'_i are pairwise inseparable.*

The equivalence result of Proposition 3 suggests a alternative route to establishing the necessity result for robust implementation in Theorem 2: by the equivalence of robust measurability and AM measurability on the union of all type spaces, we could prove the necessity by an appeal to the arguments used in Abreu and Matsushima (1992c). By contrast, our sufficiency result (Theorem 3) cannot be established using the arguments and methods in Abreu and Matsushima (1992c): as the union of all type spaces is not a finite object, the arguments in Abreu and Matsushima (1992c) — which rely on the finiteness of the type space — cannot be applied.

6.2 Interdependence and Pairwise Separability

We illustrated the notions of pairwise and mutual inseparability in Section 3 in the context of a linear model of interdependent preferences for a single object:

$$v_i(\theta_i, \theta_{-i}) = \theta_i + \gamma \sum_{j \neq i} \theta_j.$$

In this linear and symmetric model the parameter γ represented the level of interdependence in the preferences of the agents. We showed that for $\gamma < 1/(I-1)$, all payoff types of all agents are pairwise separable and suggested that pairwise separability required not too much interdependence in the preferences.

We now establish the relationship between pairwise inseparability and moderate interdependence in a substantially more general environment. We assume that the utility function of each agent i is given by a convex combination of a private value utility function v_i and an interdependent utility function w_i over the general space of lotteries Y defined in Section 2. The private value utility function v_i:

$$v_i : Y \times \Theta_i \to \mathbb{R}$$

gives rise to distinct preferences for every θ_i:

$$\theta_i \neq \theta_i' \Rightarrow v_i(\cdot, \theta_i) \quad \text{is not an affine transformation of } v_i(\cdot, \theta_i').$$

The interdependent utility function w_i:

$$w_i : Y \times \Theta_{-i} \to \mathbb{R}$$

can depend in an arbitrary way on the type profile $\theta_{-i} \in \Theta_{-i}$ of all agents except agent i. For any $\gamma_i \in [0,1]$, let $u_i^{\gamma_i}$ be the utility functions that puts weight $1 - \gamma_i$ on the private value utility v_i and weight γ_i on the interdependent utility w_i:

$$u_i^{\gamma_i}(y, \theta) \triangleq (1 - \gamma_i) v_i(y, \theta_i) + \gamma_i w_i(y, \theta_{-i}). \tag{23}$$

The interdependence in the preferences is now described by the vector of weights $\gamma = (\gamma_1, \ldots, r_I) \in [0,1]^I$. For $\gamma = (0, \ldots, 0)$ all payoff types of all agents are pairwise separable as, by assumption, the private utility function v_i gives rise to distinct preferences for all θ_i. Also, for $\gamma = (1, \ldots, 1)$, we cannot separate any pair of types for any agent. In this case, the preferences of each agent are independent of his payoff type and therefore we cannot expect to separate the payoff types of agent i on the basis of his revealed preference. We parametrize the limit set Ξ^* which by Definition 2 describes the set of pairwise inseparable types, by the vector γ, or $\Xi^*(\gamma)$.

Proposition 4 (Interdependence).

1. *The collection of sets* $\Xi_i^*(\gamma)$ *satisfies* $\Xi_i^*(\mathbf{0}) = \{\{\theta_i^1\}, \ldots, \{\theta_i^S\}\}$ *and* $\Xi_i^*(1) = 2^{\Theta_i} \backslash \emptyset$ *for all* i.
2. *If* $\hat{\gamma} \geq \gamma$, *then* $\Xi_i^*(\gamma) \subseteq \Xi_i^*(\hat{\gamma})$.

The first part of the proposition determines the structure of the pairwise separable types with minimal and maximal interdependence. The second part establishes that the sets of pairs of types θ_i and θ_i' which are inseparable are weakly increasing in the interdependence parameter γ. In particular, it shows that the separability is monotone in the parameter of interdependence. We should emphasize that as the interdependence is represented by the vector $\gamma = (\gamma_1, \ldots, \gamma_I)$, the threshold for complete separability of all types and all agents itself is a multidimensional surface in the I-dimensional hypercube.

6.3 *Intermediate Robustness Notions*

The classic Bayesian implementation literature considers implementation on a fixed type space. We believe that this approach — as usually formulated — assumes too much common knowledge (among the agents and the planner) about the environment. In relaxing these common knowledge assumptions, we take an extreme approach: we maintain the assumption that there is common knowledge of the payoff structure of the environment (i.e., the set of possible payoff types of each agent and how each agent's utility function depends on the profile of payoff types) but do not restrict agents' beliefs and higher order beliefs about other agents' types.

In a recent paper, Artemov, Kunimoto, and Serrano (2008) consider what happens to robust virtual implementation results if one imposes some restrictions on agents' beliefs in the payoff environment. In particular, call a pair $(\theta_i, \lambda_i) \in \Theta_i \times \Delta(\Theta_{-i})$ a "pseudo-type" and suppose that we add the common knowledge that agent i's pseudo-type (θ_i, λ_i) belongs to a subset $T_i \subseteq \Theta_i \times \Delta(\Theta_{-i})$. When can a social choice function be virtually implemented on all type spaces where each agent i's pseudo-type belongs to T_i? Note that an agent's pseudo-type pins down his payoff type and belief about others' payoff types, but not his higher order beliefs. Thus this assumption is intermediate between the standard approach and our robustness approach. In the special case where $T_i = \Theta_i \times \Delta(\Theta_{-i})$, this setting becomes the setting of this paper. But if T_i is a strict subset of $\Theta_i \times \Delta(\Theta_{-i})$, the conditions for robust virtual implementation will be weakened.

Now say that "pseudo-type diversity" is satisfied if

1. The set of beliefs consistent with a payoff type is a compact set, i.e., $\{\lambda_i \in \Delta(\Theta_{-i}) | (\theta_i, \lambda_i) \in T_i\}$ is a compact set for each i and $\theta_i \in \Theta_i$.
2. Two distinct payoff types cannot have the same preference over constant lotteries, i.e., $(\theta_i, \lambda_i), (\theta_i', \lambda_i') \in T_i$ and $\theta_i \neq \theta_i' \Rightarrow R_{\theta_i, \lambda_i} \neq R_{\theta_i', \lambda_i'}$.

Artemov, Kunimoto, and Serrano (2008) show that if pseudo-type diversity is satisfied, then robust virtual implementation will always be possible if the appropriate incentive compatibility conditions are satisfied (their Theorem 1). The idea is that agents' payoff types can then be identified by their preferences over constant lotteries and the Abreu and Matsushima (1992c)-style argument applied.[9]

To get a feel for the strength of the pseudo-type diversity condition, we can return to our leading example in Section 3. Recall that each $\Theta_i = [0, 1]$ and $v_i = \theta_i + \gamma \mathbb{E}_i[\sum_{j \neq i} \theta_j]$ is a sufficient statistic for agent i's preferences. Now let $\Lambda_i \subseteq \Delta([0, 1]^{I-1})$ be a compact set of beliefs over others' types that agent i may have (whatever his payoff type), so his set of possible pseudo-types is the product set $T_i = [0, 1] \times \Lambda_i$. Now if $0 < \gamma \leq \frac{1}{I-1}$, so there is not too much interdependence of preferences, pseudo-type diversity will be satisfied if and only if each Λ_i is a singleton.[10]

Artemov, Kunimoto, and Serrano (2008) also report the appropriate measurability condition required for robust virtual implementation if the pseudo-type diversity condition fails (their Definition 12 and Theorem 2).

---

[9]The version of "pseudo-type diversity" which we report is sufficient to implement the social choice functions depending just on payoff types that we study in this paper. Artemov, Kunimoto, and Serrano (2008) assume a slightly stronger version of pseudo-type diversity: they assume that each T_i is finite and that distinct pseudo-types have distinct preferences over constant lotteries even if they correspond to the same payoff type, i.e., $(\theta_i, \lambda_i), (\theta_i', \lambda_i') \in T_i$ and $(\theta_i, \lambda_i) \neq (\theta_i', \lambda_i') \Rightarrow R_{\theta_i, \lambda_i} \neq R_{\theta_i', \lambda_i'}$. This allows them to implement richer social choice functions that treat types with the same payoff types (but different beliefs over others' payoff types) differently.

[10]This example has a continuum of payoff types, so does not fit our formal framework. But we could make the same point with a finite grid of payoff types.

This will naturally be intermediate between Abreu-Matsushima measurability and our robust measurability condition. We can illustrate this also with our example. Suppose that the probability that agent i assigns to any subset of other agents' payoff types is always at least $1-\delta$ times the probability of that event under a uniform prior, so that

$$\Lambda_i = \left\{ \lambda_i \in \Delta(\Theta_{-i}) \middle| \lambda_i(E) \geq (1-\delta) \int_{\theta_{-i} \in E}, d\theta_{-i} \right.$$
$$\left. \forall \text{ measurable } E \subseteq [0,1]^{I-1} \right\}$$

and $T_i = \Theta_i \times \Lambda_i$.

Now suppose that agent i's payoff type is in Ψ_i and he knows that other agents' payoff types are in Ψ_{-i}. If agent i's beliefs are restricted to belong to Λ_i, when do there exist a pair of payoff types in Ψ_i who could not have the same expected valuation of the object? Only if

$$\max \Psi_i + \gamma \sum_{j \neq i} \left((1-\delta)\frac{1}{2} + \delta \min \Psi_j \right)$$
$$> \min \Psi_i + \gamma \sum_{j \neq i} \left((1-\delta)\frac{1}{2} + \delta \max \Psi_j \right).$$

Thus Ψ_{-i} "(δ-separates" Ψ_i if and only if

$$\max \Psi_i - \min \Psi_i \leq \gamma\delta \sum_{j \neq i} (\max \Psi_j - \min \Psi_j).$$

Now the argument of Section 3 can be adapted to show that if $\gamma\delta < \frac{1}{I-1}$, all pairs of distinct payoff types will be strategically distinguishable from each other (under δ belief restrictions) and thus incentive compatibility will be sufficient for robust virtual implementation. And if $\gamma\delta > \frac{1}{I-1}$, all pairs of payoff types will be strategically indistinguishable from each other (under δ belief restrictions) and robust virtual implementation will be impossible for any (non-constant) social choice function.

6.4 *Rationalizability and All Equilibria on All Type Spaces*

Our analysis took as given the solution concept of incomplete information rationalizability for our environment. Thus we assumed that if the agents' true payoff type profile was $\theta = (\theta_1, \ldots, \theta_I)$, they might send any message profile

$$m \triangleq (m_1, \ldots, m_I) \in \bigtimes_{i=1}^{I} S_i[\mathcal{M}](\theta_i) \triangleq S[\mathcal{M}](\theta).$$

Our motivation for employing this solution concept is that we did not want to make any assumption about agents' beliefs and higher order beliefs about other agents' payoff types. In fact, suppose one constructed a "type space" $\mathcal{T}$ specifying for each agent a set of possible epistemic types, and, for each epistemic type, a description of his (known) payoff type and his beliefs about others' epistemic types. By standard universal type space arguments, we can incorporate any beliefs and higher order beliefs about others' payoff types in such a type space. Now the type space $\mathcal{T}$ and a mechanism $\mathcal{M}$ together define a standard incomplete information game. The set of messages that can be sent by *any* type of agent i with payoff type θ_i in *any* Bayesian Nash equilibrium of the game $(\mathcal{T}, \mathcal{M})$ for *any* type space $\mathcal{T}$ is equal to $S_i[\mathcal{M}](\theta_i)$. This result is the straightforward incomplete information extension of the classic epistemic foundations result of Brandenburger and Dekel (1987), showing that the set of actions that can be played in the subjective correlated equilibria of a complete information game equals the set of actions that survive iterated deletion of strictly dominated actions in that game. Battigalli and Siniscalchi (2003) reported the incomplete information version of this result as Propositions 4.2 and 4.3. For completeness, we formally state and prove this result in the appendix of the working paper version (Bergemann and Morris (2007)).

This observation means that the gap between the solution concepts of pure strategy Bayesian Nash equilibrium (Serrano and Vohra (2001), Serrano and Vohra (2005)) and iterated deletion of (interim) strictly dominated strategies (Abreu and Matsushima (1992c)) in incomplete information virtual implementation disappears in our robust approach. We consider this to be an attraction of our approach. The intuition is that the extra bite obtained by the assumption of equilibrium is lost without complementary strong assumptions on beliefs and higher order beliefs for the implementation problem.

6.5 *Iterated Deletion of Weakly Dominated Strategies*

Our incomplete information rationalizability solution concept is equivalent to iterated deletion of strictly dominated strategies. What would happen if we looked at iterated deletion of weakly dominated strategies instead? In other words, we let $W_i^0[\mathcal{M}](\theta_i) = M_i$,

$$W_i^{k+1}[\mathcal{M}](\theta_i) = \left\{ m_i \in W_i^k[\mathcal{M}](\theta_i) \left| \begin{array}{l} \exists\, \mu_i \in \Delta_{++}\{(\theta_{-i}, m_{-i}) | m_{-i} \\ \qquad \in W_{-i}^k[\mathcal{M}](\theta_{-i})\} \quad \text{s.t.} \\ m_i \in \arg\max\limits_{m_i'} \sum\limits_{\theta_{-i}, m_{-i}} \mu_i(\theta_{-i}, m_{-i}) \\ \qquad u_i(g(m_i', m_{-i}), (\theta_i, \theta_{-i})) \end{array} \right. \right\};$$

and

$$W_i[\mathcal{M}](\theta_i) \bigcap_{k \geq 0} W_i^k[\mathcal{M}](\theta_i).$$

It is easy to see that our "negative" results would go through unchanged. If two types are pairwise inseparable $(\theta_i \sim_i')$ then the argument of Proposition 1 — unchanged — implies that they will have iteratively weakly undominated actions in common in every mechanism, or

$$W_i[\mathcal{M}](\theta_i) \cap W_i[\mathcal{M}](\theta_i') \neq \emptyset \quad \text{for all } \mathcal{M}.$$

Thus robust measurability is a necessary condition for implementation (virtual or exact) of any social choice function in iterated deletion of weakly dominated strategies in a finite (or compact) mechanism: the argument of Proposition 2 will go through unchanged in this case.

Abreu and Matsushima (1994) show that their argument for *virtual* complete information implementation in iterated deletion of *strictly* dominated strategies can be adapted to show the possibility of *exact* complete information implementation in iterated deletion of *weakly* dominated strategies, with some extra restrictions on the environment. It is a reasonable conjecture that this extension could be adapted to the standard incomplete information implementation setting of Abreu and Matsushima (1992c) and our robust incomplete information setting. However, we have not attempted this extension.

Chung and Ely (2001) have shown that in an auction environment with interdependent valuations as in section 3, the efficient outcome can be implemented in the direct mechanism under iterated deletion of weakly dominated strategies (i.e., the solution concept described above) under the assumption that $\gamma < \frac{1}{I-1}$. Our results supply a strong converse: if $\gamma > \frac{i}{I-1}$, it is not possible to implement (exactly or virtually) any non-trivial social choice function in iterated deletion of weakly dominated strategies in any finite (or compact) mechanism, direct or indirect.[11]

6.6 *Implementation in a Direct Mechanism*

We restricted attention in this paper to finite mechanisms. Thus the mechanisms here do not include any of the pathological features of "integer games" that play an important role in the full implementation literature and have been much criticized (see, e.g., Jackson (1992)). Nonetheless, the mechanisms in this paper are complex. The canonical mechanism for robust virtual implementation inherits the complexity of the mechanism of Abreu and Matsushima (1992c), on which it builds. Our maximally revealing mechanism generating strategic distinguishability is no simpler. While the mechanisms are theoretically kosher, it has been argued that their complexity and the logic of the iteration deletion in the mechanism might make them hard to use in practise. For example, Glazer and Rosenthal (1992) have made this argument about the mechanism used by Abreu and Matsushima (1992b) for complete information virtual implementation (see Abreu and Matsushima (1992a) for a response and Sefton and Yavas (1996) for later experiments inspired by the mechanism).

By requiring robustness to agents' beliefs and higher order beliefs, we reduce the amount of common knowledge about the environment that can be used by the planner in designing a mechanism. This will make it harder to achieve positive results (and our robust measurability condition is rather strong in applications). But one motivation for studying robust implementation is that we hope that robustness considerations will endogenously lead to simpler mechanisms when positive results can be achieved. By adapting results from our earlier work on exact robust implementation in direct mechanisms (Bergemann and Morris (2009)), we can report that, in at least one broad class of economic environments of interest, whenever robust virtual implementation is possible according to corollary 1, it is possible in a direct mechanism where agents simply report their payoff types. We say that preferences satisfy *aggregator single crossing* (ASC) if each agent i's preferences at type profile θ belong to a single crossing class parameterized by $h_i(\theta)$, where $h_i : \Theta \to \mathbb{R}$ is a monotonic aggregator. Bergemann and

[11]Our results are stated for a lottery space over finite outcomes, but the extension to any compact space and compact mechanisms is straightforward.

Morris (2009) established that exact robust implementation by a compact mechanism is possible if and only if the social choice function satisfies strict ex post incentive compatibility and a *contraction property* on the aggregator functions $h = (h_1, \ldots, h_I)$. In the appendix of the working paper version, we show that under the ASC assumption, robust measurability is always satisfied under the contraction property.

6.7 *Exact Implementation and Integer Games*

The first papers on incomplete information implementation focussed on exact implementation. Postlewaite and Schmeidler (1986) and Jackson (1991) identified a Bayesian monotonicity condition which (together with Bayesian incentive compatibility) was necessary and (under weak economic conditions) sufficient for exact implementation in Bayesian Nash equilibrium. Bergemann and Morris (2005a) provide a robust analogue of this result, showing that ex post incentive compatibility and a *robust monotonicity* condition are necessary and — under weak economic conditions — sufficient for exact robust implementation. All these papers follow a tradition in the implementation literature of allowing very badly behaved mechanisms, like integer games, in proving their general results. In this paper, we follow Abreu and Matsushima (1992c) in restricting attention to finite — and thus well-behaved — mechanisms. We briefly discuss the relation between these results in this section: a more complete and formal discussion in contained in the appendix of the working paper version (Bergemann and Morris (2007)).

Robust measurability and robust monotonicity turn out to be equivalent in the important class of aggregator single crossing preferences. However, in general, one can show by example that robust measurability neither implies nor is implied by robust monotonicity. Thus requiring only virtual implementation is sometimes a strict relaxation; and allowing badly-behaved mechanisms is sometimes a strict relaxation. We do not have a characterization of when exact robust implementation by a well behaved mechanism is possible (just as analogous characterizations do not exist for complete information and classical Bayesian implementation). We know only that robust measurability, robust monotonicity and strict ex post incentive compatibility will all be necessary.

We restrict attention in our analysis to social choice functions rather than social choice correspondences. Bergemann and Morris (2005b) considered the problem of *partially* robustly implementing a social choice correspondence, i.e., ensuring that whatever players' beliefs and higher order beliefs about others' types, there is *an* equilibrium leading to outcomes contained in the social choice correspondence. In the special case where the social choice correspondence is a function (and more generally in a class of separable environments), this is possible only if the function (or a selection from the correspondence in separable environments) is ex post incentive compatible. But in the general case, we do not have a satisfactory characterization of when partial robust implementation is possible. For this reason, we have not attempted a characterization of (full) robust implementation of social choice correspondences.

7 Appendix

The appendix contains omitted proofs from the main body of the paper.

Proof of Lemma 1. Suppose that

$$\sum_{\theta_{-i}\in\Theta_{-i}} \lambda_i(\theta_{-i})u_i(\bar{y}(\theta_i,\theta_{-i})) \geq \sum_{\theta_{-i}\in\Theta_{-i}} \lambda_i(\theta_{-i})u_i(x,(\theta_i,\theta_{-i})) \tag{24}$$

for all $x \in X$. If

$$\sum_{\theta_{-i}\in\Theta_{-i}} \lambda_i(\theta_{-i})u_i(\bar{y},(\theta_i,\theta_{-i})) > \sum_{\theta_{-i}\in\Theta_{-i}} \lambda_i(\theta_{-i})u_i(x',(\theta_i,\theta_{-i}))$$

for some $x' \in X$, we could conclude, that

$$\begin{aligned}\sum_{\theta_{-i}\in\Theta_{-i}} \lambda_i(\theta_{-i})u_i(\bar{y},(\theta_i,\theta_{-i})) &> \frac{1}{N}\sum_{x\in X}\sum_{\theta_{-i}\in\Theta_{-i}} \lambda_i(\theta_{-i})u_i(x,(\theta_i,\theta_{-i})) \\ &= \sum_{\theta_{-i}\in\Theta_{-i}} \lambda_i(\theta_{-i})u_i(\bar{y},(\theta_i,\theta_{-i})),\end{aligned}$$

a contradiction. So (24) implies

$$\sum_{\theta_{-i}\in\Theta_{-i}} \lambda_i(\theta_{-i})u_i(\bar{y},(\theta_i,\theta_{-i})) = \sum_{\theta_{-i}\in\Theta_{-i}} \lambda_i(\theta_{-i})u_i(x,(\theta_i,\theta_{-i})) \tag{25}$$

for all $x \in X$. But (25) implies that R_{θ_i,λ_i} is indifferent between all pure outcomes and thus all lotteries. This contradicts assumption 1 on no-complete-indifference. We conclude that the no-complete-indifference assumption implies that (24) fails for all i, i.e., that for all $i, \theta_i \in \Theta_i$ and $\lambda_i \in \Delta(\Theta_{-i})$ there exists $x \in X$ such that

$$\sum_{\theta_{-i}\in\Theta_{-i}} \lambda_i(\theta_{-i})u_i(x,(\theta_i,\theta_{-i})) > \sum_{\theta_{-i}\in\Theta_{-i}} \lambda_i(\theta_{-i})u_i(\bar{y},(\theta_i,\theta_{-i})). \quad (26)$$

Equivalently, for all $i, \theta_i \in \Theta_i$ and $\lambda_i \in \Delta(\Theta_{-i})$:

$$\max_{x\in X} \sum_{\theta_{-i}\in\Theta_{-i}} \lambda_i(\theta_{-i})[u_i(x_i,(\theta_i,\theta_{-i})) - u_i(\bar{y},(\theta_i,\theta_{-i}))] > 0$$

Now, note that for each $x \in X$ the function

$$\sum_{\theta_{-i}\in\Theta_{-i}} \lambda_i(\theta_{-i})[u_i(x,(\theta_i,\theta_{-i})) - u_i(\bar{y},(\theta_i,\theta_{-i}))]$$

is continuous in λ (in the standard topology). The conclusion follows from the compactness (in the standard topology) of $\Delta(\Theta_{-i})$ and continuity of the maximum operator. □

Proof of Lemma 4. Fix any $m_i \notin S_i[\mathcal{M}](\theta_i)$. Then there exists k such that $m_i \in S_i^k[\mathcal{M}](\theta_i)$ and $m_i \notin S_i^{k+1}[\mathcal{M}](\theta_i)$. Consider

$$\Delta_i^k = \{\mu_i \in \Delta(\Theta_{-i}) \times M_{-i} | \mu_i(\theta_{-i} \times m_i > 0 \Rightarrow m_{-i} \in S_{-i}^k[\mathcal{M}](\theta_{-i}) \text{ for each } j \neq i\}.$$

For all $\mu_i \in \Delta_i^k$, there exists $\bar{m}_i$ such that

$$\sum_{\theta_{-i},m_{-i}} \mu_i(\theta_{-i},m_{-i})u_i(g(\bar{m}_i,m_{-i}),(\theta_i,\theta_{-i})) > \sum_{\theta_{-i},m_{-i}} \mu_i(\theta_{-i},m_{-i})u_i(g(m_i,m_{-i}),(\theta_i,\theta_{-i})).$$

By compactness of Δ_i^k, there exists $\bar{\varepsilon}_i(m_i) > 0$ such that for all $\mu_i \in \Delta_i^k$ there exists $\bar{m}_i$, such that

$$\sum_{\theta_{-i},m_{-i}} \mu_i(\theta_{-i},m_{-i})u_i(g(\bar{m}_i,m_{-i}),(\theta_i,\theta_{-i})) > \sum_{\theta_{-i},m_{-i}} \mu_i(\theta_{-i},m_{-i})u_i(g(m_i,m_{-i}),(\theta_i,\theta_{-i})) + \bar{\varepsilon}_i(m_i).$$

Now let

$$\eta_{\mathcal{M}} = \min_{i,\theta_i \text{ and } m_i\notin S_i[\mathcal{M}](\theta_i)} \bar{\varepsilon}_i(m_i),$$

which establishes the desired bound. □

Proof of Lemma 5. Suppose $\pi^+ C \leq \pi^0 \eta_{\mathcal{M}^0}$. We will argue, by induction on k, that

$$(m_i^0, m_i^1) \in S_i^k[\mathcal{M}](\theta_i) \Rightarrow m_i^0 \in S_i^k[\mathcal{M}^0](\theta_i)$$

for all $k \geq 0$. This is true by definition for $k = 0$; suppose that it is true for k. Now suppose that $m_i^0 \notin S_i^{k+1}[\mathcal{M}^0](\theta_i)$ but $(m_i^0, m_i^1) \in S_i^{k+1}[\mathcal{M}](\theta_i)$ and so $(m_i^0 m_i^1) \in S_i^k[\mathcal{M}](\theta_i)$ and — by the inductive hypothesis — $m_i^0 \in S_i^k[\mathcal{M}^0](\theta_i)$. Now fix any $\mu_i \in \Delta(\Theta_{-i} \times M_{-i})$ satisfying

$$\begin{aligned} \mu_i(\theta_{-i}, (m_j^0, m_j^1)_{j \neq 1}) > 0 &\Rightarrow (m_j^0, m_j^1)_{j \neq 1} \in S_{-i}^k[\mathcal{M}](\theta_{-i}) \\ &\Rightarrow m_{-i}^0 \in S_{-i}^k[\mathcal{M}^0](\theta_{-i}). \end{aligned}$$

Let

$$\bar{\mu}_i(\theta_{-i}, m_{-i}^0) = \sum_{(m_j^1)_{j \neq 1} \in M_{-i}^1} \mu_i(\theta_{-i}, (m_j^0, m_j^1)_{j \neq 1}).$$

By Lemma 4, there exists $\bar{m}_i^0$ such that:

$$\begin{aligned} &\sum_{\theta_{-i}, m_{-i}^0} \bar{\mu}_i(\theta_{-i}, m_{-i}^0) \\ &\quad \times [u_i(g^0(\bar{m}_i^0, m_{-i}^0), (\theta_i, \theta_{-i})) - u_i(g^0(m^0, m_{-i}^0), (\theta_i, \theta_{-i}))] > \eta_{\mathcal{M}^0}. \end{aligned}$$

Thus

$$\begin{aligned} &\sum_{\theta_{-i}, m_{-i}} (u_i(\theta_{-i}, m_{-i})[u_i(g((\bar{m}_i^0, m_i^1), m_{-i}), (\theta_i, \theta_{-i})) \\ &\quad - u_i(g((m_i^0, m_i^1), m_{-i}), (\theta_i, \theta_{-i}))] > \pi^0 \eta_{\mathcal{M}^0} - \pi^+ C \geq 0. \end{aligned}$$

This contradicts our premise that $(m_i^0, m_i^1) \in S_i^{k+1}[\mathcal{M}](\theta_i)$ and we conclude that $(m_i^0, m_i^1) \in S_i^{k+1}[\mathcal{M}](\theta_i) \Rightarrow m_i^0 \in S_i^{k+1}[\mathcal{M}^0](\theta_i)$. □

The canonical mechanism asks each agent to make a series of binary choices between the central lottery y and a specific lottery y from the test set. If the test set is to be successful in eliciting the private information from agent i, then the test set should contain a sufficient number of allocations such that for every type θ_i and every belief λ_i of agent i there exists some allocation y that is strictly preferred to the central lottery $\bar{y}$.

Lemma 9 (Duality). *Type set profile Ψ_{-i} separates Ψ_i if and only if there exists $\bar{y} : \Psi_i \to Y$ such that*

$$\sum_{\theta_i \in \Psi_i} (\tilde{y}(\theta_i) - \bar{y}) = 0, \tag{27}$$

and

$$\tilde{y}(\theta_i) P_{\theta_i,\lambda_i} \bar{y}, \tag{28}$$

for all $\theta_i \in \Psi_i$ and all $\lambda_i \in \Delta(\Psi_{-i})$.

This result says that for each $\theta_i \in \Psi_i$, we can identify a direction in the lottery space, $\tilde{y}(\theta_i) - \bar{y}$, that agent i likes whatever his beliefs about Ψ_{-i}, such that the sum of those changes add up to zero. The lemma follows from the following duality result in Samet (1998):

Proposition 5 (Samet, 1998). *Let $V_1, \ldots, V_S$ be closed, convex, subsets of the N-dimensional simplex Δ^N. These sets have an empty intersection if and only if there exist $z_1, \ldots, z_S \in \mathbb{R}^N$ such that*

$$\sum_{s=1}^{S} z_S = 0,$$

and

$$v \cdot z_S > 0, \quad \text{for each } s = 1, \ldots, S \text{ and } v \in V_s.$$

This result was introduced in Samet (1998) to provide a simple proof of the observation that asymmetrically informed agents will trade against each other if and only if they do not share a common prior, i.e., their posterior beliefs could not have been derived by updating a common prior.[12] Suppose that there are N states and S agents. Each agent s observes one of a collection of signals about the true state. Each signal leads him to have a posterior $v \in \Delta^N$ over the states. Let V_s be the convex hull of his set of possible posteriors. Notice that V_s represents the set of prior beliefs he might have held over the state space before observing his signal. Thus posterior beliefs are consistent with a common prior if and only if the intersection of the V_s sets is non-empty. Now consider a multilateral bet specifying that

[12] This converse to the no trade theorem was originally proved by Morris (1994), by a more indirect duality argument.

if state n was realized, agent s will receive payment z_{sn} where the total payments sum to zero:

$$\sum_{s=1}^{S} z_{sn} = 0 \quad \text{for all } n.$$

Writing $z_s \triangleq (z_{sn})_{n=1}^{N}$, we then have

$$\sum_{s=1}^{S} z_s = 0.$$

There exists such a bet where every agent has a strictly positive expected value from accepting the bet conditional on every signal if $v \cdot z_s > 0$ for each $s = 1, \ldots, S$ and $v \in V_s$.

Proof of Lemma 9. By definition, type set profile Ψ_{-i} separates Ψ_i if, for every $R \in \mathcal{R}$, there exists $\theta_i \in \Psi_i$ such that $R_{\theta_i, \lambda_i} \neq R$ for every $\lambda_i \in \Delta(\Psi_{-i})$. Write

$$X = \{x_1, \ldots, x_n, \ldots, x_N\}, \Theta_i = \{\theta_i^1, \ldots, \theta_i^s, \ldots, \theta_i^S\}, \quad \text{and}$$
$$\Theta_{-i} = \{\theta_{-i}^1, \ldots, \theta_{-i}^w, \ldots, \theta_{-i}^W\}, \quad \text{with } W = S^{I-1}.$$

The vector

$$v_{sw} = (u_i(x_n, (\theta_{-i}^s, \theta_{-i}^w)))_{n=1}^{N},$$

is an element of $\mathbb{R}^N$. Without loss of generality (since expected utility preferences can be represented by any affine transformation), we can assume that each v_{sw} is an element of the N dimensional simplex Δ^N. Now $(v_{sw})_{w=1}^{W}$ is a collection of W elements of Δ^N, and the set of preferences

$$\{R_{\theta_i^s, \lambda_i} : \lambda_i \in \Delta(\Psi_{-i})\},$$

are represented by the convex hull of $(v_{sw})_{w=1}^{W}$, which we write as

$$V_s = \text{conv}\left((v_{sw})_{w=1}^{W}\right) \subseteq \Delta^N.$$

Thus Ψ_{-i} separates Ψ_i exactly if

$$\bigcap_{s=1}^{S} V_s = \emptyset.$$

By Proposition 5, this is true if and only if there exist $z_1, \ldots, Z_S \in \mathbb{R}^N$ such that

$$\sum_{s=1}^{S} z_s = 0, \tag{29}$$

and

$$v \cdot z_s > 0, \tag{30}$$

for all s and all $v \in V_s$. But if $(z_s)_{s=1}^S$ satisfy (29) and (30), we may choose $\in > 0$ sufficiently small such that $\tilde{y}(\theta_i^s) = \bar{y} + \varepsilon z_s \in Y$ for each s, and we have established (27) and (28).

Conversely, if (27) and (28) hold and we set $z_s = \tilde{y}(\theta_i^s) - \bar{y}$ for $s = 1, \ldots, S$, then $(z_s)_{s=1}^S$ satisfy (29) and (30). □

We now use Lemma 9 to show how, if Ψ_{-i} separates Ψ_i, we can construct a *finite* set of lotteries $\tilde{Y}_i(\Psi_i, \Psi_{-i}) \subseteq Y$ such that knowing that agent i knows that his opponent's type is in Ψ_{-i} and knowing his preferences on $\tilde{Y}_i(\Psi_i, \Psi_{-i})$ will always be enough to rule out at least one type in Ψ_i for agent i.

Lemma 10. *If Ψ_{-i} separates Ψ_i, then there exists a finite set $\tilde{Y}_i(\Psi_i, \Psi_{-i}) \subseteq Y$, such that for each $\theta_i \in \Psi_i$ and $\lambda_i \in \Delta(\Psi_{-i})$, there exists $y \in \tilde{Y}_i(\Psi_i, \Psi_{-i})$ such that*

$$\bar{y} P_{\theta_i, \lambda_i} y, \tag{31}$$

and for some $\theta_i' \in \Psi_i$,

$$y P_{\theta_i', \lambda_i'} \bar{y}, \tag{32}$$

for all $\lambda_i' \in \Delta(\Psi_{-i})$.

Proof. By Lemma 9, there exists $\tilde{y} : \Psi_i \to Y$ such that

$$\sum_{\theta_i \in \Psi_i} (\tilde{y}(\theta_i) - \bar{y}) = 0,$$

and

$$\tilde{y}(\theta_i) P_{\theta_i, \lambda_i} \bar{y} \quad \text{for all } \theta_i \in \Psi_i \quad \text{and} \quad \lambda_i \in \Delta(\Psi_{-i}).$$

Let $\tilde{Y}_i(\Psi_i, \Psi_{-i}) = \{\tilde{y}(\theta_i)\}_{\theta_i \in \Psi_i}$. Fix $\theta_i \in \Psi_i$ and $\lambda_i \in \Delta(\Psi_{-i})$. Write $\tilde{Y}_i(\Psi_i, \Psi_{-i}) = \{y^1, \ldots, y^K\}$, with $y^1 = \tilde{y}(\theta_i)$. Let $\bar{y}^0 = \bar{y}$ and

$$\bar{y}^l = \bar{y} + \varepsilon \sum_{k=1}^{l} (y^k - \bar{y}),$$

with $\varepsilon > 0$ chosen sufficiently small such that $\bar{y}^l \in Y$ for all $l = 1, \ldots, K$. We know $\bar{y}^1 P_{\theta_i, \lambda_i} \bar{y}^0$. Suppose $\bar{y}^{l+1}\ R_{\theta_i, \lambda_i} \bar{y}^l$ for all $l = 1, \ldots, K-1$. By transitivity, this would imply that $\bar{y}^K P_{\theta_i, \lambda_i} \bar{y}^0$. But $y^K = \bar{y}^0$, so we have a contradiction. We conclude that, for some $l = 1, \ldots, K-1, \bar{y}^l P_{\theta_i, \lambda_i} \bar{y}^{l+1}$. This implies that there exists θ_i' such that $\bar{y} P_{\theta_i, \lambda_i} y(\theta_i')$. Since

$$y(\theta_i') P_{\theta_i', \lambda_i'} \bar{y} \quad \text{for all } \lambda_i' \in \Delta(\Psi_{-i}),$$

the inequalities (31) and (32) are established. □

Now we will construct a large enough finite set of lotteries (the "test set") such that knowing just an agent's most preferred outcome on the test set will always reveal enough information about his preferences to separate out a type, if it is possible to do so. This will establish the proof of Lemma 7.

Proof of Lemma 7. Our proof is constructive. We first construct a set $\tilde{Y}$ consisting of the degenerate lotteries X and the sets $\tilde{Y}_i(\Psi i, \Psi_{-i})$ constructed in Lemma 10, for every triple $(i, \Psi_i, \Psi_{-i}$ with Ψ_{-i} separating Ψ_i. Knowing an agent's ranking of each element of $\tilde{Y}$ relative to the central lottery $\bar{y}$ would reveal all the information we need to extract. In order to extract this information in a single choice, we let the agent pick $f : \tilde{Y} \to \{0, 1\}$. For each $y \in \tilde{Y}, y$ is chosen with probability $1/\tilde{Y}$ if $f(y) = 1$, otherwise the central lottery $\bar{y}$ is chosen. We let Y^* be the set of all such lotteries. Now observing an agent's most preferred outcome in Y^* reveals his binary preference between $\bar{y}$ and each element of $\tilde{Y}$. Since $\tilde{Y}$ contains each $\tilde{Y}_i(\Psi_i, \Psi_{-i})$, this will ensure part (2). Since $\tilde{Y}$ contains all the lotteries which are putting probability 1 on each pure outcome, Assumption 1 (no-complete-indifference) implies that, for each θ_i and λ_i, there exist $y, y' \in \tilde{Y}$ such that $y P_{\theta_i, \lambda_i} y'$ and thus $y' \notin B_i^{Y*}(\theta_i, \lambda_i)$. This proves part (1).

Let

$$\tilde{Y} = X \cup \bigcup_{\{(i, \Psi_i, \Psi_{-i}) | \Psi_{-i} \text{ separates } \Psi_i\}} \tilde{Y}_i(\Psi_i, \Psi_{-i}).$$

Now for any $f : \tilde{Y} \to \{0,1\}$, let y_f be the lottery obtained by picking an element $y \in \tilde{Y}$ with uniform probability and then choosing lottery y if $f(y) = 1$ and $\bar{y}$ if $f(y) = 0$. Thus we define:

$$y_f \equiv \bar{y}\frac{1}{\#\tilde{Y}} \sum_{y \in \tilde{Y}} f(y)(y - \bar{y}).$$

Let Y^* be the set of such lotteries, i.e.,

$$Y^* = \{y \in Y | \exists f : \tilde{Y} \to \{0,1\} \quad \text{such that } y = y_f\}.$$

To prove part (1) of the Lemma, fix any $\theta_i \in \Theta_i$ and $\lambda_i \in \Delta(\Theta_{-i})$. By Lemma 1, there exists $x \in X \subseteq \tilde{Y}$ such that $xP_{\theta_i,\lambda_i}\bar{y}$; now let $f^0(y) = 0$, for all $y \in \tilde{Y}$, and

$$f^*(y) = \begin{cases} 0, & \text{if } y \neq x; \\ 1, & \text{if } y = x. \end{cases}$$

So we can write:

$$y_{f^0} = \bar{y}, \quad y_{f^*} = \bar{y} + \frac{1}{\#\tilde{Y}}(x - \bar{y})$$

and so $y_{f^0} \notin B_i^{Y*}(\theta_i, \lambda_i)$.

To prove part (2) of the Lemma, suppose that Ψ_{-i} separates Ψ_i. Fix $\theta_i \in \Psi_i$ and $\lambda_i \in \Delta(\Psi_{-i})$. By Lemma 10, there exists $y \in \tilde{Y}_i(\Psi_i, \Psi_{-i})$ and $\theta_i' \in \Psi_i$ such that $\bar{y}P_{\theta_i,\lambda_i}y$ and $yP_{\theta_i',\lambda_i'}\bar{y}$ for all $\lambda_i' \in \Delta(\Psi_{-i})$. So

$$y_f \in B_i^{Y*}(\theta_i, \lambda_i) \Rightarrow f(y) = 0,$$

while

$$y_f \in B_i^{Y*}(\theta_i', \Psi_i) \Rightarrow f(y) = 1,$$

and so

$$B_i^{Y*}(\theta_i, \lambda_i) \cap B_i^{Y*}(\theta_i', \Psi_i) = \emptyset,$$

which establishes the result. □

Proof of Proposition 2. Consider an arbitrary pair of types, θ_i and θ_i' such that $\theta_i \nsim \theta_i'$. Then by the definition of pairwise inseparability,

$\{\theta_i, \theta'_i\} \in \Xi^*_i$. By the construction of the inseparable sets Ξ^k_i, it follows that there is a finite stage $\bar{k}$ such that $\{\theta_i, \theta'_i\} \in \Xi^{\bar{k}}_i$ but:

$$\{\theta_i, \theta'_i\} \notin \Xi^{\bar{k}+1}_i. \tag{33}$$

By Lemma 7 we have for all i, k and m^k_i that:

$$\bar{\Theta}^k_i(m^k_i) \in \Xi^k_i, \tag{34}$$

and by Lemma 6, for each k there exists $\bar{\varepsilon} > 0$ such that

$$\{\theta_i \in \Theta_i | m^k_i \in S_i[\mathcal{M}^k_\varepsilon](\theta_i)\} \subseteq \bar{\Theta}^k_i(m^k_i), \tag{35}$$

for all $\varepsilon < \bar{\varepsilon}$ and $m^k_i \in M^k_i$. Now since Ξ^* is established in a finite number of stages, it follows that by the choosing k sufficiently large and ε sufficiently small, we obtain an augmented mechanism $\mathcal{M}^K_\varepsilon = \mathcal{M}^*$ such that if $\theta_i \not\sim \theta'_i$, then from the exclusion (33) and the inclusions (34) and (35), it follows that $S^{\mathcal{M}*}_i(\theta_i) \cap S^{\mathcal{M}*}_i(\theta'_i) = \emptyset$, which establishes the result. □

Proof of Proposition 3. (1.) Fix mutually inseparable $\Xi = (\Xi_i)^I_{i=1}$. We will use properties of Ξ to construct a type space $\mathcal{T}$. For each $\Psi_i \in \Xi_i$, there exists $\Psi^{\Psi_i}_{-i} \in \Xi_{-i}$ such that $\Psi^{\Psi_i}_{-i}$ does not separate Ψ_i. Recall that "$\Psi^{\Psi_i}_{-i}$ does not separate Ψ_i" means that there exists a preference relation R_i over uncontingent lotteries Y such that for each $\theta_i \in \Psi_i$, there exists $\lambda^{\theta_i,\Psi_i}_i \in \Delta(\Psi^{\Psi_i}_{-i})$ such that $R_{\theta_i,\lambda_i}\theta_i, \Psi_i = R_i$. Now, for each i, let

$$T_i \triangleq \{(\theta_i, \Psi_i) \in \Theta_i \times \Xi_i | \theta_i \in \Psi_i\}, \tag{36}$$

with

$$\hat{\pi}((\theta_j, \Psi_j)_{j\neq i} | (\theta_i, \Psi_i)) \triangleq \begin{cases} \lambda^{\theta_i,\Psi_i}_i(\theta_{-i}), & \text{if} \quad \Psi_{-i} = \Psi^{\Psi_i}_{-i}, \\ 0, & \text{if} \quad \text{otherwise}; \end{cases} \tag{37}$$

and

$$\hat{\theta}_i(\theta_i, \Psi_i) \triangleq \theta_i. \tag{38}$$

Now consider the partition $\mathcal{H}_i$ of the type set $\mathcal{T}_i$, as defined through (36)–(38), which is generated by the equivalence relation $(\theta_i, \Psi_i) \sim (\theta'_i, \Psi'_i)$ if $\Psi_i = \Psi'_i$. By construction, each (θ_i, Ψ_i) and (θ'_i, Ψ_i) are $(\mathcal{T}, \mathcal{H})$-indistinguishable. To see this, observe that since $\theta_i, \theta'_i \in \Psi_i$ there exists a common Ψ_{-i}, namely $\Psi^{\Psi_i}_{-i}$ such that $\lambda^{\theta_i,\Psi_i}_i(\Psi^{\Psi_i}_{-i}) = \lambda^{\theta'_i,\Psi_i}_i(\Psi^{\Psi_i}_{-i}) = 1$. Now, as the type contingent lottery $\tilde{y}_i$ has to be $\mathcal{H}$-measurable, it follow in particular that it has to be constant on $\Psi^{\Psi_i}_{-i}$ and hence is an uncontingent lottery on $\Psi^{\Psi_i}_{-i}$. But Lemma 3 shows that if any pair of payoff types

θ_i and θ_i' are pairwise inseparable, then there exists mutually inseparable $\Xi = (\Xi_i)_{i=1}^I$ and $\Psi_k \in \Xi_k$ with $\{\theta_k, \theta_k'\} \subseteq \Psi_k$.

(**2.**) For the other direction, fix a type space $\mathcal{T}$. Write $\mathcal{H}^*$ for the limit of the sequence of partitions defined above and let $\sim_{AM}^{\mathcal{T}}$ be the corresponding equivalence relation. Write $H_i^*(t_i) = \{t_i' | t_i' \sim_{AM}^{\mathcal{T}} t_i\}$ and let

$$\Xi_i \triangleq \{\Psi_i \in 2^{\Theta_i/\emptyset} | \exists t_i \in T_i \quad \text{such that} \quad \Psi_i = \{\theta_i | \exists t_i' \in H_i^*(t_i) \text{ with } \hat{\theta}_i(t_i') = \theta_i\}\}.$$

Intuitively, Ξ_i is a set of payoff types that cannot be distinguished on the particular (interim) type space $\mathcal{T}$.

Fix a player i and any $t_i \in T_i$, and let

$$\Psi_i = \{\theta_i | \exists t_i' \in H_i^*(t_i) \quad \text{with } \hat{\theta}_i'(t_i') = \theta_i\}.$$

Suppose $t_i' \sim_{AM}^{\mathcal{T}} t_i$. We know that for every $\mathcal{H}^*$-measurable $\tilde{y}_i, \tilde{y}_i \succ_{t_i}^{\mathcal{T}} \bar{y} \Rightarrow \tilde{y}_i \succeq_{t_i'}^{\mathcal{T}} \bar{y}$. Observe that each $t_i' \in H_i^*(t_i)$ must have the same support on elements of $\mathcal{H}_{-i}^*$. Pick any t_{-i}^* such that $\hat{\pi}_i(H_{-i}^*(t_{-i}^*)|t_i') > 0$ for all $t_i' \in H_i^*(t_i)$. Consider $\lambda_i^{\theta_i, \Psi_i}(\Psi_{-i}^{\Psi_i})$ which equals the uniform lottery everywhere except on t_{-i} with $t_j \sim_{AM}^{\mathcal{T}} t_j^*$ for all $j \neq i$, i.e.,

$$\tilde{y}_i(t_{-i}) = \bar{y} \quad \text{if not } t_j \sim_{AM}^{\mathcal{T}} t_j^* \text{ for some } j.$$

Note that $\tilde{y}_i$ is $\mathcal{H}^*$-measurable. Now let $\Psi_j = \{\theta_j | \exists t_j' \in H_j^*(t_j^*) \text{ with } \hat{\theta}_j(t_j') = \theta_j\}$ and observe that by construction Ψ_i is not separated by Ψ_{-i}. Thus Ξ is mutually inseparable. □

The proof of Proposition 4 will follow directly from the monotone behavior of the following auxiliary sets related to the inseparable sets. In Section 2.3 we defined a sequence of inseparable sets, $\{\Xi^k\}_{k=0}^\infty = \{(\Xi_1^k, \ldots, \Xi_I^k)\}_{k=0}^\infty$, where the $k+1$st level of sets is determined by an inductive step:

$$\Xi_i^{k+1} = \{\Psi_i \in \Xi_i^k | \Psi_{-i} \quad \text{does not separate } \Psi_i, \text{ for some } \Psi_{-i} \in \Xi_{-i}^k\}. \tag{39}$$

For our monotonicity result, it will be useful to simply fix a sequence of sets for all agents except i:

$$\{\Sigma_{-i}^k\}_{k=0}^\infty = \{(\Sigma_1^K, \ldots, \Sigma_{i-1}^K, \Sigma_{i+1}^K, \ldots, \Sigma_I^K)\}_{k=0}^\infty$$

such that the sequence satisfies the inclusion property:

$$\Sigma_j^{k+1} \subseteq \Sigma_j^k, \tag{40}$$

but without necessarily coming from the separation property as Ξ_j^{k+1} in (39). However, for agent $i, \sum_i^k$ is generated by the separation property relative to the sequence $\{\Sigma_{-i}^k\}_{k=0}^{\infty}$. In particular, $\Sigma_i^0 = 2^{\Theta_i} \backslash \emptyset$ and:

$$\Sigma_i^{k+1} \triangleq \{\Psi_i \in \Sigma_i^k | \Psi_{-i} \quad \text{does not separate } \Psi_i, \text{ for some } \Psi_{-i} \in \Sigma_{-i}^k\} \tag{41}$$

and the resulting limit set is defined by:

$$\Sigma_i^* = \bigcap_{k \geq 0} \Sigma_i^k.$$

Now we consider two sequences of sets for all agents $i, \{\hat{\Sigma}_{-i}^k\}_{k=0}^{\infty}$ and $\{\Sigma_{-i}^k\}_{k=0}^{\infty}$, such that one sequence is nested in the other, or for all $k, \hat{\Sigma}_{-i}^k \subseteq \Sigma_{-i}^k$. We then compare the resulting limit set for agent i with respect to Σ_{-i}^k and $\hat{\Sigma}_{-i}^k$ respectively. Correspondingly, we denote the respective limit sets of agent i by Σ_{-i}^* and $\hat{\Sigma}_i^*$.

Lemma 11 (Monotonicity I). *If for all $k, \hat{\Sigma}_{-i}^k \subseteq \Sigma_{-i}^k$, then $\hat{\Sigma}_i^* \subseteq \Sigma_i^*$.*

Proof. It suffices to show that for all $k, \hat{\Sigma}_i^k \subseteq \Sigma_i^k$. The proof is by induction. By construction it is true for $k = 0$. Suppose now that it holds for k and we want to establish that it holds for $k+1$. By assumption, $\hat{\Sigma}_i^k \subseteq \Sigma_i^k$ and hence consider a set $\Psi_i \in \Sigma_i^k \cap \hat{\Sigma}_i^k$. Now suppose that $\Psi_i \in \hat{\Sigma}_i^{k+1}$ and we want to show that $\Psi_i \in \Sigma_i^{k+1}$. We observe that if $\Psi_i \in \hat{\Sigma}_i^{k+1}$, then there exists some $\Psi_{-i} \in \Sigma_{-i}^k$ such that Ψ_{-i} does not separate Ψ_i. But by assumption the set $\Psi_{-i} \in \Sigma_{-i}^k$, and hence it follows that $\Psi_i \in \Sigma_i^{k+1}$ as well. □

Lemma 12 (Monotonicity II). *If $\hat{\gamma}_i > \gamma_i$, then for all $k, \Sigma_i^k \subseteq \hat{\Sigma}_i^k$.*

Proof. The proof is by induction. By construction it is true for $k = 0$. Suppose now that it holds for k and we want to establish that it holds for $k + 1$. By assumption, $\Sigma_i^k \subseteq \hat{\Sigma}_i^k$ and hence consider a set $\Psi_i \in \Sigma_i^k \cap \hat{\Sigma}_i^k$. Now suppose that $\Psi_i \in \Sigma_i^{k+1}$ and we want to show that $\Psi_i \in \hat{\Sigma}_i^{k+1}$. We observe that if $\Psi_i \in \Sigma_i^{k+1}$, then there exists some $\Psi_{-i} \in \Sigma_{-i}^k$ such that Ψ_{-i} does not separate Ψ_i. In other words, there exists for every $\theta_i \in \Psi_i$ a belief $\lambda_i(\cdot|\theta_i) \in \Delta(\Psi_{-i})$ such that for all $x \in X$ and all $\theta_i', \theta_i'' \in \Psi_i$:

$$(1 - \gamma_i)v_i(x, \theta_i') + \gamma_i \sum_{\theta_{-i} \in \Theta_{-i}} \lambda_i(\theta_{-i}|\theta_i')w_i(x, \theta_{-i})$$

$$= (1 - \gamma_i)v_i(x, \theta_i'') + \gamma_i \sum_{\theta_{-i} \in \Theta_{-i}} \lambda_i(\theta_{-i}|\theta_i'')w_i(x, \theta_{-i}).$$

As the interdependent utility $w_i(\cdot)$ does not depend on θ_i, we can rewrite the equality as:

$$(1-\gamma_i)(v_i(x,\theta_i')-v_i(x,\theta_i''))$$
$$=\gamma_i \sum_{\theta_{-i}\in\Theta_{-i}} (\lambda_i(\theta_{-i}|\theta_i'')-\lambda_i(\theta_{-i}|\theta_i'))w_i(x,\theta_{-i}). \tag{42}$$

Now we want to show that if $\hat{\gamma}_i > \gamma_i$, then we can find again associated beliefs $\hat{\lambda}_i(\theta_{-i}|\theta_i'')$ such that

$$(1-\hat{\gamma}_i)(v_i(x,\theta_i')-v_i(x,\theta_i''))$$
$$=\hat{\gamma}_i \sum_{\theta_{-i}\in\Theta_{-i}} (\hat{\lambda}_i(\theta_{-i}|\theta_i'')-\hat{\lambda}_i(\theta_{-i}|\theta_i'))w_i(x,\theta_{-i}). \tag{43}$$

We can easily verify that by letting for all $\theta_{-i}\in\Theta_{-i}$ the beliefs $\hat{\lambda}_i(\theta_{-i}|\theta_i)$ be defined by:

$$\hat{\lambda}_i(\theta_{-i}|\theta_i) \triangleq \frac{(1-\hat{\gamma}_i)\gamma_i}{\hat{\gamma}_i(1-\gamma_i)}\lambda_i(\theta_{-i}|\theta_i'') + \frac{\hat{\gamma}_i-\gamma_i}{\hat{\gamma}_i(1-\gamma_i)}\frac{1}{(I-1)S}, \tag{44}$$

we satisfy (43) if and only if we satisfy (42). Now since $\hat{\gamma}_i > \gamma_i$, it follows that

$$\frac{(1-\hat{\gamma}_i)\gamma_i}{\hat{\gamma}_i(1-\gamma_i)} < 1,$$

and hence the conditional probability distribution $\hat{\lambda}_i(\theta_{-i}|\theta_i)$ is well-defined if, as assumed, $\lambda_i(\theta_{-i}|\theta_i)$ is well-defined. But now it follows that $\Psi_i \in \hat{\Sigma}_i^{k+1}$ as well. □

Proof of Proposition 4. (1.) For $\gamma = \mathbf{0}$, we have by the definition of the private value utility function $v_i(\cdot)$ for all i and all θ_i and θ_i' : $R_i(\theta_i,\Theta_{-i}) \cap R_i(\theta_i',\Theta_{-i}) = \emptyset$. Hence it follows that we have for all $i, \Xi_i^*(\mathbf{0}) = \{\{\theta_i^1\},\ldots,\{\theta_i^S\}\}$. For $\gamma = \mathbf{1}$, we have by the definition of the interdependent value function $w_i(\cdot)$, for all i and all $\theta_i' \in \Theta_i$:

$$\bigcap_{\theta_i\in\Theta_i} R_i(\theta_i,\Theta_{-i}) = R_i(\theta_i^1,\Theta_{-i}),$$

and hence for all $i, \Xi_i^*(\mathbf{1}) = 2^{\Theta_i}\backslash\emptyset$.

(2.) It suffices to establish the result component-wise. We thus consider $\hat{\gamma} \geq \gamma$ such that $\hat{\gamma}_i > \gamma_i$ for some i and $\hat{\gamma}_j = \gamma_j$ for all $j \neq i$. Now suppose that for some agent l, we have $\Xi_l^*(\gamma) \neq \Xi_l^*(\hat{\gamma})$. Then there must be a first stage k' such that $\Xi_l^{k'}(\gamma) \neq \Xi_l^{K'}(\hat{\gamma})$, but for all $k < k'$, we have for all

$l, \Xi_l^k(\gamma) = \Xi_l^k(\hat{\gamma})$. Now since we only changed the preferences of agent i, and k' is the first stage where the sets $\Xi_l^{k'}(\gamma)$ and $\Xi_l^{k'}(\hat{\gamma})$ differ, it must be that $l = i$. But now it follows from Lemma 12 that $\Xi_i^{k'}(\gamma) \subset \Xi_i^{k'}(\hat{\gamma})$. Suppose now that there is step $k'' > k'$ such that there exists $j \neq i$ such that $\Xi_j^{k''}(\gamma) \neq \Xi_j^{k''}(\hat{\gamma})$, but for all $k < k''$, we have $\Xi_j^k(\gamma) = \Xi_j^k(\hat{\gamma})$. Now we can apply Lemma 11 to conclude that $\Xi_j^k(\gamma) \subset \Xi_j^k(\hat{\gamma})$. Now a monotonicity argument of either Lemma 11 or 12 applies at every further step along the sequence and hence we have shown that for all j, including i, we have $\Xi_j^k(\gamma) \subseteq \Xi_j^k(\hat{\gamma})$ for all k and this establishes the result. □

References

Abreu, D., and H. Matsushima (1992a): "A Response to Glazer and Rosenthal," *Econometrica*, 60, 1439–1442.

——— (1992b): "Virtual Implementation in Iteratively Undominated Strategies: Complete Information," *Econometrica*, 60, 993–1008.

——— (1992c): "Virtual Implementation in Iteratively Undominated Strategies: Incomplete Information," Discussion paper, Princeton University and University of Tokyo.

——— (1994): "Exact Implementation," *Journal of Economic Theory*, 64, 1–19.

Artemov, G., T. Kunimoto, and R. Serrano (2008): "Robust Virtual Implementation with Incomplete Information: Towards a Reinterpretation of the Wilson Doctrine," Discussion paper, Brown University and McGill University.

Battigalli, P. (1998): "Rationalizability in Incomplete Information Games," Discussion paper, Princeton University.

Battigalli, P., and M. Slniscalchi (2003): "Rationalization and Incomplete Information," *Advances in Theoretical Economics*, 3, Article 3.

Bergemann, D., and S. Morris (2005a): "Robust Implementation: The Role of Large Type Spaces," Discussion paper 1519, Cowles Foundation, Yale University.

——— (2005b): "Robust Mechanism Design," *Econometrica*, 73, 1771–1813.

——— (2007): "Strategic Distinguishability and Robust Virtual Implementation," Discussion paper 1609, Cowles Foundation, Yale University.

——— (2009): "Robust Implementation in Direct Mechanisms," *Review of Economic Studies*.

Brandenburger, A., and E. Dekel (1987): "Rationalizability and Correlated Equilibria," *Econometrica*, 55, 1391–1402.

Chung, K.-S., and J. Ely (2001): "Efficient and Dominance Solvable Auctions with Interdependent Valuations," Discussion paper, Northwestern University.

CHUNG, K.-S., and J. Ely (2007): "Foundations of Dominant Strategy Mechanisms," *Review of Economic Studies*, 74, 447–476.

Cremer, J., and R. McLean (1985): "Optimal Selling Strategies Under Uncertainty for a Discriminating Monopolist When Demands are Interdependent," *Econometrica*, 53, 345–361.

Dasgupta, P., P. Hammond, and E. Maskin (1979): "The Implementation of Social Choice Rules. Some General Results on Incentive Compatibility," *Review of Economic Studies*, 66, 185–216.

Glazer, J., and R. Rosenthal (1992): "A Note on Abreu–Matsushima Mechanisms," *Econometrica*, 60, 1435–1438.

Harsanyi, J. (1967–68): "Games with Incomplete Information Played by 'Bayesian' Players," *Management Science*, 14, 159–189, 320–334, 485–502.

Heifetz, A., and Z. Neeman (2006): "On the Generic (Im)Possibility of Full Surplus Extraction in Mechanism Design," *Econometrica,* 2006, 213–233.

Jackson, M. (1991): "Bayesian Implementation," *Econometrica*, 59, 461–477.

Jackson, M. (1992): "Implementation in Undominated Strategies: A Look at Bounded Mechanisms," *Review of Economic Studies*, 59, 757–775.

Jehiel, P., B. Moldovanu, M. Meyer-Ter-Vehn, and B. Zame (2006): "The Limits of Ex Post Implementation," *Econometrica,* 74, 585–610.

Ledyard, J. (1979): "Dominant Strategy Mechanisms and Incomplete Information," In: *Aggregation and Revelation of Preferences,* ed. by J.-J. Laffont, chap. 17, pp. 309–319. North-Holland, Amsterdam.

Mertens, J., and S. Zamir (1985): "Formalization of Bayesian Analysis for Games with Incomplete Information," *International Journal of Game Theory,* 14, 1–29.

Morris, S. (1994): "Trade with Heterogeneous Prior Beliefs and Asymmetric Information," *Econometrica*, 62, 1327–1347.

Myerson, R. (1981): "Optimal Auction Design," *Mathematics of Operations Research*, 6, 58–73.

Neeman, Z. (2004): "The Relevance of Private Information in Mechanism Design," *Journal of Economic Theory*, 117, 55–77.

Postlewaite, A., and D. Schmeidler (1986): "Implementation in Differential Information Economies," *Journal of Economic Theory*, 39, 14–33.

Samet, D. (1998): "Common Priors and Separation of Convex Sets," *Games and Economic Behavior*, 24, 172–174.

Sefton, M., and A. Yavas (1996): "Abreu-Matsushima Mechanisms: Experimental Evidence," *Games and Economic Behavior*, 16, 280–302.

Serrano, R., and R. Vohra (2001): "Some Limitations of Virtual Bayesian Implementation," *Econometrica*, 69, 785–792.

——— (2005): "A Characterization of Virtual Bayesian Implementation," *Games and Economic Behavior*, 50, 312–331.

CHAPTER 8

Multidimensional Private Value Auctions*

Hanming Fang and Stephen Morris

Abstract
We consider parametric examples of symmetric two-bidder private value auctions in which each bidder observes her own private valuation as well as noisy signals about her opponent's private valuation. We show that, in such environments, the revenue equivalence between the first and second price auctions (SPAs) breaks down and there is no definite revenue ranking; while the SPA is always efficient allocatively, the first price auction (FPA) may be inefficient; equilibria may fail to exist for the FPA. We also show that auction mechanisms provide different incentives for bidders to acquire costly information about opponents' valuation.

Keywords: Multidimensional auctions, revenue equivalence, allocative efficiency, information acquisition

1 Introduction

The paradigm of symmetric independent private value (IPV) auctions assumes that each bidder's valuation of an object is independently drawn from an identical distribution (Myerson, 1981; Riley and Samuelson, 1981). Each bidder observes her own valuation, and has no information about her opponent's valuation except for the distribution from which it is drawn. An

*We are grateful to Estelle Cantillon, David J. Cooper, Sergio Parreiras, Michael Peters, Andrew Postlewaite, seminar participants at Case Western Reserve, Ohio State, UCLA and 2004 North American Econometric Society Meetings at San Diego, two anonymous referees and especially an Associate Editor for helpful suggestions and comments. All errors are our own.

important implication of this assumption is that each bidder's belief about her opponent's valuation is independent of her own or her opponents' valuations, and it is common knowledge.

This does not seem realistic. In actual auctions, bidders may possess or may have incentives to acquire information about their opponents' valuations. Such information in most cases is *privately observed* and *noisy.* This phenomenon arises in both private and common value auction environments. For example, in highway construction procurement auctions, the capacity constraints of the bidders are an important determinant of their costs (Jofre-Bonet and Pesendorfer, 2003). While the actual cost of firm i is its private information, *other* firms may still obtain signals about firm i's cost based on their noisy observations of how much firm i is stretched in its capacity. Another example is timber auctions. Timber firms in the US Forest Service timber auctions can cruise the tract and form estimates of its characteristics (Baldwin, Marshall and Richard, 1997). A firm can obtain some noisy information about its opponents' estimates via insider rumors and industrial espionage. Our model captures two key features of both examples. First, a bidder's belief about her opponents' types is impacted by their actual types. Second, each bidder is aware that her opponents may have some signals about her type, but does not actually know the signals observed by her opponents; thus each bidder is uncertain about her opponents' belief regarding her type.

In this paper, we assume that bidders have noisy information about opponents' valuations and explore its consequences under the first-price (FPA) and second-price (SPA) auction mechanisms. Specifically, we consider parametric examples of two-bidder private value auctions in which each bidder's private valuation of the object is independently drawn from an identical distribution, and each bidder observes a noisy signal about her opponent's valuation. Thus, each bidder has a two-dimensional type that includes her own valuation (the *valuation* type) and the signal about her opponent's valuation (the *information* type). A bidder's information about her opponents' valuation is not known by her opponent. In such multidimensional auction environments, we show the following results. First, revenue equivalence between the SPA and the FPA in the standard one-dimensional symmetric IPV environments breaks down. However, our examples demonstrate that there is no general revenue ranking between the FPA and the SPA. Second, the equilibrium allocation of the object could be inefficient in the FPA but is always efficient in the SPA. Moreover, the revenue and allocative efficiency may not coincide: on the one hand, an inefficient FPA may generate a higher expected revenue for the seller; on the other hand, the seller's expected revenue could be higher in the SPA even when the object is efficiently allocated in both auctions. The inefficiency in the FPA will typically be non-monotonic in the accuracy of information, since with either complete information or no information, efficiency is obtained in the FPAs. Third, while the SPA always admits equilibrium in weakly dominant strategies, the FPA may not have any equilibrium. Finally, we show that different auction mechanisms provide different incentives for bidders to acquire costly information about opponents' private valuations. We illustrate all these results in simple examples that we can solve explicitly. However, we also argue in each case why the key features of the examples will occur more generally.

It is important to distinguish our environment from affiliated private value (APV) auctions (Wilson, 1977; Milgrom and Weber, 1982). In APV model, a bidder's belief about her opponents' valuations monotonically (in a stochastic sense) depends on her own private valuation, but does not depend on her opponents' actual valuation. That is, a bidder's private valuation at the same time provides information about opponents' valuations. Thus in APV, a bidder's belief about her opponents' valuations is no longer common knowledge; but there is still a one-to-one mapping between a bidder's belief and her own value. Our model introduces a simple but genuine *separation* between a bidder's private valuation and her signal about opponents' valuations — a bidder's belief is impacted by her opponents' realized values — thus breaking the one-to-one mapping between belief and value under the APV. It is in this sense that our model is "multidimensional." It is always possible to encode multidimensional types in a single-dimensional variable, so, as always, the content of the multidimensional signals depends on additional assumptions made about the signal space.[1]

Our examples add to the list of departures from the standard symmetric IPV auction environments in which the revenue equivalence between the FPA and SPA fails. Wilson (1977) and Milgrom and Weber (1982)

[1]Multidimensional types can always be encoded in a single-dimensional variable using the inverse Peano function (Sagan, 1984, p. 36) and other methods. The difficulty of such a one-dimensional representation of an intrinsically multidimensional problem, however, is that we could not impose reasonable restrictions on the information structure, such as types being drawn from a continuous distribution. Similar issues concerning the representation of multidimensional information with single-dimensional messages have been discussed in the mechanism design literature (see Mount and Reiter, 1974). We are grateful to the Associate Editor for bringing this issue to our attention.

showed that when bidders' valuations are symmetric and affiliated, the seller's expected revenue is (weakly) higher in the SPA than in the FPA, even though both auctions are allocatively efficient.[2] Maskin and Riley (2000) consider private value auctions in which bidders are ex ante asymmetric in the sense that different bidders' valuations are drawn from different distributions. They show that the revenue ranking between the first and the SPAs is ambiguous even though the SPA is at least as efficient as the FPA (see also (Arozamena and Cantillon, 2004; Cantillon, 2000). In their model, bidders' types are one-dimensional and the asymmetry among bidders is common knowledge. Holt (1980) and Matthews (1987) show that, when bidders are symmetrically risk averse, the seller's expected revenue in the FPA is higher than that in the SPA. Che and Gale (1998) compare the standard auctions with financially constrained bidders, and show that the seller's expected revenue in the FPA is higher than that in the SPA. The bidders in their paper are privately informed of both their valuation of the object and their financial capacity, and thus have multidimensional types. However, both the bidders' valuation and financial capacity are assumed to be independently drawn from identical distributions, hence a bidder's belief about her opponents' type remains common knowledge.

A recent paper by Kim and Che (2004) considers private value auction environments in which subgroups of bidders may *perfectly* observe the valuations of others within the group but have no information about bidders outside of their own subgroup. They show that the FPA is allocatively inefficient with positive probability and the seller's expected revenue is lower in the FPA than that in the SPA. A very nice feature of their model is that in equilibrium the competition is effectively among group leaders — bidders who have the highest valuations in their respectively subgroups — with the additional constraint that each group leader bids at least the second highest valuation in her subgroup. Their environment generates both ex ante and ex post asymmetries among group leaders. Ex ante, the leader in a larger group has stochastically higher valuation than the leader in a smaller group; and ex post, leaders may face different degrees of within-group competition. The ex post asymmetric within-group competition in particular underlies the inefficiency of the FPA. The specific forms of asymmetry in their model lead to their revenue ranking of the SPA over the FPA, rather

[2]Landsberger *et al.* (2001) showed that in asymmetric APVs auction, the FPA may generated higher expected revenue than the SPA.

than the ambiguous ranking under more general asymmetries as considered in Maskin and Riley (2000). Our model differs from Kim and Che (2004) mainly in the information structure. In our model, a bidder's information regarding her opponents' valuation is *noisy* and *private*; while in theirs, the information about rivals' types are *perfect* and *public* within a subgroup. As a result, the reasons underlying the possible allocative inefficiency of the FPA are slightly different. In their paper, the ex post asymmetry in the within-group competition faced by group leaders is the key reason: a relatively low-valued group leader facing a close competitor in her subgroup may be driven to bid more than a relatively high-valued group leader facing a distant competitor. Because the information within a subgroup is perfect and public in their model, such within-group asymmetry is publicly known. In our model, the FPA inefficiency is also related to bidder asymmetry. The contrast is that in our model the asymmetry among agents are never publicly known and are only probabilistically perceived based on bidders' noisy signals about their opponents.

We restrict the seller to two possible mechanisms for allocating the object: FPA and SPA. With more general mechanisms, in our setting, sellers could fully extract the surplus, exploiting the correlation between bidders' multidimensional types, using the type of argument employed in Cremer and McLean (1985).[3] Such mechanisms rely on very strong common knowledge assumptions among the seller and the bidders and would not work on more realistic type spaces (see (Neeman, 2004; Bergemann and Morris, 2003). For this reason, we restrict attention to simple mechanisms. Our work is an attempt to make a first step at relaxing the standard (but unfortunate) assumption in auction theory of identifying players' beliefs with their payoff types.[4] An alternative way of allowing richer beliefs into standard IPV auctions is to introduce strategic uncertainty by relaxing the solution concept from equilibrium to rationalizability. This avenue has been pursued

[3]The literature on general mechanisms with multidimensional types focusses on efficiency questions. Jehiel and Moldovanu (2001) show that, generically, there are no efficient auction mechanisms when bidders have independent multidimensional signals and interdependent valuations. McLean and Postlewaite (2004) study situations in which bidders' valuations consist of both common and idiosyncratic components. Bidders privately observe their idiosyncratic component of the valuation, and some signal regarding the common component. They show that a modification of the Vickrey auction is efficient under quite general conditions in their settings.

[4]See Feinberg and Skrzypacz (2005) for the same relaxation in the context of models of bargaining under incomplete information.

by Battigalli and Siniscalchi (2003) and Dekel and Wolinsky (2003), for first price auctions, while maintaining the assumption of no private information about others' values.

The remainder of the paper is structured as follows. Section 2 presents the parametric environment we examine; Section 3 shows the revenue non-equivalence between the FPA and SPA in our auction environment; Section 4 shows the possible inefficiency of the FPA; Section 5 shows that there may exist no equilibrium in the FPA; Section 6 provides examples that reverse the revenue ranking between the FPA and the SPA and illustrate the incentives of information acquisition under different auction mechanisms; and Section 7 concludes.

2 The Model

Two bidders, $i = 1, 2$, compete for an object. Bidders' valuations of the object are private and independently drawn from identical distributions. We assume that bidders' valuation of the object takes on three possible values $\{V_l, V_m, V_h\}$ where $V_l < V_m < V_h$.[5] The *ex ante* probability of bidder i's valuation v_i taking on value V_k is denoted by $p_k \in [0, 1)$ where $k \in \{l, m, h\}$. Of course $\sum_{k \in \{l,m,h\}} p_k = 1$. To ease exposition, we will refer to bidder 1 as "she" and bidder 2 as "he", and refer to a generic bidder as "she" when no confusion shall arise.

As in standard private value auction models, bidder i observes her private valuation $v_i \in \{V_l, V_m, V_h\}$. The novel feature of this paper is as follows: we assume that each bidder also observes a noisy signal about her opponent's valuation. For tractability, we assume that the noisy signal takes on two possible qualitative categories $\{L, H\}$. Bidder i's signal $s_i \in \{L, H\}$ about j's valuation v_j is generated as follows. For $k \in \{l, m, h\}$, and $i, j \in \{l, 2\}$, $i \neq j$,

$$\Pr(s_i = L | v_j = V_k) = q_k, \quad \Pr(s_i = H | v_j = V_k) = 1 - q_k, \tag{1}$$

where $q_k \in [0, 1]$. We assume that $q_l \geq q_m \geq q_h$. Note that when $q_l = q_m = q_h$, the signals are completely uninformative about the opponent's

[5] Wang (1991) and Campbell and Levin (2000) studied common value auctions with discrete valuations.

valuation.[6] We assume that bidders' signals s_1 and s_2 are independent. To summarize, each bidder has a two-dimensional type $(v_i, s_i) \in \{V_l, V_m, V_h\} \times \{L, H\}$ where v_i is called bidder i's *valuation* type and s_i her *information* type. The primitives of our model are a tuple of nine parameters as follows:

$$\varepsilon = \left\{ \langle V_k, p_k, q_k \rangle_{k \in \{l,m,h\}} : V_l < V_m < V_h, p_k \in [0,1], \right.$$

$$\left. \sum_k p_k = 1, q_k \in [0,1], q_l \geq q_m \geq q_h \right\}.$$

Any element $e \in \varepsilon$ is called an *auction environment.*

We first compare the seller's expected revenue and the allocative efficiency of the standard auctions. Since we are in a two-person private value environment, Dutch and English auctions are strategically equivalent to the FPA and the SPA, respectively. Thus we will only analyze the FPA and the SPA: Bidders simultaneously submit bids; the high bidder wins the object. In the event of a tie, we assume that the bidder with higher valuation wins the object if the bidders' valuations are different; and the tie-breaking can be arbitrary if the bidders' valuations are the same.[7]

As usual, we will analyze the auctions as a Bayesian game of incomplete information between two bidders in which the type space for each bidder is $T \equiv \{V_l, V_m, V_h\} \times \{L, H\}$. Bidder i's generic type is $t_i = (v_i, s_i) \in T$. Given her information type s_i, bidder i updates her belief about j's valuation type v_j according to Bayes' rule as follows. For $s_i \in \{L, H\}$, and $k \in \{l, m, h\}$,

$$\Pr(v_j = V_k | s_i = L) = \frac{p_k q_k}{\sum_{k' \in \{l,m,h\}} p_{k'} q_{k'}},$$

$$\Pr(v_j = V_k | s_i = H) = \frac{p_k (1 - q_k)}{\sum_{k' \in \{l,m,h\}} p_{k'} (1 - q_{k'})}. \quad (2)$$

Analogously, given her valuation type v_i, bidder i updates her belief about j's information type s_j according to the signal technology specified by (1).

[6]Because completely uninformative signals are the same as no signals at all, this special case corresponds to the standard one-dimensional IPV model.

[7]It is now well known that tie-breaking rules are important in guaranteeing equilibrium existence in FPAs. This tie-breaking rule is endogenous yet incentive compatible in the sense that bidders with tying bids in equilibrium would truthfully reveal their values if asked. See Jackson (2002) for more general discussions of endogenous sharing rules. Kim and Che (2004) and Maskin and Riley (2000) used a similar assumption.

For any $(t_1, t_2) = ((v_1, s_1), (v_2, s_2)) \in T^2$, the joint probability mass is

$$\Pr(t_1, t_2) = \Pr(v_1)\Pr(s_1|v_2) \times \Pr(v_2)\Pr(s_2|v_1)$$

and the conditional probability is

$$\Pr(t_i|t_j) = \Pr(v_i|s_j)\Pr(s_i|v_j) \quad \text{where } i \neq j. \tag{3}$$

3 Seller's Expected Revenue

We first show that the celebrated revenue equivalence result for the standard one-dimensional IPV auctions breaks down in our multidimensional setting. To demonstrate this result in the simplest possible fashion, we consider a special case of the above model:

- $p_m = 0$, $p_l \in (0,1)$, $p_h \in (0,1)$. That is, the bidders' valuations are only of two possible types, $\{V_l, V_h\}$.
- $q_l = 1 - q_h = q \in [1/2, 1]$. That is, signal L is equally indicative of value V_l as signal H is of value V_h. The parameter q measures the accuracy of the signal: when $q = 1/2$, the signals are completely uninformative; and when $q = 1$, the signals are perfectly informative.

3.1 *Second-Price Auction*

In the SPA, it is routine to show that the unique equilibrium in weakly dominant strategies in this multidimensional setting is for a bidder of type (v_i, s_i) to bid her private value v_i regardless of her information type. That is, the equilibrium bidding strategy of bidder i in the SPA, denoted by B_i^{SPA}, is

$$B_i^{\text{SPA}}(v_i, s_i) = v_i \quad \text{for all } (v_i, s_i) \in \{V_l, V_h\} \times \{L, H\}. \tag{4}$$

In fact, this equilibrium characterization for the SPA is completely general to any private value auction environment and does not depend on the number of bidders, discrete valuation and signal types. We thus conclude that the multidimensional SPA is efficient; and the seller's expected revenue is independent of accuracy of the signals, hence equal to that in the standard environment where bidders only observe their own valuations.

3.2 *First-Price Auction*

The unique equilibrium of the FPA for this special case is characterized as follows:

Proposition 1. *If* $p_m = 0, q_l = 1 - q_h = q \in [1/2, 1]$, *then the unique equilibrium of the FPA is symmetric and is described as follows: for* $i = 1, 2$,

1. $B_i^{\mathrm{FPA}}(V_l, s) = V_l$ *for* $s \in \{L, H\}$.
2. *Type-*(V_h, L) *bidder* i *mixes over* $[V_l, \bar{b}_{(V_h,L)}]$ *according to* CDF $G_{(V_h,L)}(\cdot)$ *specified by*

$$G_{(V_h,L)}(b) = \frac{p_l q\,(b - V_l)}{p_h(1-q)^2(v_h - b)}, \tag{5}$$

where

$$\bar{b}_{(v_h,L)} = \frac{p_h\,(1-q)^2 V_h + p_l q V_l}{p_h(1-q)^2 + p_l q}. \tag{6}$$

3. *Type-*(V_h, H) *bidder* i *mixes over* $[\bar{b}_{(V_h,L)}, \bar{b}_{(V_h,H)}]$ *according to CDF* $G_{(V_h,H)}(\cdot)$ *specified by*

$$G_{(V_h,H)}(b) = \frac{(p_l + p_h q)(1-q)(b - \bar{b}_{(V_h,L)})}{p_h q^2 (V_h - b)}, \tag{7}$$

where

$$\bar{b}_{(V_h,H)} = \frac{p_h q^2 V_h + (p_l + p_h q)(1-q)\bar{b}_{(V_h,L)}}{p_h q^2 + (p_l + p_h q)(1-q)}. \tag{8}$$

The proof is relegated to the appendix. Figure 1 illustrates the equilibrium. The reason that a bidder with valuation V_l will bid V_l regardless of her signal about her opponent's value is similar to that in Bertrand competition between two firms with identical costs. Type-(V_h, L) bidder will play a mixed strategy over a support $[V_l, \bar{b}_{(V_h,L)}]$. The CDF $G_{(V_h,L)}(\cdot)$ is chosen

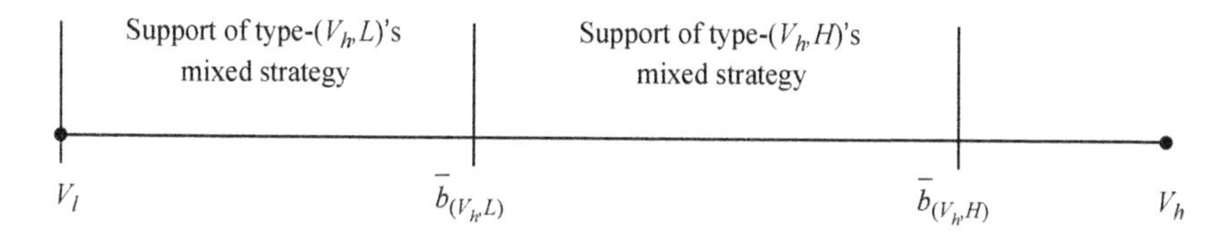

Fig. 1. A graphic illustration of the equilibrium of the FPA in Proposition 1.

to ensure that each bid in the support generates the same constant expected surplus. Type-(V_h, H) bidder will mix over a higher support $[\bar{b}_{(V_h,L)}, \bar{b}_{(V_h,H)}]$ because she perceives her opponent to more likely have valuation V_h, thus bidding more aggressively.

3.3 *Revenue Non-Equivalence*

Now we compare the seller's expected revenue from the SPA and the FPA. Because bidders bid their own private valuations in the SPA, the seller receives V_h if and only if both bidders have valuation type V_h (an event that occurs with probability p_h^2) and the seller receives V_l otherwise. Hence the seller's expected revenue from the SPA, denoted by R^{SPA}, is

$$R^{\text{SPA}} = (1 - p_h^2)V_l + p_h^2 V_h. \tag{9}$$

Since in the SPA a bidder obtains positive surplus $(V_h - V_l)$ only when her valuation is V_h and her opponent's valuation is V_l, an event that occurs with probability $p_h p_l$, each bidder's *ex ante* expected surplus from the SPA, denoted by M^{SPA}, is

$$M^{\text{SPA}} = p_h p_l (V_h - V_l).$$

In the unique equilibrium of the FPA characterized in Proposition 1, the object is always efficiently allocated. Thus, the expected social welfare is $p_l^2 V_l + (l - p_l^2) V_h$. In equilibrium, bidders with valuation type V_l obtains zero expected surplus; and type-(V_h, L) and type-(V_h, H) bidders respectively obtain expected surplus $K_{(V_h,L)}$ and $K_{(V_h,H)}$ as described by (A.2) and (A.5) in the appendix. The *ex ante* probabilities that bidder i is of type (V_h, L) and (V_h, H) are, respectively, $\Pr[t_i = (V_h, L)] = p_h[p_h(1-q) + p_l q]$ and $\Pr[t_i = (V_h, H)] = p_h[p_h q + p_l(1-q)]$. Thus, the *ex ante* expected surplus of each bidder from the FPA, denoted by M^{FPA}, is

$$\begin{aligned} M^{\text{FPA}}(q) &= \Pr[t_i = (V_h, L)] K_{(V_h,L)} + \Pr[t_i = (V_h, H)] K_{(V_h,H)} \\ &= \frac{p_h q(1-q) + p_l q}{p_h (1-q)^2 + p_l q} p_h p_l (V_h - V_l). \end{aligned}$$

We have the following observations. First, M^{FPA} depends on q and M^{SPA} is independent of q. The intuition is simply that bidders strategically use their information about opponent's valuation only in the FPA. Second, $M^{\text{FPA}}(q) > M^{\text{SPA}}$ for all $q \in (1/2, 1)$ and $M^{\text{FPA}}(1/2) = M^{\text{FPA}}(1) = M^{\text{SPA}}$. That is, a bidder's expected surplus is strictly higher in the FPA than that in the SPA except for the completely uninformative and completely informative signal cases. When $q = 1/2$, the signals are completely uninformative, and bidders would simply disregard their information type. We can see from Lemma A.1 that the probability densities of $G_{(V_h,L)}$ and $G_{(V_h,H)}$ can be smoothly pasted at $\bar{b}_{(V_h,L)}$ when $q = 1/2$, which implies that effectively, when $q = 1/2$, bidders of valuation type V_h are simply playing a mixed strategy on the whole support of $[V_l, \bar{b}_{(V_h,H)}]$. When $q = 1$, the FPA becomes a complete information auction, and it is well known that it is revenue equivalent to the SPA.

The seller's expected revenue in the FPA, denoted by R^{FPA}, is simply the difference between the expected social welfare and the sum of the bidders' expected surplus. That is,

$$\begin{aligned}R^{\text{FPA}}(q) &= [p_l^2 V_l + (1 - p_l^2)V_h] - 2M^{\text{FPA}}(q)\\ &= (1 - p_h^2)V_l + p_h^2 V_h - \frac{2p_h^2 p_l(2q-1)(1-q)}{p_h(1-q)^2 + p_l q}(V_h - V_l)\\ &= R^{\text{SPA}} - \frac{2p_h^2 p_l(2q-1)(1-q)}{p_h(1-q)^2 + p_l q}(V_h - V_l). \qquad (10)\end{aligned}$$

The following proposition summarizes the comparison between $R^{\text{FPA}}(q)$ and R^{SPA}:

Proposition 2 (Revenue non-equivalence). *Let $p_m = 0$ and $q_l = 1 - q_h = q$. For any $q \in (1/2, 1)$, $R^{\text{FPA}}(q) < R^{\text{SPA}}$; and $R^{\text{FPA}}(1/2) = R^{\text{FPA}}(1) = R^{\text{SPA}}$; moreover, $R^{\text{FPA}}(q)$ has a unique minimizer.*

That $R^{\text{FPA}}(q)$ has a unique minimizer in q follows from simple algebra. Figure 2 depicts the seller's expected revenues as a function of $q \in [1/2, 1]$

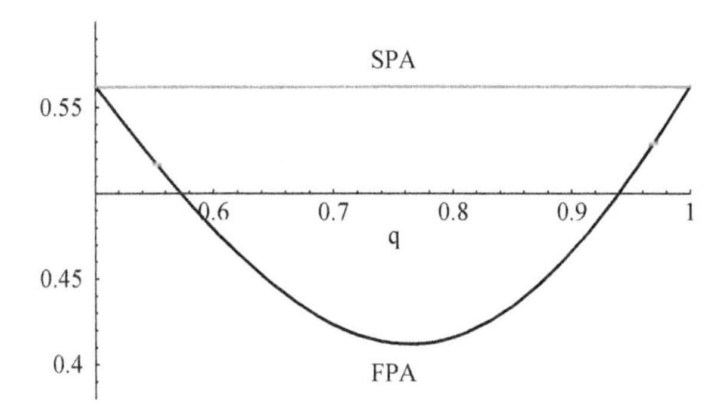

Fig. 2. Seller's expected revenues in the SPA and FPA: $p_h = 0.75$, $V_h = 1$, $V_l = 0$.

from the two auction mechanisms for an example where $p_h = 0.75$, $V_l = 0$, $V_h = 1$.

The standard revenue equivalence theorem (Riley and Samuelson, 1981; Myseron, 1981) crucially relies on the bidders' types being single-dimensional and on the valuations being drawn from continuous distributions. Both are important in the argument that bidders' expected payoffs are completely determined from the winning probabilities. One can show, however, that the revenue equivalence between the FPA and SPA still holds in a single dimensional private value auction environment with finite number of valuation types (see Appendix B for a proof). Therefore, the revenue non-equivalence we show in Proposition 2 is not due to the discreteness of the valuation space, rather it is due to the unique information structure.

We also note that in this two-valuation example (since $p_m = 0$), the SPA generates a higher expected revenue for the seller than the FPA despite the fact that both auction mechanisms are allocatively efficient. This is similar to the one-dimensional APV auctions: the objects are allocated efficiently in both the first and second-price APV auctions, but the SPA generates weakly higher expected revenue for the seller (Milgrom and Weber, 1982). However, in a simple one-dimensional correlated-value analog of our discrete-value example, the FPA and the SPA are actually revenue equivalent. To see this, suppose that bidders' valuations are correlated as follows: $\Pr(v_i = V_l | v_j = V_l) = \Pr(v_i = V_h | v_j = V_h) = \rho \in [1/2, 1]$. Such symmetric correlation requires that, ex ante, v_i takes on values V_l and V_h with probability 1/2, i.e., $\rho_l = \rho_h = 1/2$. Each bidder privately observes her valuation, and *no* additional information about her opponent's value. The seller's expected revenue under the SPA is

$$\frac{\rho V_h}{2} + \left(1 - \frac{\rho}{2}\right) V_l,$$

since with probability $\rho/2$ both bidders receive V_h and hence bid V_h; and with complementary probability at least one bidder receives V_l and the resulting second highest bid is V_l. It can be shown, analogous to the proof of Proposition 1 in the appendix, that the unique symmetric equilibrium of the FPA is as follows: type-V_l bidder bids her valuation V_l, while a type-V_h bidder bids according to a mixed strategy on the support $[V_l, \rho V_h + (1-\rho)V_l]$ with CDF $G^{CV}(\cdot)$ given by

$$G^{CV}(b) = \frac{(1-\rho)(b - V_l)}{\rho(V_h - b)} \quad \text{if } b \in [V_l, \rho V_h + (1-\rho)V_l].$$

Type-V_h bidder's expected surplus under the mixed strategy is simply $(1-\rho)(V_h - V_l)$. The seller's expected revenue under the FPA can be calculated as follows. The expected social surplus under the FPA is

$$\frac{\rho}{2}V_l + \left(1 - \frac{\rho}{2}\right) V_h$$

since the object is always efficiently allocated. The bidders' expected surplus is 0 for type-V_l bidder; and $(1-\rho)(V_h - V_l)$ for type-V_h bidder. Since ex ante, bidders are of type V_l and V_h with equal probability, the seller's expected revenue under the FPA is

$$\frac{\rho}{2}V_l + \left(1 - \frac{\rho}{2}\right) V_h - (1-\rho)(V_h - V_l) = \left(1 - \frac{\rho}{2}\right) V_l + \frac{\rho}{2}V_h.$$

Thus a one-dimensional correlated-value analog of our two-value example would yield revenue equivalence between the FPA and the SPA regardless of the degree of correlation. Note that this result is not robust, as (Milgrom and Weber, 1982) showed that in continuous affiliated value environments, the expected revenue from the SPA is in general at least weakly higher than that from the FPA.

4 Efficiency

While the SPA is always allocatively efficient in equilibrium, we argue in this section that the allocative efficiency of the FPA in Section 3 is an artifact of the two-valuation example. When we make all three valuations occur with positive probability, i.e., $p_k > 0$ for $k \in \{l, m, h\}$, the unique symmetric equilibrium of the FPA may be allocatively inefficient. Allocative inefficiency may arise in the FPA if a type-(V_m, H) bidder infers that her opponent is mostly likely of valuation type V_h and hence bid more aggressively than a type-(V_h, L) bidder, who perceives her opponent to be weak and is willing to sacrifice the probability of winning in exchange for a bigger surplus when winning against an opponent with valuation V_l. The subtle point is that this intuition works only if the following conditions are met: (1) Type-(V_m, H) bidder's posterior belief about her opponent puts a small weight on (V_h, L), and big weight on (V_h, H). This requires that q_m be sufficiently small and $\Pr(v_j = V_h | s_i = H)$ be sufficiently large; (2). Type-(V_h, L) bidder's posterior belief about her opponent puts a big weight on V_l. This requires that q_l be sufficiently large; (3). V_h cannot be too large relative to V_m since otherwise, type-(V_h, L) bidder is not willing to lower her probability of winning by bidding conservatively.

Arguments similar to those in Section 3 can be used to establish that, first, in any symmetric equilibrium of the FPA, bidders with valuation V_l must bid V_l in pure strategy regardless of their information type; second, other types of bidders must bid in mixed strategies with contiguous and non-overlapping supports; third, the support of type-(V_m, L) bidder's mixed strategy must be lower than that of type-(V_m, H); the support of type-(V_h, L) bidder's mixed strategy must be lower than that of type-(V_h, H); the support of type-(V_m, H) bidder's mixed strategy must be lower than that of type-(V_h, H), and the support of type-(V_m, L) bidder's mixed strategy must be lower than that of type-(V_h, L). Thus the symmetric equilibrium of the FPA in this section takes only two possible forms depending on the order of the mixed strategy supports of type-(V_m, H) and type-(V_h, L) bidders. A symmetric equilibrium is *efficient* if the equilibrium mixed strategy support of type-(V_m, H) bidder is lower than that of type-(V_h, L) bidder; and it is *inefficient* if the equilibrium mixed strategy support of type-(V_h, L) bidder is lower than that of type-(V_m, H) bidder. Our first interesting result is.

Proposition 3. *(Efficient and inefficient equilibria cannot coexist in the FPA). Any auction environment $e \in \varepsilon$ cannot simultaneously have both an efficient and an inefficient symmetric equilibrium in the FPA.*

We now show that, in contrast to the SPA, the FPA may be allocatively inefficient. The following result is proved by constructing an explicit auction environment with inefficient equilibrium. Proposition 3 then guarantees that it does not admit any efficient symmetric equilibrium. That there is an open set of auction environments in ε with inefficient equilibrium in the FPA follows from continuity.

Proposition 4 (Inefficiency of FPA). *There exists an open set of auction environments in ε in which the unique symmetric equilibrium of the FPA is inefficient.*

Our model has an interesting implication regarding the impact of more information on efficiency. The probability of the object being inefficiently allocated in an inefficient equilibrium is the probability that the two bidders' types are (V_m, H) and (V_h, L) respectively, which is given by

$$2\Pr\{t_i = (V_m, H), t_j = (V_h, L), i \neq j\} = 2p_m p_h (1 - q_h) q_m.$$

Recall that $1 - q_h$ is the probability of bidder i obtaining $s_i = H$ when her opponent's valuation $v_j = V_h$. Thus the higher $1 - q_h$ is, the more informative the signal H is about V_h, and also of course, the more informative the signal L is about V_l. Thus, the probability of inefficient allocation in the FPA may be increasing in $1 - q_h$ *locally* in the set of auction environments with inefficient equilibrium. We want to emphasize, however, that this is only a local result: as signal becomes more informative, the inefficient equilibrium may cease to exist, and efficiency may be restored.

Finally, it can also be verified that the seller's expected revenue in the inefficient equilibrium in the FPA is again smaller than that in the SPA except for some knife-edge cases with measure zero.

The key observations of this section would clearly continue to hold in more general settings. For example, suppose that bidders' private values were independently drawn from a continuous distribution and each bidder observed a continuous signal, correlated with the value of the other bidder. In a continuous setting, efficiency would require that each bidder's strategy depend only on his valuation and not his signal. This would be impossible if the signal was informative. To the extent that equilibrium exists in the FPA in such environments, the probability of inefficiency would therefore always be non-monotonic, since we have inefficiency with intermediate informativeness of signals, but we have efficiency with either no information about others' values (efficiency is a well known property of the FPA with symmetric distributions and IPVs) or full information about others' values (there is efficiency in the FPA with complete information).

5 Equilibrium Existence

Up to now, we have assumed that bidders' information about the opponent's private valuation is of the same accuracy. In this section, we show that the existence of equilibrium in the FPA is contingent on this assumption in our model. For this purpose, we consider again the example we used in Section 3, with the exception that the accuracy of bidder i's signal regarding bidder j's valuation is $q_i \in [1/2, 1)$ and we let $q_1 > q_2$. Recall that $q_i = \Pr(s_i = L|v_j = V_l) = \Pr(s_i = H|v_j = V_h)$. Our main result in this section can be stated as follows.

Proposition 5. *If $p_m = 0$ and $1 \geq q_1 > q_2 \geq 1/2$, then the FPA does not admit any equilibrium for generic values of $\langle V_l, V_h, p_l, q_1, q_2 \rangle$.*

Examples of non-existence of equilibrium with multidimensional types are also presented [Jackson] in the context of auctions with both private and common value components, and the example of this section has a similar flavor.

The non-existence problem is surely not an artifact of the discrete type assumption. Consider again the case where bidders' private values were independently drawn from a continuous distribution and each bidder observed a continuous signal, correlated with the value of the other bidder. Even with strong assumptions of the signals (e.g., the monotone likelihood ratio property), if bidder 1 knows that bidder 2 is following a strategy that is monotonic in his valuation and his signal of bidder 1's valuation, bidder 1 will, in some cases, not have a best response that is monotonic (in the same sense). To see why, suppose that bidder 1 has a bimodal distribution on bidder 2's valuation, and thus on bidder 2's bid. Suppose that improvements in bidder 1's signal translate up her beliefs about bidder 2's bids. For low values of the signal, it will be optimal for bidder 1 to bid such that she wins against both modal bids. However, as her signal improves, there will be a point where she will give up on winning against 2's high modal bid and her bid will jump down to just above the 2's low modal bid. Thus her bid will jump downwards as her signal improves. It is hard to think of a primitive assumption on the signal structure that will prevent this type of non-monotonicity. This lack of monotonicity implies that the existence arguments such as those of Athey (2001); Reny and Zamir (2004) will not help in this problem.[8]

6 Discussion: Revenue and Information Acquisition

In this section, we first present two examples of multidimensional private value auctions in which the revenue ranking of the FPA and the SPA are reversed; and then discuss the incentives of information acquisition.

6.1 *Revenue*

We have observed that the SPA is efficient and the FPA is not in general. This may suggest the possibility that the SPA will generate more revenue (as suggested by the three valuation example of Section 4). We have also seen an example where the seller's revenue in the SPA is higher than in

[8]It is possible that equilibrium existence may be restored by introducing communication and more complicated endogenous tie-breaking rules a la Jackson *et al.* (2002). However, it is not at all clear what would be the endogenous tie-breaking rule that would be compatible with equilibrium in this asymmetric environment.

the FPA, even though the there is efficient allocation of the object under both auctions (the two valuation example in Section 3). What can be said in general about the revenue ranking?

There is an easy way to see that a general revenue ranking is not possible. For some special information structures, each bidder will know what private signal the other bidder has observed. This will be true if each bidder observes a partition of the other bidder's valuations. Now even though we start with a model that is completely symmetric across bidders, conditional on the observed signals, bidders are playing in an IPVs environment with asymmetric distributions. But from the work of Maskin and Riley (2000), we already know that revenue ranking may go either way. We can use this insight to construct the following example where revenue in the FPA is higher than that in the SPA. Presumably, this revenue ranking would continue to hold in nearby models where private signals were not common knowledge among the players.

Example 1. Consider a private value auction with two bidders, $i = 1, 2$. Suppose that v_1 and v_2 are independent and both drawn from Uniform [0, 1]. Bidders also observe a noisy signal about their opponent's valuation. Suppose that the signal is generated as follows: for $i \neq j$,

$$s_i = \begin{cases} L & \text{if } v_j \in \left[0, \dfrac{1}{2}\right], \\ H & \text{if } v_j \in \left(\dfrac{1}{2}, 1\right]. \end{cases}$$

That is, a bidder observes a signal that tells her if her opponent's value is higher or lower than 1/2; and this information structure is common knowledge.

In the equilibrium of the SPA for the auction environment described in Example 1, each bidder will bid their own private valuation. In the FPA, however, we have to consider three cases: (i) $v_i \in [0, 1/2]$ for $i = 1, 2$; (ii) $v_i \in (1/2, 1]$ for $i = 1, 2$; and (iii) $v_i \in [0, 1/2]$ and $v_j \in (1/2, 1]$ where $i \neq j$. In case (i), both bidders effectively compete in an auction environment in which it is common knowledge that the valuations are both drawn from Uniform [0, 1/2] distributions. In case (ii), both bidders effectively compete in an auction environment in which it is common knowledge that the valuations are both drawn from Uniform [1/2, 1] distributions. In case (iii), however, the bidders are asymmetric in their valuation distributions and it is common knowledge. Clearly the FPA and the SPA are revenue equivalent in case (i) and (ii) events. In case (iii) events, however, the bidder asymmetry breaks the revenue equivalence. It can be easily verified that case (iii) events satisfy the conditions for a theorem of Maskin and Riley [(2000), Proposition 4.3] which shows that the FPA would generate a higher expected revenue for the seller than the SPA under this type of asymmetry. Thus, we reach the conclusion that overall in this example, the seller's expected revenue is higher in the FPA than that in the SPA. It can also be numerically verified that in case (iii) events, the FPA may be allocatively inefficient. Thus we have an example that the FPA generates higher expected revenue for the seller than the SPA despite its possible allocative inefficiency relative to the SPA.

6.2 *Information Acquisition*

So far we have assumed that bidder's information about her opponent's valuation or valuation distribution is provided by nature without incurring any cost. In reality, of course, such information is be costly to acquire.[9] Now we argue that if bidders have to costly acquire such information, then different auction mechanisms provide vastly different incentives for such information acquisition. This, together with the difference in revenue and allocative efficiency between the FPA and the SPA we documented earlier, provides yet another reason for the auction designer to prefer one auction mechanism over another even in private value auction environments.

In the SPA, bidders do not strategically use information about their opponents' valuation, thus there is no incentives at all to acquire such information if it is costly. This observation is completely general for any private value auction environments. The lack of incentives to acquire information about one's opponents in the SPA is related to the fact that bidding one's private valuation is an *ex post* equilibrium in the SPA.

In the FPA, however, information about the opponent's valuation does have strategic consequences in the bidding, thus bidders do have incentives to acquire such information if the cost is sufficiently small. We illustrate such incentives using an extension of Example 1 above. Suppose that bidder i can, at a cost $c_i > 0$, purchase a signal about her opponent's valuation that reveals whether her opponent's valuation is below or above $1/2$. Assume that a bidder's signal purchase decision is observable to her opponent.

[9]Most of the existing literature in information acquisition in auctions are concerned with common value auctions (for example Mathews, 1984; Persico, 2000).

Suppose that the timing of the game is as follows: first, bidders decide whether to purchase such a signal technology at cost c_i; second, nature draws private valuations from Uniform $[0, 1]$ for each bidder; third, a bidder observes whether her opponent's private valuation is below or above 1/2 if and only if she purchased the signal technology; and finally, bidders compete for the object in the FPA.

The equilibrium bidding strategies in the FPA depend on the signal purchase decisions:

- If neither bidder purchased the signal technology in stage 1, then both bidders will play the symmetric FPA in which the opponent's valuation is drawn from Uniform $[0, 1]$ distribution. Thus bidders bid $v/2$ and the expected surplus for each bidder is given by $\left(\int_0^1 v^2\, dv\right)/2 = 1/6 \approx 0.16667$.
- If both bidders purchase the signal technology, then bidder i's ex ante expected surplus from the subsequent FPA is calculated as follows. (1) With probability 1/4, both bidder valuations will be below 1/2. In this case, bidders will bid $v/2$ in equilibrium and the expected surplus for each bidder is $(\int_0^{1/2} v^2\, dv)/2 = 1/48$. (2) With probability 1/4, both valuations will be above 1/2. In this case, bidders will bid $v/2$ again in equilibrium and the expected surplus for each bidder is $(\int_{1/2}^{1} v^2\, dv)/2 = 7/48$. (3) With probability 1/4, bidder i's valuation is below 1/2 and bidder j's valuation is above 1/2, where $j \neq i$. (4) With probability 1/4, bidder i's valuation is above 1/2 and bidder j's valuation is below 1/2. The equilibria of the FPA in the events of case(3) and (4) cannot be analytically solved, but numerical calculation shows that bidder i's expected surplus in case (3) and (4) are 0.01848 and 0.34808 respectively.[10] Thus bidder i's *ex ante* expected surplus if both bidders purchased the signal technology is

$$\frac{1/48 + 7/48 + 0.01848 + 0.34808}{4} \approx 0.13331.$$

- If bidder i does not purchases the signal technology but bidder j does, then bidder i and j's *ex ante* expected surplus from the subsequent FPA can be calculated as follows. (1) With probability 1/2, bidder i's valuation

[10]John Riley and Estelle Cantillon graciously provided various versions of BIDCOMP2 fortran codes that are used in calculating the bidders' ex ante expected payoffs in the asymmetric auctions.

is below 1/2. In this case, bidder i's belief about j's valuation is Uniform [0, 1] while bidder j's belief about i's valuation is Uniform [0, 1/2] and this is common knowledge. Hence, bidder i's expected surplus is that of a "weak" bidder (in the terminology of Maskin and Riley (2000)) with Uniform [0, 1/2] valuation distribution against a "strong" bidder with Uniform [0, 1] distribution in the FPA, which can be analytically calculated to be approximately 0.0242334.[11] Likewise, bidder j (the "strong" bidder in this case)'s ex ante expected payoff is approximately 0.253449. (2) With probability 1/2, bidder i's valuation is above 1/2. In this case, bidder i's belief about bidder j's valuation is Uniform [0, 1] while bidder j's belief about j's valuation is Uniform [1/2, 1] and this is common knowledge. Hence, bidder j's expected surplus is that of a "strong" bidder with Uniform [1/2, 1] valuation distribution against a "weak" bidder with Uniform [0, 1] distribution in the FPA, which can be numerically calculated to be 0.218465. Likewise, bidder j (the "weak" bidder in this case)'s expected surplus is approximately 0.104104. Thus bidder i (the non-purchaser)'s *ex ante* expected payoff is approximately

$$\frac{0.0242334 + 0.218465}{2} = 0.12135$$

and bidder j (the purchaser)'s *ex ante* expected payoff is approximately

$$\frac{0.253449 + 0.104104}{2} = 0.17878.$$

Table 1 lists the ex ante expected payoff matrix for the two bidders taking into account the information acquisition cost c_i and c_j. When c_i and c_j are sufficiently small, the unique equilibrium in the information acquisition stage is that both bidders purchase the signals. Both bidders are made worse off through two channels. First, they incur the information acquisition cost; second, in the subsequent FPA, they will be engaged in more fierce competition and the seller will be able to extract a higher revenue. The social

[11] The unique equilibrium of a two-bidder asymmetric FPA with valuation distributions Uniform $[0, h_1]$ and Uniform $[0, h_2]$ respectively where $h_1 > 0$, $h_2 > 0$ and $h_1 \neq h_2$ are given by

$$b_1(v) = \frac{\sqrt{1 + mv^2} - 1}{mv}, \quad b_2(v) = \frac{1 - \sqrt{1 - mv^2}}{mv},$$

where $m = (h_1^2 - h_2^2)/(h_1 h_2)^2$ is a constant. Appendix B in Fang and Morris (2003) provides an elementary derivation of the above equilibrium. See also Griesmer *et al.* (1967); Plum (1992).

Table 1. The expected payoff matrix.

Bidder i\Bidder j	No purchase	Purchase
No Purchase	0.16667, 0.16667	0.12135, $0.17878 - c_j$
Purchase	$0.17878 - c_i$, 0.12135	$0.13331 - c_i$, $0.13331 - c_j$

welfare is also decreased for two reasons. First, the information acquisition cost is dissipative; second, the object will be allocated inefficiently with positive probability.

This example also illustrates the possibility that a decrease in the cost of information acquisition may increase allocative inefficiency in the FPAs. Imagine that initially the information acquisition cost c_i are sufficiently high that in equilibrium neither bidder purchases the signal technology. Thus we know that the subsequent FPA is allocatively efficient. However, as c_i is sufficiently low, both bidders will purchase information in equilibrium and the subsequent FPA is allocatively inefficient with positive probability.

7 Conclusion

This paper presents examples of two-bidder private value auctions in which each bidder observes her own private valuation as well as noisy signals about her opponent's private valuation. This departs from the one-dimensional symmetric IPV paradigm and provides a simple but genuine separation between a bidder's private valuation and her signal about opponents' valuations, unlike the one-dimensional APV model. We partially characterize the equilibrium of the FPA when each bidder's signal about her opponent's valuation is drawn from the same distribution, and show that the revenue-equivalence between standard auctions fails. Our examples demonstrate that, first, the revenue ranking between the FPA and the SPA is ambiguous; second, the equilibrium allocation of the object could be inefficient in the FPA but is always efficient in the SPA, but the revenue and allocative efficiency may not coincide: an inefficient FPA may generate a higher expected revenue for the seller; but it is also possible that the seller's expected revenue is higher in the SPA even when the object is efficiently allocated in both auctions. We also show that the equilibrium existence of the FPA may be problematic in multidimensional type environments. Finally, we show that different auction mechanisms provide different incentives for bidders to acquire cost information about opponents' private valuations. We also provide examples that the allocative inefficiency in the FPA may increase as the signal becomes more informative; and the allocative inefficiency may increase in the FPA as the information acquisition costs are decreased. While the results in our paper are derived in examples, we have explained how the underlying intuitions are general.

Appendix A. Proofs

Proof of Proposition 1. Proposition 1 follows from the following intermediate lemmas:

Lemma A.1. *In any equilibrium of the FPA, type-(V_l, s) bidders bid V_l in pure strategies for $s \in \{L, H\}$. That is, for $i = 1, 2$,*

$$B_i^{\mathrm{FPA}}(V_l, s) = V_l \quad \textit{for } s \in \{L, H\}.$$

Proof. We first argue that bidders with valuation V_l must bid in pure strategies in equilibrium. Suppose that type-(V_l, H) bidders plays a mixed strategy equilibrium on support $[\underline{b}, \bar{b}]$ with $\underline{b} < \bar{b}$. (The lower limit of the interval may be open, but this is not important for the argument.) Clearly $\bar{b} \leq V_l$. Since the bid $(\underline{b} + \bar{b})/2$ wins positive probability, it yields a positive surplus for type-(V_l, H) bidder. However, bids close to $\underline{b}$ will win with probability almost zero, hence the expected surplus will approach zero. A contradiction for the indifference condition required for the mixed strategy. Hence type-(V_l, H) bidders must bid in pure strategies. Identical arguments show that type-(V_l, L) bidders must also bid in pure strategy. Now we argue that, if type-(V_l, L) and (V_l, H) bidders must bid their valuation V_l in pure strategy. To see this, suppose that type-(V_l, L) and (V_l, H) bidder 2 bids less than V_l. Then bidder 1 of these types can deviate by bidding ε more than bidder 2, which will be a profitable deviation if ε is made arbitrarily close to zero. A contradiction. □

Lemma A.2. *Together with the strategies specified in Lemma* A.1 *for bidders with valuation V_l, the following constitute a symmetric equilibrium:*

1. *Type-(V_h, L) bidders play a mixed strategy on $[V_l, \bar{b}_{(V_h,L)}]$ according to* CDF *$G_{(V_h,L)}(\cdot)$ given by* (5) *where $\bar{b}_{(V_h,L)}$ is given by* (6).
2. *Type-(V_h, H) bidders play a mixed strategy on $[\bar{b}_{(V_h,L)}, \bar{b}_{(V_h,H)}]$ according to* CDF *$G_{(V_h,H)}(\cdot)$ given by* (7) *where $\bar{b}_{(V_h,H)}$ is given by* (8).

Proof. Suppose that bidder 2 bids according to the postulated strategies.

First, consider type-(V_h, L) bidder 1. Her expected payoff from submitting a bid $b \in [V_l, \bar{b}_{(V_h,L)}]$ is

$$(V_h - b)\left\{\frac{p_l q}{p_l q + p_h(1-q)} + \frac{p_h(1-q)^2}{p_l q + p_h(1-q)} G_{(V_h,H)}(b)\right\}, \tag{A.1}$$

where

- The term $p_l q/[p_l q + p_h(1-q)]$ is the probability that bidder 2 has a valuation type V_l conditional on bidder 1's own information type L [recall formula (2)]. By Lemma A.1, bidder 2 with valuation type V_l bids V_l with probability one. Thus bidder 1 wins with probability 1 against such an opponent with any bid in the interval $[V_l, \bar{b}_{(V_h,L)}]$ (note that the tie-breaking rule is applied at the bid V_l).
- The term $p_h(1-q)^2/[p_l q + p_h(1-q)]$ is the probability that bidder 2 is of type (V_h, L) conditional on bidder 1's own type (V_h, L) [recall formula (3)]. Since type-(V_h, L) bidder 2 is postulated to bid in mixed strategies according to $G_{(V_h,L)}(\cdot)$, bidder 1's bid of b wins against such an opponent with probability $G_{(V_h,L)}(b)$.

Plugging $G_{(V_h,L)}(\cdot)$ as described by (5) into (A.1) yields a positive constant, denoted by $K(V_h, L)$,given by

$$K_{(V_h,L)} = \frac{p_l q}{p_h(1-q) + p_l q}(V_h - V_l), \tag{A.2}$$

which is type-(V_h, L) bidder's expected surplus. Therefore type-(V_h, L) bidder 1 indeed is indifferent between any bids in the interval $[V_l, \bar{b}_{(V_h,L)}]$ provided that bidder 2 follows the postulated strategy.

Now we check that type-(V_h, L) bidder 1 does not have incentive to deviate to other bids. First, she clearly does not have incentive to deviate to bids lower than or equal to V_l, since it would have yielded her a zero surplus instead of a positive $K_{(V_h,L)}$. Now suppose that she deviates to $\bar{b}_{(V_h,L)} < b \leq \bar{b}_{(V_h,H)}$, her expected payoff would be

$$(V_h - b)\left\{\frac{p_l q}{p_l q + p_h(1-q)} + \frac{p_h(1-q)^2}{p_l q + p_h(1-q)} + \frac{p_h(1-q)q}{p_l q + p_h(1-q)} G_{(V_h,H)}(b)\right\}, \tag{A.3}$$

where the term $p_h(1-q)q/[p_h(1-q)+p_lq]$ the probability that bidder 2 is of type (V_h, H) conditional on bidder 1's own type (V_h, L); and $G_{(V_h,H)}(b)$ is the probability that a bid $b \in (\bar{b}_{(V_h,L)}, \bar{b}_{(V_h,H)}]$ wins against such an opponent. Plugging $G_{(V_h,H)}(\cdot)$ as described by (7) into (A.3), we obtain

$$\frac{p_lq + p_h(1-q)^2}{p_lq + p_h(1-q)}(V_h - b) + \frac{p_h(1-q)q}{p_lq + p_h(1-q)} \frac{(p_l + p_hq)(1-q)(b - \bar{b}_{(V_h,L)})}{p_hq^2}$$
$$= \frac{\{[p_lq + p_h(1-q)^2]V_h - (p_l + p_hq)(1-q)^2\bar{b}_{(V_h,L)}\} + p_l[(1-q)^2 - q^2]b/q}{p_lq + p_h(1-q)},$$

which is non-increasing in b since $q \geq l/2$. Hence type-(V_h, L) bidder 1 does not have incentive to deviate to bids in the interval $(\bar{b}_{(V_h,L)}, \bar{b}_{(V_h,H)}]$; which also implies that her expected payoff would be even smaller if she bids more than $\bar{b}_{(V_h,H)}$.

Now consider type-(V_h, H) bidder 1. Given that bidder 2 plays according to the postulated strategies, her expected payoff from bidding $b \in [\bar{b}_{(V_h,L)}, \bar{b}_{(V_h,H)}]$ is given by

$$(V_H - b)\left\{\frac{p_l(1-q)}{p_hq + p_l(1-q)} + \frac{p_hq(1-q)}{p_hq + p_l(1-q)} + \frac{p_hq^2}{p_hq + p_l(1-q)}G_{(V_h,H)}(b)\right\} \tag{A.4}$$

where

- The term $p_l(1-q)/[p_hq + p_l(1-q)]$ is the probability that bidder 2 has valuation V_l conditional on bidder 1's signal H; and the term $[p_hq(1-q)]/[p_hq + p_l(1-q)]$ is the probability that bidder 2 is of type (V_h, L). In both events, a bid $b \in [\bar{b}_{(V_h,L)}, \bar{b}_{(V_h,H)}]$ wins against such opponents with probability one under the postulated strategies by bidder 2.
- The term $p_hq^2/[p_hq + p_l(1-q)]$ is probability that bidder 2 is of type (V_h, H) conditional on bidder 1's own type (V_h, H). In this case, a bid $b \in [\bar{b}_{(V_h,L)}, \bar{b}_{(V_h,H)}]$ wins with probability $G_{(V_h,H)}(b)$.

Plugging $G_{(V_h,H)}(\cdot)$ as described by (7) into (A.4), we obtain a positive constant, denoted by $K_{(V_h,H)}$, given by

$$K_{(V_h,H)} = \frac{(p_l + p_hq)(1-q)}{p_hq + p_l(1-q)}(V_h - \bar{b}_{(V_h,L)}). \tag{A.5}$$

Hence type-(V_h, H) bidder 1 is indeed indifferent between any bids in the interval $[\bar{b}_{(V_h,L)}, \bar{b}_{(V_h,H)}]$.

Now we check that type-(V_h, H) bidder 1 does not have incentive to deviate to other bids. First, she does not have incentive to bid more than $\bar{b}_{(V_h,H)}$, since bidding $\bar{b}_{(V_h,H)}$ strictly dominates any higher bid given bidder 2's strategies; second, she does not have incentive to bid less than or equal to V_l since such bids will yield a zero surplus. Now we show that she does not have incentive to bid in the interval $(V_l, \bar{b}_{(V_h,L)})$. Her expected payoff from a bid $b \in (V_l, \bar{b}_{(V_h,L)})$ is given by

$$(V_h - b)\left\{\frac{p_l(1-q)}{p_h q + p_l(1-q)} + \frac{p_h q(1-q)}{p_h q + p_l(1-q)} G_{(V_h,L)}(b)\right\} \tag{A.6}$$

since such a bid loses to type-(V_h, H) opponent with probability one and win against a type-(V_h, L) opponent with probability $G_{(V_h,L)}(b)$. Plugging $G_{(V_h,L)}(\cdot)$ as described by (5) into (A.6), we get

$$\begin{aligned}\frac{p_l(1-q)}{p_h q + p_l(1-q)}&(V_h - b) + \frac{p_h q(1-q)}{p_h q + p_l(1-q)} \frac{p_l q(b - V_l)}{p_h(1-q)^2} \\ &= \frac{[p_l(1-q)^2 V_h - p_l q^2 V_l] + p_l[q^2 - (1-q)^2]b}{[p_h q + p_l(1-q)](1-q)},\end{aligned}$$

which is non-decreasing in b since $q \geq 1/2$. Hence type-(V_h, H) bidder 1 does not have incentive to deviate to bids in the interval $(V_l, \bar{b}_{(V_h,L)})$.

Finally, note that the expressions for $\bar{b}_{(V_h,L)}$ and $\bar{b}_{(V_h,H)}$ respectively satisfy $G_{(V_h,L)}(\bar{b}_{(V_h,L)}) = 1$ and $G_{(V_h,H)}(\bar{b}_{(V_h,H)}) = 1$. This concludes the proof that the postulated bidding strategies constitute a symmetric equilibrium. □

Lemma A.3. *The symmetric equilibrium described in Lemma* A.2 *is the unique symmetric equilibrium of the FPA.*

Proof. The argument proceeds in three steps.

Step 1: We show that in any symmetric equilibrium type-(V_h, L) and type-(V_h, H) bidders must bid in mixed strategies. For example, suppose to the contrary that, say, a type-(V_h, L) bidder 2 bids in pure strategy an amount $\tilde{b} < V_h$, then type-(V_h, L) bidder 1 can profitably deviate by bidding $\tilde{b} + \varepsilon$ where $\varepsilon > 0$ is arbitrarily small. Such a deviation will provide a discrete positive jump in type-(V_h, L) bidder 1's probability of winning, hence it is profitable. The argument for type-(V_h, H) bidders is analogous.

Step 2: We show that in any symmetric mixed strategy equilibrium, the supports of $G_{(V_h,L)}(\cdot)$ and $G_{(V_h,H)}(\cdot)$ are contiguous and non-overlapping.

That the supports should be contiguous follows from the same ε-deviation argument as the one to rule out pure strategies. Now suppose that the supports of $G_{(V_h,L)}(\cdot)$ and $G_{(V_h,H)}(\cdot)$ overlap in an interval $[b_1, b_2]$ with $b_2 > b_1$. To be consistent with mixed strategies, it must be the case that, the expected surplus for both types from any bid $b \in [b_1, b_2]$ is constant. That is, for some constants $\tilde{K}_{(V_h,L)}$ and $\tilde{K}_{(V_h,H)}$,

$$(V_h - b)\left\{\frac{p_l(1-q)}{p_h q + p_l(1-q)} + \frac{p_h q^2 G_{(V_h,H)}(b)}{p_h q + p_l(1-q)} + \frac{p_h q(1-q) G_{(V_h,H)}(b)}{p_h q + p_l(1-q)}\right\} = \tilde{K}_{(V_h,H)}, \tag{A.7}$$

$$(V_h - b)\left\{\frac{p_l q}{p_h(1-q) + p_l q} + \frac{p_h(1-q) q G_{(V_h,H)}(b)}{p_h(1-q) + p_l q} + \frac{p_h(1-q)^2 G_{(V_h,H)}(b)}{p_h(1-q) + p_l q}\right\} = \tilde{K}_{(V_h,L)}. \tag{A.8}$$

Multiplying Eq. (A.7) by $(1-q)[p_h q + p_l(1-q)]$, and Eq. (A.8) by $q[p_h (1-q) + p_l q]$, and summing up, we obtain

$$\tilde{K}_{(V_h,H)}(1-q)[p_h q + p_l(1-q)] = -\tilde{K}_{(V_h,L)} q[p_h(1-q) + p_l q] = (V_h - b)p_l[(1-q)^2 - q^2]. \tag{A.9}$$

Because the left-hand side of Eq. (A.9) is a constant, this equation holds only for a single value of b unless $q = 1/2$. Therefore, the supports of the symmetric equilibrium mixed strategies of type-(V_h, L) and (V_h, H) bidders must be non-overlapping. The same argument also shows that the supports of the symmetric equilibrium mixed strategies of type-(V_h, L) and type-(V_h, H) bidders cannot overlap at more than one point.

Step 3: We show that the support of type-(V_h, L) bidders' mixed strategy must be lower than that of type-(V_h, H) bidders. Suppose to the contrary. Let $[V_l, \tilde{b}]$ be the support of type-(V_h, H) bidder and $[\tilde{b}, \hat{b}]$ be the support of type-(V_h, L) bidder, for some $\hat{b} > \tilde{b}$. For type-(V_h, L) bidders to randomize on $[\tilde{b}, \hat{b}]$, it must be the case that

$$(V_h - b)\left\{\frac{p_l q}{p_h(1-q) + p_l q} + \frac{p_h(1-q)q}{p_h(1-q) + p_l q} + \frac{p_h(1-q)^2}{p_h(1-q)^2 + p_l q}\tilde{G}_{(V_h,L)}(b)\right\} = (V_h - \tilde{b})\frac{p_l q + p_h q(1-q)}{p_h(1-q) + p_l q},$$

from which, after solving for $\tilde{G}_{(V_h,L)}(b)$, we obtain

$$\tilde{G}_{(V_h,L)}(b) = \frac{q(1-p_h q)(b-\tilde{b})}{p_h(1-q)^2(V_h-b)}. \tag{A.10}$$

Suppose that type-(V_h, H) bidder 2 mixes over $[V_l, \tilde{b}]$, and type-(V_h, L) bidder 2 mixes over $[\tilde{b}, \hat{b}]$ according to $\tilde{G}(V_h, L)(\cdot)$ as described by (A.10). Then the expected surplus for type-(V_h, H) bidder 1 from bidding $b \in (\tilde{b}, \hat{b})$ is given by

$$\begin{aligned}(V_h - b)&\left\{\frac{p_l(1-q)}{p_h q + p_l(1-q)} + \frac{p_h q^2}{p_h q + p_l(1-q)} + \frac{p_h q(1-q)}{p_h q + p_l(1-q)}\tilde{G}_{(V_h,L)}(b)\right\}\\ &= \frac{(1-q)[p_l(1-q)+p_h q^2]V_h - q^2(1-p_h q)\tilde{b} + p_l[q^2-(1-q)^2]b}{(1-q)\,[p_h q + p_l(1-q)]},\end{aligned}$$

which is non-decreasing in b. Therefore, type-(V_h, H) bidder will have an incentive to bid higher than $\tilde{b}$ if her opponent follows the prescribed strategies, a contradiction.

Combining Steps 1–3 and Lemma A.1, we know that the equilibrium described in Lemma A.2 is the only symmetric equilibrium. □

Lemma A.4. *There is no asymmetric equilibrium.*

Proof. First, arguments similar to step 3 in the proof of Lemma A.3 can be used to show that in an asymmetric equilibrium, the support of type-(V_h, L) bidders must be lower than that of type-(V_h, H) bidders.

Now suppose that type-(V_h, L) bidder 1 and bidder 2 respectively play a mixed strategy on the support $[V_l, \tilde{b}_1]$ and $[V_l, \tilde{b}_2]$, and without loss of generality, suppose that $\tilde{b}_1 > \tilde{b}_2$. Since type-$(V_h, L)$ bidder 1 must be indifferent between any bids in $(V_l, \tilde{b}_2]$, type-(V_h, L) bidder 2's mixed strategy, denoted by $\tilde{G}_{2(V_h,L)}$, must satisfy

$$\begin{aligned}(V_h - b)&\left\{\frac{p_l q}{p_l q + p_h(1-q)} + \frac{p_h(1-q)}{p_l q + p_h(1-q)}\tilde{G}_{2(V_h,L)}(b)\right\}\\ &= (V_h - V_l)\frac{p_l q}{p_l q + p_h(1-q)}\end{aligned}$$

from which we obtain that $\tilde{b}_2 = \bar{b}_{(V_h,L)}$ where $\bar{b}_{(V_h,L)}$ is specified by formula (6). Now since type-(V_h, L) bidder 2 is indifferent between any bids in $[V_l, \tilde{b}_2]$, type-(V_h, L) bidder 1's mixed strategy CDF, denoted by

$\tilde{G}_{1(V_h,L)}$, in the interval $[V_l, \tilde{b}_2]$ must satisfy

$$(V_h - b)\left\{\frac{p_l q}{p_l q + p_h(1-q)} + \frac{p_h(1-q)}{p_l q + p_h(1-q)}\tilde{G}_{1(V_h,L)}(b)\right\}$$
$$= (V_h - V_l)\frac{p_l q}{p_l q + p_h(1-q)}$$

from which we obtain that

$$\tilde{G}_{1(V_h,L)}(b) = \frac{p_l q(b - V_l)}{p_h(1-q)^2(V_h - b)}.$$

But then $\tilde{G}_{1(V_h,L)}(\tilde{b}_2) = \tilde{G}_{1(V_h,L)}(\bar{b}_{(V_h,L)}) = 1$. Hence $\tilde{b}_1 = \tilde{b}_2$, a contradiction. □

Proof of Proposition 3. Suppose to the contrary that there is an auction environment that admits both types of symmetric equilibrium in the FPA. First, since the support of type-(V_m, L) bidders must be lower than those of type-(V_m, H), (V_h, L), and (V_h, H) bidders in both equilibria, the upper limit of type-(V_m, L) bidders' mixed strategies in both equilibria must be the same, which we denote by $\bar{b}_{(V_m,L)}$.

Let $\bar{b}^{\text{eff}}_{(V_h,L)}$ be the upper limit of the mixed strategy support of type-(V_h, L) bidder in the efficient equilibrium and let $\bar{b}^{\text{ineff}}_{(V_m,H)}$. be the upper limit of the mixed strategy support of type-(V_m, H) bidder in the inefficient equilibrium. We then consider two possible cases:

Case 1: $\bar{b}^{\text{eff}}_{(V_h,L)} \geq \bar{b}^{\text{ineff}}_{(V_m,H)}$ illustrated in Fig. 3. Since in the inefficient equilibrium type-(V_h, L) bidder is indifferent between any bids in $[\bar{b}_{(V_m,L)} \geq \bar{b}^{\text{ineff}}_{(V_h,L)}]$, her expected surplus in the inefficient equilibrium is the same as

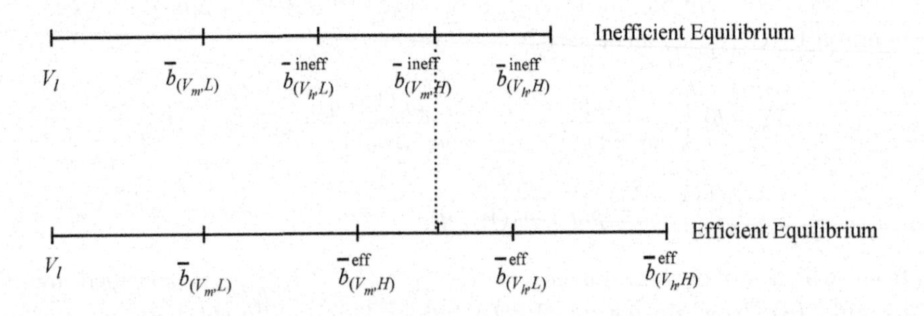

Fig. 3. Case 1 in the Proof of Proposition 3.

that when she bids $\bar{b}_{(V_m,L)}$ (recall our tie-breaking rule), which is simply

$$Z_1 = (V_h - \bar{b}_{(V_m,L)}) \left\{ \frac{p_l q_l}{\sum_{k\in\{l,m,h\}} p_k q_k} + \frac{p_m q_m (1 - q_h)}{\sum_{k\in\{l,m,h\}} p_k q_k} \right\}, \quad \text{(A.11)}$$

where the term in the bracket is the expected probability of winning against bidders with V_l valuation and type-(V_m, L) bidders. Given that her opponent follows the prescribed strategy in the inefficient equilibrium, her expected payoff from deviating to a bid of $\bar{b}^{\text{ineff}}_{(V_m,H)}$ is

$$Z_2 = (V_h - \bar{b}^{\text{ineff}}_{(V_m,H)}) \left\{ \frac{p_l q_l}{\sum_{k\in\{l,m,h\}} p_k q_k} + \frac{p_m q_m}{\sum_{k\in\{l,m,h\}} p_k q_k} + \frac{p_h q_h (1 - q_h)}{\sum_{k\in\{l,m,h\}} p_k q_k} \right\}, \quad \text{(A.12)}$$

where the term in the bracket is the expected probability of winning against bidders with V_l and V_m valuations and type-(V_h, L) bidders. By the requirement of the inefficient equilibrium, we have $Z_1 > Z_2$.[12] Moreover, since in this case $V_h > \bar{b}^{\text{eff}}_{(V_h,L)} \geq \bar{b}^{\text{ineff}}_{(V_m,H)}$ by assumption, we immediately have

$$Z_2 \geq (V_h - \bar{b}^{\text{eff}}_{(V_h,L)}) \left\{ \frac{p_l q_l}{\sum_{k\in\{l,m,h\}} p_k q_k} + \frac{p_m q_m}{\sum_{k\in\{l,m,h\}} p_k q_k} + \frac{p_h q_h (1 - q_h)}{\sum_{k\in\{l,m,h\}} p_k q_k} \right\}. \quad \text{(A.13)}$$

But the right-hand side of inequality (A.13) is exactly type-(V_h, L) bidder's expected surplus in the efficient equilibrium, which by the definition of the efficient equilibrium is required to be larger than Z_1 [as given by expression (A.11)]. This is so because Z_1 is also type-(V_h, L) bidder's expected surplus from deviating to a bid of $\bar{b}_{(V_m,L)}$ in the efficient equilibrium. Thus we have $Z_2 \geq Z_1$, which is a contradiction to our earlier conclusion that $Z_2 > Z_1$.

Case 2: $\bar{b}^{\text{eff}}_{(V_h,L)} < \bar{b}^{\text{ineff}}_{(V_m,H)}$. A contradiction can be derived for type-(V_m, H) bidder using argu ments analogous to Case 1. □

[12]The strict, rather than the usual weak, inequality, is valid because, tedious algebra shows that the deviation surplus function is strictly decreasing in the whole interval $[\bar{b}^{\text{ineff}}_{(V_h,L)}, \bar{b}^{\text{ineff}}_{(V_m,H)}]$ in order to be consistent with the inefficient equilibrium.

Proof of Proposition 4. Consider an example of the model presented in Section 2 as follows:

- $V_l = 0$, $V_m = 1$ and $V_h = 2$. We can set $V_l = 0$ and $V_m = 1$ by normalization and scaling with no loss of generality. For the inefficient equilibrium described below to exist, V_h cannot be too high relative to V_m.
- $p_l = p_m = p_h = 1/3$. That is, ex ante bidders' valuations of the object take on the three values with equal probability. This assumption is purely for computational ease in Bayesian updating;
- $q_l = 0.9, q_m = 0.1, q_h = 0.05$. As we described in the beginning of this section, the intuition for inefficient equilibrium requires that q_l be big, q_m to be small, and q_h even smaller.

As in the example in Section 3, we know that bidders with valuation type V_l will bid V_l regardless of their information types. Now we show that the following mixed strategies for the other types of bidders constitute the unique symmetric equilibrium:

- Type-(V_m, L) bidders bid according to a mixed strategy on the support $[V_l, \bar{b}_{(V_m,L)}]$ with CDF $G_{(V_m,L)}(\cdot)$ where

$$\bar{b}_{(V_m,L)} = \frac{q_m^2 V_m + q_l V_l}{q_l + q_m^2},$$

$$G_{(V_m,L)}(b) = \frac{q_l}{q_m^2}\left(\frac{b - V_l}{V_m - b}\right).$$

- Type-(V_h, L) bidders bid according to a mixed strategy on the support $[\bar{b}_{(V_m,L)}, \bar{b}_{(V_h,L)}]$ with CDF $G_{(V_h,L)}(\cdot)$ where

$$\bar{b}_{(V_h,L)} = \frac{[q_l + q_m q_h]\bar{b}_{(V_m,L)} + q_h^2 V_h}{q_l + q_m q_h + q_h^2},$$

$$G_{(V_h,L)}(b) = \frac{q_l + q_m q_h}{q_h^2}\frac{b - \bar{b}_{(V_m,L)}}{V_h - b}.$$

- Type-(V_m, H) bidders bid according to a mixed strategy on the support $[\bar{b}_{(V_h,L)}, \bar{b}_{(V_m,H)}]$ with CDF $G_{(V_m,H)}(\cdot)$ where

$$\bar{b}_{(V_m,H)} = \frac{[1 - q_l + (1 - q_m)q_m + (1 - q_h)q_m]\bar{b}_{(V_h,L)} + (1 - q_m)^2 V_m}{1 - q_l + (1 - q_m)q_m + (1 - q_h)q_h + (1 - q_m)^2},$$

$$G_{(V_m,H)}(b) = \frac{1 - q_l + (1 - q_m)q_m + (1 - q_h)q_m}{(1 - q_m)^2}\frac{b - \bar{b}_{(V_h,L)}}{V_m - b}.$$

- Type-(V_h, H) bidders bid according to a mixed strategy on the support $[\bar{b}_{(V_m,H)}, \bar{b}_{(V_h,H)}]$ with CDF $G_{(V_h,H)}(\cdot)$ where

$$\bar{b}_{(V_h,H)} = \frac{(1-q_h)^2 V_h + [3 - q_l - q_m - q_h - (1-q_h)^2]\bar{b}_{(V_m,H)}}{3 - q_l - q_m - q_h},$$

$$G_{(V_h,H)}(b) = \left[\frac{3 - q_l - q_m - q_h - (1-q_h)^2}{(1-q_h)^2}\right] \frac{b - \bar{b}_{(V_m,H)}}{V_h - b}.$$

Under the above parameterization,

$$\bar{b}_{(V_h,H)} \approx 1.32531 > \bar{b}_{(V_m,H)} \approx 0.744012$$
$$> \bar{b}_{(V_h,L)} \approx 0.0164684 > \bar{b}_{(V_m,L)} \approx 0.010989 > V_l = 0.$$

To show that the above strategy profile constitutes an equilibrium, we need to demonstrate that, given that the opponent follows the postulated strategies, each type-(v, s) bidder, where $v \in \{V_m, V_h\}$ and $s \in \{L, H\}$, obtains a constant expected surplus from any bids in the support of the CDF $G_{(v,s)}(\cdot)$, which is in turn higher than the expected surplus from any other deviation bids. The details of the verifications are straightforward but arithmetically tedious, and can be found at Appendix A of Fang and Morris (2003). □

Proof of Proposition 5. Using standard ε-deviation arguments, we can show that (1) bidders with valuation V_l must bid V_l in pure strategy in any equilibrium; (2) each bidder of type-(V_h, L) and type-(V_h, H) must bid in mixed strategies with bids higher than V_l; (3) the highest bid that may be submitted by each bidder must be the same; (4) there is no gap in the bids submitted in equilibrium. We denote the mixed strategy CDF of type-(V_h, L) and type-(V_h, H) bidder i by $G_{i(V_h,L)}$ and $G_{i(V_h,H)}$ respectively, where $i = 1, 2$.

Next, we show that for each bidder i, the supports of $G_{i(V_h,L)}$ and $G_{i(V_h,H)}$ cannot overlap at more than one point. Without loss of generality, consider bidder 1. Let B_1 be the set of points in which the supports of $G_{i(V_h,L)}$ and $G_{i(V_h,H)}$ overlap. For any overlap bid $b \in B_1$, the following must be true:

$$(V_h - b)\left[\frac{p_l q_1}{p_l q_1 + p_h(1-q_1)} + \frac{p_h(1-q_1)(1-q_2)G_{2(V_h,L)}(b)}{p_l q_1 + p_h(1-q_1)} + \frac{p_h(1-q_1)q_2 G_{2(V_h,H)}(b)}{p_l q_1 + p_h(1-q_1)}\right] = \tilde{K}_{1(V_h,L)},$$

$$(V_h - b)\left[\frac{p_l(1-q_1)}{p_l(1-q_1)+p_h q_1} + \frac{p_h q_1(1-q_2)G_{2(V_h,L)}(b)}{p_l(1-q_1)+p_h q_1}\right.$$
$$\left. + \frac{p_h q_1 q_2 G_{2(V_h,H)}(b)}{p_l(1-q_1)+p_h q_1}\right] = \tilde{K}_{1(V_h,H)}.$$

Similar to the arguments in step 2 of the proof of Lemma A.3, the above system equations can hold for at most one value of b.

Therefore, we are left with four possible cases to consider depending on the order of the supports of type-(V_h, L) and (V_h, H) mixed strategies for each bidder. We will derive a contradiction for one of the cases, and the other cases can be dealt with analogously.

We consider the following case: The support of $G_{i(V_h,L)}$ is $[V_l, \bar{b}_{i(V_h,L)}]$ and the support of $G_{i(V_h,H)}$ is $[\bar{b}_{i(V_h,L)}, \bar{b}_{i(V_h,H)}]$. From discussions above, $\bar{b}_{1(V_h,H)} = \bar{b}_{2(V_h,H)} = \bar{b}_{(V_h,H)}$.

Step 1: Simple calculation shows that it must be the case that $\bar{b}_{1(V_h,L)} > \bar{b}_{2(V_h,L)}$.

Step 2: From the necessary indifference condition of type-(V_h, L) bidder 1 in the interval $[V_l, \bar{b}_{2(V_h,L)}]$, we can obtain $G_{2(V_h,L)}$:

$$(V_h - b)\left[\frac{p_l q_1}{p_l q_1 + p_h(1-q_1)} + \frac{p_h(1-q_1)(1-q_2)G_{2(V_h,L)}(b)}{p_l q_1 + p_h(1-q_1)}\right]$$
$$= (V_h - V_l)\frac{p_l q_1}{p_l q_1 + p_h(1-q_1)}$$
$$\Rightarrow G_{2(V_h,L)}(b) = \frac{p_l q_1}{p_h(1-q_1)(1-q_2)}\frac{b - V_l}{V_h - b},$$
$$\bar{b}_{2(V_h,L)} = \frac{p_h(1-q_1)(1-q_2)V_h + p_l q_1 V_l}{p_h(1-q_1)(1-q_2) + p_l q_1}.$$

Step 3: The indifference condition for type-(V_h, L) bidder 2 requires that $G_{1(V_h,L)}(b)$ must satisfy, for $b \in [V_l, \bar{b}_{2(V_h,L)}]$,

$$(V_h - b)\left[\frac{p_l q_2}{p_l q_2 + p_h(1-q_2)} + \frac{p_h(1-q_2)(1-q_1)G_{1(V_h,L)}(b)}{p_l q_1 + p_h(1-q_2)}\right]$$
$$= (V_h - V_l)\frac{p_l q_2}{p_l q_2 + p_h(1-q_2)}$$

from which we can obtain $G_{1(V_h,L)}(b)$ for $b \in (V_l, \bar{b}_{2(V_h,L)}]$ as

$$G_{1(V_h,L)}(b) = \frac{p_l q_2}{p_h(1-q_1)(1-q_2)}\frac{b - V_l}{V_h - b}.$$

Step 4: To obtain the $G_{1(V_h,L)}(b)$ for $b \in [\bar{b}_{2(V_h,L)}, \bar{b}_{1(V_h,L)}]$, we make use of the indifference condition of type-(V_h, H) bidder 2, which is given by

$$(V_h - b)\left[\frac{p_l(1-q_2)}{p_l(1-q_2)+p_h q_2} + \frac{p_h q_2(1-q_1)}{p_l(1-q_2)+p_h q_2} G_{1(V_h,L)}(b)\right]$$
$$= (V_h - \bar{b}_{2(V_h,L)})\left[\frac{p_l(1-q_2)}{p_l(1-q_2)+p_h q_2} + \frac{p_h q_2(1-q_1)}{p_l(1-q_2)+p_h q_2} \times G_{1(V_h,L)}(\bar{b}_{2(V_h,L)})\right]$$

hence, for $b \in [\bar{b}_{2(V_h,L)}, \bar{b}_{1(V_h,L)}]$

$$G_{1(V_h,L)}(b) = \frac{p_l(1-q_2)[b - \bar{b}_{2(V_h,L)}] + p_h q_2(1-q_1)G_{1(V_h,L)}(\bar{b}_{2(V_h,L)})[V_h - \bar{b}_{2(V_h,L)}]}{p_h q_2(1-q_1)(V_h - b)}.$$

Setting $G_{1(V_h,L)}(b) = 1$, we obtain

$$\bar{b}_{1(V_h,L)}) = \frac{p_h q_2(1-q_1)\{[1 - G_{1(V_h,L)}(\bar{b}_{2(V_h,L)})]V_h + G_{1(V_h,L)}(\bar{b}_{2(V_h,L)}\bar{b}_{2(V_h,L)}\} + p_l(1-q_2)\bar{b}_{2(V_h,L)}}{p_h q_2(1-q_1) + p_l(1-q_2)}.$$

Step 5: The indifference condition of type-(V_h, L) bidder 1 for the bids in the interval $[\bar{b}_{2(V_h,L)}, \bar{b}_{1(V_h,L)}]$ requires that $G_{2(V_h,H)}(b)$ for $b \in [\bar{b}_{2(V_h,L)}, \bar{b}_{1(V_h,L)}]$ must satisfy

$$(V_h - b)\left[\frac{p_l q_1}{p_l q_1 + p_h(1-q_1)} + \frac{p_h(1-q_1)(1-q_2)}{p_l q_1 + p_h(1-q_1)} + \frac{p_h(1-q_1)q_2}{p_l q_1 + p_h(1-q_1)} G_{2(V_h,H)}(b)\right] = (V_h - V_l)\frac{p_l q_1}{p_l q_1 + p_h(1-q_1)}$$

thus,

$$G_{2(V_h,H)}(b) = \frac{b - V_l}{V_h - b}\frac{p_l q_1}{p_h(1-q_1)q_2} - \frac{p_h(1-q_1)(1-q_2)}{p_h(1-q_1)q_2}$$

from which can obtain $G_{2(V_h,H)}(\bar{b}_{1(V_h,L)})$.

Step 6: The indifference condition of type-(V_h, H) bidder 2 requires that $G_{1(V_h,H)}(b)$ satisfy

$$(V_h - b)\left[\frac{p_l(1-q_2)}{p_l(1-q_2)+p_hq_2} + \frac{p_hq_2(1-q_1)}{p_l(1-q_2)+p_hq_2} + \frac{p_hq_2q_1}{p_l(1-q_2)+p_hq_2}G_{1(V_h,H)}(b)\right]$$
$$= (V_h - \bar{b}_{2(V_h,L)})\left[\frac{p_l(1-q_2)}{p_l(1-q_2)+p_hq_2} + \frac{p_hq_2(1-q_1)}{p_l(1-q_2)+p_hq_2}G_{1(V_h,L)}(\bar{b}_{2(V_h,L)})\right],$$

which implies a value for $\bar{b}_{1(V_h,H)}$.

Step 7: Likewise, the indifference condition of type-(V_h, H) bidder 1 requires that $G_{2(V_h,H)}(b)$ satisfy, for $b \in [\bar{b}_{1(V_h,L)}, \bar{b}_{2(V_h,H)}]$,

$$(V_h - b)\left[\frac{p_l(1-q_1)}{p_l(1-q_1)+p_hq_1} + \frac{p_hq_1}{p_l(1-q_1)+p_hq_1}G_{2(V_h,H)}(b)\right]$$
$$= (V_h - \bar{b}_{1(V_h,L)})\left[\frac{p_l(1-q_1)}{p_l(1-q_1)+p_hq_1} + \frac{p_hq_1}{p_l(1-q_1)+p_hq_1}G_{2(V_h,H)}(\bar{b}_{1(V_h,L)})\right],$$

which implies a value for $\bar{b}_{2(V_h,H)}$.

Step 8: For generic values of $\langle V_l, V_h, p_l, q_1, q_2\rangle$, $\bar{b}_{1(V_h,H)}$ and $\bar{b}_{2(V_h,H)}$ are not equal, which contradicts the equilibrium requirement by the standard ε-deviation argument. (see Fig. 4 for a graphic illustration of the above steps). □

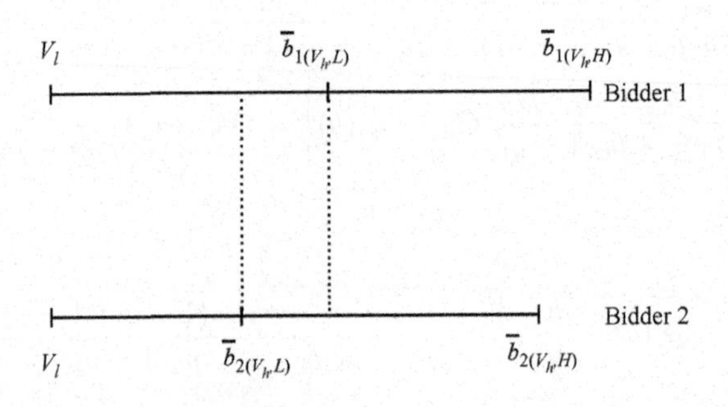

Fig. 4. A graphic illustration of the Proof of Proposition 5.

Appendix B.

In this appendix, we provide a proof that the seller receives the same expected revenue from the FPA and the SPA in an IPV environment where the valuations are drawn from a discrete distribution. This extends Maskin and Riley (1985) who showed the revenue equivalence for an environment with two bidders and two valuations. There are n agents bidding for a single object. The bidders' valuations are independent drawn from a discrete distribution with values $0 < V_1 < V_2 < \cdots < V_m$, and the probability of V_j is written as $p_j \in (0, 1)$. Assume that ties are broken as in Section 2.

Proposition B.1. *The seller's expected revenue is the same under the SPA and the FPA in the above environment.*

Proof. The weakly dominant strategy equilibrium in the SPA is for each bidders to bid her own valuation. Maskin and Riley (1996) showed that the FPA will admit a unique symmetric equilibrium. In this equilibrium, type-V_1 bids V_1 with probability 1; type-V_2 bids in a mixed strategy on $[V_1, \bar{b}_2]$; and in general type-V_k bids in a mixed strategy on $[\bar{b}_{k-1}, \bar{b}_k]$ where $\bar{b}_k > \bar{b}_{k-1}$.

Due to the monotonicity of the equilibrium bidding strategy, the expected probabilities of winning for a type-V_k bidder under both FPA and SPA are the same and they are $\sum_{j=1}^{k-1} p_j + p_k/n$. Since a type-$V_k$ bidder's expected payoff is given by her expected probability of winning $\times$ V_k minus her expected payment, her expected payment must be the same under the FPA and the SPA if her expected payoffs are the same, which we show below. Clearly, the expected payoffs for type-V_1 bidders are 0 under both the SPA and the FPA.

For $k = 2, \ldots, m$, a type-V_k bidder's equilibrium expected payoff in the SPA is given by

$$(p_1)^{n-1}(V_k - V_1) + [(p_1 + p_2)^{n-1} - p_1^{n-1}](V_k - V_2) \\ + \cdots + \left[\left(\sum_{j=1}^{k-1} p_j\right)^{n-1} - \left(\sum_{j=1}^{k-2} p_j\right)^{n-1}\right](V_k - V_{k-1}), \quad \text{(B.1)}$$

where $(p_1)^{n-1}$ and $\left(\sum_{j=1}^{k-1} p_j\right)^{n-1} - \left(\sum_{j=1}^{k-2} p_j\right)^{n-1}$ are respectively the probability that the highest value among the $n-1$ opponents is V_1 and V_k.

In the unique symmetric equilibrium of the FPA, for $k \geq 2$, a type-V_k bidder will bid in a mixed strategy on a support $[\bar{b}_{k-1}, \bar{b}_k]$; thus her expected payoff is equal to that from bidding $\bar{b}_{k-1}$. Given the tie-breaking rule, a bid of $\bar{b}_{k-1}$ by a type-V_k bidder will win when all her $n-1$ opponents have valuation lower than V_k, which occurs with probability $\left(\sum_{j=1}^{k-1} p_j\right)^{n-1}$. Thus her expected payoff from bidding $\bar{b}_{k-1}$ is

$$\left(\sum_{j=1}^{k-1} p_j\right)^{n-1} (V_k - \bar{b}_{k-1}). \tag{B.2}$$

It can be verified, by induction, that $\bar{b}_{k-1}$, for $k \geq 2$, must satisfy

$$\left(\sum_{j=1}^{k-1} p_j\right)^{n-1} \bar{b}_{k-1} = (p_1)^{n-1} V_1 + [(p_1 + p_2)^{n-1} - (p_1)^{n-1}] V_2 + \cdots + \left[\left(\sum_{j=1}^{k-1} p_j\right)^{n-1} - \left(\sum_{j=1}^{k-2} p_j\right)^{n-1}\right] V_k. \tag{B.3}$$

Plugging (B.3) into (B.2), we immediately show that type-V_k bidder's expected payoff in the FPA is identical to that in the SPA given by expression (B.1). Thus all types of bidders' expected payment must be the same under the FPA and the SPA. Hence the seller's expected revenue must be the same. □

References

Arozamena, L., and E. Cantillon (2004): "Investment incentives in procurement auctions," *Rev. Econ. Stud.*, 71, 1–18.

Athey, S. (2001): "Single crossing properties and the existence of pure strategy equilibria in games with incomplete information," *Econometrica*, 69, 861–890.

Baldwin, L., R. Marshall, and J. F. Richard (1997): "Bidder collusion at forest service timber sales," *J. Polit. Economy* 105, 657–699.

Battigalli, P., and M. Siniscalchi (2003): "Rationalizable bidding in first-price auctions," *Games Econ. Behav.* 45, 38–72.

Bergemann, D., and S. Morris (2003): "Robust Mechanism Design," Cowles Foundation Discussion Paper No. 1421.

Campbell, C. M., and D. Levin (2000): "Can the seller benefit from an insider in common-value auctions?," *J. Econ. Theory*, 91, 106–120.

Cantillon, E. (2000): "The effect of bidders' asymmetries on expected revenue in auctions," Cowles Foundation Discussion paper 1279.

Che, Y.-K., and I. Gale (1998): "Standard auctions with financially constrained bidders," *Rev. Econ. Stud.*, 65, 1–22.

Cremer, J., and R. McLean (1985): "Optimal selling strategies under uncertainty for a discriminating monopolist when demands are interdependent," *Econometrica*, 53, 345–361.

Dekel, E., and A. Wolinsky (2003): "Rationalizable outcomes of large private-value first-price discrete auctions," *Games Econ. Behav.*, 43, 175–188.

Fang, H., and S. Morris (2003): "Multidimensional Private Value Auctions," Cowles Foundation Discussion paper no. 1423.

Feinberg, Y., and A. Skrzypacz (2005): "Uncertainty about uncertainty and delay in bargaining," *Econometrica*, 73, 69–91.

Griesmer, J. H., R. E. Levitan, and M. Shubik (1967): "Toward a study of bidding processes: part IV: games with unknown costs," *Naval Res. Logistics Quart.*, 14, 415–433.

Holt, C. (1980): "Competitive bidding for contracts under alternative auction procedures," *J. Polit. Economy*, 88, 433–445.

Jackson, M. The non-existence of equilibrium in auctions with two dimensional types, Mimeo, California Institute of Technology.

Jackson, M., L. K. Simon, J. M. Swinkels, and W. R. Zame (2002): "Communication and equilibriumin discontinuous games of incomplete information," *Econometrica*, 70, 1711–1740.

Jehiel, P., and B. Moldovanu (2001): "Efficient design with interdependent valuations," *Econometrica*, 69, 1237–1259.

Jofre-Bonet, M., and M. Pesendorfer (2003): "Estimation of a dynamic auction game," *Econometrica*, 71, 1443–1489.

Kim, J., and Y.-K. Che (2004): "Asymmetric information about rivals' types in standard auctions," *Games Econ. Behav.*, 46, 383–397.

Landsberger, M., J. Rubinstein, E. Wolfstetter, and S. Zamir (2001): "First-price auctions when the ranking of valuations is common knowledge," *Rev. Econ. Design*, 6, 461–480.

Maskin, E., and J. Riley (1985): "Auction theory with private values," *Amer. Econ. Rev. Papers Proc.*, 75, 150–155.

Maskin, E., and J. Riley (1996): "Uniqueness in sealed high bid auctions," Mimeo, UCLA.

Maskin, E., and J. Riley (2000): "Asymmetric auctions," *Rev. Econ. Stud.*, 67, 413–438.

Maskin, E., and J. Riley (2000): "Equilibrium in sealed high bid auctions," *Rev. Econ. Stud.*, 67, 439–454.

Matthews, S. (1984): "Information acquisition in discriminatory auctions." In: *Bayesian Models in Economic Theory*, ed. by M. Boyer, R. E. Kihlstrom. North-Holland, Amsterdam.

Mathews, S. (1987): "Comparing auctions for risk averse buyers: a buyer's point of view," *Econometrica*, 55, 633–646.

McLean, R., and A. Postlewaite (2004): "Informational size and efficient auctions," *Rev. Econ. Stud.*, 71, 809–827.

Milgrom, P., and R. Weber (1982): "A theory of auctions and competitive bidding," *Econometrica*, 70, 1089–1122.

Mount, K., and S. Reiter (1974): "The information size of message spaces," *J. Econ. Theory*, 8, 161–192.

Myerson, R. (1981): "Optimal auction design," *Math. Oper. Res.*, 6, 58–73.

Neeman, Z. (2004): "The relevance of private information in mechanism design," *J. Econ. Theory*, 117, 55–77.

Persico, N. (2000): "Information acquisition in auctions," *Econometrica*, 68, 135–148.

Plum, M. (1992): "Characterization and computation of Nash-equilibria for auctions with incomplete information," *Int. J. Game Theory*, 20, 393–418.

Reny, P., and S. Zamir (2004): "On the existence of pure strategy monotone equilibria in asymmetric first-price auctions," *Econometrica*, 72, 1105–1125.

Riley, J., and W. F. Samuelson (1981): "Optimal auctions," *Amer. Econ. Rev.*, 71, 381–392.

Sagan, H. (1984): "Space-Filling Curves," Springer, Berlin.

Wang, R. (1991): "Common-value auctions with discrete private information," *J. Econ. Theory*, 54, 429–447.

Wilson, R. (1977): "A bidding model of perfect competition," *Rev. Econ. Stud.*, 44, 511–518.

CHAPTER 9

The Robustness of Robust Implementation*

Moritz Meyer-ter-Vehn and Stephen Morris

Abstract

We show that a mechanism that robustly implements optimal outcomes in a one-dimensional supermodular environment continues to robustly implement ε-optimal outcomes in all close-by environments. Robust implementation of ε-optimal outcomes is thus robust to small perturbations of the environment. This is in contrast to ex-post implementation which is not robust in this sense as only trivial social choice functions are ex-post implementable in generic environments.

Keywords: Robust implementation, ex-post implementation, social choice correspondence, belief-dependent outcomes

1 Introduction

Fix a mechanism design environment where each agent has a payoff type and agents' payoffs from different allocations depend on the profile of payoff types. If we come up with a mechanism where, whatever the agents' beliefs and higher order beliefs, there is an equilibrium where outcomes are consistent with a social choice correspondence, we say that the mechanism partially robustly implements the social choice correspondence.

If we restrict attention to 'direct mechanisms' where agents truthfully report their payoff types but not their beliefs, a social choice correspondence can be partially robustly implemented only if it can be implemented in

*This research is supported by NSF grant SES-0850718. We would like to thank the associate editor at JET and three anonymous referees for their thoughtful advice that improved the exposition of this note, and seminar audiences at ASSA 2010, the Princeton Conference on Mechanism Design, Toronto, and UCLA for helpful comments.

ex-post equilibrium.[1] Bergemann and Morris (2005) show that this equivalence between robust mechanism design and ex-post mechanism design continues to hold for some environments and social choice correspondences, even if mechanisms eliciting and using beliefs are allowed. In particular, this is true in a quasi-linear environment when there is a unique acceptable social allocation as a function of the payoff types, but transfers can depend on beliefs and are not restricted by budget balance.

In many contexts this observation is negative since ex-post incentive compatibility can be very restrictive. Jehiel, Meyer-ter-Vehn, Moldovanu and Zame (2006) show that in a generic class of multi-dimensional quasi-linear environments, only trivial allocation functions can be ex-post implemented. This implies a disturbing discontinuity. In a one-dimensional single-crossing environment, efficient allocations are implementable in ex-post equilibrium and thus robustly. But if the single-crossing environment is embedded in a larger, multi-dimensional payoff environment where single-crossing is perturbed generically, only trivial allocation functions are ex-post implementable even when the mechanism can be tailored to the perturbation.

Here, we show that this negative result and the discontinuity around single-crossing are properties of ex-post implementation, but not properties of robust implementation. When we allow the entire social choice to depend on beliefs and strengthen the solution concept to full implementation in rationalizable strategies, continuity is restored. More precisely, Theorem 1 shows that if a mechanism $\mathcal{M}$ fully implements an optimal social choice function in a one-dimensional, supermodular, baseline environment, then $\mathcal{M}$ also fully implements an almost optimal social choice correspondence whenever agents have approximate common knowledge that the baseline environment is close in a sense we make precise. This observation highlights that robust implementation is more flexible than ex-post implementation in economically important environments. The intuition for the result is that rationalizable strategies vary continuously in underlying payoffs, so if agents have strict incentives to report truthfully in the baseline environment, then reports can be only slightly off in a perturbed environment.

[1]This observation goes back to Ledyard (1978) and Dasgupta, Hammond and Maskin (1979) in the case of private values; the extension to interdependent values is straightforward.

The following example from Jehiel, Meyer-ter-Vehn, Moldovanu and Zame (2006) illustrates the main idea. Consider a quasi-linear environment where an object must be allocated to one of two agents. Agent i has a payoff type $\theta_i = (\theta_i^p, \theta_i^c) \in [0,1]^2$, with the interpretation that θ_i^p is a private value component for agent i, while θ_i^c is a component that enters both agents' valuations of the object. The value of the object to agent i is $\theta_i^p + \varepsilon\theta_i^c\theta_{-i}^c$. In this environment, only trivial allocation functions are ex-post implementable. However, if the designer ignores the interdependence term $\varepsilon\theta_i^c\theta_{-i}^c$ and just conducts a second-price auction, it is weakly dominated for type (θ_i^p, θ_i^c) to bid outside $[\theta_i^p, \theta_i^p + \varepsilon]$. The resulting allocation depends on beliefs and is generally not efficient but the welfare loss is less than ε. Thus the social choice correspondence consisting of ε-efficient allocations is robustly implementable.[2,3] One attractive robustness feature that carries over from this example to our general result is that our mechanism $\mathcal{M}$ is defined only with reference to the baseline environment, rather than being tailored to the true type space. Consequently, if $\mathcal{M}$ is efficient and detail-free, in the spirit of Dasgupta and Maskin (2000), and Perry and Reny (2002), then we can achieve almost efficient, detail-free implementation on all close-by type spaces.

In this paper, we generalize the logic of this example by considering as baseline a class of payoff environments from Bergemann and Morris (2009), where payoffs are interdependent and not necessarily quasi-linear, but satisfy a supermodularity condition and a contraction property that limits the interdependence. Theorem 1 shows that if a direct mechanism implements an optimal social choice function in a baseline environment, then the same mechanism implements the social choice correspondence of almost optimal outcomes in any close-by environment.

1.1 *Literature*

Examples in Bergemann and Morris (2005) illustrate the point that robust implementation can occur when ex-post implementation is impossible, using

[2]While similar in spirit, our notion of implementing ε-efficient allocations is strictly weaker than virtual efficient implementation because it allows for the possibility that the efficient allocation is chosen with probability zero.

[3]Bikhchandani (2006) addresses the impossibility theorem of Jehiel, Meyer-ter-Vehn, Moldovanu and Zame (2006) in an alternative way. He shows that non-trivial allocation functions are ex-post implementable when the object can be allocated to neither agent, and there are no allocative externalities.

mechanisms that elicit and respond to beliefs. By showing here that close to optimal SCCs are robustly implementable, we highlight the economic significance of this point. Börgers (1991), Smith (2010), and Yamashita (2011) show how mechanisms that elicit and respond to beliefs can robustly improve on ex-post incentive-compatible mechanisms, respectively in a general social choice setting, a public good setting with private values, and a bilateral trade setting.[4] A subtle difference to our note is that Börgers (1991) and Smith (2010) show that any ex-post mechanism is outperformed by some robust mechanism for some beliefs, while we show that there is a social choice correspondence that is robustly implementable but not ex-post implementable.[5]

Our result that belief-dependent robust mechanisms may improve on ex-post mechanisms contrasts with work by Chung and Ely (2007) who show that an ex-post mechanism is the optimal robust mechanism for revenue maximization in a class of single-object allocation settings with quasi-linear payoffs.

2 Setup

Basics: There is a finite set of agents, $i = 1, \ldots, I$, and a compact outcome space X with associated lottery space $Y = \Delta(X)$. A mechanism $\mathcal{M} = (M_1, \ldots, M_I, g)$ is given by measurable message sets M_i and a measurable outcome function $g : M \to Y$.

Types: We represents agents' beliefs and preferences, and higher order beliefs and preferences, with a type space $\mathcal{T} = (T_i, \pi_i, \tilde{u}_i)_{i=1}^{I}$, where each agent i has a measurable set of types T_i, a measurable function describing the agent's beliefs about others' types $\pi_i : T_i \to \Delta(T_{-i})$, and a measurable, bounded Bernoulli utility function $\tilde{u}_i : X \times T \to [\underline{u}, \bar{u}]$. This utility function extends to lotteries in the standard expected utility way. We put no restrictions on the agents' beliefs and preferences, so this type space can

[4]Thus, these papers relate to the negative results about ex-post implementation by Gibbard (1973), Satterthwaite (1975) and Hagerty and Rogerson (1987) in a similar way that this note relates to Jehiel, Meyer-ter-Vehn, Moldovanu and Zame (2006).

[5]While Yamashita (2011) compares mechanisms by their expected welfare, his two-price mechanism can be shown to also robustly implement a social choice correspondence that is not ex-post implementable.

be understood as the universal type space of Harsanyi, and Mertens and Zamir (1985) over all possible bounded expected utility functions.

Interim rationalizability: We ask when the planner can design a mechanism such that rational strategies of the agents lead to acceptable outcomes for the planner. We use the solution concept of interim (correlated) rationalizability (Dekel, Fudenberg and Morris (2007). Agent i is assumed to send some message that survives iterated deletion of messages that are not best responses to any belief over message-type profiles that (a) is consistent with π_i, and (b) puts probability one on profiles that have not yet been deleted. A formal description of the set of *interim rationalizable messages* $\tilde{R}_i^{\mathcal{M}}(t_i)$ for type t_i of agent i playing mechanism $\mathcal{M}$ is as follows:

$$\tilde{R}_{i,0}^{\mathcal{M}} = M_i,$$

$$\tilde{R}_{i,k+1}^{\mathcal{M}}(t_i) = \left\{ m_i \in M_i \,\middle|\, \begin{array}{l} \text{there exists } \tilde{\mu}_i \in \Delta(M_{-i} \times T_{-i}) \text{ such that} \\ (1) \;\; \tilde{\mu}_i(\{(m_{-i}, t_{-i}) | m_j \in \tilde{R}_{j,k}^{\mathcal{M}}(t_j) \\ \qquad \text{for all } j \neq i\}) = 1 \\ (2) \;\; \mathrm{marg}_{T_{-i}} \tilde{\mu}(\cdot) = \pi_i(\cdot | t_i) \\ (3) \;\; \int \tilde{u}_i(g(m_i, m_{-i}), (t_i, t_{-i})) d\tilde{\mu}_i(m_{-i}, t_{-i}) \\ \qquad \geq \sup_{m_i'} \int \tilde{u}_i(g(m_i', m_{-i}), (t_i, t_{-i})) \\ \qquad \times \, d\tilde{\mu}_i(m_{-i}, t_{-i}) \end{array} \right\},$$

$$\tilde{R}_i^{\mathcal{M}}(t_i) = \bigcap_{k \geq 1} \tilde{R}_{i,k}^{\mathcal{M}}(t_i).$$

Dekel, Fudenberg and Morris (2007) argue that interim rationalizable messages are those that are consistent with common knowledge of rationality among the agents and that are played in some subjective correlated equilibrium. We use rationalizability rather than equilibrium as solution concept as it is more natural and strengthens our results.

Implementation: The outcomes acceptable to the planner are represented by a social choice correspondence (SCC) $F : T \to Y$.

Definition 1. Mechanism $\mathcal{M}$ robustly implements SCC F at type profile $t \in T$ if

$$m \in \tilde{R}^{\mathcal{M}}(t) \Rightarrow g(m) \in F(t).$$

This definition does not require existence of rationalizable messages. We argue in Section 4 that this does not matter for the interpretation of Theorem 1.[6]

Approximate optimality: We will be particularly interested in approximately optimal SCCs. Label a planner as agent 0 with utility $\tilde{u}_0 : Y \times T \to \mathbb{R}$. A SCC $F : T \to Y$ is λ-optimal at type profile $t \in T$ if

$$\tilde{u}_0(y', t) \geq \sup_y\{\tilde{u}_0(y, t)\} - \lambda \quad \forall y' \in F(t).$$

If the planner maximizes the sum of the agents' utilities, so $\tilde{u}_0(y, t) = \sum \tilde{u}_i(y, t)$, and in addition the environment is quasi-linear, then SCC F is λ-*optimal* if and only if allocations are λ-efficient.

The purpose of the paper is to give new sufficient conditions for the implementation of approximately optimal SCCs.

3 Baseline payoff environments

The payoff (type) environment $(\Theta, u) = ((\Theta_i)_{i=1}^I, (u_i)_{i=0}^I)$ is given by compact one-dimensional payoff type spaces $\Theta_i \subseteq \mathbb{R}$ for agents $i = 1, \ldots, I$ and continuous Bernoulli utility functions $\mu_i : Y \times \Theta \to \mathbb{R}$ for $i = 0, 1, \ldots, I$.

Our sufficient condition relies on approximate common knowledge that payoffs are close to those generated by such a payoff environment that satisfies additional properties. A payoff environment does not specify agents' beliefs and higher order beliefs: we will be allowing them to have any beliefs and higher order beliefs.

We first describe the notion of closeness capturing approximate common knowledge of a payoff environment. Then we describe an alternative solution concept defined directly on the payoff environment that constrains behavior of types close to the payoff environment. Then we report the additional properties of the payoff environment that we use in our positive results.

3.1 *Approximate common knowledge*

We say that the payoff environment (Θ, u) is γ-*approximate common knowledge at type profile* t, if there is an event $E = (E_i)_{i=1}^I \subseteq T$ with $t \in E$ and

[6] To ensure existence of rationalizable messages we would need to assume compactness of type spaces T_i and message spaces M_i, as well as continuity of utility functions $\tilde{u}_i$, belief functions π_i, and the outcome function g of the mechanism.

measurable mappings to payoff types $\tilde{\theta}_i : E_i \to \Theta_i$ such that:

1. Baseline payoffs and true payoffs are γ-close on E:

$$|u_i(y, \tilde{\theta}(t)) - \tilde{u}_i(y, t)| \leq \gamma \quad \forall i, y \in Y, t \in E.$$

2. The event E is common $(1 - \gamma)$-belief among agents:

$$\pi_i(E_{-i}|t_i) \geq 1 - \gamma \quad \forall i, t_i \in E_i.$$

This definition allows for two perturbations of the payoff environment. Condition 1 allows payoffs to be misspecified by up to γ at every type profile. Outside the set E payoffs can be completely misspecified, but condition 2 requires the set E to be $(1 - \gamma)$-belief closed. For an illustration, if T is any type space over Θ with perturbed utilities $u_i(\cdot)$ satisfying $|u_i(y, \theta) - \tilde{u}_i(y, \theta)| \leq \gamma$, then the payoff environment (Θ, u) is γ-approximate common knowledge on all of T.

3.2 *Payoff environment solution concept*

Instead of applying the solution concept of Definition 1 directly, we connect it to an approximate version of rationalizability defined directly on the payoff environment (Θ, u). For a fixed mechanism $\mathcal{M}, \delta > 0$ and each payoff type θ_i of each agent i, we iteratively delete messages that are not within δ of a best response for any belief over other agents' remaining payoff types and messages. Formally, we set

$$R_{i,0}^{\mathcal{M},\delta}(\theta_i) = M_i,$$

$$R_{i,k+1}^{\mathcal{M},\delta}(\theta_i) = \left\{ m_i \in M_i \;\middle|\; \begin{array}{l} \text{there exists } \mu_i \in \Delta(M_{-i} \times \Theta_{-i}) \text{ such that} \\ (1) \;\; \mu_i(\{(m_{-i}, \theta_{-i}) | m_j \in \tilde{R}_{j,k}^{\mathcal{M},\delta}(\theta_j) \\ \qquad \text{for all } j \neq i\}) = 1 \\ (2) \;\; \displaystyle\int u_i(g(m_i, m_{-i}), (\theta_i, \theta_{-i})) d\mu_i(m_{-i}, \theta_{-i}) \\ \qquad \geq \sup_{m_i'} \displaystyle\int u_i(g(m_i', m_{-i}), (\theta_i, \theta_{-i})) \\ \qquad \times d\mu_i(m_{-i}, \theta_{-i}) - \delta \end{array} \right\},$$

$$R_i^{\mathcal{M},\delta}(\theta_i) = \bigcap_{k \geq 1} R_{i,k}^{\mathcal{M},\delta}(\theta_i)$$

and say that *message* m_i *is* δ*-rationalizable for payoff type* θ_i in mechanism $\mathcal{M}$, if $m_i \in R_i^{\mathcal{M},\delta}(\theta_i)$. If message m_i is interim rationalizable for a type t_i and the payoff environment (Θ, u) is approximate common knowledge at t_i, then m_i is δ-rationalizable for the corresponding payoff type θ_i in mechanism $\mathcal{M}$:

Lemma 1. *Fix a payoff environment* (Θ, u) *and a mechanism* $\mathcal{M} = (M_1, \ldots, M_I, g(\cdot))$. *For any* $\delta > 0$ *there exists* $\gamma > 0$, *such that whenever* (Θ, u) *is* γ*-approximate common knowledge at* $t = (t_i, t_{-i})$, *then any interim rationalizable message* m_i *of type* t_i *is also* δ*-rationalizable for the payoff type* $\tilde{\theta}_i(t_i)$. *Formally,*

$$\tilde{R}_i^{\mathcal{M}}(t_i) \subseteq R_i^{\mathcal{M},\delta}(\tilde{\theta}_i(t_i)).$$

Proof. In Appendix A. □

This result is a consequence of the upper hemicontinuity of interim rationalizable outcomes (Dekel, Fudenberg and Morris, 2007). It is an incomplete information analogue of a result of Ely (2001) on complete information rationalizability. Battigalli and Siniscalchi (2003) define a class of rationalizability solution concepts (called 'Δ-rationalizability') for incomplete information environments where common knowledge of certain first-order beliefs is built; here we are analogously building approximate common knowledge of the payoff environment.

As a corollary of this lemma we have:

Corollary 1. *For any* $\delta > 0$ *and mechanism* $\mathcal{M}$ *there exists* $\gamma > 0$, *such that* $\mathcal{M}$ *implements SCC* F *at type profile* $t \in T$ *whenever*

1. *payoff environment* (Θ, u) *is* γ*-approximate common knowledge at* t, *and*
2. $m \in R^{\mathcal{M},\delta}(\tilde{\theta}(t)) \Rightarrow g(m) \in F(t)$.

3.3 *One-dimensional, contractive, supermodular payoff type environments*

We now introduce the restrictions on the payoff environment that our main result appeals to. Recall that each Θ_i is a compact subset of $\mathbb{R}$ and that payoff functions $u_i(y, \theta), u_0(y, \theta)$ are continuous in θ. We consider slightly stronger assumptions than those in Bergemann and Morris (2009).

A1 Assumption — Monotone aggregator.

The type profile θ affects payoffs u_i via (across-agents) aggregators $h_i : \Theta \to \mathbb{R}$, so that

$$u_i(y, \theta) = v_i(y, h_i(\theta_i, \theta_{-i})).$$

The aggregator h_i is continuous, and strictly increasing in θ_i, and payoffs $v_i : Y \times \mathbb{R} \to \mathbb{R}$ are continuous in aggregated types.

A2 Assumption — Supermodularity.

There is an order $\prec_i$ on Y such that the payoff function $v_i(y, \phi)$ is weakly supermodular.

A3 Assumption — Contraction property.

There exists a strictly increasing function $\alpha_{CP} : \mathbb{R}^+ \to \mathbb{R}^+$, such that for all $\xi > 0$

$$h_i(\theta_i, \theta_{-i}) - h_i(\theta_i', \theta_{-i}') > \alpha_{CP}(\xi)$$

if $\theta_i - \theta_i' \geq \xi$ and $|\theta_j - \theta_j'| < \xi$ for all $j \neq i$.[7]

We will also consider a uniform version of ex-post incentive compatibility. Social choice function $f : \Theta \to Y$ is *uniformly ex-post incentive compatible*, if there exists a strictly increasing function $\alpha_{IC} : \mathbb{R}^+ \to \mathbb{R}^+$, such that for all i, $\theta_i \neq \theta_i'$ and θ_{-i} the payoff-loss when type θ_i reports θ_i' in the direct revelation mechanism $f : \Theta \to Y$ is bounded below by α_{IC}:

$$u_i(f(\theta_i, \theta_{-i}), (\theta_i, \theta_{-i})) - u_i(f(\theta_i', \theta_{-i}), (\theta_i, \theta_{-i})) \geq \alpha_{IC}(|\theta_i - \theta_i'|).$$

Instead of justifying these assumptions directly, we show that they are satisfied in the two following examples from Bergemann and Morris (2009), and assert that they are also satisfied in the further examples 'Binary allocation' and 'Information aggregation' in that paper.

Single-object allocation: Quasi-linear utilities with type spaces $\Theta_i = [0, 1]$, valuations $v_i(\theta) = \theta_i + \gamma \sum_{j \neq i} \theta_j$, and $\gamma < 1/(I-1)$ satisfy assumptions A1–A3. Writing $\hat{\theta}_i$ for the report of agent i in the direct mechanism, the generalized Vickrey mechanism allocates to $i \in$arg $\max_j \{\hat{\theta}_j\}$ and charges the winner $pi = \max_{j \neq i}\{\hat{\theta}_j\} + \gamma \sum_{j \neq i} \hat{\theta}_j$. Truthtelling is ex-post incentive compatible and gives rise to an efficient allocation. To render this mechanism uniformly ex-post incentive compatible, it is played with probability $1 - \varepsilon$, while with probability $\varepsilon \hat{\theta}_i / I$ the object is allocated to i for a

[7]We are grateful to an anonymous referee for pointing out this simplified definition.

payment of $\hat{\theta}_i/2+\gamma\hat{\theta}_i\sum_{j\neq i}\hat{\theta}_j$, and with probability $\varepsilon(I-\sum_j\hat{\theta}j)$ the object is not allocated.

Public good provision: Quasi-linear utilities with valuations $v_i(x,\theta) = (\theta_i+\gamma\sum_{j\neq i}\theta_j)x$, cost functions $\frac{1}{2}x^2$, and $\gamma < 1/(I-1)$ satisfy assumptions A1–A3. The generalized Vickrey mechanism is ex-post efficient and uniformly ex-post incentive compatible.

4 Main Result

Theorem 1. *Assume that (Θ, u) satisfies A1–A3, mechanism $\mathcal{M} = (\Theta_1,\ldots,\Theta_N, f)$ is uniformly ex-post incentive compatible, and the social choice function f is λ-optimal at every θ. Fix $\varepsilon > 0$.*

Then there exists $\gamma > 0$, such that the same mechanism $\mathcal{M}$ robustly implements a social choice correspondence F that is $(\lambda+\varepsilon)$-optimal at any type profile t with γ-approximate common knowledge of the payoff environment (Θ, u).

In the introductory auction example the mechanism that performs well in the perturbed environment is just the second-price auction. Similarly here, the $(\lambda+\varepsilon)$-optimal mechanism $\mathcal{M}$ does not depend on the perturbed true environment. This independence is important, because it implies that Theorem 1 does not exploit Definition 1 by constructing a mechanism $\mathcal{M}$ in which rationalizable messages do not exist.

We prove Theorem 1 in two steps. Lemma 2 shows that rationalizable reports of t_i must be close to $\tilde{\theta}_i(ti)$. Then it essentially suffices to show that the objective function is continuous in the reports.

Lemma 2. *Assume that the payoff environment (Θ, u) satisfies A1–A3, and that $\mathcal{M} = (\Theta_1,\ldots,\Theta_N, f)$ is uniformly ex-post incentive compatible. Then for every $\xi > 0$, there exists $\delta > 0$, such that for all types θ_i only reports $\theta'_i \in [\theta_i-\xi, \theta_i+\xi]$ are δ-rationalizable.*

Proof. If this is not the case, then there is $\xi > 0$ such that for any $\delta > 0$

$$\sup\{|\theta'_i-\theta_i| : i, \theta_i, \theta'_i \in R_i^{\mathcal{M},\delta}(\theta_i)\} \geq \xi.$$

Let the LHS be maximal for agent i. Applying the contraction property, there exist θ_i and $\theta'_i \in R_i^{\mathcal{M},\delta}(\theta_i)$ with $\theta_i \geq \theta'_i+\xi$ such that

$$h_i(\theta_i,\theta_{-i}) - h_i(\theta'_i,\theta'_{-i}) > \alpha_{CP}(\xi) \quad \forall\theta_{-i}\in\Theta_{-i}, \theta'_{-i}\in R_{-i}^{\mathcal{M},\delta}(\theta_{-i}).$$

The aggregator h_i is uniformly continuous, so we can choose $\zeta \in (0, \xi)$ with

$$h_i(\theta_i, \theta_{-i}) \geq h_i(\theta'_i, +\zeta, \theta'_{-i}) \quad \forall \theta_{-i} \in \Theta_{-i}, \theta'_{-i} \in R^{\mathcal{M},\delta}_{-i}(\theta_{-i}).$$

Fix any $\theta'_{-i} \in R^{\mathcal{M},\delta}_{-i}(\theta_{-i})$. The payoff-loss of type $\theta'_i + \zeta$, when he reports θ'_i instead of $\theta'_i + \zeta$ and others truthfully report θ'_{-i}, is at least $\alpha_{IC}(\zeta)$ by uniform ex-post incentive compatibility:

$$u_i(f(\theta'_i + \zeta, \theta'_{-i}), (\theta'_i + \zeta, \theta'_{-i})) - u_i(f(\theta'_i, \theta'_{-i}), (\theta'_i + \zeta, \theta'_{-i})) \geq \alpha_{IC}(\zeta).$$

This payoff-loss is weakly greater for type θ_i and any true type of others θ_{-i}

$$\begin{aligned} &u_i(f(\theta'_i + \zeta, \theta'_{-i}), (\theta_i, \theta_{-i})) - u_i(f(\theta'_i, \theta'_{-i}), (\theta_i, \theta_{-i})) \\ &\quad \geq u_i(f(\theta'_i + \zeta, \theta'_{-i}), (\theta'_i + \zeta, \theta'_{-i})) - u_i(f(\theta'_i, \theta'_{-i}), (\theta'_i + \zeta, \theta'_{-i})) \geq \alpha_{IC}(\zeta) \end{aligned}$$

because (1) the aggregated type is weakly greater $h_i(\theta_i, \theta_{-i}) \geq h_i(\theta'_i + \zeta, \theta'_{-i})$ by the above, (2) the aggregated value function v_i is supermodular in outcome and aggregator, and (3) outcomes are increasing $f(\theta'_i + \zeta, \theta'_{-i}) \succcurlyeq_i f(\theta'_i, \theta'_{-i})$ by supermodularity and uniform ex-post incentive compatibility.

Since this argument holds for any $\theta'_{-i} \in R^{\mathcal{M},\delta}_{-i}(\theta_{-i})$, we obtain the desired contradiction $\theta'_i \notin R^{\mathcal{M},\delta}_i(\theta_i)$ for any $\delta < \alpha_{IC}(\zeta)$. □

Proof of Theorem 1. We first show that there exists $\delta > 0$ such that every δ-rationalizable reports θ' of types θ lead to $(\lambda + \varepsilon/2)$-optimal outcomes in the payoff environment. For any $\varepsilon > 0$, fix $\xi > 0$ such that $|u_0(y, \theta') - u_0(y, \theta)| \leq \varepsilon/4$ for all $\theta' \in \prod[\theta_i - \xi, \theta_i + \xi]$. By Lemma 2 there exists $\delta > 0$, such that only reports $\theta'_i \in [\theta_i - \xi, \theta_i + \xi]$ are δ-rationalizable. So, for type profile θ, δ-rationalizable reports $\theta' \in R^{\mathcal{M},\delta}(\theta)$, and any outcome $y \in Y$ obtain

$$u_0(y, \theta) - u_0(f(\theta'), \theta) \leq u_0(y, \theta') - u_0(f(\theta'), \theta') + \varepsilon/2 \leq \lambda + \varepsilon/2,$$

proving that outcome $f(\theta')$ is $(\lambda + \varepsilon/2)$ optimal for types θ in the payoff environment.

To finish the proof, choose $\gamma < \varepsilon/4$ such that all rationalizable messages θ' of types $t \in E$ are δ-rationalizable for the γ-close payoff types $\theta = \tilde{\theta}(t)$. Then

$$\tilde{u}_0(y, t) - \tilde{u}_0(f(\theta'), t) \leq u_0(y, \theta) - u_0(f(\theta'), \theta) + \varepsilon/2 \leq \lambda + \varepsilon$$

finishing the proof. □

5 Discussion

We have reported sufficient conditions for implementing almost optimal social choice correspondences. Theorem 1 implies that for a fixed type space Θ and an open set of quasi-linear payoff functions, ε-efficient social choice correspondences are robustly implementable.

To what extent can these sufficient conditions be weakened? Bergemann and Morris (2009) show that in many payoff environments without the contraction property A3, it is impossible to fully implement efficient outcomes. Then, a fortiori, efficient outcomes cannot be approximated in close-by environments, and arguments by Oury and Tercieux (2009) suggest that it may be impossible to even partially robustly implement an almost efficient correspondence. However, there may be baseline payoff environments that violate our sufficient conditions but that still allow for interesting mechanisms with unique rationalizable messages. In such environments — that assume the result of Bergemann and Morris (2009 rather than its conditions — some regularity assumptions ensure that only close-by messages are δ-rationalizable as in Lemma 2. The proof of Theorem 1 then bounds the payoff-loss when true types have almost common knowledge of the baseline environment. These regularity assumptions are satisfied if either (1) the baseline environment is finite, or (2) it is compact and the mechanism is continuous.[8]

Abandoning the assumptions on the payoff environment more broadly can lead to a more dramatic failure of robust implementation. We illustrate this with a concluding example of a quasi-linear environment where values are not close to supermodular, and a mechanism designer with no information about agents' first-order beliefs weakly prefers a default choice over designing a mechanism that tries to take into account agents' information.

Example 1.[9] There are two agents, Rowena and Colin, with binary payoff types $\Theta_R = \{u, d\}$, $\Theta_C = \{l, r\}$ and binary allocations $x \in X = \{0, 1\}$ with lotteries $\Delta(X) = [0, 1]$. Utilities are quasi-linear $u_i(x, \theta, p) = xv_i(\theta) + p_i$ and values $v_i(\theta)$ are given by

[8] We are grateful to an anonymous referee for pointing this out to us. The argument that the rationalizability correspondence is upper hemicontinuous is analogue to Theorem 3 in Oury and Tercieux (2009).

[9] This environment is taken from a working paper version of Jehiel, Meyer-ter-Vehn, Moldovanu, Zame (2006).

values	l	r
u	$2, -1$	$-2, 1$
d	$-2, 1$	$2, -1$

Consider type spaces $T_i = \Theta_i \times \Delta(\Theta_{-i})$ with typical element $t_i(\theta_i, \pi_i)$; we extend beliefs over payoff types $\pi_i \in \Delta(\Theta_{-i})$ to beliefs over types $\Delta(T_{-i})$ by assuming beliefs over first-order beliefs to be uniform.

In Appendix A we prove that for any incentive-compatible mechanism $f : T \to [0, 1]$ and any planner preferences over allocations $u_0(x, \theta)$, there is a default allocation $y_0 \in [0, 1]$ that beats the worst-case performance of $\mathcal{M}$: For all payoff types θ there exist belief-types π such that[10]

$$u_0(f(\theta, \pi), \theta) \leq u_0(y_0, \theta).$$

This is in stark contrast to the mechanisms that robustly lead to ε-optimal outcomes in close to supermodular environments. The reason for this negative result is that, say, Rowena does not know whether her type u values allocation $x = 1$ higher than her type d; that depends on her belief about Colin's type. The circularity of the monotonicity constraints in the payoff matrix implies that no mechanism can robustly separate the payoff types, i.e., ensure allocations $f(\theta, \pi) < y_0 < f(\theta', \pi)$ for some θ, θ', y_0 and all π.

The contrast with the introductory auction example is instructive: There, the order of types depends on beliefs only for types with close-by private value components. However, type $\theta_i = (1, 1)$ is unambiguously higher than type $\theta_i = (0, 0)$. This partial order ensures that ε-efficient, allocation correspondences are robustly implementable.

Appendix A

Proof of Lemma 1. For $\delta > 0$, let $\gamma = \min\{\delta/(2 + 2(\bar{u} - \underline{u})), 1/2\}$. We argue by induction and suppose that $\tilde{R}^{\mathcal{M}}_{i,k-1}(ti) \subseteq R^{\mathcal{M},\delta}_{i,k-1}(\tilde{\theta}_i(t_i))$ for $(k - 1)$th level rationalizable actions of all types $t_i \in E_i$.

[10]The argument extends immediately to implementation in rationalizable strategies: In any mechanism $\mathcal{M} = (M_R, M_C, g)$ where every type has a rationalizable message, for every payoff types θ there exist belief types π and rationalizable messages $m \in R^{\mathcal{M}}(\theta, \pi)$ such that $u_0(g(m), \theta) \leq u_0(y_0, \theta)$.

We fix $m_i \in \tilde{R}^{\mathcal{M}}_{i,k}$ and will show that $m_i \in R^{\mathcal{M},\delta}_{i,k}(\tilde{\theta}_i(t_i))$. By the definition, there exists a belief $\tilde{\mu}_i \in \Delta(M_{-i} \times T_{-i})$ that rationalizes message m_i for type t_i:

(1) $\tilde{\mu}_i(\{(m_{-i}, t_{-i})|m_j \in \tilde{R}^{\mathcal{M}}_{j,k-1}(t_j) \text{ for all } j \neq i\}) = 1$,
(2) $\text{marg}_{T_{-i}}\tilde{\mu}_i = \pi_i(t_i)$,

$$(3) \int \tilde{u}_i(g(m_i, m_{-i}), (t_i, t_{-i}))d\tilde{\mu}_i(m_{-i}, t_{-i})$$
$$\geq \sup_{m'_i} \int \tilde{u}_i(g(m'_i, m_{-i}), (t_i, t_{-i}))d\tilde{\mu}_i(m_{-i}, t_{-i}).$$

We now define a belief $\mu_i \in \Delta(M_{-i} \times \Theta_{-i})$ that rationalizes message m_i for type $\theta_i = \tilde{\theta}_i(t_i)$. Let μ_i be the image of the measure $\tilde{\mu}_i \in \Delta(M_{-i} \times E_{-i})$ (restricted to E_{-i}) under the mapping $\tilde{\theta}_{-i} : E_{-i} \to \Theta_{-i}$

$$\mu_i(X_{-i}) = \frac{\tilde{\mu}_i((m_{-i}, t_{-i}) : (m_{-i}, \tilde{\theta}_{-i}(t_{-i})) \in X_{-i})}{\tilde{\mu}_i(M_{-i}, E_{-i})}.$$

Now

$$\mu_i(\{(m_{-i}, \theta_{-i})|m_j \in R^{\mathcal{M},\delta}_{j,k-1}(\theta_j) \text{ for all } j \neq i\}) = 1$$

because $\tilde{\mu}_i(\{(m_{-i}, t_{-i})|m_j \in \tilde{R}^{\mathcal{M}}_{j,k-1}(t_j) \text{ for all } j \neq 1\})$ by assumption, and $\tilde{R}^{\mathcal{M}}_{j,k-1}(t_j) \subseteq \tilde{R}^{\mathcal{M},\delta}_{j,k-1}(\theta_j)$ by the induction hypothesis.

Secondly, for all m'_i

$$\tilde{\mu}_i(M_{-i}, E_{-i}) \int_{M_{-i}\times\Theta_{-i}} u_i(g(m_i, m_{-i}), (\tilde{\theta}_i(t_i), \theta_{-i}))d\mu_i(m_{-i}, \theta_{-i})$$
$$\geq \int_{M_{-i}\times E_{-i}} (\tilde{u}_i(g(m_i, m_{-i}), (t_i, t_{-i})) - \gamma)d\tilde{\mu}_i(m_{-i}, t_{-i})$$
$$\geq \int_{M_{-i}\times T_{-i}} (\tilde{u}_i(g(m_i, m_{-i}), (t_i, t_{-i})) - \gamma)d\tilde{\mu}_i(m_{-i}, t_{-i}) - \gamma\overline{u}$$
$$\geq \int_{M_{-i}\times T_{-i}} (\tilde{u}_i(g(m'_i, m_{-i}), (t_i, t_{-i})) - \gamma)d\tilde{\mu}_i(m_{-i}, t_{-i}) - \gamma\overline{u}$$
$$\geq \int_{M_{-i}\times E_{-i}} (\tilde{u}_i(g(m'_i, m_{-i}), (t_i, t_{-i})) - \gamma)d\tilde{\mu}_i(m_{-i}, t_{-i}) - \gamma(\overline{u} - \underline{u})$$

$$\geq \tilde{\mu}_i(M_{-i}, E_{-i}) \int\limits_{M_{-i}\times\Theta_{-i}} (u_i(g(m_i', m_{-i}), (\tilde{\theta}_i(t_i), \theta_{-i})) - 2\gamma)$$
$$\times\, d\mu_i(m_{-i}, \theta_{-i}) - \gamma(\overline{u} - \underline{u}).$$

As $t_i \in E_i$ it must be that $\tilde{\mu}_i(M_{-i}, E_{-i}) = \pi_i(E_i) \geq 1 - \gamma$, and so m_i is a $\gamma(2 + (\overline{u} - \underline{u})/(1-\gamma))$-best reply to the assessment μ_i. This establishes $m_i \in R_{j,k}^{\mathcal{M},\delta}(\tilde{\theta}_i(t_i))$. □

Proof of concluding example. Fix planner's preferences $u_0(x,\theta)$ and an IC direct mechanism $f : T \to [0,1]$; depending on the context we write type profiles as $t = (\theta_R, \theta_C, \pi)$ or $t = (\theta, \pi)$. For fixed beliefs $\pi = (\pi_R, \pi_C)$ the standard IC constraints not to misreport one's payoff type imply the following monotonicity constraints:

$$\sum_{\theta_C} \pi_R(\theta_C)(v_R(\theta_R', \theta_C) - v_R(\theta_R, \theta_C))(f(\theta_R', \theta_C, \pi)$$
$$- f(\theta_R, \theta_C, \pi)) \geq 0, \qquad \text{(Mon-R)}$$
$$\sum_{\theta_R} \pi_C(\theta_R)(v_C(\theta_R, \theta_C') - v_C(\theta_R, \theta_C))(f(\theta_R, \theta_C', \pi)$$
$$- f(\theta_R, \theta_C, \pi)) \geq 0. \qquad \text{(Mon-C)}$$

We now show that for all payoff types $\theta, \theta' \in \{0,1\}^2$ there exist beliefs π such that the monotonicity constraints imply $f(\theta,\pi) \leq f(\theta',\pi)$. This implies that there exists $y_0 \in [0,1]$ such that for every θ there exist beliefs $\underline{\pi}_\theta$ and $\overline{\pi}_\theta$ with $f(\theta, \underline{\pi}_\theta) \leq y_0 \leq f(\theta, \overline{\pi}_\theta)$ and thus finishes the proof.

To do so fix $\theta = (u, l)$, say. To show $f(u,l,\pi) \leq f(u,r,\pi)$ and $f(u,l,\pi) \leq f(d,r,\pi)$, we consider beliefs π with $\pi_C(u) = 1$ so that l is the 'low type' for Colin who values allocation $x = 1$ at -1, and $\pi_R(r) = 1$ so that u is the 'low type' for Rowena. Then, first (Mon-C) implies that

$$(v_C(u,l) - v_C(u,r))(f(u,l,\pi) - f(u,r,\pi))$$
$$= (-1-1)(f(u,l,\pi) - f(u,r,\pi)) \geq 0$$

so $f(u,l,\pi) \leq f(u,r,\pi)$; second (Mon-R) implies that

$$(v_R(d,r) - v_R(u,r))(f(d,r,\pi) - f(u,r,\pi)) = (1+1)(f(d,r,\pi) - f(u,r,\pi)) \geq 0$$

so $f(u,r,\pi) \leq f(d,r,\pi)$ and thus $f(u,l,\pi) \leq f(d,r,\pi)$.

The last case, i.e. showing that $f(u,l,\pi') \leq f(d,l,\pi')$ for some π', needs a different argument: When Rowena knows $\pi_R(l) = 1$ her incentive constraints goes into the wrong direction and (Mon-R) implies $f(u,l,\pi) \geq f(d,l,\pi)$. Rather we fix π' with $\pi'_R(r) = 1$ and $\pi'_C(u) = \pi'_C(d) = 1/2$. By the above, we know $f(u,r,\pi) \leq f(d,r,\pi)$. By (Mon-C)

$$\begin{aligned}
0 &\leq (v_C(u,r) - v_C(u,l))(f(u,r,\pi) - f(u,l,\pi))/2 \\
&\quad + (v_C(d,r) - v_C(d,l))(f(d,r,\pi) - f(d,l,\pi))/2 \\
&= (f(u,r,\pi) - f(u,l,\pi)) - (f(d,r,\pi) - f(d,l,\pi)) \\
&\leq -f(u,l,\pi) + f(d,l,\pi)
\end{aligned}$$

and hence $f(u,l,\pi) \leq f(d,l,\pi)$. One way to understand this argument is that the utility (net of transfers) of Colin's type r is weakly negative when he reports truthfully, because $f(u,r,\pi) \leq f(d,r,\pi)$. If $f(u,l,\pi) > f(d,l,\pi)$ he could get a strictly positive net utility from misreporting his type as l. □

References

BATTIGALLI, P., AND M. SINISCALCHI (2003): "Rationalization and incomplete information," *Adv. Theor. Econ.*, 3, Article 3.

BERGEMANN, D., AND S. MORRIS (2005): "Robust mechanism design," *Econometrica*, 73, 1771–1813.

BERGEMANN, D., AND S. MORRIS (2009): "Robust implementation in direct mechanisms," *Rev. Econ. Stud.*, 76, 1175–1206.

BIKHCHANDANI, S. (2006): "Ex post implementation in environments with private goods," *Theoretical Econ.*, 1, 369–393.

BÖRGERS, T. (1991): "Undominated strategies and coordination in normal-form games," *Soc. Choice Welfare*, 8, 65–78.

CHUNG, K., AND J. ELY (2007): "Foundations of dominant-strategy mechanisms," *Rev. Econ. Stud.*, 74, 447–476.

DASGUPTA, P., P. HAMMOND, AND E. MASKIN (1979): "The implementation of social choice rules: Some results on incentive compatibility," *Rev. Econ. Stud.*, 46, 185–216.

DASGUPTA, P., AND E. MASKIN (2000): "Efficient auctions," *Quart. J. Econ.*, 115, 341–388.

DEKEL, E., D. FUDENBERG, AND S. MORRIS (2007): "Interim correlated rationalizability," *Theoretical Econ.*, 2, 15–40.

Ely, J. (2001): "Rationalizability and approximate common knowledge," http://www.kellogg.northwestern.edu/research/math/papers/1324.pdf.

Gibbard, A. (1973): "Manipulation of voting schemes," *Econometrica*, 41, 587–601.

Hagerty, K., and W. Rogerson (1987): "Robust trading mechanisms," *J. Econ. Theory*, 42, 94–107.

Jehiel, P., M. Meyer-ter-Vehn, B. Moldovanu, and W. Zame (2006): "The limits of ex post implementation," *Econometrica*, 74, 585–610.

Ledyard, J. (1978): "Incentive compatibility and incomplete information," *J. Econ. Theory*, 18, 171–189.

Mertens, J.-F., and S. Zamir (1985): "Formulation of Bayesian analysis for games of incomplete information," *Int. J. Game Theory*, 14, 1–29.

Oury, M., and O. Tercieux (2009): "Continuous implementation," IAS Working Paper 90.

Perry, M., and P. Reny (2002): "An efficient auction," *Econometrica*, 70, 1199–1213.

Satterthwaite, M. (1975): "Strategy-proofness and arrow's conditions: Existence and correspondence theorems for voting procedures and social welfare functions," *J. Econ. Theory*, 10, 187–217.

Smith, D. (2010): "A prior free efficiency comparison of mechanisms for the public good problem," http://www-personal.umich.edu/~dougecon/Doug_Smith_JMP.pdf.

Yamashita, T. (2011): "Robust welfare guarantees in bilateral trading mechanisms," http://www.stanford.edu/~takuroy/files/wi.pdf.

CHAPTER 10

Rationalizable Implementation*

Dirk Bergemann, Stephen Morris and Olivier Tercieux

Abstract
We consider the implementation of social choice functions under complete information in rationalizable strategies. A strict version of the monotonicity condition introduced by Maskin is necessary under the solution concept of rationalizability. Assuming the social choice function is responsive, i.e., in distinct states it selects distinct outcomes, we show that strict Maskin monotonicity is also sufficient under a mild "no worst alternative" condition. In particular, no economic condition is required. We discuss how our results extend when the social choice function is not responsive.

Keywords: Implementation, complete information, rationalizability, Maskin monotonicity, robust mechanism design, robust implementation

1 Introduction

We consider the implementation of social choice functions under complete information in rationalizable strategies. We say that a social choice function f is rationalizably implemented if there exists a mechanism such that every rationalizable strategy profile leads to the realization of the social choice function f. A priori, implementation in rationalizable strategies does not require the existence of a (pure or mixed) Nash equilibrium that leads to the realization of f, and hence this implementation notion is neither stronger nor weaker than that of Nash implementation. However, we establish that

*We benefited from the suggestions of the guest editor and four anonymous referees. We are grateful to Navin Kartik, Ludovic Renou and Roberto Serrano for comments on an earlier draft. This research is supported by NSF Grant #SES-0851200.

a strict (and thus stronger) version of the monotonicity condition shown by Maskin (1999) to be necessary for Nash implementation is necessary under the more stringent solution concept of rationalizability. Assuming the social choice function is responsive (i.e., it never picks the same outcome in two distinct states), we show that it is also sufficient under a "no worst alternative" (NWA) condition. In particular, no economic condition is required.

We are able to obtain this strong result because — like much of the classical implementation literature — we allow infinite mechanisms (including "integer games"); and — unlike the classical implementation literature — we allow for stochastic mechanisms.

In earlier works (see Bergemann and Morris (2009, 2011)), two of us established necessary and sufficient conditions for "robust implementation" in incomplete information environments. There we showed that a social choice function f can be Bayesian equilibrium implemented for all possible beliefs and higher order beliefs if and only if f is implementable under an incomplete information version of rationalizability. The results here are obtained by refining and further developing the rationalizability arguments for the complete information environment. We can establish stronger necessary and sufficient conditions than in the incomplete information environment. We can also dispense with an economic condition on the environment. In turn, we establish necessary conditions and sufficient conditions almost equivalent to Nash equilibrium implementation when the social choice function is responsive. The augmented mechanism which establishes the sufficiency result permits each agent to propose a menu of allocations. This construction already appeared in Maskin (1999) and Maskin and Sjöström (2004) to establish complete information implementation in the presence of mixed strategies. The sufficiency arguments for Nash equilibrium implementation typically rely on a no-veto property of the social choice function. In contrast, we use a weak condition, introduced as "no worst alternative" by Cabrales and Serrano (2009), to establish the sufficiency argument. This condition requires that in state θ and for every agent i, the social choice $f(\theta)$ is not the worst alternative among all possible allocations. The no worst alternative property plays a role in our proof that is quite distinct from the no veto property in the classic Nash equilibrium results. The no worst alternative property guarantees that in the augmented mechanism, any report in state θ in which an agent expresses his disagreement with the remaining agents cannot be a rationalizable report. By contrast, the no veto property guaranteed that if an agent were to express his disagreement, then further disagreement by other agents would only be possible in equilibrium if it would lead to the same equilibrium allocation as prescribed by $f(\theta)$.

These results narrow an open question in the literature. The existing literature shows that Maskin monotonicity is necessary for Nash implementation in any mechanism (even if stochastic mechanisms are allowed[1]). Abreu and Matsushima (1992) shows that if implementation is made easier by (i) requiring only virtual implementation; and (ii) imposing a weak domain restriction ruling out identical preferences; then implementation is always possible even if it is made harder by (iii) requiring finite mechanisms; and (iv) requiring the stronger solution concept of rationalizability. In addition, Abreu and Matsushima (1994) extend their earlier argument from virtual to exact implementation, but now using the iterative elimination of weakly dominated strategies rather than strictly dominated strategies.[2] Our result shows that it is possible to exactly implement a social choice function, in rationalizable strategies, even if domain restriction (ii) fails, as long as infinite, stochastic, mechanisms are allowed.

2 Setup

The *environment* consists of a collection of I agents (we write $\mathcal{I}$ for the set of agents); a finite set of possible states Θ; a countable set of pure allocations Z (we write $Y \equiv \Delta(Z)$ for the set of lotteries on Z); and, to each state, we associate for each player i a von Neumann–Morgenstern utility function $u_i : Z \times \Theta \to \mathbb{R}$, extended to lotteries as $u_i : Y \times \Theta \to \mathbb{R}$ with

$$u_i(y, \theta) = \sum_{z \in Z} y_z u_i(z, \theta).$$

We sometimes abuse notations and identify an outcome z in Z with the element of Y that assigns probability one to the correspondin gcoordinate.

[1]In such a case, Maskin monotonicity (that is usually defined on the set of pure allocations) has to be stated on the set of lotteries on pure allocations.

[2]Abreu and Matsushima (1994) actually note that their permissive implementation result holds under one round of removal of weakly dominated strategies followed by iterative removal of strictly dominated strategies. Hence, our result (that a stronger version of Maskin monotonicity is necessary for rationalizable implementation) underlines a well-known fact that replacing removal of weakly dominated strategies by removal of strictly dominated strategies can yield very different results.

Thus at two distinct states θ and θ', all agents can have the same ordinal preferences; this contrasts with some of the literature that associates a state with a profile of ordinal preferences (e.g. Maskin (1999)). We discuss alternative formulations in the final section. A *mechanism* $\mathcal{M}$ is given by $\mathcal{M} = ((M_i)_{i=l}^I, g)$, where each M_i is countable, $M = M_1 \times \cdots \times M_I$ and $g : M \to Y$.

The environment and the mechanism together describe a game of complete information for each $\theta \in \Theta$. We will use (correlated) rationalizability as a solution concept.[3] Our formal definition will coincide with the standard definition with finite or compact message spaces. But we will also allow infinite, non-compact, message spaces; in this case, our definition is equivalent to one introduced in Lipman (1994). Let a message set profile $S = (S_1, \ldots, S_I)$, where each $S_i \in 2^{M_i}$, and we write $\mathcal{S}$ for the collection of message set profiles. The collection $\mathcal{S}$ is a lattice with the natural ordering of set inclusion: $S \leq S'$ if $S_i \subseteq S'_i$ for all i. The largest element is $\bar{S} = (M_1, \ldots, M_I)$. The smallest element is $\underline{S} = (\varnothing, \varnothing, \ldots, \varnothing)$.

We define an operator $b^\theta : \mathcal{S} \to \mathcal{S}$ to iteratively eliminate never best responses with $b^\theta = (b_1^\theta, \ldots b_i^\theta, \ldots, b_I^\theta)$ and b_i^θ is defined by

$$b_i^\theta(S) = \left\{ m_i \in M_i \;\middle|\; \begin{array}{l} \text{there exists } \lambda_i \in \Delta(M_{-i}) \text{ such that} \\ (1)\ \lambda_i(m_{-i}) > 0 \Rightarrow m_j \in S_j \text{ for each } j \neq i, \\ (2)\ m_i \in \arg\max_{m'_i \in M_i} \sum_{m_{-i} \in M_{-i}} \\ \qquad \lambda_i(m_{-i}) u_i(g(m'_i, m_{-i}), \theta), \end{array} \right\}.$$

We observe that b^θ is increasing by definition: i.e., $S \leq S' \Rightarrow b^\theta(S) \leq b^\theta(S')$. By Tarski's fixed point theorem, there is a largest fixed point of b^θ, which we label $S^{\mathcal{M},\theta}$. Thus (i) $b^\theta(S^{\mathcal{M},\theta}) = S^{\mathcal{M},\theta}$ and (ii) $b^\theta(S) = S \Rightarrow S \leq S^{\mathcal{M},\theta}$. If $m_i \in S_i^{\mathcal{M},\theta}$, we say that message m_i is rationalizable in (the complete information game parameterized by) state θ.

We can also construct the fixed point $S^{\mathcal{M},\theta}$ by starting with $\bar{S}$ — the largest element of the lattice — and iteratively applying the operator b^θ.

---

[3]The original definition of rationalizability of Bernheim (1984) and Pearce (1984) required agents' conjectures over their opponents' play to be independent. We follow the convention of some of the recent literature (e.g., Osborne and Rubinstein (1994)) in using "rationalizability" for the correlated version of rationalizability (see Brandenburger and Dekel (1987) for an early definition and discussion). Our results do not rely on the use of the correlated version of rationalizability.

If the message sets are finite, we have

$$S_i^{\mathcal{M},\theta} \triangleq \bigcap_{n\geq 0} b_i^\theta([b^\theta]^n(\bar{S})).$$

In this case, the solution concept is equivalent to iterated deletion of strictly dominated strategies (see Brandenburger and Dekel (1987)). But if the mechanism $\mathcal{M}$ is infinite, transfinite induction may be necessary to reach the fixed point.[4] We will also sometimes use the following notation

$$S_{i,k}^{\mathcal{M},\theta} \triangleq b_i^\theta([b^\theta]^{k-1}(\bar{S})).$$

again using transfinite induction if necessary. Thus $S_i^{\mathcal{M},\theta}$ is the set of messages surviving (transfinite) iterated deletion of never best responses. It is possible to show formally that $S_i^{\mathcal{M},\theta}$ is the set of messages that agent i might send consistent with common certainty of rationality and the fact that payoffs are given by θ (see Lipman (1994). Finally, we will say that a message set profile $S = (S_1, \ldots, S_I)$ has the best-response property in state θ if $S \subseteq b^\theta(S)$, or equivalently, if for each player i and message $m_i \in S_i$, there exists $\lambda_i \in \Delta(M_{-i})$ such that $\lambda_i(m_{-i}) > 0 \Rightarrow m_j \in S_j$ for each $j \neq i$, and

$$m_i \in \arg\max_{m_i' \in M_i} \sum_{m_{-i}\in M_{-i}} \lambda_i(m_{-i})u_i(g(m_i', m_{-i}), \theta).$$

It is easy to check that if S has the best-response property in state θ, then $S \subseteq S^{\mathcal{M},\theta}$.

Now a *social choice function* (SCF) f is given by $f : \Theta \to Y$. Mechanism $\mathcal{M}$ *implements* f *in rationalizable strategies* if there exists $\mathcal{M}$ such that, for all $\theta, S^{\mathcal{M},\theta} \neq \varnothing$ and $m \in S^{\mathcal{M},\theta} \Rightarrow g(m) = f(\theta)$. SCF f is *implementable in rationalizable strategies* if there exists $\mathcal{M}$ such that $\mathcal{M}$ implements f in rationalizable strategies. The definition of rationalizable implementation does not require the existence of a (pure or mixed) Nash equilibrium that leads the realization of the social choice function f. Hence, a priori, rationalizable implementation need not be stronger (neither weaker) than Nash implementation. However, in the next section, we provide necessary conditions and sufficient conditions for rationalizable implementation almost equivalent to Nash implementation.

[4]Lipman (1994) contains a formal description of the transfinite induction required. As he notes "we remove strategies which are never a best reply, taking limits where needed".

3 Main Result

We first recall the definition of Maskin monotonicity restricted to social choice functions:

Definition 1. (*Maskin monotonicity*). Social choice function f satisfies Maskin *monotonicity* if:

1. $f(\theta) = f(\theta')$ whenever

$$u_i(f(\theta),\theta) \geq u_i(y,\theta) \Rightarrow u_i(f(\theta),\theta') \geq u_i(y,\theta') \quad \text{for all } i \text{ and } y;$$

or, equivalently,

2. $f(\theta) \neq f(\theta')$ implies

$$u_i(f(\theta),\theta) \geq u_i(y,\theta) \quad \text{and} \quad u_i(y,\theta') > u_i(f(\theta),\theta') \quad \text{for some } i \text{ and } y.$$

The latter condition states that in case the desired alternative differs at state θ and θ', there must exist at least one agent who, if the true state were θ' and she expected other agents to claim the state is θ, could be offered a reward y that would give her a strict incentive to "report" the deviation of other agents, where the reward y would not tempt her if the true state was in fact θ i.e., she would have a (weak) incentive to "report truthfully". The strengthening of Maskin monotonicity we will use, reinforces the latter statement, requiring that the reward y gives a strict incentive to "report truthfully" if the true state were θ.

Definition 2. (*Strict Maskin monotonicity*). Social choice function f satisfies strict Maskin *monotonicity* if:

1. $f(\theta) = f(\theta')$ whenever

$$u_i(f(\theta),\theta) > u_i(y,\theta) \Rightarrow u_i(f(\theta),\theta') \geq u_i(y,\theta') \quad \text{for all } i \text{ and } y; \tag{1}$$

or, equivalently,

2. $f(\theta) \neq f(\theta')$ implies

$$u_i(f(\theta),\theta) > u_i(y,\theta) \quad \text{and} \quad u_i(y,\theta') > u_i(f(\theta),\theta') \quad \text{for some } i \text{ and } y. \tag{2}$$

Maskin monotonicity, which is necessary for Nash implementation, is weaker than strict Maskin monotonicity. We show in the following proposition that strict Maskin monotonicity (and hence Maskin monotonicity) is necessary for rationalizable implementation.

Proposition 1. (*Necessary conditions*). *If f is implementable in rationalizable strategies, then f satisfies strict Maskin monotonicity.*

Proposition 1 is a consequence of the following lemma. In words, it states that, given a social choice function f, if θ and θ' satisfy condition (1) in the definition of strict Maskin monotonicity and, in addition, f is implementable by a mechanism $\mathcal{M}$, then the set of rationalizable message profiles must be the same in state θ and θ'.

Lemma 1. *Pick θ and θ' satisfying condition* (1). *If mechanism $\mathcal{M} = ((M_i)_{i=1}^I, g)$ implements f in rationalizable strategies, then we have $S^{\mathcal{M},\theta} = S^{\mathcal{M},\theta'}$.*

Proof. Pick θ and θ' satisfying condition (1) and fix any mechanism $\mathcal{M} = ((M_i)_{i=1}^I, g)$ that implements f in rationalizable strategies.

We first show that $S^{\mathcal{M},\theta} \subseteq S^{\mathcal{M},\theta'}$. Because $b^\theta(S^{\mathcal{M},\theta}) = S^{\mathcal{M},\theta}$, $S^{\mathcal{M},\theta}$ has the best-response property in state θ (i.e., for each player i and all $m_i \in S_i^{\mathcal{M},\theta}$, there exists $\lambda_i^{m_i,\theta} \in \Delta(M_{-i})$ such that $\lambda_i^{m_i,\theta}(m_{-i}) > 0 \Rightarrow m_j \in S_j^{\mathcal{M},\theta}$ for each $j \neq i$), and

$$\sum_{m_{-i} \in M_{-i}} \lambda_i^{m_i,\theta}(m_{-i}) u_i(g(m_i, m_{-i}), \theta) \geq \sum_{m_{-i} \in M_{-i}} \lambda_i^{m_i,\theta}(m_{-i}) u_i(g(m_i', m_{-i}), \theta) \tag{3}$$

for all $m_i' \in M_i$. We want to show that m_i is also a best response against $\lambda_i^{m_i,\theta}$ in state θ'. Since i and $m_i \in S_i^{\mathcal{M},\theta}$ have been fixed arbitrarily, this will prove that $S^{\mathcal{M},\theta}$ has the best-response property in state θ' and so that $S^{\mathcal{M},\theta} \subseteq S^{\mathcal{M},\theta'}$ as claimed. Note first that for any m_{-i} such that $\lambda_i^{m_i,\theta}(m_{-i}) > 0$, $m_{-i} \in S_{-i}^{\mathcal{M},\theta}$ and so because $m_i \in S_i^{\mathcal{M},\theta}$, we have $g(m_i, m_{-i}) = f(\theta)$. Thus,

$$\sum_{m_{-i} \in M_{-i}} \lambda_i^{m_i,\theta}(m_{-i}) u_i(g(m_i, m_{-i}), \theta') = u_i(f(\theta), \theta') = \sum_{m_{-i} \in M_{-i}} \lambda_i^{m_i,\theta}(m_{-i}) u_i(g(m_i', m_{-i}), \theta') \tag{4}$$

for all $m_i' \in S_i^{\mathcal{M},\theta}$. In addition, we claim that

$$\sum_{m_{-i} \in M_{-i}} \lambda_i^{m_i,\theta}(m_{-i}) u_i(g(m_i, m_{-i}), \theta) = u_i(f(\theta), \theta) > \sum_{m_{-i} \in M_{-i}} \lambda_i^{m_i,\theta}(m_{-i}) u_i(g(m_i', m_{-i}), \theta)$$

for any $m_i' \notin S_i^{\mathcal{M},\theta}$. By (3), the above is true with a weak inequality. Now if an equality were to hold, some $m_i' \notin S_i^{\mathcal{M},\theta}$ would be a best response against $\lambda_i^{m_i,\theta}$ in state θ and so $(\{m_i'\} \cup S_i^{\mathcal{M},\theta}) \times S_{-i}^{\mathcal{M},\theta}$ would have the best-response property in state θ implying that $m_i' \in S_i^{\mathcal{M},\theta}$ which is false by assumption. Now we know that

$$u_i(f(\theta),\theta) > u_i(y,\theta) \Rightarrow u_i(f(\theta),\theta') \geq u_i(y,\theta') \quad \text{for all } i \text{ and } y,$$

and so applying this to the lotteries $y \triangleq \sum_{m_{-i}} \lambda_i^{m_i,\theta}(m_{-i})g(m_i', m_{-i})$, we get that

$$\sum_{m_{-i}\in M_{-i}} \lambda_i^{m_i,\theta}(m_{-i})u_i(g(m_i,m_{-i}),\theta')$$
$$= u_i(f(\theta),\theta') \geq \sum_{m_{-i}\in M_{-i}} \lambda_i^{m_i,\theta}(m_{-i})u_i(g(m_i',m_{-i}),\theta') \tag{5}$$

for any $m_i' \notin S_i^{\mathcal{M},\theta}$. Finally, (4) and (5) ensure that m_i is also a best response against $\lambda_i^{m_i,\theta}$ in state θ'.

Now, let us show that $S^{\mathcal{M},\theta} \supseteq S^{\mathcal{M},\theta'}$. Since f is implementable in rationalizable strategies by $\mathcal{M}$ and $S^{\mathcal{M},\theta} \subseteq S^{\mathcal{M},\theta'}$, we have $f(\theta) = f(\theta')$ Take any player i and any $m_i^* \in S_i^{\mathcal{M},\theta'}$ we show that $m_i^* \in S_i^{\mathcal{M},\theta}$. Pick any $m_i \in S_i^{\mathcal{M},\theta}$ and $\lambda_i^{m_i,\theta} \in \Delta(M_{-i})$ satisfying $\lambda_i^{m_i,\theta}(m_{-i}) > 0 \Rightarrow m_j \in S_j^{\mathcal{M},\theta}$ for all $j \neq i$, and

$$u_i(f(\theta),\theta) = \sum_{m_{-i}} \lambda_i^{m_i,\theta}(m_{-i})u_i(g(m_i,m_{-i}),\theta)$$
$$\geq \sum_{m_{-i}} \lambda_i^{m_i,\theta}(m_{-i})u_i(g(m_i',m_{-i}),\theta) \tag{6}$$

for any $m_i' \in M_i$. Note that since $S^{\mathcal{M},\theta} \subseteq S^{\mathcal{M},\theta'}$, we have that $\lambda_i^{m_i,\theta}(m_{-i}) > 0 \Rightarrow m_j \in S_j^{\mathcal{M},\theta'}$ for all $j \neq i$; in addition, $m_i^* \in S_i^{\mathcal{M},\theta'}$ and so $g(m_i^*, m_{-i}) = f(\theta') = f(\theta)$ for any m_{-i} such that $\lambda_i^{m_i,\theta}(m_{-i}) > 0$. Hence

$$\sum_{m_{-i}} \lambda_i^{m_i,\theta}(m_{-i})u_i(g(m_i^*,m_{-i}),\theta)$$
$$= u_i(f(\theta),\theta) = \sum_{m_{-i}} \lambda_i^{m_i,\theta}(m_{-i})u_i(g(m_i,m_{-i}),\theta) \tag{7}$$

and thus (6) and (7) together show that m_i^* is a best response against $\lambda_i^{m_i,\theta}$ in state θ which proves that $m_i^* \in S_i^{\mathcal{M},\theta}$, as claimed. □

It is clear that Proposition 1 is obtained as a corollary of Lemma 1.

Oury and Tercieux (2009) have shown that Maskin monotonicity is a necessary condition for "continuous" partial implementation of a social choice function, where "continuous" means that the direct mechanism itself must work for types that are close to the complete information types in the product topology. They also show that full implementation in rationalizable strategies is necessary. Hence, an alternative way to prove the necessity of Maskin monotonicity would be to use this latter result and Proposition 1.

We need two extra conditions for the sufficiency result.

Definition 3. (*Responsive social choice function*). Social choice function f is responsive if $\theta \neq \theta' \Rightarrow f(\theta) \neq f(\theta')$.

The notion of responsiveness requires that the social choice function "responds" to a change in the state with a change in the social allocation.

Definition 4. (*No worst alternative*). Social choice function f satisfies "no worst alternative" (NWA) if, for each i and θ, there exists $\underline{y}_i(\theta)$ such that

$$u_i(f(\theta), \theta) > u_i(\underline{y}_i(\theta), \theta). \tag{8}$$

Property NWA requires that an agent never gets his worst outcome under the social choice function. The NWA property appears in Cabrales and Serrano (2009) as a sufficient condition to guarantee implementation in best-response dynamics. Given the set of allocations $\{\underline{y}_i(\theta)\}_{\theta \in \Theta}$, we define the average allocation $\underline{y}_i$ of this set by setting

$$\underline{y}_i \triangleq \frac{1}{\#\Theta} \sum_{\theta \in \Theta} \underline{y}_i(\theta). \tag{9}$$

Note that under NWA, for all θ and all i, there exists $y_i(\theta)$ such that

$$u_i(y_i(\theta), \theta) > u_i(\underline{y}_i, \theta); \tag{10}$$

this can be established by defining $y_i(\theta)$ as follows:

$$y_i(\theta) \triangleq \frac{1}{\#\Theta} \sum_{\hat{\theta} \neq \theta} \underline{y}_i(\hat{\theta}) + \frac{1}{\#\theta} f(\theta).$$

We also define the average allocation $\underline{y}$ of the set $\{\underline{y}_i\}_{i \in \mathcal{I}}$ by setting

$$\underline{y} \triangleq \frac{1}{I} \sum_{i \in \mathcal{I}} \underline{y}_i.$$

Here again, we note that under NWA, for all θ and all i, there exists $y_i^*(\theta)$ such that

$$u_i(y_i^*(\theta), \theta) > u_i(\underline{y}, \theta), \tag{11}$$

where the above inequality clearly holds after defining $y_i^*(\theta)$ as follows:

$$y_i^*(\theta) \triangleq \frac{1}{I}\sum_{j \neq i} \underline{y}_j + \frac{1}{I} y_i(\theta).$$

We now construct an auxiliary set of allocations, denoted by $\{z_i(\theta, \theta')\}_{\theta,\theta'}$, which uses the existence of the allocations $\{\underline{y}_i(\theta)\}_{\theta\in\Theta}$. The allocations $\{z_i(\theta, \theta')\}_{\theta,\theta'}$ are going to appear in the canonical mechanism to be defined shortly where they guarantee the existence of better response for agent i should the remaining agents choose to misreport the true state. In particular, the following lemma establishes that for agent i the allocation $z_i(\theta, \theta')$ represents an improvement if the true state is θ but the other agents misreport it to be θ'. It also establishes that $z_i(\theta, \theta')$ would not constitute an improvement relative to $f(\theta')$ if the true state were indeed θ'.

Lemma 2. *If social choice function f satisfies "no worst alternative" (NWA) then for each player i, there exists a collection of lotteries $\{zi(\theta, \theta')\}_{\theta,\theta'}$ such that for all θ, θ' :*

$$u_i(f(\theta'), \theta') > u_i(z_i(\theta, \theta'), \theta'), \tag{12}$$

and for $\theta \neq \theta'$:

$$u_i(z_i(\theta, \theta'), \theta) > u_i(z_i(\theta', \theta'), \theta). \tag{13}$$

Proof. Based on the allocations $\{\underline{y}_i(\theta)\}_{\theta\in\Theta}$ from Definition 4, we define our collection of lotteries as follows. First, for all θ':

$$z_i(\theta', \theta') \triangleq (1 - \varepsilon)\underline{y}_i(\theta') + \varepsilon \underline{y}_i,$$

with $\underline{y}_i$ as defined in (9), and for all θ, θ' with $\theta \neq \theta'$:

$$z_i(\theta, \theta') \triangleq (1 - \varepsilon)\underline{y}_i(\theta') + \frac{\varepsilon}{\#\Theta}\left(\sum_{\hat{\theta} \neq \theta} \underline{y}_i(\hat{\theta}) + f(\theta)\right).$$

By NWA and the finiteness of the state space Θ. we can find a sufficiently small, but positive, $\varepsilon > 0$ such that for all θ and θ' : $u_i(f(\theta'), \theta') > u_i(z_i(\theta, \theta'), \theta')$ which establishes inequality (12). Now we observe that the only difference between $z_i(\theta', \theta')$ and $z_i(\theta, \theta')$ is the fact that the lottery

$\underline{y}_i(\theta)$ is replaced by the lottery $f(\theta)$. But now by NWA, this is clearly increasing the expected utility of agent i in state θ, and hence we have for θ, θ' with $\theta \neq \theta'$:

$$u_i(z_i(\theta, \theta'), \theta) > u_i(z_i(\theta', \theta'), \theta),$$

which establishes the strict inequality (13). □

We establish the sufficient conditions for implementation in rationalizable strategies by means of a canonical mechanism. The canonical mechanism shares many basic features with the implementation mechanism suggested by Maskin and Sjöström (2004) to establish complete information implementation in the presence of mixed strategies, and is a modification of the original mechanism suggested by Maskin (1999). The aforementioned allocations $\{z_i(\theta, \theta')\}_{\theta, \theta'}$ appear in the mechanism if agent i reports a state θ different from the reported state θ' by all the other agents. In this case, the allocation $z_i(\theta', \theta')$ is chosen with positive probability, yet this probability can be lowered by a suitable message of agent i and be replaced by a more favorable allocation $z_i(\theta, \theta')$. In the proposition below, we show that Maskin monotonicity together with NWA are sufficient for rationalizable implementation. The fact that we do not refer to strict Maskin monotonicity in this statement may seem surprising given that in Proposition 1 we showed that strict Maskin monotonicity is a necessary condition for rationalizable implementation. This is due to the simple fact that under NWA, strict Maskin monotonicity and Maskin monotonicity are equivalent.[5]

Proposition 2. (*Sufficient conditions*). *If $I \geq 3$, f is responsive, satisfies Maskin monotonicity and NWA, then f is implementable in rationalizable strategies.*

Proof. We establish the result by constructing an implementing mechanism $\mathcal{M} = (M, g)$. First, recall that by definition of Maskin monotonicity, for all θ and θ' such that $f(\theta) \neq f(\theta')$, there exist i and $y(\theta, \theta') \in Y$ with

[5]To see this just note that if f is Maskin monotonic then $f(\theta) \neq f(\theta')$ implies the existence of some i and y satisfying $u_i(f(\theta), \theta) \geq u_i(y, \theta)$ and $u_i(y, \theta') > u_i(f(\theta), \theta')$. Now, under NWA, there exists $\underline{y}_i(\theta)$ such that $u_i(f(\theta), \theta) > u_i(\underline{y}_i(0), \theta)$. Now if one sets $\tilde{y} = \varepsilon \underline{y}_i(\theta) + (1 - \varepsilon) y$, for ε small enough we get $u_i(f(\theta), \theta) > u_i(\tilde{y}, \theta)$ and $u_i(\tilde{y}, \theta') > u_i(f(\theta), \theta')$, showing that f is strict Maskin monotonic.

$u_i(f(\theta),\theta) \geq u_i(y(\theta,\theta'),\theta)$ and $u_i(y(\theta,\theta'),\theta') > u_i(f(\theta),\theta')$. We define the following finite set of lotteries:

$$\mathcal{Y} = \{z_i(\theta,\theta')\}_{i,\theta,\theta'} \cup \{y(\theta,\theta')\}_{\{\theta,\theta'|f(\theta)\neq f(\theta')\}} \cup \{y_i^*(\theta)\}_{i,\theta},$$

where the collection $\{z_i(\theta,\theta')\}_{i,\theta,\theta'}$ has been defined in Lemma 2 while the collection $\{y_i^*(\theta)\}_{i,\theta}$ has been established in (11).

Each agent i sends a message $m_i = (m_i^1, m_i^2, m_i^3, m_i^4)$, where $m_i^1 \in \Theta$, $m_i^2 \in \mathbb{Z}_+, m_i^3 : \Theta \to \mathcal{Y}, m_i^4 \in \mathcal{Y}$. The third component of the message profile will allow agent i to suggest an allocation $m_i^3(\theta)$ contingent on all the other agents $j \neq i$ reporting $m_j^1 = \theta$. The outcome function will make use of the "uniformly worse outcome" defined earlier by $\underline{y}$. Now the outcome $g(m)$ is determined by the following rules:

Rule 1. If $m_i^1 = \theta$ and $m_i^2 = 1$ for all i, pick $f(\theta)$.

Rule 2. If there exists $i \in I$ — called the deviating player — such that $(m_j^1, m_j^2) = (\theta, 1)$ for all $j \neq i$ and $(m_i^1, m_i^2) \neq (\theta, 1)$, then we go to two subrules:

(i) if $u_i(f(\theta),\theta) \geq u_i(m_i^3(\theta),\theta)$, pick $m_i^3(\theta)$ with probability $1 - 1/(m_i^2+1)$ and $z_i(\theta,\theta)$ with probability $1/(m_i^2+1)$;

(ii) if $u_i(f(\theta),\theta) < u_i(m_i^3(\theta),\theta)$, pick $z_i(\theta,\theta)$ with probability 1.

Rule 3. In all other cases, we identify a pivotal agent i by requiring that $m_i^2 \geq m_j^2$ for all $j \in I$ and that if for $j \neq i$, $m_i^2 = m_j^2$, then $i < j$. The rule then requires that with probability $1 - 1/(m_i^2+1)$ we pick m_i^4, and with probability $1/(m_i^2+1)$ we pick $\underline{y}$.

Claim 1. *It is never a best reply for agent i to send a message with $m_i^2 > 1$ (i.e., $m_i \in b_i^\theta(\bar{S}) \Rightarrow m_i^2 = 1$).*

Proof of Claim 1. We proceed by contradiction and suppose that $m_i = (m_i^1, m_i^2, m_i^3, m_i^4) \in S_i^{\mathcal{M},\theta}$ and $m_i^2 > 1$. Then for any profile of messages m_{-i} that player i's opponents may play, (m_i, m_{-i}) will trigger either Rule 2 or Rule 3. But in this case, whatever agent i's beliefs $\lambda_i \in \Delta(M_{-i})$ about the other agents' messages, his payoff can be increased by modifying m_i appropriately, in particular by increasing the integer choice from m_i^2. To see this, denote the set of messages of all agents excluding i in which Rule 2 is triggered by

$$M_{-i}^2 \triangleq \{m_{-i} \in M_{-i} | m_j^1 = \theta' \text{ and } m_j^2 = 1 \text{ for some } \theta' \text{ for all } j \neq i\}, \tag{14}$$

and the set of messages of all agents excluding i in which Rule 3 is triggered as the complement set

$$M_{-i}^3 \triangleq M_{-i} \backslash M_{-i}^2. \tag{15}$$

Suppose first that agent i has a belief $\lambda_i \in \Delta(M_{-i})$ under which Rule 3 is triggered with positive probability, so that $\lambda_i(M_{-i}^3) > 0$. Note that if agent i plays m_i, with strictly positive probability $\underline{y}$ is provided. Hence, because from (11), $y_i^*(\theta) \in \mathcal{Y}$ is such that $u_i(y^*(\theta), \theta) > u_i(\underline{y}, \theta)$, i's expected utility conditional on Rule 3 satisfies

$$\sum_{m_{-i} \in M_{-i}^3} \lambda_i(m_{-i}) u_i(g(m_i, m_{-i}), \theta) < \sum_{m_{-i} \in M_{-i}^3} \lambda_i(m_{-i}) \max_{y \in \mathcal{Y}} u_i(y, \theta).$$

Now, if i deviates to $\hat{m}_i = (\hat{m}_i^1, \hat{m}_i^2, \hat{m}_i^3, \hat{m}_i^4)$ where $\hat{m}_i^4 \in \arg\max_{y \in \mathcal{Y}} u_i(y, \theta)$, it is easily checked that i's expected utility conditional on Rule 3 tends to

$$\sum_{m_{-i} \in M_{-i}^3} \lambda_i(m_{-i}) \max_{y \in \mathcal{Y}} u_i(y, \theta)$$

as $\hat{m}_i^2$ tends to infinity. Thus, player i can always improve his expected payoff conditional on Rule 3 by deviating from m_i to $\hat{m}_i$ and announcing $\hat{m}_i^2$ large enough.

Now suppose that agent i believes that Rule 2 will be triggered with positive probability, so that $\lambda_i(M_{-i}^2) > 0$. We again consider a deviation to $\hat{m}_i = (\hat{m}_i^1, \hat{m}_i^2, \hat{m}_i^3, \hat{m}_i^4)$ and observe that the choice of $\hat{m}_i^4$ does not affect the outcome of the mechanism conditional on Rule 2. We also note that for any $m_{-i} \in M_{-i}^2$ such that $\lambda_i(m_{-i}) > 0$, (m_i, m_{-i}) does not trigger Rule 2(ii). Indeed, if it were the case, we would have $u_i(g(m_i, m_{-i}), \theta) = u_i(z_i(m_{-i}^1, m_{-i}^1), \theta)$. We have to distinguish two cases: whether players $j \neq i$ send message θ or not. First, consider the case where $m_{-i}^1 \neq \theta$.[6] Now, player i could change m_i to $\hat{m}_i$ having $\hat{m}_i^3(m_{-i}^1) = z_i(\theta, m_{-i}^1)$ and keeping m_i unchanged otherwise. By Lemma 2, $u_i(f(m_{-i}^1), m_{-i}^1) > u_i(z_i(\theta, m_{-i}^1), m_{-i}^1) = u_i(\hat{m}_i^3(m_{-i}^1), m_{-i}^1)$, and so by construction of the mechanism, $(\hat{m}_i, m_{-i})$ now triggers Rule 2(i). Again using Lemma 2 and

[6] We sometimes abuse notations and write $m_{-i}^1 = \theta$ whenever $m_j^1 = \theta$ for all $j \neq i$.

the fact that $m^1_{-i} \neq \theta$, we get

$$\begin{aligned} u_i(g(m_i, m_{-i}), \theta) = u_i(z_i(m^1_{-i}, m^1_{-i}), \theta) &< (1 - 1/(m_i^2 + 1)) \\ &\times u_i(z_i(\theta, m^1_{-i}), \theta) + (1/(m_i^2 + 1)) \\ &\times u_i(z_i(m^1_{-i}, m^1_{-i}), \theta) = u_i(g(\hat{m}_i, m_{-i}), \theta). \end{aligned}$$

Hence, the expected utility of player i would strictly increase, which yields the contradiction. Consider the second case where $m^1_{-i} = \theta$, player i could change m_i to $\hat{m}_i$ having $\hat{m}_i(m^1_{-i}) = f(\theta)$ and keeping m_i unchanged otherwise. It is clear that by construction of the mechanism, $(\hat{m}_i, m_{-i})$ now triggers Rule 2(i). Since Lemma 2 gives us

$$\begin{aligned} u_i(g(m_i, m_{-i}), \theta) = u_i(z_i(m^1_{-i}, m^1_{-i}), \theta) &< (1 - 1/(m_i^2 + 1))u_i(f(\theta), \theta) \\ &+ (1/(m_i^2 + 1))u_i(z_i(\theta, \theta), \theta) \\ &= u_i(g(\hat{m}_i, m_{-i}, \theta), \end{aligned}$$

the expected utility of player i would strictly increase, which here again yields a contradiction. So now we know that for any $m_{-i} \in M^2_{-i}$ such that $\lambda_i(m_{-i}) > 0$, (m_i, m_{-i}) does not trigger Rule 2(ii). Using a similar reasoning, it is easily shown that for any $m_{-i} \in M^2_{-i}$ such that $\lambda_i(m_{-i}) > 0$, we must have $u_i(m_i^3(m^1_{-i}), \theta) > u_i(z_i(m^1_{-i}, m^1_{-i}), \theta)$, hence the expected payoff conditional on Rule 2 from m_i:

$$\begin{aligned} \sum_{m_{-i} \in M^2_{-i}} &\lambda_i(m_{-i})[(1 - 1/(m_i^2 + 1))u_i(m_i^3(m^1_{-i}), \theta) \\ &+ (1/(m_i^2 + 1))u_i(z_i(m^1_{-i}, m^1_{-i}), \theta)] \end{aligned}$$

is strictly increasing in m_i^2. It follows that the choice of $\hat{m}_i$ with $\hat{m}_i^2$ large and strictly larger than m_i^2 strictly improves the expected utility of agent i if either Rule 2 or 3 is triggered, which yields the desired contradiction. □

Claim 2. *$(\theta, 1, m_i^3, m_i^4) \in S_i^{\mathcal{M}, \theta}$ for all i, θ, m_i^3, m_i^4.*

Proof of Claim 2. Suppose that player i in state θ puts probability 1 on each other agent j sending a message of the form $(\theta, 1, m_j^3, m_j^4)$. If player i announces a message of the form $(\theta, 1, m_i^3, m_i^4)$, he gets payoff $u_i(f(\theta), \theta)$. If he announces a message not of this form, the outcome is determined by Rule 2. Since by Lemma 2, $u_i(z_i(\theta, \theta), \theta) < u_i(f(\theta), \theta)$, it is clear that by construction of the mechanism, his payoff from invoking Rule 2 is bounded above by $u_i(f(\theta), \theta)$. □

Claim 3. *If* $m_i = (\theta', 1, m_i^3, m_i^4) \in S_i^{\mathcal{M},\theta}$, *then* $\theta' = \theta$.

Proof of Claim 3. Suppose $m_i = (\theta', 1, m_i^3, m_i^4) \in S_i^{\mathcal{M},\theta}$. Given the message m_i, we can define the set of messages of the remaining agents which trigger Rule 1, 2 or 3, respectively. In particular, we define M_{-i}^1 as the set of $m_{-i} \in M_{-i}$ such that (m_i, m_{-i}) triggers Rule 1. Similarly, $M_{-i}^{2,i}$ is defined as the set of $m_{-i} \in M_{-i}$ such that (m_i, m_{-i}) triggers Rule 2 where player i is the deviating player. Now consider a given belief λ_i of agent i. If $\lambda_i(\{m_{-i} \in M_{-i}^1\}) = 0$, then Rule 2 or 3 will be triggered with probability one. Although, Rule 2 can now be triggered with a "deviating player" being different of i, it is easily checked that a similar argument as in Claim 1 applies and so the message m_i cannot be a best reply by agent i. Suppose now that the belief λ_i of agent i is such that:

$$0 < \lambda_i(\{m_{-i} \in M_{-i}^1\}) < 1. \tag{16}$$

While we still argue that agent i can strictly increase his expected utility by selecting an integer $\hat{m}_i^2 > 1$, we observe that a complication arises as with λ_i given by (16), a choice of $\hat{m}_i^2 > 1$ leads from an allocation determined by Rule 1 to an allocation determined by Rule 2, and hence the realization of an unfavorable allocation $\underline{y}$ with positive probability. But now we observe that by selecting $\hat{m}_i$ such that:

$$\hat{m}_i^3(\hat{\theta}) = \begin{cases} f(\theta'), & \text{if } \hat{\theta} = \theta', \\ m_i^3(\hat{\theta}), & \text{if otherwise,} \end{cases}$$

$\hat{m}_i^4 \in \arg\max_{y \in \mathcal{Y}} u_i(y, \theta)$ and by choosing an integer $\hat{m}_i^2$ sufficiently large, the small loss in Rule 2 can always be offset by a gain in Rule 3 relative to the allocation achieved under $g(m_i, m_{-i})$. More formally, for $m_i = (\theta', 1, m_i^3, m_i^4)$, since $0 < \lambda_i(\{m_{-i} \in M_{-i}^1\}) < 1$ and since — as claimed before — for all $m_{-i} \in M_{-i}^{2,i}$ such that $\lambda_i(m_{-i}) > 0$, $u_i(m_i^3(m_{-i}^1), \theta) > u_i(z_i(m_{-i}^1, m_{-i}^1), \theta)$, i's expected payoff from playing m_i is strictly lower than

$$\sum_{m_{-i} \in M_{-i}^1} \lambda_i(m_{-i}) u_i(f(\theta'), \theta) + \sum_{m_{-i} \in M_{-i}^{2,i}} \lambda_i(m_{-i}) u_i(m_i^3(m_{-i}^1), \theta)$$
$$+ \sum_{m_{-i} \notin M_{-i}^1 \cup M_{-i}^{2,i}} \lambda_i(m_{-i}) \max_{y \in \mathcal{Y}} u_i(y, \theta)$$

while for $\hat{m}_i = (\theta', \hat{m}_i^2, \hat{m}_i^3, \hat{m}_i^4)$, it is easily checked that as $\hat{m}_i^2$ tends to infinity, i's expected payoffs tend toward the expression above. Hence, choosing

$\hat{m}_i^2$ large enough, $\hat{m}_i$ is a better response against λ_i for player i than m_i, a contradiction.

So if $m_i = (\theta', 1, m_i^3, m_i^4) \in S_i^{\mathcal{M},\theta}$, it follows player i must be convinced that each other player must be choosing a message of the form $(\theta', 1, m_j^3, m_j^4)$, and hence $\lambda_i(\{m_{-i} \in M_{-i}^1\}) = 1$. Thus there must exist a message of the form $m_j = (\theta', 1, m_j^3, m_j^4) \in S_j^{\mathcal{M},\theta}$ for all j. Now, proceed by contradiction and assume that $f(\theta') \neq f(\theta)$. By Maskin monotonicity, we know that there exist j with $u_j(f(\theta'), \theta') \geq u_j(y(\theta', \theta), \theta')$ and $u_j(y(\theta', \theta), \theta') > u_j(f(\theta'), \theta)$. By the above argument, we know that player j's belief against which $m_j = (\theta', 1, m_j^3, m_j^4)$ is a best reply assigns probability one to each player $l \neq j$ sending a message of the form $m_l = (\theta', 1, m_l^3, m_l^4)$. Hence, player j's expected payoff from playing m_j is $u_j(f(\theta'), \theta)$, while if j deviates to $\hat{m}_j = (\theta', \hat{m}_j^2, \hat{m}_j^3, \hat{m}_j^4)$, where $\hat{m}_j^2 > 1$ and

$$\hat{m}_j^3(\hat{\theta}) = \begin{cases} y(\theta', \theta), & \text{if } \hat{\theta} = \theta', \\ m_i^3(\hat{\theta}), & \text{if otherwise,} \end{cases}$$

player j believes with probability one that Rule 2(i) will be triggered. Hence, player j's expected payoff would be

$$(1 - 1/(\hat{m}_j^2 + 1))u_j(y(\theta', \theta), \theta) + (1/(\hat{m}_j^2 + 1))u_j(z_j(\theta', \theta'), \theta).$$

Note that as $\hat{m}_j^2$ tends to infinity, this expression tends to $u_j(y(\theta', \theta), \theta)$ which is strictly larger than $u_j(f(\theta'), \theta)$. Hence for $\hat{m}_j^2$ large enough, $\hat{m}_j$ is better response for player j than $(\theta', 1, m_j^3, m_j^4)$, a contradiction. Thus $f(\theta') = f(\theta)$. Since the social choice function has been assumed to be responsive, we get $\theta' = \theta$ as claimed. □

Completion of proof. Claims 1–3 together imply that for each $\theta : S_i^{\mathcal{M},\theta} \neq \varnothing$ and $m_i \in S_i^{\mathcal{M},\theta} \Rightarrow m_i^2 = 1$ and $m_i^1 = 0$. Thus $S^{\mathcal{M},\theta} \neq \varnothing$ and $m \in S^{\mathcal{M},\theta} \Rightarrow g(m) = f(\theta)$. □

The mechanism $\mathcal{M}$ used here allows each agent to propose a menu of choices $m_i^3 = \{m_i^3(\theta)\}_{\theta \in \Theta}$. The menu m_i^3 gives agent i the opportunity to select an appropriate allocation in case that Rule 2 is triggered. In our sufficiency argument, the NWA property replaces the no veto property which commonly appears in the sufficiency argument for implementation in Nash equilibrium. Yet, in terms of the proof, the role of the NWA property is quite distinct from the no veto property. The NWA property guarantees that in the augmented mechanism, any report in state θ in which an agent expresses his disagreement with the remaining agents (i.e., $m_i^2 > 1$) cannot be a rationalizable report. By contrast, the no veto property guaranteed that if an agent were to express his disagreement, then further disagreement by other agents would only be possible in equilibrium if it would lead to the same equilibrium allocation as prescribed $f(\theta)$.

We note that our mechanism not only implements in rationalizable messages but also implements in Nash equilibrium (the proof of Claim 2 above indeed establishes the existence of a pure Nash equilibrium at each state). In recent work, Bochet (2007) and Benoit and Ok (2008) report sufficient conditions for implementation in Nash equilibrium strategies using stochastic mechanisms. In both contribution, the principal uses randomized allocation only out of equilibrium while pure alternatives are always chosen in equilibrium. Their conditions, the *top strict difference* condition and the *top coincidence* condition, respectively, do not imply nor are they implied by the NWA property required for sufficiency. In related work, Serrano and Vohra (2007) have used stochastic implementing mechanisms to provide weak sufficient conditions for Bayesian implementation in *mixed strategy* Bayes Nash equilibrium.

4 The Non-responsive Case

In this section, we discuss extensions of our results to the cases when the social choice function is not responsive. We will provide a strengthening of strict Maskin monotonicity that can be shown to be sufficient (together with a strengthening of the NWA) even if the social choice function is not responsive. We also show that the strengthening of strict Maskin monotonicity is actually necessary for rationalizable implementation given a weak condition on the class of mechanisms to be considered. This weak condition is trivially satisfied when the social choice function is responsive.

Now, given a social choice function f, let us consider the unique partition of $\Theta : P_f = \{\Theta_z\}_{z \in f(\Theta)}$ such that

$$\Theta_z = \{\theta \in \Theta | f(\theta) = z\}. \tag{17}$$

We now introduce the following notion which reduces to strict Maskin monotonicity in case f is responsive.

Definition 5. (*Strict Maskin monotonicity**). Social choice function f satisfies strict Maskin *monotonicity** if there exists a partition $\mathcal{P}$ of Θ finer than $\mathcal{P}_f$ such that for any θ:

1. $\theta' \in \mathcal{P}(\theta)$ whenever for all i and y

$$[\text{for all } \hat{\theta} \in \mathcal{P}(\theta) : u_i(f(\theta), \hat{\theta})] > u_i(y, \hat{\theta})] \Rightarrow [u_i(f(\theta), \theta') \geq u_i(y, \theta')]; \tag{18}$$

   or, equivalently,
2. $\theta' \notin \mathcal{P}(\theta)$ implies for some i and y

$$u_i(y, \theta') > u_i(f(\theta), \theta') \quad \text{and} \quad u_i(f(\theta), \hat{\theta}) > u_i(y, \hat{\theta}) \text{ for all } \hat{\theta} \in \mathcal{P}(\theta). \tag{19}$$

Before we establish the necessary and sufficient conditions, we briefly describe the complications that arise with a non-responsive social choice function. By definition, under a non-responsive social choice function there are at least two states, θ and θ', that lead to the same social choice: $f(\theta) = f(\theta') = z$. Now, a priori, the principal would not need to know whether it is the state θ or θ' which leads to the realization of the social choice z. In fact, it would appear that it would be sufficient to learn that the realized state belongs to the set Θ_z of states which lead to the social choice z. Now, such a coarse reporting protocol as suggested by the above partition $\mathcal{P}_f$ would be sufficient if the agents were known to report truthfully, yet a problem arises if they might not report truthfully. For, if an agent now alleges collusive behavior of the remaining agents, the principal may lack the information to verify whether the whistle-blower himself is behaving in an incentive compatible manner. After all, the principal would merely know that the reported state is in some set Θ_z but would not know the identity of the state itself. Thus, while it might not be useful to distinguish between any two states $\theta, \theta' \in \Theta_z$ if the agents were to report truthfully, it might be critical to distinguish between θ and θ' in order to fend off undesirable equilibrium play by the agents. This discussion might therefore suggest that the inequalities (18), or alternatively (19), should be satisfied for the finest possible partition of states. But, as we argue next, such a condition would (i) require too much to constitute a necessary condition, and (ii) be impossible to satisfy by any implementing mechanism.

The first observation is straightforward to establish. Consider for the moment the strict Maskin monotonicity* condition in the version of (19), which we might refer to as the whistle-blower inequality. Now suppose that the social choice problem is such that the inequalities (19) are satisfied even for the coarse partition $\mathcal{P}_f$ itself. In this case, we would find that the principal would not need to distinguish between any two states $\theta, \theta' \subset \Theta_z$, either for truthtelling or, by condition (19), for whistle-blowing behavior.

The second observation stems from an earlier result. Lemma 1 gave a sufficient condition under which the set of rationalizable actions for any pair of states, θ and θ', have to be identical for all agents. For the purpose here we can restrict attention to any two states with $\theta, \theta' \in \Theta_z$. In this case, the condition (1) reads as follows:

$$u_i(z, \theta) > u_i(y, \theta) \Rightarrow u_i(z, \theta') \geq u_i(y, \theta') \quad \text{for all } i \text{ and } y.$$

In words, if for every agent i, the upper contour set (relative to the allocation $f(\theta) = f(\theta') = z$), at one state, say θ', is included in the upper contour set of the other state, say θ, then the sets of rationalizable actions have to coincide. But of course, once the sets of rationalizable actions have to agree, it will be impossible to distinguish behavior in state θ from behavior in state θ'. The inclusion property of the upper contour sets, given by condition (1), thus imposes an upper bound on how fine the partition $\mathcal{P}$ can be chosen while remaining compatible with rationalizable behavior. We finally observe that the partition $\mathcal{P}$ may yet have to be coarser than is indicated by the pairwise inclusion property. To see this, consider $\theta, \theta', \theta'' \in \Theta_z$, and suppose that the upper contour sets (relative to the allocation z) in state θ' as well as in state θ'' are included in the upper contour sets in state θ, but that the upper contour sets in the state θ' and θ'' themselves do not display an inclusive relationship. Now, Lemma 1 tells us that $S^{\mathcal{M},\theta} = S^{\mathcal{M},\theta'}$ and that $S^{\mathcal{M},\theta} = S^{\mathcal{M},\theta''}$ which of course implies that $S^{\mathcal{M},\theta'} = S^{\mathcal{M},\theta''}$ even though the condition (1) does not apply to the states θ' and θ'' themselves.

As we already stated, we can prove that strict Maskin monotonicity* is necessary under a weak condition on the class of mechanisms we consider. This condition states that for any state θ and any rationalizable message m_i of any player i in this state, the message m_i is also best-response to some belief with support in the set of rationalizable actions of the other players and for any state $\hat{\theta}$ such that $S^{\mathcal{M},\hat{\theta}} = S^{\mathcal{M},\theta}$, best responses against this belief are non-empty.

Definition 6. (*The best-response property*). Given a social choice function f, a mechanism $\mathcal{M}$ has the *best-response property* if for all θ and all $m_i \in S_i^{\mathcal{M},\theta}$, there exists $\lambda_i^{(m_i,\theta)} \in \Delta(M_{-i})$ satisfying $\lambda_i^{(m_i,\theta)} > 0 \Rightarrow m_j \in S_j^{\mathcal{M},\theta}$

for each $j \neq i$, and such that m_i is a best response against $\lambda_i^{m_i,\theta}$ in state θ and

$$\arg\max_{m_i'} \sum_{m_{-i}} \lambda_i^{m_i,\theta}(m_{-i}) u_i(g(m_i', m_{-i}), \hat{\theta}) \neq \emptyset$$

for all $\hat{\theta}$ such that $S^{\mathcal{M},\hat{\theta}} = S^{\mathcal{M},\theta}$.

Note that if f is responsive then any implementing mechanism must satisfy $S^{\mathcal{M},\hat{\theta}} = S^{\mathcal{M},\theta} \Rightarrow \hat{\theta} = \theta$ and so any implementing mechanism must have the best-response property. Moreover, the best-response property also holds for any pair θ, θ' which are directly related through the inclusion property (1). The best-response property then secures that it applies also to profiles which are indirectly related as in the example of $\theta, \theta', \theta'' \in \Theta_z$ discussed above. Hence, the subsequent Proposition 3 generalizes Proposition 1 above.

Proposition 3. (*Necessary conditions*). *If f is implementable in rationalizable strategies by a mechanism $\mathcal{M}$ having the best-response property, then f satisfies strict Maskin monotonicity**.

In order to show this, we prove the following lemma that generalizes Lemma 1.

Lemma 3. *Assume the existence of a mechanism $\mathcal{M} = ((M_i)_{i=1}^I, g)$, that has the best-response property and that implements f in rationalizable strategies. Pick θ and θ' satisfying condition (18) where the partition $\mathcal{P}$ is assumed to be $\mathcal{P}(\theta'') = \{\tilde{\theta} \in \Theta | S^{\mathcal{M},\theta''} = S^{\mathcal{M},\tilde{\theta}}\}$ for any θ''. We have $S^{\mathcal{M},\theta} = S^{\mathcal{M},\theta'}$.*

Proof. Fix any mechanism $\mathcal{M} = ((M_i)_{i=I}^I, g)$ that has the best-response property and that implements f and pick θ and θ' satisfying condition (18) for $\mathcal{P}(\theta'') = \{\tilde{\theta} \in \Theta | S^{\mathcal{M},\theta''} = S^{\mathcal{M},\tilde{\theta}}\}$ i.e., for all i and y

$$[\text{for all } \hat{\theta} \in \mathcal{P}(\theta) : u_i(f(\theta), \hat{\theta}) > u_i(y, \hat{\theta})] \Rightarrow [u_i(f(\theta), \theta') \geq u_i(y, \theta')].$$

Note that by construction, $\hat{\theta} \in \mathcal{P}(\theta) \Rightarrow S^{\mathcal{M},\hat{\theta}} = S^{\mathcal{M},\theta}$. In addition, since $\mathcal{M}$ implements f in rationalizable strategies, for any state in $\mathcal{P}(\theta'')$, f picks the outcome $f(\theta'')$ and so $\mathcal{P}$ is finer than $\mathcal{P}_f$.

We first show that $S^{\mathcal{M},\theta} \subseteq S^{\mathcal{M},\theta'}$. Because $b^{\theta}(S^{\mathcal{M},\theta}) = S^{\mathcal{M},\theta}$, $S^{\mathcal{M},\theta}$ has the best-response property in state θ i.e., for all player i and all $m_i \in S_i^{\mathcal{M},\theta}$, there exists $\lambda_i^{m_i,\theta} \in \Delta(M_{-i})$ such that $\lambda_i^{m_i,\theta}(m_{-i}) > 0 \Rightarrow m_j \in S_j^{\mathcal{M},\theta}$ for each $j \neq i$, and

$$\sum_{m_{-i} \in M_{-i}} \lambda_i^{m_i,\theta}(m_{-i}) u_i(g(m_i, m_{-i}), \theta) \geq \sum_{m_{-i} \in M_{-i}} \lambda_i^{m_i,\theta}(m_{-i}) u_i(g(m_i', m_{-i}), \theta)$$

for all $m_i' \in M_i$. In addition, since $\mathcal{M}$ has the best-response property, $\lambda_i^{m_i,\theta}$ can be chosen so that:

$$\arg\max_{m_i'} \sum_{m_{-i}} \lambda_i^{m_i,\theta}(m_{-i}) u_i(g(m_i', m_{-i}), \hat{\theta}) \neq \emptyset,$$

for all $\hat{\theta} \in \mathcal{P}(\theta)$. This in turn implies that for all $\hat{\theta} \in \mathcal{P}(\theta)$:

$$\sum_{m_{-i} \in M_{-i}} \lambda_i^{m_i,\theta}(m_{-i}) u_i(g(m_i, m_{-i}), \hat{\theta}) \geq \sum_{m_{-i} \in M_{-i}} \lambda_i^{m_i,\theta}(m_{-i}) u_i(g(m_i', m_{-i}), \hat{\theta}) \tag{20}$$

for all $m_i' \in M_i$. To see this, observe that if it were not true, we would have for some $\hat{\theta} \in \mathcal{P}(\theta)$:

$$\sum_{m_{-i}} \lambda_i^{m_i,\theta}(m_{-i}) u_i(g(m_i^*, m_{-i}), \theta) > \sum_{m_{-i} \in M_{-i}} \lambda_i^{m_i,\theta}(m_{-i}) u_i(g(m_i, m_{-i}), \hat{\theta}) = u_i(f(\theta), \hat{\theta})$$

where m_i^* denotes a best response to $\lambda_i^{m_i,\theta}$ in state $\hat{\theta}$ and so $g(m_i^*, m_{-i}) \neq f(\theta)$ for some $(m_i^*, m_{-i}) \in S^{\mathcal{M},\hat{\theta}}$ which contradicts the fact that $\mathcal{M} = ((M_i)_{i=1}^I, g)$ implements f.

Now, we want to show that m_i is also a best response against $\lambda_i^{m_i,\theta}$ in state θ'. Since i and $m_i \in S_i^{\mathcal{M},\theta}$ have been fixed arbitrarily, this will prove that $S^{\mathcal{M},\theta}$ has thc best-response property in state θ' and so that $S^{\mathcal{M},\theta} \subseteq S^{\mathcal{M},\theta'}$ as claimed. Note first that for any m_{-i} such that $\lambda_i^{m_i,\theta}(m_{-i}) > 0$,

$m_{-i} \in S_{-i}^{\mathcal{M},\theta}$ and so because $m_i \in S_i^{\mathcal{M},\theta}$, we have $g(m_i, m_{-i}) = f(\theta)$. Thus,

$$\sum_{m_{-i} \in M_{-i}} \lambda_i^{m_i,\theta}(m_{-i}) u_i(g(m_i, m_{-i}), \theta')$$
$$= u_i(f(\theta), \theta') = \sum_{m_{-i} \in M_{-i}} \lambda_i^{m_i,\theta}(m_{-i}) u_i(g(m_i', m_{-i}), \theta') \quad (21)$$

for all $m_i' \in S_i^{\mathcal{M},\theta}$. In addition, we claim that for all $\hat{\theta} \in \mathcal{P}(\theta)$:

$$\sum_{m_{-i} \in M_{-i}} \lambda_i^{m_i,\theta}(m_{-i}) u_i(g(m_i, m_{-i}), \hat{\theta})$$
$$= u_i(f(\theta), \hat{\theta}) > \sum_{m_{-i} \in M_{-i}} \lambda_i^{m_i,\theta}(m_{-i}) u_i(g(m_i', m_{-i}), \theta') \quad (22)$$

for any $m_i' \notin S_i^{\mathcal{M},\theta}$. Indeed, by (20), the above is true with a weak inequality. Now if an equality were to hold, some $m_i' \notin S_i^{\mathcal{M},\theta}$ would be a best response against $\lambda_i^{m_i,\theta}$ in some state $\hat{\theta}$. Thus the set $(\{m_i'\} \cup S_i^{\mathcal{M},\hat{\theta}}) \times S_{-i}^{\mathcal{M},\hat{\theta}} = (\{m_i'\} \cup S_i^{\mathcal{M},\theta}) \times S_{-i}^{\mathcal{M},\theta}$ would have the best-response property in state $\hat{\theta}$ implying that $m_i' \in S_i^{\mathcal{M},\hat{\theta}} = S_i^{\mathcal{M},\theta}$ which is false by assumption.

Now, by assumption, we know that

$$[\hat{\theta} \in \mathcal{P}(\theta) : u_i(f(\theta), \hat{\theta}) > u_i(y, \hat{\theta})]$$
$$\Rightarrow [u_i(f(\theta), \theta') \geq u_i(y, \theta')] \quad \text{for all } i \text{ and } y$$

and so applying this to the lotteries $y \triangleq \sum_{m_{-i}} \lambda_i^{m_i,\theta}(m_{-i}) g(m_i', m_{-i})$, Eq. (22) yields

$$\sum_{m_{-i} \in M_{-i}} \lambda_i^{m_i,\theta}(m_{-i}) u_i(g(m_i, m_{-i}), \theta')$$
$$= u_i(f(\theta), \theta') \geq \sum_{m_{-i} \in M_{-i}} \lambda_i^{m_i,\theta}(m_{-i}) u_i(g(m_i', m_{-i}), \theta') \quad (23)$$

for any $m_i' \notin S_i^{\mathcal{M},\theta}$. Finally, (21) and (23) ensure that m_i is also a best response against $\lambda_i^{m_i,\theta}$ in state θ'. Hence, $S^{\mathcal{M},\theta} \subseteq S^{\mathcal{M},\theta'}$ as claimed.

To complete the proof, we have to show that $S^{\mathcal{M},\theta} \supseteq S^{\mathcal{M},\theta'}$. The argument is the same as in Lemma 1. □

Note that if f is implementable by a mechanism $\mathcal{M}$ that has the best-response property, then if one were to pick the partition $\mathcal{P}$ given by $\mathcal{P}(\theta'') = \{\tilde{\theta} \in \Theta | S^{\mathcal{M},\theta''} = S^{\mathcal{M},\tilde{\theta}}\}$ for any θ'', then whenever θ and θ' satisfy condition (18), by Lemma 3, we must have $S^{\mathcal{M},\theta} = S^{\mathcal{M},\theta'}$ and so, by definition, $\theta' \in \mathcal{P}(\theta)$. Hence, Proposition 3 is obtained as a corollary of Lemma 3.

As mentioned earlier, it is easily checked that our sufficiency argument can be extended to this setting provided that a strengthening of NWA is used. To be more specific, if one assumes that f is (strict) Maskin monotonic and that for any state θ, there exists some outcome that is worse than the outcome selected by f at any state in the partition cell $\mathcal{P}(\theta)$, then we can build a mechanism similar to the one built in the proof of Proposition 2.[7] In the revised mechanism each player is asked to report a partition cell $\boldsymbol{P}$ in $\mathcal{P}$, an integer, a mapping from $\mathcal{P}$ to $\mathcal{Y}$ and a lottery in $\mathcal{Y}$. Essentially, everything would go as if we were replacing each state θ by the partition cell containing θ. In particular, as in the responsive case, we can show that for any rationalizable message, using condition (19) in the definition of Maskin monotonicity*, each agent will report truthfully, i.e., will report $\mathcal{P}(\theta)$ whenever the true state is θ and announce an integer equal to 1. The modified notions of strict Maskin monotonicity and NWA as well as the sufficiency argument itself are presented in detail in the working paper version (see Serrano and Vohra (2010)). The sufficiency theorem is stated and established as Proposition 4 of the working paper.

5 Concluding Remarks

Ordinal and cardinal preferences. The canonical approach (e.g. Maskin (1999) of implementation with complete information associates to each state a profile of ordinal preferences and does not introduce any cardinal representation. The main reason for this is the use of (pure) Nash equilibrium as the solution concept. For this solution concept, there is no need to specify cardinal representations. However, once we change the solution concept to rationalizability, we need to specify profiles of von Neumann–Morgenstern utility functions. As already mentioned, we adopted a cardinal approach i.e., we associate to each state a single cardinal representation. Another way to proceed would be to maintain an ordinal approach by associating to each state ordinal preferences and require rationalizable implementation to be achieved irrespective of the cardinal representation used at this state. This is for instance the approach followed in Mezzetti and Renou (2010). Let us now be more formal and identify a state θ with a profile of ordinal preferences $(\succsim_i^\theta)_i$ over Z. We say that $\mathbf{u} = (u_i(\cdot,\theta))_{i,\theta}$ is a (cardinal) representation of $(\succsim_i^\theta)_{i,\theta}$ if $[u_i(z,\theta) \geq u_i(z',\theta) \Leftrightarrow z \succsim_i^\theta z']$ for each player

[7] As for Proposition 2, the proof would go through if we just considered Maskin monotonicity* instead of strict Maskin monotonicity*.

i and state θ. In the sequel, we write $\max_i^\theta Z$ for $\{z \in Z | z \succsim_i^\theta z' \text{ for all } z' \in Z\}$, in the same way $\min_i^\theta Z$ stands for $\{z \in Z | z' \succsim_i^\theta z \text{ for all } z' \in Z\}$. We then say that a social choice function f is *implementable in rationalizable strategies independently of the cardinal representation* if there exists a mechanism $\mathcal{M}$ such that, for any cardinal representation $\mathbf{u}$ of $(\succsim_i^\theta)_{i,\theta}$, for all θ, $S^{\mathcal{M},\theta} \neq \varnothing$ and $m \in S^{\mathcal{M},\theta} \Rightarrow g(m) = f(\theta)$. In the cardinal case, we know by Proposition 1 that going from the solution concept of Nash equilibrium to rationalizability implies that the usual necessary condition (Maskin monotonicity) is strengthened to its strict version. The following result is then a straightforward corollary of Proposition 1.

Corollary 1. (*Necessary condition for all cardinal representations*). *If f is* implementable in rationalizable strategies independently of the cardinal representation, *then f satisfies strict Maskin monotonicity for any cardinal representation.*

We now present a monotonicity notion in the ordinal preference setting, namely strict set monotonicity.

Definition 7. (*Strict set monotonicity*). A social choice function $f : \Theta \to Z$ satisfies strict set *monotonicity* if $f(\theta) = f(\theta')$ whenever for all i, either $f(\theta) \in \min_i^\theta Z$ or $f(\theta) \in \max_i^{\theta'} Z$ or the following two conditions are satisfied:

$$f(\theta) \succ_i^\theta z \Rightarrow f(\theta) \succ_i^{\theta'} z \quad \text{for all } z \in Z,$$

and

$$f(\theta) \succsim_i^\theta z \Rightarrow f(\theta) \succsim_i^{\theta'} z \quad \text{for all } z \in Z.$$

The next result establishes that strict Maskin monotonicity for any cardinal representation has a close analogue when we consider the ordinal setting, namely strict set monotonicity. The central condition of our characterization can hence be mapped into the purely ordinal setting in a natural way.

Proposition 4. (*Strict Maskin and strict set monotonicity*). *Assume Z is finite. A social choice function $f : \Theta \to Z$ satisfies strict Maskin monotonicity for any cardinal representation if and only if it is strict set monotone.*

Proof. Assume f is strict Maskin monotone for any cardinal representation. Pick θ and θ' s.t. for all i : $f(\theta) \in \min_i^\theta Z$ or $f(\theta) \in \max_i^{\theta'} Z$ or the following two conditions are satisfied:

$$f(\theta) \succ_i^\theta z \Rightarrow f(\theta) \succ_i^{\theta'} z \quad \text{for all } z \in Z, \tag{24}$$

and

$$f(\theta) \succsim_i^\theta z \Rightarrow f(\theta) \succsim_i^{\theta'} z \quad \text{for all } z \in Z. \tag{25}$$

We have to show that $f(\theta) = f(\theta')$. In order to prove this, we will show that there is a cardinal representation $\mathbf{u} = (u_i(\cdot, \tilde{\theta}))_{i,\tilde{\theta}}$ of $(\succsim_i^{\tilde{\theta}})_{i,\tilde{\theta}}$ s.t.

$$u_i(f(\theta), \theta) > \sum_{z\in Z} y(z) u_i(z, \theta)$$
$$\Rightarrow u_i(f(\theta), \theta') \geq \sum_{z\in Z} y(z) u_i(z, \theta') \quad \text{for all} \quad i \quad \text{and} \quad y \in Y.$$

Pick an arbitrary player i. In the case of $f(\theta) \in \min_i^\theta Z$ or $f(\theta) \in \max_i^{\theta'} Z$, it is clear that any representation satisfies the implication. Now, consider the case where the conditions (24) and (25) both hold. By the proof of Theorem 1 in Mezzetti and Renou (2010), there exists a cardinal representation such that $u_i(f(\theta), \theta) = u_i(f(\theta), \theta')$ and $u_i(z, \theta) \leq u_i(z, \theta')$ for all $z \in Z$. The proof then follows directly.

Suppose that f is strict set monotonic. Pick θ and θ' and a cardinal representation $\mathbf{u} = (u_i(\cdot, \tilde{\theta}))_{i,\tilde{\theta}}$ of $(\succsim_i^{\tilde{\theta}})_{i,\tilde{\theta}}$ such that

$$u_i(f(\theta), \theta) > \sum_{z\in Z} y(z) u_i(z, \theta) \Rightarrow u_i(f(\theta), \theta') \geq \sum_{z\in Z} y(z) u_i(z, \theta'). \tag{26}$$

We have to show that $f(\theta) = f(\theta')$. Now, pick an arbitrary player i. If $f(\theta) \in \min_i^\theta Z$ or $f(\theta) \in \max_i^{\theta'} Z$, strict set monotonicity of f completes the proof. So assume that $f(\theta) \notin \min_i^\theta Z$ or $f(\theta) \notin \max_i^{\theta'} Z$. We show that (i) $f(\theta) \succsim_i^\theta z \Rightarrow f(\theta) \succsim_i^{\theta'} z$ for all $z \in Z$ and (ii) $f(\theta) \succ_i^\theta z \Rightarrow f(\theta) \succ_i^{\theta'} z$ for all $z \in Z$. Note first that condition (26) clearly implies that $f(\theta) \succ_i^\theta z \Rightarrow f(\theta) \succsim_i^{\theta'} z$ for all $z \in Z$. If (i) were not true, there would exist $z \sim_i^\theta f(\theta)$ and $z \succ_i^{\theta'} f(\theta)$. Choose $\underline{z} \in \min_i^\theta Z$ and consider the lottery $(1-\varepsilon)z + \varepsilon\underline{z}$. For any $\varepsilon > 0$, this lottery is strictly dominated by $f(\theta)$ at θ since $f(\theta) \notin \min_i^\theta Z$. However, at θ', for $\varepsilon > 0$ small enough, $(1-\varepsilon)z + \varepsilon\underline{z}$ is strictly preferred to $f(\theta)$ at θ', this contradicts (26). So, (i) must be true. Secondly, if (ii) were not true, there would exist $f(\theta) \succ_i^\theta z$ and $z \succ_i^{\theta'} f(\theta)$. Choose $\bar{z} \in \max_i^{\theta'} Z$ and consider the lottery $(1-\varepsilon)z + \varepsilon\bar{z}$. Clearly, for any $\varepsilon > 0$, the lottery $(1-\varepsilon)z + \varepsilon\bar{z}$ is strictly preferred to $f(\theta)$ at θ' since $f(\theta) \notin \max_i^{\theta'} Z$. But now for $\varepsilon > 0$ small enough, $f(\theta)$ is strictly preferred to $(1-\varepsilon)z + \varepsilon\bar{z}$ at state θ, this contradicts (26). So, (ii) must be true. □

We note that the necessary (and sufficient under additional noveto requirements) condition for implementation in *mixed Nash equilibrium* by Mezzetti and Renou (2010) is strictly weaker than our necessary condition.[8]

Social choice sets and social choice correspondences. So far, we restricted our attention to the analysis to social choice functions. We now discuss extension of the results to social choice correspondences and social choice sets. A social choice correspondence defines a set of permissible allocations and rationalizability is a set-based solution concept. Thus, there are a number of plausible extensions of the definition of rationalizable implementation to social choice correspondences. The extensions basically vary to the extent that one wishes to restrict attention to selections in the set of outcome profiles. We first note that the concept of social choice correspondence prevalent in the complete information literature (see Maskin (1999) differs from the concept of a social choice set, prevalent in the incomplete information literature (see Palfrey and Srivastava (1989) and Jacksonm (1991). Our results for social choice functions extend immediately to social choice sets. Indeed, say that a social choice set $F = \{f | f : \Theta \to Y\}$ is implementable if and only if for each element f of F, there is a mechanism that implements f.[9] Then the above necessary and sufficient conditions for social choice functions can be directly applied to provide necessary as well as sufficient conditions for the implementation of a social choice set.

In contrast, the analysis would change substantially if we sought to obtain necessary or sufficient conditions for rationalizable implementation

---

[8]The central concept in Mezzetti and Renou (2010) is the notion of weak set monotonicity. A social choice function f is weak set monotonic if $f(\theta) = f(\theta')$ whenever for all i: either $f(\theta) \in \max_i^{\theta'} X$ or the following two conditions are satisfied:

$$f(\theta) \succ_i^{\theta} z \Rightarrow f(\theta) \succ_i^{\theta'} z \quad \text{for all } z \in Z,$$

and

$$f(\theta) \succsim_i^{\theta} z \Rightarrow f(\theta) \succsim_i^{\theta'} z \quad \text{for all } z \in Z.$$

It can be shown, along the lines of the proof of Proposition 4, that weak set monotonicity is equivalent to Maskin monotonicity for any cardinal representation. We are grateful to Ludovic Renou for discussions clarifying the relation between these monotonicity conditions.

[9]This notion of implementation of the social choice set permits the implementing mechanism to depend on the function selected from the social choice set, this is slightly different from the one given in Palfrey and Srivastava (1989) and Jackson (1991).

Table 1.

$u(\cdot,\cdot)$	a	b	c	d
α	$1+\varepsilon, 0$	$0, 1+\varepsilon$	1,1	$1+2\varepsilon, 1+2\varepsilon$
β	$1+\varepsilon, 0$	$0, \varepsilon$	1,1	$1+2\varepsilon, 1+2\varepsilon$

of social choice correspondences. To illustrate this point, we now show that Maskin monotonicity may not even be a necessary condition for implementation in rationalizable strategies (according to at least one natural definition of these terms).[10] We describe the difficulty of social choice correspondences with the following approach (and subsequent example). A *social choice correspondence* (SCC) is a mapping $F : \Theta \to 2^Y/\varnothing$. A social choice correspondence F is implementable in rationalizable strategies if there exists a mechanism $\mathcal{M}$ with $g[S^{\mathcal{M},\theta}] = F(\theta)$ for all $\theta \in \Theta$. A SCC F is *Maskin monotonic* if: whenever $y^* \in F(\theta)$ and $u_i(y^*,\theta) \geq u_i(y,\theta) \Rightarrow u_i(y^*,\theta') \geq u_i(y,\theta')$ for all i and y; then $y^* \in F(\theta')$. Now consider the following example. There are 2 agents; $\Theta = \{\alpha,\beta\}$; $Z = \{a,b,c,d\}$; payoffs are given by Table 1.

The social choice correspondence is $F^*(\alpha) = \{a,b,c,d\}$ and $F^*(\beta) = \{d\}$. Now we demonstrate that F^* is not Maskin monotonic. To see why, pick $a \in F^*(\alpha)$. Observe that player 1's preferences are state independent and so $u_1(a,\alpha) \geq u_1(y,\alpha) \Rightarrow u_1(a,\beta) \geq u_1(y,\beta)$. In addition, a is the (unique) worst outcome for player 2 at state θ and so if $u_2(a,\alpha) \geq u_2(y,\alpha)$ then $y = a$ and so we must trivially have $u_2(a,\beta) \geq u_2(y,\beta)$. Thus, Maskin monotonicity would require $a \in F^*(\beta)$ which is false. But nonetheless F^* is implementable in rationalizable strategies. Consider the following mechanism $\mathcal{M}$ with $M_i = \{m_i^1, m_i^2, m_i^3\}$ and deterministic g given by the following matrix:

$g(\cdot)$	m_2^1	m_2^2	m_2^3
m_1^1	a	b	c
m_1^2	b	a	c
m_1^3	c	c	d

[10]As shown in Mezzetti and Renou (2010), a similar issue arises when one considers implementation in mixed Nash equilibrium where — contrary to the usual requirement — implementation does not ask for each alternative in the set of desired alternatives to be the outcome of a *pure* Nash equilibrium.

Now, for each i, $S_{i,k}^{\mathcal{M},\alpha} = M_i$ for all k and thus $S_i^{\mathcal{M},\alpha} = M_i$. Thus $g[S^{\mathcal{M},\alpha}] = \{a,b,c,d\} = F^*(\alpha)$. But in state β, we have in subsequent rounds of elimination:

$$S_{1,0}^{\mathcal{M},\beta} = \{m_1^1, m_1^2, m_1^3\} \quad \text{and} \quad S_{2,0}^{\mathcal{M},\beta} = \{m_2^1, m_2^2, m_2^3\},$$
$$S_{1,1}^{\mathcal{M},\beta} = \{m_1^1, m_1^2, m_1^3\} \quad \text{and} \quad S_{2,1}^{\mathcal{M},\beta} = \{m_2^3\},$$
$$S_{1,2}^{\mathcal{M},\beta} = \{m_1^3\} \quad \text{and} \quad S_{2,2}^{\mathcal{M},\beta} = \{m_2^3\},$$

and thus $g[S^{\mathcal{M},\beta}] = \{d\} = F^*(\beta)$. We thus showed that F^* is implementable in rationalizable strategies, yet did not satisfy Maskin monotonicity. The example, in particular the failure of Maskin monotonicity to remain a necessary condition, suggests that novel, but currently unidentified, necessary conditions would arise with social choice correspondences. On the other hand, the ordinal versions of the necessary conditions for social choice functions, described in Corollary 1 and Proposition 4 suggest that the insights in Mezzetti and Renou (2010) might be extended to social choice correspondences. In particular, Mezzetti and Renou (2010) obtain necessary and sufficient conditions for mixed strategy Nash implementation of correspondence using the ordinal approach. It is a natural but open conjecture that strict versions of their conditions would be appropriate for rationalizable implementation.

Strict Maskin monotonicity. We identified a necessary condition for rationalizable implementation, strict Maskin monotonicity, that is strictly stronger than the usual one for Nash implementation. Here we provide an example of a social choice function that is not rationalizable implementable but which is Nash implementable. There are 3 agents; the states are given by $\Theta = \{\alpha, \beta\}$; and the deterministic allocations are $Z = \{a,b,c,d\}$; the resulting payoffs are given by Table 2.

The social choice correspondence is $f(\alpha) = a$ and $f(\beta) = b$. It is easily checked that f is Maskin monotonic ($u_1(f(\alpha),\alpha) = u_1(b,\alpha)$ but $u_1(f(\alpha),\beta) < u_1(b,\beta)$; similarly, $u_1(f(\beta),\beta) > u_1(c,\beta)$ but $u_1(f(\beta),\beta) <$

Table 2.

$u(\cdot,\cdot)$	a	b	c	d
α	0,0,0	0,1,0	1,0,0	0,0,1
β	0,0,0	1,1,1	0,0,0	0,0,0

$u_1(c, \beta)$) and satisfies no-veto-power. Hence, standard arguments (see Maskin (1999) and Maskin and Sjöström (2004) show that f is implementable in (pure or mixed) Nash equilibrium. However, for any player i and $y \in \Delta(Z)$: $u_i(f(\alpha), \alpha) \leq u_i(y, \alpha)$ and so this social choice function cannot be strict Maskin monotonic, and so it is not implementable in rationalizable strategies.

Finally, from a purely game-theoretic point of view, the results presented in Propositions 1 and 2 may appear surprisingly strong. Given that we are investigating a social choice function, the notion of full implementation is akin to requiring that the game has a unique equilibrium (outcome). The present implementation results then say that necessary conditions to get a unique rationalizable outcome are similar to those required for a unique Nash equilibrium outcome. In addition, provided that the social choice function is responsive, under mild additional conditions, a unique rationalizable outcome arises under the same condition as a unique Nash equilibrium outcome. This is noteworthy as the necessary and almost sufficient condition of Maskin monotonicity is much weaker than the well-known conditions under which there are close connections between Nash equilibrium and rationalizability, such as supermodular or concave games. The Nash equilibrium results indicate the strength of the implementation approach to reduce the number of equilibria. The arguments presented here complement and extend these results. By using infinite message spaces and stochastic allocations, we strengthen the positive implementation results to the weaker solution concept of rationalizability.

References

ABREU, D., AND H. MATSUSHIMA (1992): "Virtual Implementation in Iteratively Undominated Strategies: Complete Information," *Econometrica*, 60, 993–1008.

ABREU, D., AND H. MATSUSHIMA (1994): "Exact implementation," *J. Econ. Theory*, 64, 1–19.

BENOIT, J.-P., AND E. OK (2008): "Nash implementation without no-veto power," Games Econ. Behav., 64, 51–67.

BERGEMANN, D., AND S. MORRIS (2009): "Robust Implementation in Direct Mechanisms," *Rev. Econ. Stud.*, 76, 1175–1206.

BERGEMANN, D., AND S. MORRIS (2011): "Robust Implementation in General Mechanisms," *Games Econ. Behav.*, doi:10.1016/j.geb.2010.05.001, forthcoming.

BERGEMANN, D., S. MORRIS, AND O. TERCIEUX (2010): "Rationalizable implementation," Cowles Foundation, Yale University, Tech. Rep. CFDP1697R.

BERNHEIM, D. (1984): "Rationalizable strategic behavior," *Econometrica*, 52, 1007–1028.

BOCHET, O. (2007): "Nash implementation with lottery mechanisms," *Soc. Choice Welfare*, 28, 111–125.

BRANDENBURGER, A., AND E. DEKEL (1987): "Rationalizability and correlated equilibria," *Econometrica*, 55, 1391–1402.

CABRALES, A., AND R. SERRANO (2009): "Implementation in adaptive better-response dynamics: Towards a general theory of bounded rationality in mechanisms," Universitat Carlos III de Madrid and Brown University, Tech. Rep.

JACKSON, M. (1991): "Bayesian implementation," *Econometrica*, 59, 461–477.

MASKIN, E. (1999): "Nash Equilibrium and Welfare Optimality," *Rev. Econ. Stud.*, 66, 23–38.

MASKIN, E., AND T. SJÖSTRÖM (2004): "Implementation theory," In: *Handbook of Social Choice and Welfare*, ed. by K. Arrow, A. Sen and K. Suzumura, vol. 1, North-Holland, Amsterdam.

MEZZETTI, C., AND L. RENOU (2010): "Implementation in mixed Nash equilibrium," http://ssrn.com/abstract=1396629.

LIPMAN, B. (1994): "A note on the implications of common knowledge of rationality," *Games Econ. Behav.*, 6, 114–129.

OSBORNE, M., AND A. RUBINSTEIN (1994): "A Course in Game Theory," MIT Press, Cambridge.

OURY, M., AND O. TERCIEUX (2009): "Continuous implementation," Paris School of Economics, Tech. Rep.

PALFREY, T., AND S. SRIVASTAVA (1989): "Implementation with incomplete information in exchange economies," *Econometrica*, 57, 115–134.

PEARCE, D. (1984): "Rationalizable strategic behavior and the problem of perfection," *Econometrica*, 52, 1029–1050.

SERRANO, R., AND R. VOHRA (2007): "Multiplicity of mixed equilibria in mechanisms: A unified approach to exact and approximate implementation," Brown University, Tech. Rep.

CHAPTER 11

Pricing Without Priors*

Dirk Bergemann and Karl H. Schlag

Abstract
We consider the problem of pricing a single object when the seller has only minimal information about the true valuation of the buyer. Specifically, the seller only knows the support of the possible valuations and has no further distributional information. The seller is solving this choice problem under uncertainty by minimizing her regret. The pricing policy hedges against uncertainty by randomizing over a range of prices. The support of the pricing policy is bounded away from zero. Buyers with low valuations cannot generate substantial regret and are priced out of the market. We generalize the pricing policy without priors to encompass many buyers and many qualities.

1 Introduction

We consider the problem of seller who has to price a given product with minimal information about the willingness to pay of the buyer. We offer a solution to the pricing problem of the seller by analyzing the pricing policy under regret minimization.

"There always is a first time" — with growing and globalizing markets the number of situations in which market participants have little information about their environment appears to be increasing. Market surveys can be costly and time consuming. Unless stakes are high, with marketplaces evolving and trading partners changing, it is useful to know how to

*This research is partially supported by NSF grants SES-0518929 and CNS-0428422. We would like to thank Stephen Morris and an anonymous referee for helpful comments.

set a price without the need to gather additional information. Traditional decision theory determines the optimal price according to the prior belief. Yet there is little guidance as how to form these initial beliefs.

We formally model the problem of optimal pricing with minimal information and build on the axiomatic literature on decision-making under uncertainty. The objective function of the seller is to minimize the regret from a given pricing policy. The regret of the seller is the difference between the profit under complete information and the realized profit under incomplete information. The regret of the seller can be positive for two reasons: (i) the buyer has a low valuation relative to the price and hence does not purchase the object, or (ii) he has a high valuation relative to the price and hence the seller could have obtained a higher revenue. The notion of regret contains a benchmark against which the realized profit is measured and offers a trade-off which determines the optimal policy.[1]

The idea of a minimax regret rule was first suggested by Savage (1951) in his reading of Wald (1950). A decision theoretic axiomatization of regret was provided by Milnor (1954) and more recently by Stoye (2007). It is noteworthy that the axiomatic foundations for the *minimax regret criterion* do not refer to regret at all; rather, they relax the axiom of independence of irrelevant alternatives. Namely, the irrelevance of the alternative is only maintained if it would not change the choice outcome under complete information. In this way, the minimax regret criterion captures the idea of a decision maker who is concerned about foregone opportunities, and hence the term "regret." We wish to emphasize that the concern for regret arises from the axioms and not from any emotional or behavioral considerations. In particular, there is no need for the decision maker to learn the true state of the world after making her decision.

How should the seller price under the minimax regret criterion? The pricing policy has to resolve the conflict between the regret which arises with low prices against the regret associated with high prices. If the seller offers a low price, regret can arise through the arrival of a high valuation buyer. On the other hand, if the seller offers a high price, regret can be caused by a

[1]The notion of regret shares features with the notion of competitiveness which is central in optimal design problems analyzed in computer science (see the recent survey to online design problems by Borodin and El-Yaniv, 1998). The competitiveness of a policy is the ratio (rather than the difference) of realized profit against maximal profit under complete information. Neeman (2003) analyzes the competitiveness of the second price auction and Bergemann and Valimaki (2006) survey robust models in mechanism design.

valuation just below the offered price. It then becomes evident that a single price will always expose the seller to substantial regret. Consequently, the seller can decrease her exposure by offering many prices in the form of a random pricing policy. With a random pricing policy, the seller diminishes the likelihood of large regret.

The intuition regarding the regret-minimizing policy is easy to establish in comparison to the optimal revenue maximizing policy for a given distribution. An optimal policy for a given distribution of valuations is always to offer the object at a deterministic price. In contrast a regret minimizing policy will offer many prices (with varying probability). With a single price, the risk of missing a trade at a valuation just below the given price is substantial. On the other hand, if the seller were simply to lower the price, she would miss the chance of extracting revenue from higher valuation customers. She resolves this conflict by offering low prices to the low valuation customers with positive probability.

We shall contrast the policy under the minimax regret criterion with the *maximin utility criterion* which seeks to maximize the worst case outcome. In the setting here, the worst outcome arises when the buyer has a valuation below the offered price. The maximin utility criterion forces the seller to set the price equal to the lowest possible valuation (provided it generates positive profits). This conservative point view fails to provide a trade-off in terms of foregone opportunities by focussing exclusively on the worst case scenario from the perspective of profits.

The current analysis complements our earlier work on robust monopoly pricing in Bergemann and Schlag (2007). There we considered a robust version of the classic problem of optimal monopoly pricing with incomplete information. In the robust version of the problem the seller only knows that demand will be in a small neighborhood of a given model distribution. We characterized the optimal pricing policy under two distinct, but related, decision criteria with multiple priors: (i) maximin expected utility and (ii) minimax expected regret. The resulting optimal pricing policy under either criterion depends on the model distribution and the size of the neighborhood. In the current contribution we do not allow for any prior information about the valuation of the buyer nor do we allow for variation in the uncertainty faced by the seller. In particular, we cannot say how the seller would be responding to an increase in uncertainty. The absence of prior information then allows us to focus on the trade-offs inherent to an environment without information.

A recent paper by Eren and van Ryzin (2006) considers a product differentiation problem without prior information and under regret minimization. They consider a market with differentiated products (either horizontal or vertical) and determine the optimal product positioning without market information. Perakis and Roels (2006) consider the inventory problem of the news-vendor model with partial information under regret minimization.

2 Model

Consider a seller of a good who faces a single potential buyer. The seller sets a price p for a unit of the good. The buyer wishes to buy at most one unit of the good and has a value v, his willingness to pay, belonging to a closed interval such that $v \in [\underline{v}, 1]$ where $\underline{v} \geq 0$.[2] The net utility of the buyer of purchasing the product at price p is given by

$$u(v,p) = v - p.$$

The marginal cost of the seller is constant and equal to $c \in [0,1)$, and the cost c is incurred only if the good is sold. The profit of the seller equals

$$\pi(p,v) \triangleq (p-c)\mathbf{1}_{\{v \geq p\}}, \tag{1}$$

where $\mathbf{1}_{\{v \geq p\}}$ is the indicator function specifying

$$\mathbf{1}_{\{v \geq p\}} = \begin{cases} 0 & \text{if } v < p, \\ 1 & \text{if } v \geq p. \end{cases}$$

The value v of the good is private information to the buyer and unknown to the seller. The only information the seller has is that $v \in [\underline{v}, 1]$. Clearly, the buyer purchases the good if $v \geq p$ and does not purchase if $v < p$.

We solve the problem in which the seller seeks to minimize the maximal expected regret. The *regret* of the seller charging price p is determined as the difference between the maximal profit the seller could make if she knew the value v and the profit she makes by setting p. The maximal profit when knowing v is given by

$$\max_p \pi(p,v) = \max\{v-c, 0\},$$

[2] The normalization to 1 is without loss of generality and the value v can interpreted as the relative value in relation to the maximum possible value.

and we obtain the following formula for regret:

$$r(p, v) \triangleq \max\{v - c, 0\} - (p - c)\mathbf{1}_{\{v \geq p\}}. \tag{2}$$

The regret is equal to the foregone profits of the seller due to not knowing the true value of the buyer. The regret is non-negative and can only vanish if $p = v$ or if $v \leq c$. The seller experiences strictly positive regret in two different cases: (i) the good is sold but the buyer would have been willing to pay more, so $p < v$ and $r(p, v) = v - p$, or (ii) the good was not sold but the willingness to pay of the buyer exceeded the cost or $p > v > c$ and $r(p, v) = v - c$. An upper bound on the valuation of the buyer is needed to ensure that the regret is finite.

The pricing policy with regret can be determined as an equilibrium strategy of a zero-sum game between the seller and adversarial nature. In the zero-sum game, the payoff to the seller is equal to $-r(p, v)$, to nature it is equal to $r(p, v)$ for a given realization of price p and valuation v. (The equilibrium behavior of the buyer is incorporated in the definition of regret given in definition [2].) The seller may use a mixed pricing strategy $\Phi \in \Delta\mathbb{R}$ and nature may choose a distribution over valuations, denoted by $F \in \Delta[\underline{v}, 1]$. The regret of the seller choosing a mixed pricing policy $\Phi \in \Delta\mathbb{R}$ given a valuation v is defined by the expected regret, so

$$r(\Phi, v) = \int r(p, v) d\Phi(p),$$

and by extension the expected regret given Φ and F is given by

$$r(\Phi, F) = \iint r(p, v) d\Phi(p) dF(v).$$

A pair of strategies (Φ^*, F^*) is a Nash equilibrium of the zero-sum game if it forms a saddle point:

$$r(\Phi^*, F) \leq r(\Phi^*, F^*) \leq r(\Phi, F^*), \quad \forall \Phi, \ \forall F. \tag{3}$$

The pricing strategy Φ^* is said to *attain minimax regret* and the equilibrium strategy of nature F^* is called a *least favorable demand.* The value $r^* \triangleq r(\Phi^*, F^*)$ is referred to as the value of the minimax regret.

The behavior in the minimax regret problem has a well-known relationship to Bayesian decision making. The pricing policy Φ^* that attains minimax regret also maximizes the expected profits of a Bayesian decision maker who is endowed with a least favorable demand F^* as prior. In this sense it is as if the minimax regret approach selects a specific prior.

3 Pricing Without Priors

The regret of the seller arises from two, qualitatively different kind of exposures. If the valuation of the buyer is very high, then the regret may arise from having offered a price too low relative to the valuation. We refer to this as the *upward exposure*. On the other hand, by having offered a price too high, the buyer risks to have a valuation below the price and the regret of the seller arises from not selling at all. Correspondingly, we refer to this as the *downward exposure*. At every given price p, the seller faces both a downward and an upward exposure. In this context, a deterministic price policy will always leave the seller exposed to substantial regret and the regret can be significantly reduced by offering a probabilistic pricing policy. We observe that a buyer with a low valuation cannot generate substantial regret and hence we may expect that the seller will never offer a price to sell to a customer with a low valuation. Consequently, the lower bound on the valuations given by $\underline{v}$ will only play a role in the determination of the equilibrium if it is not too low. A critical value for the lower bound $\underline{v}$ is given by $c + (1 - c)/e$ and we define

$$k \triangleq \max\{\underline{v}, c + (1 - c)/e\}.$$

The seller may "hedge" against regret and resolve the dilemma of facing both downward and upward exposure by "trying her luck" in a well-calibrated manner. If the seller is to be indifferent in her pricing policy against the least favorable demand, then the marginal profit must be zero over the range of prices which the seller offers. In the language of optimal monopoly pricing this means that the virtual utility of different prices has to be constant and equal to zero:

$$p - c - \frac{1 - F^*(p)}{f^*(p)} = 0. \tag{4}$$

In turn for nature to be indifferent between different valuations, it must be that the regret,

$$r(v, \Phi^*(p)) = v - c - \int_{p \leq v} (p - c) d\Phi^*(p),$$

is constant for those valuations (which satisfy $v \geq c$). By differentiating with respect to v we obtain

$$1 - (p - c)\phi^*(p) = 0,$$

or

$$\phi^*(p) = \frac{1}{p - c}. \tag{5}$$

It is now reasonable to guess that the distributions of seller and of nature share the same support over some interval $[a, b] \subseteq [\max\{\underline{v}, c\}, 1]$. We observe that the upper bound of the interval has to be $b = 1$ as an increase in the valuation from $v = b$ to $v = 1$ could otherwise strictly increase the regret of the seller. On the other hand, given the interval $[a, 1]$, nature may always choose a valuation just below a. This choice of valuation would yield a regret arbitrarily close to $a - c$ as the seller would fail to sell the good with prices $p \geq a$. In consequence the regret will be equal to $a - c$. The value of a is lowest if the distribution Φ^* of prices does not display a mass point and is obtained at $a = c \cdot 1 + (1 - c) \cdot (1/e)$ as we have

$$\int_{c+(1-c)/e}^{1} \frac{1}{p - c} dp = 1.$$

The equilibrium strategies are then identified by the lowest possible a subject to the indifference conditions (4) and (5), the requirement that Φ^* and F^* are well-defined distributions, and $a \geq \underline{v}$. The latter conditions imply that the least favorable demand F^* has a mass point at the upper end of the interval and that the pricing policy has a mass point at the lower end of the interval if $\underline{v} > c + (1 - c)/e$.

Proposition 1 (Pricing without Priors). *The unique minimax regret strategy is given by* Φ^*:

$$\Phi^*(p) = \begin{cases} 0 & \text{if } 0 \leq p < \kappa, \\ 1 + \ln \dfrac{p - c}{1 - c} & \text{if } \kappa \leq p \leq 1, \end{cases} \tag{6}$$

and Φ^* *has a point mass at* $p = \underline{v}$ *if and only if* $\underline{v} > c + (1 - c)/e$.

Proof. A least favorable demand is given by F^* with

$$F^*(v) = \begin{cases} 0 & \text{if } 0 \leq v < \kappa, \\ 1 - \dfrac{\kappa - c}{v - c} & \text{if } \kappa \leq v \leq 1, \\ 1 & \text{if } v = 1, \end{cases} \tag{7}$$

Given the pair (Φ^*, F^*) we need to verify the saddlepoint condition (3). The expected regret for a given price p is

$$\begin{aligned} r(p, F^*) &= \kappa - c + \int_\kappa^1 \frac{\kappa - c}{v - c} dv - (\kappa - c) \\ &= (\kappa - c) \ln \frac{1 - c}{\kappa - c}, \quad \text{for } p \in [\kappa, 1], \end{aligned}$$

and

$$\begin{aligned} r(p, F^*) &= \kappa - c + \int_\kappa^1 \frac{\kappa - c}{v - c} dv - (p - c) \\ &> (\kappa - c) \ln \frac{1 - c}{\kappa - c} \quad \text{for } 0 \leq p < \kappa. \end{aligned}$$

Similarly, the expected regret from given valuation v is

$$\begin{aligned} r(\Phi^*, v) &= v - c - \int_\kappa^v dp - (\kappa - c)\left(1 + \ln \frac{\kappa - c}{1 - c}\right) \\ &= -(\kappa - c) \ln \frac{\kappa - c}{1 - c}. \end{aligned} \tag{8}$$

for $v \in [\kappa, 1]$, and

$$r(\Phi^*, v) = \max\{v - c, 0\} < -(\kappa - c) \ln \frac{\kappa - c}{1 - c}$$

for $\underline{v} \leq v < \kappa$. We have thus verified that (Φ^*, F^*) satisfies condition (3). The uniqueness of Φ^* follows as nature has to be indifferent over all $v \in (\kappa, 1]$. □

The solution Φ^* of the regret minimization problem simultaneously determines a least favorable demand F^* given by (7) and a performance guarantee for the seller in terms of the maximal regret given by (8).

We observe that if the seller were restricted to choose a deterministic price policy, then the regret minimizing price would have to balance the upside exposure $1 - p$ and the downside exposure $p - c$ in a single price p.

Corollary 1 (Deterministic Pricing). *If the seller is constrained to pure strategies, then*

$$p^* = \begin{cases} \frac{1}{2}(1+c) & \text{if } \underline{v} < \frac{1}{2}(1+c), \\ \underline{v} & \text{if } \underline{v} \geq \frac{1}{2}(1+c). \end{cases}$$

The associated regret r^* for the seller is naturally higher under the restriction to pure strategies. At this point, it may be instructive to briefly consider a possible alternative objective in the presence of large uncertainty, namely, to choose a price that maximizes the minimum profit. Here the seller chooses a price (distribution) Φ^* such that

$$\Phi^* \in \arg\min_{\Phi} \sup_{v} \pi(\Phi, v).$$

With the maximin criterion, the seller chooses a price policy Φ^* that puts all the mass on $p = \underline{v}$ if $\underline{v} > c$ and is indifferent over all prices in $[c, 1]$ if $\underline{v} \leq c$. Under the minimax criterion, the seller is exclusively concerned with missing sales at valuations above marginal cost and hence she sets the price equal to the lowest possible valuation provided $\underline{v} > c$. If however $\underline{v} \leq c$ then all prices achieve the same minimal profit equal to 0 and every price above c is a solution to the maximin problem.

4 Discussion

Robustness. In this article we considered the optimal pricing of a single object with minimal information about the nature of the demand. Specifically, the information of the seller consisted of the interval of possible valuations without any additional distributional information. As the seller minimized her regret, randomization over prices played an important role. It is used to protect the seller against suffering from foregone opportunities. We argued that the optimal price policy under minimax regret can be understood in the classic expected utility (profit) framework as an optimal pricing rule under a specific prior. Yet the randomization over many prices would never emerge as the unique optimal pricing policy in the expected utility setting as there is always an optimal price which is deterministic.

In Bergemann and Schlag (2007) we consider the problem of optimal pricing when the seller has some prior information given by a model distribution and by a specified neighborhood around the model distribution in which the true demand distribution is known to be. The resulting model can be interpreted as a robust version of the classic problem of optimal monopoly pricing.

This article and Bergemann and Schlag (2007) make distinct extreme assumptions about multiple priors. Here, the set of multiple priors is the set of all demand distributions, there it is a small neighborhood around a model distribution. Many intermediate scenarios are interesting for future research. In particular, it seems natural to analyze a dynamic version of the robust pricing problem in which the uncertainty decreases over time due to the sampling of information.

Many buyers. We defined the pricing problem of the seller as offering a single product for a single buyer with an unknown valuation. The model and the results allow a further interpretation, namely, as offering the same product simultaneously to a finite number or a continuum of buyers. The notion of regret is subadditive with equality holding when all buyers have the same valuation and hence the problem of minimizing (average) regret when facing many small buyers or a single large buyer leads to the same solution as outlined in Proposition 1.

Product differentiation. In the current model, the buyer has a binary choice between accepting or rejecting a single product. A natural generalization of the model would allow for many different qualities of the same product class as in Mussa and Rosen (1978). There, the marginal willingness to pay for quality is constant and given by v and the cost of providing quality q is given by a convex cost function $c(q)$. Without prior information, the seller would now like to offer a menu of qualities to as to minimize her regret. The optimal menu $(q^*(v), p^*(v))$ would offer a combination of qualities $q^*(v)$ and prices $p^*(v)$ such that the buyers would self-select and such that the regret is minimized. With complete information, the seller would choose for every value v, the first best quantity $q_{FB}(v)$ which maximizes the social surplus $v \cdot q - c(q)$. The regret of the seller is the difference between the maximal net revenue and the realized net revenue. The regret minimization again requires that the regret is constant across all types which receive offers from the seller, that is, for all payoff types v with $q^*(v) > 0$ and the solution of the regret minimization problem is given by the following differential equation in v:

$$v \cdot q^{*\prime}(v) - c'(q^*(v)) = 0,$$

which can be solved after imposing the relevant boundary conditions.

References

BERGEMANN, D., AND K. SCHLAG (2007): "Robust Monopoly Pricing," Discussion paper No. 1527R, Cowles Foundation for Research in Economics, Yale University.

BERGEMANN, D., AND J. VALIMAKI (2006): "Information in Mechanism Design," In: *Advances in Economics and Econometrics*, ed. by R. Blundell, W. Newey, and T. Persson. Cambridge University Press, Cambridge, UK.

BORODIN, A., AND R. EL-YANIV (1998): *Online Computation and Competitive Analysis.* Cambridge University Press, Cambridge, UK.

EREN, S., AND G. VAN RYZIN (2006): "Product Line Positioning without Market Information," Working paper, Columbia Business School, Columbia University.

MILNOR, J. (1954): "Games against Nature," In: *Decision Processes*, ed. by R. Thrall, C. Coombs, and R. Davis. Wiley, New York.

Mussa, M., and S. Rosen (1978): "Monopoly and Product Quality," *Journal of Economic Theory*, 18, 301–317.

NEEMAN, Z. (2003): "The Effectiveness of English Auctions," *Games and Economic Behavior*, 43, 214–238.

PERAKIS, G., AND G. ROELS (2006): "Regret in the Newsvendor Model with Partial Information," Working paper, MIT.

SAVAGE, L. (1951): "The Theory of Statistical Decision," *Journal of the American Statistical Association*, 46, 55–67.

STOYE, J. (2007): "Axioms for Minimax Regret Choice Correspondences," Working paper, New York University.

WALD, A. (1950): *Statistical Decision Functions.* Wiley, New York.

CHAPTER 12

Robust Monopoly Pricing*

Dirk Bergemann and Karl Schlag

Abstract
We consider a robust version of the classic problem of optimal monopoly pricing with incomplete information. In the robust version, the seller faces model uncertainty and only knows that the true demand distribution is in the neighborhood of a given model distribution. We characterize the pricing policies under two distinct decision criteria with multiple priors: (i) maximin utility and (ii) minimax regret. The equilibrium price under either criterion is lower then in the absence of uncertainty. The concern for robustness leads the seller to concede a larger information rent to all buyers with values below the optimal price without uncertainty.

Keywords: Monopoly, robustness, multiple priors, maximin utility, minimax regret, robust mechanism design

1 Introduction

In the past decade, the theory of mechanism design has found increasingly widespread applications in the real world, favored partly by the growth of the electronic marketplace and trading on the Internet. Many trading platforms, such as auctions and exchanges, implement key insights of the theoretical literature. With an increase in the use of optimal mechanisms,

*The first author acknowledges financial support by NSF Grants #SES-0518929 and #CNS-0428422. The second author acknowledges financial support from the Spanish Ministerio de Educacion y Ciencia, Grant MEC-SEJ2006-09993. We thank the editor, Christian Hellwig, and an associate editor for their helpful suggestions. We are grateful for constructive comments from Rahul Deb, Peter Klibanoff, Stephen Morris, David Pollard, Phil Reny, John Riley and Thomas Sargent.

the robustness of these mechanisms with respect to the model specification becomes an important issue. In this note, we investigate a robust version of the classic monopoly problem under incomplete information. The determination of the optimal monopoly price is the most elementary instance of a revenue maximization problem in mechanism design.

We analyze the robustness of the optimal selling policy by enriching the standard model to account for model uncertainty. In the classic model, the valuation of the buyer is drawn from a given prior distribution. In contrast, in the robust version, the seller only knows that the true distribution is in the neighborhood of a given model distribution. The size of the neighborhood represents the extent of the model uncertainty faced by the seller.

The optimal pricing policy of the seller in the presence of model uncertainty is an instance of decision-making with multiple priors. We therefore build on the axiomatic decision theory with multiple priors and obtain interesting new insights for monopoly pricing. The methodological insight is that robustness can be guaranteed by considering decision making under multiple priors. The strategic insight is that we are able predict how an increase in uncertainty effects the pricing policy by using exclusively the data of the model distribution.

There are two leading approaches to incorporate multiple priors into axiomatic decision making: maximin utility and minimax regret. In the maximin utility approach with multiple priors, due to Gilboa and Schmeidler (1989), the decision maker evaluates each action by its minimum utility across all priors. The decision maker selects the action that maximizes the minimum utility. The minimax regret approach was axiomatized by Milnor (1954) and recently adapted to multiple priors by Hayashi (2008) and Stoye (2008). Here, the decision maker evaluates foregone opportunities using regret and chooses an action that minimizes the maximum expected regret among the set of priors.

The analysis of the optimal pricing under the two decision criteria reveals that either criterion leads to a robust policy in the sense of statistical decision theory. A family of policies, indexed by the size of the uncertainty, is said to be *robust*, if for any demand sufficiently close to the model distribution, the difference between the expected profit under the optimal policy for this demand and the expected profit under the candidate policy is arbitrarily small. While the optimal policies under *maximin utility* and *minimax regret* share the robustness property, the exact response to the uncertainty leads to distinct qualitative features under these two criteria.

The pricing policy of the seller is obtained as the equilibrium strategy of a zero-sum game between the seller and adversarial nature. In this construction, nature selects a *least favorable demand* given the objective of the seller. The choice by nature attempts to exploit the sensitivity of the objective function of the seller to the information regarding the demand. In consequence, the strategy of the seller is to minimize the sensitivity of his objective function with respect to the demand information. The sensitivity of the objective function to the private information shapes the equilibrium under either criterion. The central role of the information sensitivity is most immediate in the case of the maximin utility criterion, where the seller maximizes the minimal profit across a set of demand distributions. Consider for a moment the profit function of the seller at a candidate price p. The expected profit depends on the distribution of valuations *only* through the upper cumulative probability at price p, namely the probability that the valuation of the buyer is equal to or exceeds p. In particular, any variation of the distribution function which does not affect the upper cumulative probability at p, does not affect the value of the profit function. Given the sensitivity of the profit function, nature then seeks to minimize the upper cumulative probability. In turn, the seller minimizes the sensitivity to the information by choosing his optimal price as if nature would choose the lowest possible upper cumulative probability in the neighborhood of the model distribution. In consequence the equilibrium choice of the seller always consists in lowering her price relative to the optimal price in the absence of uncertainty.

The logic of the equilibrium is identical under the minimax regret criterion, the modifications that arise are due to the distinct informational sensitivity of the objective function. When we consider the minimax regret criterion, the notion of regret modifies the trade-off for the seller and for nature. The regret of the seller is the difference between the realized valuation of the buyer and the realized profit obtained by the seller. The regret of the seller can therefore be positive for two reasons: (i) a buyer has a low valuation relative to the price and hence fails to purchase the object, or (ii) he has a high valuation relative to the price and hence the seller could have realized a higher price. In turn, the expected regret is the difference between the expected valuation of the buyer and the expected profit, where the expected valuation represents the natural upper bound on the profits of the seller. Given this additive form of the objective function, and given a candidate price p, the expected regret therefore depends on the mean of the valuation and the upper cumulative probability at the candidate price p.

The later element appears as in the maximin criterion, but the sensitivity to the mean of the demand distribution newly appears in the regret minimization problem. In equilibrium, the pricing policy of the seller has to minimize the exposure to these two different statistics of the demand distribution simultaneously. In particular, if the seller were to concern herself exclusively with the upper cumulative probability, and hence as in the profit maximization lower the price too much, then nature would take advantage by increasing the mean of the valuation and hence increase the regret from this new, second, source. The seller resolves the conflict between these two statistics by a random pricing policy which offers trades at a range of prices. The range of the prices, i.e. the support of the equilibrium price distribution is chosen so that the expected regret is equalized across all prices, and the frequency of the prices is chosen such that no other demand distribution can lead to a larger regret. The resulting randomized pricing policy still has the feature that, relative to the optimal price in the absence of uncertainty, the expected price paid by almost all buyers with valuations within the support of the mixed pricing policy decreases when uncertainty increases. Yet, the upper segment of the buyers see higher prices with positive probability.

This brief description of the equilibrium policies emphasizes the common determinants of the policies under maximin utility and minimax regret, and traces the divergent aspects to differences in the objective functions. We will return to these differences and their axiomatic foundations in the final section. The common concern for robustness leads to many shared features in the equilibrium policies. First, and most importantly, the equilibrium price is lower (at least with positive probability) then it would be in the absence of uncertainty. With maximin utility, the hedging concern is so strong that the lower price is quoted with probability one. With minimax regret, the hedging concern leads to a range of offers, below and above, the price in the absence of uncertainty. Second, in terms of the information rent, the concern for robustness leads the seller to concede a larger information rent to all buyers with value below the optimal price without uncertainty. In the conclusion we discuss the extent to which these arguments may carry over to more general mechanism design settings.

We conclude the introduction with a brief discussion of the directly related literature. A recent paper by Bose, Ozdenoren, and Pape (2006) determines the optimal auction in the presence of an uncertainty averse seller and bidders. Lopomo, Rigotti, and Shannon (2009) consider a general mechanism design setting when the agents, but not the principal, have incomplete preferences due to Knightian uncertainty. In related work,

Bergemann and Schlag (2008) consider the optimal monopoly problem under regret without *any* priors. There, the analysis is concerned with optimal policies in the absence of information rather than robustness and responsiveness to uncertainty as in the current contribution. Linhart and Radner (1989) analyzed bilateral trade under minimax regret. A related notion of regret was considered by Engelbrecht-Wiggans (1989) in the context of auctions, and recently, Engelbrecht-Wiggans and Katok (2007) and Filiz-Ozbay and Ozbay (2007) present experimental evidence indicating concern for regret in first price auctions. In a complete information environment, Renou and Schlag (2010) use minimax regret to analyze strategic uncertainty.

2 Model

Monopoly. The seller faces a single potential buyer with value $v \in [0,1]$ for a unit of the object. The value v is private information to the buyer and unknown to the seller. The buyer wishes to buy at most one unit of the object. The marginal cost of production is constant and normalized to zero. The net utility of the buyer with value v of purchasing a unit of the object at price p is $v - p$. The *profit* of selling a unit of the object at a deterministic price $p \in \mathbb{R}_+$ if the valuation of the buyer is v is:

$$\pi(p, v) \triangleq p\mathbb{I}_{\{v \geq p\}},$$

where $\mathbb{I}_{\{v \geq p\}}$ is the indicator function specifying:

$$\mathbb{I}_{\{v \geq p\}} = \begin{cases} 0, & \text{if } v < p, \\ 1, & \text{if } v \geq p. \end{cases}$$

By extension, if the seller chooses a randomized pricing policy, represented by a probability distribution $\Phi \in \Delta\mathbb{R}_+$, then the expected profit when facing a buyer with value v equals:

$$\pi(\Phi, v) \triangleq \int \pi(p, v) d\Phi(p).$$

In the standard version of the monopoly with incomplete information, the seller maximizes the expected profit for a given prior$F$ over valuations.

For a given distribution F and deterministic price p the expected profit is:

$$\pi(p, F) \triangleq \int \pi(p, v) dF(v).$$

We note that the demand generated by the distribution F can either represent a single large buyer or many small buyers. Here we phrase the results in terms of a single large buyer, but the results generalize naturally to the case of many small buyers. With a random pricing policy Φ, the expected profit is given by:

$$\pi(\Phi, F) \triangleq \int \int \pi(p, v) dF(v) d\Phi(p).$$

A random pricing policy $\Phi^*(F)$ that maximizes the profit for a given distribution F solves:

$$\Phi^*(F) \in \arg \max_{\Phi \in \Delta \mathbb{R}_+} \pi(\Phi, F)$$

A well-known result by Riley and Zeckhauser (1983) states that for every distribution F, there exists a deterministic price $p^*(F)$ that maximizes profits.

Uncertainty. In the robust version the seller faces uncertainty (or ambiguity) in the sense of Ellsberg (1961). The uncertainty is represented by a set of possible distributions. The set is described by a model distribution F_0 and includes all distributions in a neighborhood of size ε of the model distribution F_0. The magnitude of the uncertainty is quantified by the size of the neighborhood around the model distribution. Given the model distribution F_0 we denote by p_0 a profit maximizing price at $F_0 : p_0 \triangleq p^*(F_0)$. For the remainder of the paper we shall assume that at the model distribution F_0: (i) p_0 is the unique maximizer of the profit function $\pi(p, F_0)$ and (ii) the density f_0 is continuously differentiable near p_0. These regularity assumptions enable us to use the implicit function theorem for the local analysis.

We consider two different decision criteria that allow for multiple priors: maximin utility and minimax regret. In either approach, the unknown state of the world is identified with the value v of the buyer.

Neighborhoods. We consider the neighborhoods induced by the Prohorov metric, the standard metric in robust statistical decision theory (see Huber and Robust (1981)). Given the model distribution F_0, the ε neighborhood under the Prohorov metric, denoted by $\mathcal{P}_\varepsilon(F_0)$, is:

$$\mathcal{P}_\varepsilon(F_0) \triangleq \{F|F(A) \leq F_0(A^\varepsilon + \varepsilon, \ \forall \text{ measurable } A \subseteq [0,1]\}, \tag{1}$$

where the set A^ε denotes the closed ε neighborhood of any measurable set A:

$$A^\varepsilon \triangleq \left\{ x \in [0,1] \min_{y \in A} d(x,y) \leq \varepsilon \right\},$$

where $d(x,y) = |x - y|$ is the distance on the real line. We shall use the language of small neighborhood and ε-neighborhood in the following interchangeably.

The Prohorov metric has evidently two components. The additive term ε in (1) allows for a *small* probability of *large* changes in the valuations relative to the model distribution whereas the larger set A^ε permits *large* probabilities of *small* changes in the valuations. The Prohorov metric is a metric for weak convergence of probability measures. In the context of our demand model, the Prohorov metric gives a literal description of the two relevant sources of model uncertainty. With a large probability, the seller could misperceive the willingness to pay by a small margin, and with a small probability, the seller could be mistaken about the market parameters by a large margin.

Maximin utility. Under maximin utility, the seller maximizes the minimum utility, where the utility of the seller is simply the profit, by searching for

$$\Phi_m \in \arg\max_{\Phi \in \Delta\mathbb{R}_+} \min_{F \in \mathcal{P}_\varepsilon(F_0)} \pi(\Phi, F).$$

Accordingly, we say that Φ_m *attains maximin utility*. We refer to F_m as a *least favorable demand* (for maximin utility) if

$$F_m \in \arg\min_{F \in \mathcal{P}_\varepsilon(F_0)} \max_{\Phi \in \Delta\mathbb{R}_+} \pi(\Phi, F).$$

The least favorable demand F_m minimizes across all profit maximizing pricing policies. Occasionally, it is useful to explicitly state the dependence of the optimal policies Φ_m and F_m on the size ε of the neighborhood, in which case we write $\Phi_{m,\varepsilon}$ and $F_{m,\varepsilon}$.

Minimax regret. The *regret* of the monopolist at a given price p and valuation v is:

$$r(p,v) \triangleq v - p\mathbb{I}_{\{v \geq p\}} = v - \pi(p,v). \qquad (2)$$

The regret of the monopolist charging price p facing a buyer with value v is the difference between the profit the monopolist could make if she were to know the value v of the buyer before setting her price and the profit she makes without this information. The regret is non-negative and can only vanish if $p = v$. The regret of the monopolist is strictly positive in either of two cases: (i) the value v exceeds the price p, the indicator function is $\mathbb{I}_{\{v \geq p\}} = 1$; or (ii) the value v is below the price p, the indicator function is $\mathbb{I}_{\{v \geq p\}} = 0$.

The expected regret of a random pricing policy Φ given a demand distribution F is:

$$r(\Phi, F) \triangleq \int\int r(p,v)d\Phi(p)dF(v) = \int vdF(v) - \int \pi(p,F)d\Phi(p). \qquad (3)$$

In the final expression of the expected regret in (3), we see that the expected regret is, as mentioned in the introduction, the difference between the expected valuation and the expected profit. It follows that the probabilistic pricing policy Φ is profit maximizing at F if and only if Φ minimizes (expected) regret when facing F. The pricing policy Φ_r *attains minimax regret* if it minimizes the maximum regret over all distributions F in the neighborhood of a model distribution F_0:

$$\Phi_r \in \arg\min_{\Phi \in \Delta\mathbb{R}_+} \max_{F \in \mathcal{P}_\varepsilon(F_0)} r(\Phi, F).$$

F_r is called a *least favorable demand* if

$$F_r \in \arg\min_{F \in \mathcal{P}_\varepsilon(F_0)} \max_{\Phi \in \Delta\mathbb{R}_+} r(\Phi, F) = \arg\min_{F \in \mathcal{P}_\varepsilon(F_0)} \left(\int vdF(v) - \max_{\Phi} \pi(\Phi, F) \right).$$

Thus, a least favorable demand maximizes the regret of a profit maximizing seller who knows the true demand. While the regret criterion seems to relate to foregone opportunities when the information is revealed ex post, this particular interpretation is solely an additional feature of the minimax regret model. In particular, the decision maker does not need the information to become available ex post to evaluate his expected regret.[1]

Robust policy. For a given model distribution F_0, we define a robust family of random pricing policies, $\{\Phi_\varepsilon\}_{\varepsilon>0}$, which are indexed by the size of the neighborhood ε as follows.

Definition 1. (*Robust pricing policy*). A family of pricing policies $\{\Phi_\varepsilon\}_{\varepsilon>0}$ is called *robust* if, for each $\gamma > 0$, there is $\varepsilon > 0$ such that $F \in \mathcal{P}_\varepsilon(F_0) \Rightarrow \pi(\Phi^*(F), F) - \pi(\Phi_\varepsilon, F) < \gamma$.

The above notion requires that for every, arbitrarily small, upper bound γ, on the difference in the profits between the optimal policy $\Phi^*(F)$ without uncertainty and an element of the robust family of policies $\{\Phi_\varepsilon\}$, we can find a sufficiently small neighborhood ε so that the robust policy Φ_ε meets the upper bound γ for all distributions in the neighborhood. An ideal candidate for a robust policy is the optimal policy $\Phi^*(F)$ itself. In other words, we would require that for each $\gamma > 0$, there is $\varepsilon > 0$ such that:

$$F \in \mathcal{P}_\varepsilon(F_0) \Rightarrow \pi(\Phi^*(F), F) - \pi(\Phi^*(F_0), F) < \gamma. \tag{4}$$

This notion of robustness, applied directly to the optimal policy $\Phi^*(F)$, constitutes the definition of α robustness in (2003) where it is shown that the profit maximizing price in the optimal monopoly problem is not robust to model misspecification.[2] One of the objectives here is to identify robust policies by considering decision making under multiple priors that do not suffer from such discontinuity in the profits.

3 Maximin Utility

We consider a monopolist who maximize the minimum profit for all distributions in the neighborhood of the model distribution F_0. The pricing rule that attains maximin utility is the equilibrium strategy in a game between

[1] The axiomatic approach is distinct from the ex-post measure of regret due to Hannan (1957) in the context of repeated games and from the behavioral approaches to regret due to Bell (1982) and Loomes and Sugden (1982).

[2] The non-robustness is demonstrated in Prasad (2003) by the following example: consider a Dirac distribution which puts probability one on valuation v. The optimal monopoly price p is equal to v. This policy is not robust to model misspecification, since if the true model puts probability one on a value arbitrarily close, but strictly below v, then the revenue is 0 rather than v.

the seller and adversarial nature. The seller chooses a probabilistic pricing policy, a distribution $\Phi \in \Delta\mathbb{R}_+$, and nature chooses a demand distribution $F \in \mathcal{P}_\varepsilon(F_0)$. In this game, the payoff of the seller is the expected profit while the payoff of nature is the negative of the expected profit. A Nash equilibrium of this zero-sum game is a solution (Φ_m, F_m) to the saddle point problem:

$$\pi(\Phi, F_m) \leq \pi(\Phi_m, F_m) \leq \pi(\Phi_m, F), \quad \forall \Phi \in \Delta\mathbb{R}_+, \ \forall F \in \mathcal{P}_\varepsilon(F_0). \qquad (\mathrm{SP}_m)$$

The objective of adversarial nature is to lower the expected profit of the seller. For a given price p, the expected profit of the seller is

$$\pi(p, F) = \int \pi(p, v) dF(v) = p(1 - F(p)).$$

The profit minimizing demand, given p, is then achieved by decreasing the cumulative probability of valuations equal or larger than p by as much as possible within the neighborhood $\mathcal{P}_\varepsilon(F_0)$. The profit minimizing demand thus minimizes the probability of sale, the upper cumulative probability $1 - F(p)$. Given the model distribution F_0 and the size ε of the neighborhood, there is a unique distribution, which minimizes the probability, $1 - F(p)$, for all p in the unit interval *simultaneously*. We obtain this least favorable demand explicitly by shifting the probabilities as far down as possible, given the constraints imposed by the model distribution F_0 and the size ε of the neighborhood. We shift, for every v, the cumulative probability of the model distribution F_0 at the point $v + \varepsilon$ downwards to be the cumulative probability at the point v. In addition, we transfer the very highest valuations with probability ε to the lowest valuation, namely $v = 0$. This results in the distribution $F_{m,\varepsilon}$ given by:

$$F_{m,\varepsilon}(v) \triangleq \min\{F_0(v + \varepsilon) + \varepsilon, 1\}, \qquad (5)$$

that is within the ε neighborhood of F_0. The first shift, generated by $v + \varepsilon$, represents small changes in valuations that occur with large probability. The second shift, generated by $F_0(\cdot) + \varepsilon$, represents large changes that occur with small probability. It is easily verified that $F_{m,\varepsilon}$ is a profit minimizing demand for *any* price p given the constraint imposed by the size of the neighborhood. In other words, the profit minimizing demand does not depend on the, possibly probabilistic, price p of the seller. Given that the profit minimizing demand $F_{m,\varepsilon}$ does not depend on the offered prices, the monopolist acts as if the demand is given by $F_{m,\varepsilon}$. In consequence, the seller maximizes profits at $F_{m,\varepsilon}$ by choosing a deterministic price $p_{m,\varepsilon}$ where $p_{m,\varepsilon} \triangleq p^*(F_{m,\varepsilon})$.

Proposition 1. (*Maximin utility*). *For every $\varepsilon > 0$, there exists a pair $(p_{m,\varepsilon}, F_{m,\varepsilon})$, such that $P_{m,\varepsilon} \in [0,1]$ attains maximin utility and $F_{m,\varepsilon}$ is a least favorable demand.*

An important aspect of the above result is that the construction of the profit minimizing demand does not require a local argument, and hence the above equilibrium characterization is valid for arbitrarily large neighborhoods around the model distribution. The construction of the least favorite demand, given by (5), also reveals that $F_{m,\varepsilon}$ is first-order stochastically dominated by any other distribution in the neighborhood $\mathcal{P}_\varepsilon(F_0)$. By this constructive argument, the result of Proposition 1 then extends to any notion of neighborhood (and/or generating metric) which forms a lattice (strictly speaking, we only need the semi-lattice property) with respect to first order stochastic dominance.[3] The optimal pricing policy of the seller can easily be recast as canonical mechanism design problem, using the language of virtual utility, as shown by Myerson (1981) and Bulow and Roberts (1989). By using the incentive constraints to replace the transfers, the maximization problem of the seller can be represented as:

$$\int \left(\max_{x \in [0,1]} x(v) \left(v - \frac{1 - F(v)}{f(v)} \right) \right) f(v) dv. \tag{6}$$

For a given distribution F, the pointwise optimal solution $x^*(v) \in \{0,1\}$ is to assign the object, $x^*(v) = 1$, if the virtual utility $v - (1 - F(v))/f(v)$ is positive, and to not assign the object, $x^*(v) = 0$, if it is negative. We can rewrite the above integral after disregarding the valuations which have negative virtual utility as they receive zero weight, $x^*(v) = 0$, in the optimal solution:

$$\int_{\{v | v - (1 - F(v))/f(v) \geq 0\}} (v f(v) - (1 - F(v))) dv. \tag{7}$$

In this reformulation of the objective function of the seller, we see that the least favorable demand, as established in Proposition 1, depresses the mean valuation, conditional on allocating the good, and simultaneously depresses the information cost, or sensitivity to the private information,

[3] We thank the editor for pointing out the relationship to the lattice property of the neighborhoods.

$1 - F(v)$. Thus, the least favorable demand generates the lowest feasible mean valuation, but the resulting allocation improves the *ex-post* efficiency as there are some intermediate types with willingness-to-pay v which will receive the object under $F_{m,\varepsilon}$, but would not receive it under any other distribution $F \in \mathcal{P}_\varepsilon(F_0)$.

How does the optimal price change with an increase in uncertainty? The rate of the change in the price depends on the curvature of the profit function at the model distribution F_0. By the assumption of concavity, we know that the curvature is negative. We can apply the implicit function theorem to the optimal price p_0 at the model distribution F_0 and obtain the following comparative static result.

Proposition 2. (*Pricing under maximin utility*). *The price $p_{m,\varepsilon}$ responds to an increase in uncertainty at $\varepsilon = 0$ by:*

$$\left.\frac{dp_{m,\varepsilon}}{d\varepsilon}\right|_{\varepsilon=0} = -1 + \frac{1 - f_0(p_0)}{\partial\pi^2(p_0, F_0)/\partial p^2} < -\frac{1}{2}.$$

Accordingly, the price that attains maximin utility responds to an increase in uncertainty with a lower price. Marginally, this response is equal to -1 if the objective function is infinitely concave.

Consider now the profits realized by the price $p_{m,\varepsilon}$ — which attains maximin utility within the neighborhood $\mathcal{P}_\varepsilon(F_0)$ — at a given distribution $F \in \mathcal{P}_\varepsilon(F_0)$. By construction, these profits are at least as high as those obtained when facing the least favorable demand $F_{m,\varepsilon}$. We use the lower bound on the profits supported by F_m to show that the optimal profits are continuous in the demand distribution F. This implies that profits achieved by $p_{m,\varepsilon}$ when facing F are close to those achieved by $p^*(F)$ when facing F. The family of pricing rules that attain maximin utility thus qualify as robust.

Proposition 3. (*Robustness*). *The family of pricing policies $\{p_{m,\varepsilon}\}_{\varepsilon>0}$ is a robust family of pricing policies.*

4 Minimax Regret

Next we consider the minimax regret problem of the seller, where a (probabilistic) pricing policy Φ_r and a least favorable demand F_r are the equilibrium policies of a zero-sum game. In this zero-sum game, the payoff of the seller is the negative of the regret while the payoff to nature is regret itself.

A Nash equilibrium (Φ_r, F_r) is a solution to the saddle point problem:

$$r(\Phi_r, F) \leq r(\Phi_r, F_r) \leq r(\Phi, F_r), \quad \forall \Phi \in \Delta\mathbb{R}_+, \ \forall F \in \mathcal{P}_\varepsilon(F_0). \qquad (\mathrm{SP}_r)$$

The saddlepoint result permits us to link minimax regret behavior to payoff maximizing behavior under a prior as follows. If the minimax regret is derived from the equilibrium characterization in (SP_r) then any price chosen by a monopolist who minimizes maximal regret, is at the same time a price which maximizes expected profit against a particular demand, namely, the least favorable demand. In fact, the saddle point condition requires that Φ_r is a probabilistic price that maximizes profits given F_r and F_r is a regret maximizing demand given Φ_r.

In the equilibrium of the zero-sum game, the probabilistic price has to resolve the conflict between the regret which arises with low prices, against the regret associated with high prices. If she offers a low price, nature can cause regret with a distribution which puts substantial probability on high valuation buyers. On the other hand, if she offers a high price, nature can cause regret with a distribution which puts substantial probability at valuations just below the offered price. If we consider the formula of the expected regret at a deterministic price p, rather than a general random pricing policy Φ, as in (3),

$$r(p, F) = \int v dF(v) - \int \pi(p, v) dF(v) = \int v dF(v) - p(1 - F(p)) \qquad (8)$$

we see this tension in terms of the expected valuation, the first term, and the expected profit, the second term. The first term is controlled by the mean valuation, whereas the second is controlled by the upper cumulative probability, $1 - F(p)$, the probability of a sale at price p. Now, the analysis of the maximin utility problem showed that a distribution which minimizes $1 - F(p)$ can be determined independently of p. The unique solution, given by the distribution $F_{m,\varepsilon}$ in (5) has the property that it is first order stochastically dominated by all other distributions in the neighborhood $\mathcal{P}_\varepsilon(F_0)$. An immediate consequence of the first order stochastic dominance is that the distribution $F_{m,\varepsilon}$ achieves the lowest mean valuation among all distributions of $F \in \mathcal{P}_\varepsilon(F_0)$. Now, as nature is seeking to maximize regret, the first term would suggest for nature to choose the distribution with highest mean in $\mathcal{P}_\varepsilon(F_0)$ whereas the second term would suggest to choose the distribution with the lowest upper cumulative probability, namely $F_{m,\varepsilon}$, which by the above argument is also the distribution with the lowest mean. The relative importance of these two terms depends on the choice of the price p

by the seller, and hence, in contrast to the case of maximin utility, the least favorable demand is the result of an equilibrium argument and cannot be constructed independently of the strategy of the seller (as it was the case with maximin utility). It also suggests that a deterministic pricing policy in the form of single price p exposes the seller to substantial regret, and that the seller can decrease her exposure by offering many prices in the form of a random pricing policy. We prove the existence of a solution to the saddlepoint problem (SP_r) using results from Reny (1999).

Proposition 4. *(Existence of minimax regret). A solution* (Φ_r, F_r) *to the saddlepoint condition* (SP_r) *exists.*

We should emphasize that the above existence result does not use local arguments, and establishes the existence of a Nash equilibrium for arbitrary neighborhoods, small or large. By contrast, the following explicit characterization of the equilibrium pricing strategy of the seller uses a local argument, namely the implicit function theorem, and hence is valid only for small neighborhoods.

The tension between the mean valuation and the upper cumulative probability changes the structure of the least favorable demand and the equilibrium pricing policy relative to the maximin utility analysis. Intuitively, nature seeks to accomplish two conflicting goals. First she tries to maximize the expected valuation, which represents the upper bound for the profit of the seller, and hence also the maximal value of regret, while, second, she attempts to minimize the expected profits. These goals are conflicting as an increase in the expected valuation ought to lead eventually to an increase in the surplus the seller can extract. Now for a given candidate price p, the expected profit only depends on the upper cumulative probability, $1 - F(p)$, at p. Now, to the extent that the upper cumulative probability is held constant at p (and so is p itself), nature would seek to maximize the expected valuation. But by the logic of the first order stochastic dominance, this means to maximize the upper cumulative probability everywhere but at p. This means, that in contrast to the least favorable demand, under maximin utility, the least favorable demand in the minimax regret will maximize rather than minimize $1 - F(p')$ for all $p' \neq p$. But this thought experiment suggests a discontinuity in the form of a downward jump of the upper cumulative probability to $1 - F(p)$ (or correspondingly an upward jump to $F(p)$) from the left, and a constant upper cumulative probability to the right of p, as long as permitted by the size of the neighborhood. But now observe, that if there were a flat segment in the probability distribution to the right of p, then the seller would a profitable deviation. By increasing the price from p to the largest price $p'' > p$ where the equality $F(p'') = F(p)$ would still prevail, the seller could raise his price without losing sales, clearly an improvement. It follows that in equilibrium, nature has to suspend the maximization of the upper cumulative probability precisely in the support of the prices offered by the seller, denoted by $[a, c]$ in the proposition below. In this interval, a real trade-off arises between the maximization of the expected valuation and the minimization of the profits. In particular, nature is attempting to maintain the prices offered sufficiently low and by the logic familiar from maximin utility this involves lowering the upper cumulative probability as much as feasible within the constraint imposed by the size of the neighborhood $\mathcal{P}_\varepsilon(F_0)$. The constraint on the choice set is going to be binding at some point $b \in [a, c]$, where nature cannot lower the upper probability, $1 - F(b)$, any further, and at this insensitive point b where "undercutting" is infeasible, the random pricing strategy of the seller can place a single atom. Everywhere else, the random pricing strategy Φ_r keeps nature exactly indifferent with respect to local changes of the demand distribution.

Proposition 5. (*Minimax regret*).

1. *Given* $\delta > 0$, *for every* ε *sufficiently small, there exist* a, b *and* c *with* $0 < a < b < c < 1$ *and* $p_0 - \delta < a < p_0 < c < p_0 + \delta$ *such that a minimax regret probabilistic price* Φ_r *is given by:*

$$\Phi_r(p) = \begin{cases} 0 & \text{if } 0 \le p < a, \\ \ln\dfrac{p}{a} & \text{if } a \le p < b, \\ 1 - \ln\dfrac{c}{p} & \text{if } b \le p < c, \\ 1 & \text{if } c < p \le 1. \end{cases}$$

2. *The boundary points* a, b *and* c *respond to an increase in uncertainty at* $\varepsilon = 0$:

   (i) $\lim_{\varepsilon\to 0} a'(0) = -\infty$;
   (ii) $\lim_{\varepsilon\to 0} b'(0)$ *is finite*; *and*
   (iii) $\lim_{\varepsilon\to 0} c'(0) = \infty$.

The proof of Proposition 5 relies on a straightforward but lengthy application of the implicit function theorem and is provided in Proposition 5 of (Bergermann and Schlag, 2008). The least favorable demand makes the seller indifferent among all prices $p \in [a, c]$. As uncertainty increases, the interval over which the seller randomizes increases rapidly in order to protect against nature either undercutting or moving mass towards higher valuations. At the same time, the mass point b does not change drastically.

We now investigate the comparative static behavior of the random price $p_{r,\varepsilon}$ governed by the distribution $\Phi_{r,\varepsilon}$. The response of the expected price, $\mathbb{E}[p_{r,\varepsilon}]$, to a marginal increase in uncertainty can be explained by the first order effects. For a small level of uncertainty, we may represent the regret through a linear approximation $r^* = r_0 + \varepsilon \cdot \partial r^* / \partial \varepsilon$, where r_0 is the regret at the model distribution. The optimal response of the seller to an increase in uncertainty is now to find a probabilistic price which minimizes the additional regret $\varepsilon \cdot \partial r^* / \partial \varepsilon$ coming from the increase in uncertainty. Locally, the cost of moving the price away from the optimum is given by the second derivative of the objective function. With small uncertainty, the curvature of the regret is identical to the curvature of the profit function. The rate at which the minimax regret price responses to an increase in uncertainty is then simply the ratio of the response of the marginal regret to a change in price divided by the curvature of the profit function.

Corollary 1. (*Comparative statics with minimax regret*).

1. *The expected price* $\mathbb{E}[p_{r,\varepsilon}]$ *responds to an increase in uncertainty at* $\varepsilon = 0$ *by:*

$$\frac{d}{d\varepsilon}\mathbb{E}[p_{r,\varepsilon}]|_{\varepsilon=0} = \begin{cases} -1 - \dfrac{f_0(p_0)+1}{\partial \pi^2(p_0, F_0)/\partial p^2} > -1 & \text{if } p_0 \leq \dfrac{1}{2}, \\ -1 - \dfrac{f_0(p_0)-1}{\partial \pi^2(p_0, F_0)/\partial p^2} < -\dfrac{1}{2} & \text{if } p_0 > \frac{1}{2}. \end{cases} \tag{9}$$

2. *If* ε *is sufficiently small, then for any* $v \in (a, c)\backslash b$,

$$\frac{d}{d\varepsilon}\mathbb{E}[p_{r,\varepsilon} | p_{r,\varepsilon} \leq v] < 0.$$

By comparing Corollary 1 and Proposition 2, we find that the marginal response of the expected price $\mathbb{E}[p_{r,\varepsilon}]$ to an increase in uncertainty is identical under minimax regret and maximin profit for $p_0 > \frac{1}{2}$. In both cases, the (expected) equilibrium price is lower than under the model distribution F_0 without uncertainty. The difference between the two criteria arises at a low level of p_0 at which the seller is less aggressive in lowering her price under minimax regret. When the optimal price in the absence of uncertainty is low, $p_0 < \frac{1}{2}$, then the trade-off that nature faces in her two conflicting goals, namely to maximize the expected valuation, while, second, to minimize the expected profits, is resolved in favor of the former, namely to maximize the expected valuation. We discussed this trade-off above following Proposition 4. Now, if $p_0 < \frac{1}{2}$, and hence in the lower half of the support $[0, 1]$, an increase in the expected valuation is guaranteed to increase regret by more, namely $1 - p_0 > \frac{1}{2}$ than a lost sale which would increase regret only by $p_0 < \frac{1}{2}$. This explains the change in the derivative of the expected price $\mathbb{E}[p_{r,\varepsilon}]$ at the midpoint of the support [0,1]. In fact, for the case of $p_0 < \frac{1}{2}$, it turns out that the expected price can be strictly increasing in ε. As nature finds it to her advantage to increase the mean value of the distribution, and hence also to increase the upper cumulative probability, the seller is responding to the increase in the demand distribution with a raise in the expected equilibrium price. But since the expected equilibrium price is close to $p_0 < \frac{1}{2}$, the resulting least favorable demand still leads to an increase in regret relative to the model distribution. For example, the increase in the price occurs if the model density is linear and strictly decreasing. This response of the equilibrium price to an increase in uncertainty stands in stark contrast to the maximin behavior where any increase in the size of the uncertainty has a downward effect on prices, regardless of the model distribution.

The change in the expected price, as given by Corollary 1(1) also represent the change in welfare to a buyer who purchases with probability one, or $v > c$. The impact of the uncertainty on a buyer whose purchase occurs with probability less than one, or $v < c$ is stated in Corollary 1(2). The derivative of the conditional price is defined everywhere but at b, where it has a discontinuity as the mass point at b changes its location with an increase in ε. It shows that the price *conditional* on a purchase decreases when the uncertainty increases. The decrease in the price *conditional on purchase* does not contradict the possible increase in the expected price. The increase in the unconditional price is driven by the changes in the support of Φ_r — in particular the increase of c — and increases (in the sense of first order stochastic dominance) of the unconditional distribution of prices. In Bergemann and Schlag (2008) we established this result in terms of menu of prices, where we showed that for every type $v \in (a, c)$, the price paid per unit of the object is decreasing with an increase in uncertainty, see Proposition 7 in Bergemann and Schlag (2008).[4]

We conclude by showing that the solution to the minimax regret problem also generates a robust family of policies in the sense of Definition 1.

Proposition 6. (*Robustness*). *If* $\{\Phi_{r,\varepsilon}\}_{\varepsilon>0}$ *attains minimax regret at* F_0 *for all sufficiently small* ε, *then* $\{\Phi_{r,\varepsilon}\}_{\varepsilon>0}$ *is a robust family of pricing policies.*

5 Discussion

We conclude by relating the pricing behavior of the seller in the incomplete information monopoly to the axiomatic foundations. Finally, we spell out how the insights from the specific model here might relate to more general models of mechanism design and uncertainty.

Axioms and behavioral implications. From an axiomatic perspective, the maximin utility and minimax regret criteria represent different departures from the standard model of Anscombe and Aumann (1963) by allowing for multiple priors. The maximin utility criterion emerges when imposing the independence axiom only when mixing with constant actions. The minimax regret criterion allows the choice to be menu dependent by requiring independence of irrelevant alternatives only when the changes in the menu do not change the best outcome in any of the states (see Stoye (2008)). Both criteria include an additional axiom to capture aversion to ambiguity by postulating that the decision maker will hedge against uncertainty by mixing whenever indifferent. This hedging leads the decision maker under either criterion to sometimes offer the object for sale at lower prices than he would at the model distribution F_0, but in the absence of uncertainty. Interestingly, the difference between maximin utility and minimax regret then arises in the strength of the hedging motive. In the absence of the axiom of independence of irrelevant alternatives, the choice under regret does depend on the opportunities, both in terms of missed sales and missed revenues, the respective downward and upward opportunities. In contrast,

[4]In the menu representation, a buyer with willingness-to-pay v, pays a transfer $t_r(v)$ to receive the object with probability $q_r(v)$. The conditional expected price with random pricing here is equal to the per unit price in the menu representation there, or: $\mathbb{E}[p_{r,\varepsilon} \leq v] = t_r(v)/q_r(v)$.

the maximin utility maintains the independence of irrelevant alternatives, and this leads the decision-maker to act as if the lowest distribution (in terms of first-order dominance) were the true distribution.

The choice of metric (and neighborhoods). We investigated the robust policies with respect to neighborhoods generated by the Prohorov metric. The question then arises as to how sensitive the results are to the choice of the metric. In particular, there are a number of other distances, such as the Levy metric or the bounded Lipschitz metric which also metrize the weak topology. Of course, these distances define different neighborhoods and hence different choice sets for nature. However, these distinct notions differ only in the support sets over which the distributions are evaluated. Therefore the comparative static results near $\varepsilon = 0$ are unaffected by the specific notion for the metric.

Beyond small neighborhoods. We analyzed the pricing policies when robustness is required with respect to small neighborhoods. But we required the assumption of small neighborhoods only in the use of the implicit function theorem for the comparative static results and in the explicit construction of the random pricing under minimax regret. In related work, Bergemann and Schlag (2008) we considered the monopoly problem under minimax regret in the absence of *any* restrictions about the uncertainty, in other words, *very large* neighborhoods. The analysis there was notably easier as there were no constraints on the choice of strategy by nature. But interestingly, the distinct features of minimax regret strategy were preserved, namely the logarithmic distribution of the prices with a single mass point. This suggests that the intermediate case of large neighborhoods would support similar results for minimax regret. The associated analysis, however, is beyond the scope of this note, as it requires the use of general optimal control techniques to keep track of the multitude of constraints imposed by the neighborhood on the least favorable demand. But as we know that the minimax regret policy has to remain a random pricing policy, and as we have established the general form of the maximin policy, we already know that the distinct features of the minimax regret and maximin utility are preserved beyond small neighborhoods.

Beyond monopoly pricing. We analyzed robust policies in a simple mechanism design environment, namely the monopolistic sale of a single unit under incomplete information. The robust pricing policies displayed less sensitivity to private information and hedged against the uncertainty by offering sales at lower prices relative to the policy without uncertainty.

We expect these insights to extend to mechanism design problems beyond the single good monopoly pricing as the logic of the equilibrium construction indicates. The common element of revenue maximizing mechanism, whether it pertains to single good, multi-unit goods (nonlinear pricing) or multi-person problems (auctions) is that the principal maximizes the virtual utility rather than the social utility. In all of the above problems, the virtual utility takes the form, $v - (1 - F(v))/f(v)$, where the later term represents the cost of private information to the seller. Now, we saw in the maximin utility that nature lowers the revenue by lowering the entire social surplus by minimizing the upper cumulative probability. But, as we saw then, this means that the virtual utility of each type v is increased, and hence that low types v will have positive virtual utility in the presence of uncertainty, whereas the would not have in the absence of uncertainty. It follows that agents with lower valuation v now receive the object or receive an assignment more generally. But as the participation constraint still binds for these types, it means that the prices will be lower, and overall the cost of the private information, represented by $(1 - F(v))/f(v)$ will be lower. It is this general aspect of the robust policy, namely lower prices through lower inverse hazard rates, that we expect to emerge in general allocation environments as well. In other words, the robust revenue maximizing policy is closer to the socially efficient allocation as the information cost, $(1 - F(v))/f(v)$, carries less weight, and the resulting prices are closer to the externality prices imposed by the efficient Groves mechanism. Given the prominent role of the notion of first-order stochastic dominance and the immediate link to the information cost $(1 - F(v))/f(v)$ in the argument presented here, it is conceivable that the present argument extends directly to the above mentioned, more general, allocation problems.

By contrast, in the minimax regret problem, we can expect the general downward trend of information cost and allocation policy to be attenuated relative to the maximin problem as adversarial nature attempts to maximize the potential for regret, namely the social value. But the exact determination of the robust policy appears to be much more difficult to establish in general environments, where we cannot exploit the specific structure of the allocation problem as in this note.

Appendix A

The appendix contains the proofs for the results in the main body of the text.

Proof of Proposition 1. As F_m is given by (5), we have that $\pi(p, F_m) \leq \pi(p, F)$ for all $F \in \mathcal{P}_\varepsilon(F_0)$. On the other hand, if $p_m = p^*(F_m)$, then $\pi(p_m, F_m) \geq \pi(p, F_m)$ holds for all p by definition of p_m. Jointly this implies that (p_m, F_m) is a saddle point of (SP_m) and p_m attains maximin payoff and F_m is a least favorable demand. □

Proof of Proposition 2. For sufficiently small ε our assumptions on F_0 imply that F_m is differentiable near p_m. Since p_m is optimal given demand F_m, we find that p_m satisfies the associated first order conditions:

$$\frac{d}{dp}(p(1 - F_m(p)))|_{p=p_m} = 0.$$

The earlier strict concavity assumption on $\pi(p, F_0)$ implies that we can apply the implicit function theorem at $\varepsilon = 0$ to the above equation to obtain

$$\left.\frac{dp_m}{d\varepsilon}\right|_{\varepsilon=0} = -1 + \frac{1 - f_0(p_0)}{-2f_0(p_0) - p_0 f_0'(p_0)} = \frac{f_0(p_0) + p_0 f_0'(p_0) + 1}{-2f_0(p_0) - p_0 f_0'(p_0)}.$$

Since $-2f_0(p_0) - P_0 f_0'(P_0) < 0$, we observe that the l.h.s. of the above equation as a function of $f_0(p_0)$ is increasing in $f_0(p_0)$ and hence by taking the limit as $f_0(p_0)$ tends to infinity it follows that this expression is bounded above by $-1/2$. □

Proof of Proposition 3. We show that for any $\gamma > 0$, there exists $\varepsilon > 0$ such that $F \in \mathcal{P}_\varepsilon(F_0)$ implies $\pi(p^*(F), F) - \pi(p_m, F) < \gamma$. Note that $\pi(p_m, F) \geq \pi(p_m, F_m)$ and thus $\pi(p^*(F), F) - \pi(p_m, F) \leq \pi(p^*(F), F) - \pi(p_m, F_m)$. Since $\pi(p_m, F_m) = \pi(p^*(F_m), F_m)$ the proof is complete once we show that $\pi(p^*(F), F)$ is a continuous function of F with respect to the weak topology. Consider F, G such that $G \in \mathcal{P}_\varepsilon(F)$. Using the fact that $G(p) \leq F(p + \varepsilon) + \varepsilon$, we obtain

$$\begin{aligned}\pi(p^*(G), G) &\geq \pi(p^*(F) - \varepsilon, G) = (p^*(F) - \varepsilon)(1 - G(p^*(F) - \varepsilon)) \\ &\geq (p^*(F) - \varepsilon) - (1 - F(p^*(F)) - \varepsilon) \geq \pi(p^*(F), F) - 2\varepsilon.\end{aligned}$$

Since the Prohorov norm is symmetric and thus $F \in \mathcal{P}_\varepsilon(G)$, it follows that

$$\pi(p^*(F), F) + 2\varepsilon \geq \pi(p^*(G), G) \geq \pi(p^*(F), F) - 2\varepsilon,$$

and hence we have proven that $\pi(p^*(F), F)$ is continuous in F. □

Proof of Proposition 4. We apply Corollary 5.2 in Reny (1999) to show that a saddle point exists. For this we need to verify that the zero-sum game between the seller and nature is a compact Hausdorff game for which the mixed extension is both reciprocally upper semi continuous and payoff secure.

Clearly we have a compact Hausdorff game. Reciprocal upper semi continuity follows directly as we are investigating a zero-sum game. So all we have to ensure is *payoff security.* Payoff security for the monopolist means that we have to show for each (F_r, Φ_r) with $F_r \in \mathcal{P}_\varepsilon(F_0)$ and for every $\delta > 0$ that there exists $\gamma > 0$ and $\overline{\Phi}$ such that $F \in \mathcal{P}_\gamma(F_r)$ implies $r(\overline{\Phi}, F) \leq r(\Phi_r, F_r) + \delta$.

Let $\gamma \triangleq \delta/4$ and let $\overline{\Phi}$ be such that $\overline{\Phi}(p) \triangleq \Phi_r(p+\gamma)$. Then using the fact that $F(v) \geq F_r(v-\gamma) - \gamma$ we obtain

$$\int_0^1 v dF(v) \leq 2\gamma + \int_0^1 v dF_r(v).$$

Using the fact that $F(v) \leq F_r(v+\gamma) + \gamma$ we obtain

$$\pi(\overline{\Phi}, F) \geq \pi(\Phi_r(p+\gamma), \min\{F_r(v+\gamma)+\gamma, 1\}) \geq \pi(\Phi_r, F_r) - 2\gamma,$$

and hence $r(\overline{\Phi}, F) \leq r(\Phi_r, F_r) + \delta$. To show payoff security for nature we have to show for each (Φ_r, F_r) with $F_r \in \mathcal{P}_\varepsilon(F_0)$ and for every $\delta > 0$ that there exists $\gamma > 0$ and $\overline{F} \in \mathcal{P}_\varepsilon(F_0)$ such that $\Phi \in \mathcal{P}_\gamma(\Phi_r)$ implies $r(\Phi, \overline{F}) \geq r(\Phi_r, F_r) - \delta$.

Here we set $\overline{F} \triangleq F_r$. Given $\gamma > 0$ consider any $\Phi \in \mathcal{P}_\gamma(\Phi_r)$. All we have to show is that $\pi(\Phi, F_r) \leq \pi(\Phi_r, F_r) + \delta$ for sufficiently small γ. Note that $\Phi(p) \leq \Phi_r(p+\gamma) + \gamma$ implies

$$\begin{aligned}
\pi(\Phi, F_r) &\leq \gamma + \int (p+\gamma) \int_p^1 dF_r(v) d\Phi_r(p+\gamma) \\
&= \gamma + \int p \int_{p-\gamma}^1 dF_r(v) d\Phi_r(p) \\
&= \gamma + \pi(\Phi_r, F_r) + \int p \int_{[p-\gamma,p)} dF_r(v) d\Phi_r(p) \\
&\leq \gamma + \pi(\Phi_r, F_r) + \int \int_{[p-\gamma,p)} dF_r(v) d\Phi_r(p).
\end{aligned}$$

Given the continuity of the last integral term above in the boundary point γ, the claim is established. □

Proof of Proposition 6. Assume that Φ_r attains minimax regret but is not robust. So there exists $\gamma > 0$, such that for all $\varepsilon > 0$, there exists F_ε

such that $F_\varepsilon \in \mathcal{P}_\varepsilon(F_0)$ but

$$\pi(p^*(F_\varepsilon), F_\varepsilon) - \pi(\Phi_r, F_\varepsilon) \geq \gamma. \tag{10}$$

Assume that (Φ_r, F_r) is a saddle point of the regret problem (SP_r) given $\varepsilon > 0$. Then $\pi(\Phi_r, F_r) = \pi(p^*(F_r), F_r)$ and we can rewrite the left-hand side of (10) as follows:

$$\begin{aligned}\pi(p^*(F_\varepsilon), F_\varepsilon) - \pi(\Phi_r, F_\varepsilon) = \pi(p^*(F_\varepsilon), F_\varepsilon) - \pi(p^*(F_r), F_r)\\ + \pi(\Phi_r, F_r) - \pi(\Phi_r, F_\varepsilon.)\end{aligned} \tag{11}$$

Using (SP_r) we also obtain

$$\begin{aligned}0 \leq r(\Phi_r, F_r) - r(\Phi_r, F_\varepsilon) = \int v dF_r(v) - \int v dF_\varepsilon(v)\\ + \pi(\Phi_r, F_\varepsilon) - \pi(\Phi_r, F_r),\end{aligned}$$

so that:

$$\pi(\Phi_r, F_r) - \pi(\Phi_r, F_\varepsilon) \leq \int v dF_r(v) - \int v dF_\varepsilon(v).$$

Entering this into (11) we obtain from (10) that:

$$\pi(p^*(F_\varepsilon), F_\varepsilon) - \pi(p^*(F_r), F_r) + \int v dF_r(v) - \int v dF_\varepsilon(v) \geq \gamma. \tag{12}$$

Since $F_\varepsilon, F_r \in \mathcal{P}_\varepsilon(F_0)$ and since $h(v) = v$ is a continuous function and the Prohorov norm metrizes the weak topology we obtain, if ε is sufficiently small, that

$$\int v dF_r(v) - \int v dF_\varepsilon(v) < \gamma/2. \tag{13}$$

In the proof of Proposition 3 we showed that $\pi(p * (F), F)$ as a function of F is continuous with respect to the weak topology. Hence

$$\pi(p^*(F_\varepsilon), F_\varepsilon) - \pi(p^*(F_r), F_r) < \gamma/2. \tag{14}$$

if ε is sufficiently small. Comparing (12) to (13) and (14) yields the desired contradiction. □

References

Anscombe, F., and R. Aumann (1963): "A definition of subjective probability," *Ann. Math. Statist.*, 34, 199–205.

Bell, D. (1982): "Regret in decision making under uncertainty," *Oper. Res.*, 30, 961–981.

Bergemann, D., and K. Schlag (2008): "Pricing without priors," *J. Europ. Econ. Assoc. Papers Proc.*, 6, 560–569.

Bergemann, D., and K. Schlag (2008): "Robust monopoly pricing," Discussion paper 1527RR, Cowles Foundation for Research in Economics, Yale University.

Bose, S., E. Ozdenoren, and A. Pape (2006): "Optimal auctions with ambiguity," *Theoret. Econ.*, 1, 411–438.

Bulow, J., and J. Roberts (1989): "The simple economics of optimal auctions," *J. Polit. Economy*, 97, 1060–1090.

Ellsberg, D. (1961): "Risk, ambiguity and the savage axioms," *Quart. J. Econ.*, 75, 643–669.

Engelbrecht-Wiggans, R. (1989): "The effect of regret on optimal bidding in auctions," *Management Sci.*, 35, 685–692.

Engelbrecht-Wiggans, R., and E. Katok (2007): "Regret in auctions: Theory and evidence," *Econ. Theory*, 33, 81–101.

Filiz-Ozbay, E., and E. Ozbay (2007): "Auctions with anticipated regret: Theory and experiment," *Amer. Econ. Rev.*, 97, 1407–1418.

Gilboa, I., and D. Schmeidler (1989): "Maxmin expected utility with non-unique prior," *J. Math. Econ.*, 18, 141–153.

Hannan, J. (1957): "Approximation to Bayes risk in repeated play," In: *Contributions to the Theory of Games*, ed. by M. Dresher, A. Tucker and P. Wolfe. Princeton University Press, Princeton, pp. 97–139.

Hayashi, T. (2008): "Regret aversion and opportunity dependence," *J. Econ. Theory*, 139, 242–268.

Huber, P. J. (1981): "Robust Statistics," John Wiley and Sons, New York.

Linhart, P., and R. Radner (1989): "Minimax — Regret strategies for bargaining over several variables," *J. Econ. Theory*, 48, 152–178.

Loomes, G., and R. Sugden (1982): "Regret theory: An alternative theory of rational choice under uncertainty," *Econ. J.*, 92, 805–824.

Lopomo, G., L. Rigotti, and C. Shannon (2009): "Uncertainty in mechanism design," Discussion paper.

Milnor, J. (1954): "Games against nature," in: *Decision Processes*, ed. by R. Thrall, C. Coombs and R. Davis. Wiley, New York.

Myerson, R. (1981): "Optimal auction design," *Math. Oper. Res.*, 6, 58–73.

Prasad, K. (2003): "Non-robustness of some economic models," *Top. Theor. Econ.*, 3, 1–7.

Renou, L., and K. Schlag (2010): "Minimax regret and strategic uncertainty," *J. Econ. Theory*, 145, 264–286.

Reny, P. J. (1999): "On the existence of pure and mixed strategy Nash equilibria in discontinuous games," *Econometrica*, 67, 1029–1056.

Riley, J., and R. Zeckhauser (1983): "Optimal selling strategies: When to haggle, when to hold firm," *Quart. J. Econ.*, 98, 267–290.

Stoye, J. (2008): "Axioms for minimax regret choice correspondences," Discussion paper, New York University.

Author Index

Subject Index

CPSIA information can be obtained
at www.ICGtesting.com
Printed in the USA
BVOW06*2212251116
468845BV00003B/9/P